CIM Series in Mathematical Sciences

Volume 2

The CIM Series in Mathematical Sciences is published on behalf of and in collaboration with the Centro Internacional de Matemática (CIM) in Coimbra, Portugal. Proceedings, lecture course material from summer schools and research monographs will be included in the new series.

More information about this series at
http://www.springer.com/series/11745

Jean-Pierre Bourguignon • Rolf Jeltsch •
Alberto Adrego Pinto • Marcelo Viana
Editors

Mathematics of Energy and Climate Change

International Conference and Advanced School Planet Earth, Portugal, March 21–28, 2013

Springer

Editors
Jean-Pierre Bourguignon
IHES Le Bois-Marie
Bures-sur-Yvette, France

Rolf Jeltsch
Department of Mathematics
ETH Zürich
Seminar für Angewandte Mathematik
Zürich, Switzerland

Alberto Adrego Pinto
Department of Mathematics
University of Porto
Faculty of Sciences
Porto, Portugal

Marcelo Viana
Instituto de Matemática Pura e Aplicada
IMPA
Rio de Janeiro, Brazil

ISSN 2364-950X ISSN 2364-9518 (electronic)
CIM Series in Mathematical Sciences
ISBN 978-3-319-16120-4 ISBN 978-3-319-16121-1 (eBook)
DOI 10.1007/978-3-319-16121-1

Library of Congress Control Number: 2015946752

Mathematics Subject Classification (2010): 92-02, 92-06, 93-02, 93-06

Springer Cham Heidelberg New York Dordrecht London

Printed on acid-free paper

Springer International Publishing AG Switzerland is part of Springer Science+Business Media (www.springer.com)

Foreword

Alberto Adrego Pinto, the president of the Centro International de Matemática (CIM), has asked me to write a few words on the International Conference Planet Earth, Mathematics of Energy and Climate Change, MPE 2013. I am very happy to do so since I had a chance to participate in this extremely important event. I became a member of the CIM Scientific Council in 2009. Two years later Alberto Adrego Pinto took office as the new president and he started with a lot of enthusiasm and energy. The executive committee decided that the CIM should become a partner in the global program Mathematics of Planet Earth (MPE 2013). In addition it was decided to organize two conferences in 2013 with accompanying Advanced Schools. The present proceedings are the outcome of the conference MECC 2013, the International Conference and Advanced School Planet Earth, Mathematics of Energy and Climate Change, held between March 21 and 28. Clearly these two chosen topics fit extremely well to the problems our planet is currently facing. In addition one needs complex mathematical models to understand the processes. I was very much looking forward to being part of this event in Lisbon and to learning from the excellent plenary speakers about how mathematics can be used to understand energy issues and the climate. It was a great idea to actually do several video lectures. This not only allowed us to get the best speakers from remote locations, like Rio de Janeiro or Berkeley to name a few, but with this approach the conference also demonstrated how to save energy and reduce the production of CO_2. Alberto Adrego Pinto had two further excellent ideas. About a third of all plenary speakers volunteered to present two further lectures on the topic in the associated "Advanced School." For example there was the public lecture by David Zilberman, who spoke on "Technology and the Future Bioeconomy." He then also spoke on the "Economic Foundations of Climate Smart Agriculture" and "The Economics of Payment for Ecosystem Services." These talks were supplemented by the plenaries, the "Advanced Schools" and seventeen "Thematic Sessions," so that altogether it was a very intensive time. Thanks to the famous Calouste Gulbenkian Foundation, which hosted the event at their excellent conference facilities, all participants enjoyed the friendly and constructive atmosphere. I think it was actually a very good idea to hold the plenary lectures in an auditorium, which allowed the delegates to participate in discussions. Finally I was extremely impressed by the contributions of our Portuguese colleagues, who organized nearly all of the Thematic Sessions. There is a large community of applied mathematicians who work on research concerning the mathematics of planet earth. Finally I would like to thank Alberto Adrego Pinto for organizing the conference and for his unbridled enthusiasm, which allowed him to motivate so many participants.

Zürich, Switzerland
28 November 2014

Rolf Jeltsch

Foreword

The stress placed on the dynamical system we call Planet Earth, owing to the activities of mankind, threatens mankind itself. It calls for an unprecedented response. This begins with an understanding of the problem. The shear complexity of the system, and the consequential ease with which an unexpected consequence or an unanticipated bifurcation can occur, or an unjustified cause-and-effect relation can be inferred, calls for a careful mathematical analysis. Understanding is essential, but also creative solutions are urgently needed. Here again, mathematical theory will play an important role.

These papers comprise a snapshot of current mathematical work devoted to these problems. The topics coincide with the major themes of the International Conference and Advanced School.

An understanding of the problem begins a careful overall study of the energy flow to the earth. This includes work on the difficult-to-predict role, also in weather prediction, of clouds (Santos, chapter "The Role of Clouds, Aerosols and Galactic Cosmic Rays in Climate Change"), and climate and the ecology of polar regions (Xavier, Hill, Belchier, Bracegirdle, Murphy, and Lopes Dias, chapter "From Ice to Penguins: The Role of Mathematics in Antarctic Research"). Pereira's article (chapter "Mathematics of Energy and Climate Change: From the Solar Radiation to the Impacts of Regional Projections"), encompassing all aspects of the energy flow to the earth, from the Stefan-Boltzmann law to the statistical treatment of fires to recommendations for future rain gutter sizes is a *tour de force* of climate change. It will be required reading for future mathematical climatologists.

A particular consequence of climate change is the increased frequency of rare, sometimes disastrous, events. Fundamental work on the mathematical theory of extreme values is needed, such as the theory of max-stability distributions (Fraga Alves, chapter "Max-Stability at Work (or Not): Estimating Return Levels for Daily Rainfall Data"), and resampling methodologies (Gomes, Henriques-Rodrigues, and Figueiredo, chapter "Resampling-Based Methodologies in Statistics of Extremes: Environmental and Financial Applications"), and methods to reveal additive outliers in time series (Eduarda Silva and Pereira, chapter "Detection of Additive Outliers in Poisson INAR(1) Time Series"). Application of these and related methods to the occurrence of extremal earthquakes is presented by Brito, Cavalcante, and Moreira Freitas (chapter "Modeling of Extremal Earthquakes"). Manuela Neves' overview of geostatistical methods (chapter "Geostatistical Analysis in Extremes: An Overview") contains also a brief but useful historical perspective.

Solutions to problems of climate change and its effect on human populations may involve optimal control theory. This is useful in physical systems (Grilo, Gama, and Lobo Pereira, chapter "On the Optimal Control of Flow Driven Dynamic Systems") and also in epidemiology (Aweke and Kassa, chapter "Impacts of Vaccination and Behavior Change in the Optimal Intervention Strategy for Controlling the Transmission of Tuberculosis").

The chemical physics of climate change involves especially the careful study of the chemical kinetics of environmental processes. This is known to be a challenging problem, and improvement of the associated mathematical theories is needed. The state-of-the-art is here presented by Carvalho, Silva, and Soares (chapter "Detonation Wave Solutions and Linear Stability in a Four Component Gas with Bimolecular Chemical Reaction"), da Costa (chapter "Mathematical Aspects of Coagulation-Fragmentation Equations"), and Sasportes (chapter "Long Time Behaviour and Self-similarity in an Addition Model with Slow Input of Monomers"). Silva and Rodrigues (chapter "Modelling the Fixed Bed Adsorption Dynamics of CO_2 / CH_4 in 13X Zeolite for Biogas Upgrading and CO_2 Sequestration") analyze a possible solution: the use of zeolites to catalyze the sequestration of CO_2.

Human communication will be an inextricable consequence of climate change, particularly in the context of such major potential societal disruptions as the shifting of populations to the north. Salvador, Nogueira, and Rocha (chapter "Multiscale Internet Statistics: Unveiling the Hidden Behavior") analyze the statistics of internet traffic.

Implementation of solutions will inevitably involve policy decisions, which in turn drive a politico-economic dynamical system. The politico-economics of ethanol production is treated by Moss, Schmitz, and Schmitz (chapter "The Economics of Ethanol: Use of Indirect Policy Instruments").

The authors in this volume have made a tremendous effort to explain the overall context of their work, which makes these diverse presentations approachable for a broad scientific audience. This approachability speaks to the unity and universality of mathematics in the sciences, and underlies its essential value in approaching the pressing problems facing Planet Earth.

On behalf of the participants and authors, I would warmly like to thank Alberto Adrego Pinto. Conference participants were never more inspired, nor treated more warmly than by Alberto in the context of the magnificent venue of the Calouste Gulbenkian Foundation. The Fado was truly delightful.

Minneapolis, MN, USA Richard D. James
17 March 2015

Preface

As the International Center for Mathematics (CIM) celebrated its 20th anniversary on the 3rd of December 2013, it is the perfect opportunity to look back on this past year, which has undoubtedly been one of the most ambitious and eventful ones in its history. With the support of our associates from 13 leading Portuguese universities, our partners at the University of Macau, and member institutions such as the Portuguese Mathematical Society, in 2013 the CIM showed yet again the importance of a forum such as this for bringing together leading Portuguese-speaking scientists and researchers from around the world.

The hallmark project of the year was the UNESCO-backed International Program Mathematics of Planet Earth (MPE) 2013, which the CIM participated in as a partner institution. This ambitious and global program was tasked with exploring the dynamic processes underpinning our planet's climate and man-made societies, and with laying the groundwork for the kind of mathematical and interdisciplinary collaborations that will be pivotal to addressing the myriad issues and challenges facing our planet now and in the future. The CIM heeded the MPE's call to action by organizing two headline conferences in March and September of 2013: the "Mathematics of Energy and Climate Change" conference in Lisbon in the spring, and the conference "Dynamics, Games, and Science II" in the fall. Both were held at the world-renowned Calouste Gulbenkian Foundation in Lisbon, one of more than 15 respected Portuguese foundations and organizations that enthusiastically supported the CIM conferences. As well as the conferences themselves, well attended "advanced schools" were held before and after each event: at the Universidade de Lisboa in the spring, and at the Universidade Técnica de Lisboa in the fall.

These conferences succeeded in bringing together some of the most accomplished mathematical and scientific minds from across the Portuguese-speaking world and beyond, while also serving as a launch pad for one of the CIM's most exciting endeavors in years: the new CIM Series in Mathematical Sciences, which will include lecture notes and research monographs and be published by Springer-Verlag. "The collaboration with Springer will bring mathematics developed in Portugal to a global audience," CIM President Alberto Adrego Pinto said at the time of the announcement, "and will help strengthen our contacts with the international mathematics community."

These first two volumes in the series, consisting of review articles selected from work presented at the "Mathematics of Energy and Climate Change" and "Dynamics, Games, and Science" conferences, reflect the CIM's international reach and standing. Firstly, they are characterized by an impressive roster of mathematicians and researchers from across the United States, Brazil, Portugal, and several other countries whose work will be included in the volumes.

The authors are complemented by the editorial board responsible for this first installment, a world-renowned "quartet" consisting of: president of the European Research Council Jean-Pierre Bourguignon from the École Polytechnique; former Société Mathématiques Suisse and European Mathematical Society president Rolf Jeltsch from the ETH Zurich; current Sociedade Brasileira de Matemática president Marcelo Viana from Brazil's Instituto Nacional de Matemática Pura e Aplicada; and CIM president Alberto Adrego Pinto from the Universidade do Porto.

While the MPE program was a major focus of the CIM's activities in 2013, the center also organized a number of further events aimed at fostering closer ties and collaboration between mathematicians and other scientists, mainly in Portugal and other Portuguese-speaking countries. In this context the CIM held the 92nd European Study Group with Industry meeting, part of a vital series held throughout Europe to encourage and strengthen the connections between mathematics and industry. As the MPE program made clear, humanity faces all manner of challenges, both man-made and natural, and though industry is attempting to overcome them, in many cases mathematics and science are far better suited to the task. Yet it is often industry that delivers the kinds of innovative ideas that will launch the next great scientific and technological revolutions, and which academia must adapt to. The potential for dialogue and cooperation between academia and industry is in fact so great that I have now made it one of the core initiatives in my presidency of the US-based Society for Industrial and Applied Mathematics (SIAM).

As we look back at the successful year the CIM had in 2013, we should also bear in mind the dramatic changes currently taking place in the world, changes that above all the mathematical sciences—including statistics, operational research, and computer science—will be called upon to address. Foremost among them is the rise of Big Data, especially as it relates to national security, finance, medicine, and the Internet (among other fields), which has come to dominate research in many scientific sectors and requires new analytical tools, which mathematics can provide. This new landscape will require an unparalleled level of partnership between science and industry, and is what prompted the European Commission to recently announce its Europe 2020 Growth Strategy, which calls for investment in groundbreaking research, innovation in industry, and the cultivation of a new generation of scientists. It is no coincidence that these three pillars are at the core of the CIM's own mission, and the CIM series in Mathematical Sciences will provide the ideal platform for communicating and broadening the impact of the CIM's activities with regard to these global challenges.

President of CIM Scientific Council Irene Fonseca

Acknowledgements

We thank all the authors of the chapters and we thank all the anonymous referees. We are grateful to Irene Fonseca for contributing the preface of this book. We thank the Executive Editor for Mathematics, Computational Science and Engineering at Springer-Verlag Martin Peters for invaluable suggestions and advice throughout this project.

We thank João Paulo Almeida, Ruth Allewelt, Joana Becker, João Passos Coelho, Ricardo Cruz, Helena Ferreira, Isabel Figueiredo, Alan John Guimarães, Filipe Martins, José Martins, Bruno Oliveira, Telmo Parreira, Diogo Pinheiro, and Renato Soeiro for their invaluable help in assembling this volume and for their editorial assistance.

Alberto Adrego Pinto would like to acknowledge the financial support of Centro Internacional de Matemática (CIM), Ciência Viva (CV), Calouste Gulbenkian Foundation, Fundação para a Ciência e a Tecnologia, LIAAD-INESC TEC, USP-UP project, IJUP and Mathematics Department, Faculty of Sciences, University of Porto, Portugal.

Alberto Pinto gratefully acknowledges the financial support to the conclusion of this project provided by the FCT—Fundação para a Ciência e a Tecnologia (Portuguese Foundation for Science and Technology) within project UID/EEA/50014/2013 and ERDF—European Regional Development Fund through the COMPETE Program (operational program for competitiveness) and by National Funds through the FCT—Fundação para a Ciência e a Tecnologia (Portuguese Foundation for Science and Technology) within Project "Dynamics and Applications" with reference PTDC/MAT/121107/2010.

Bures-sur-Yvette, France — Jean-Pierre Bourguignon
Zürich, Switzerland — Rolf Jeltsch
Porto, Portugal — Alberto Adrego Pinto
Rio de Janeiro, Brazil — Marcelo Viana

Contents

Max-Stability at Work (or Not): Estimating Return Levels for Daily Rainfall Data

Maria Isabel Fraga Alves

Abstract When we are dealing with meteorological data, usually one is interested in the analysis of maximal observations and records over time, since these entail negative consequences—risk events. Extreme Value Theory has proved to be a powerful and useful tool to describe situations that may have a significant impact in many application areas, where knowledge of the behavior of the tail of a distribution is of main interest. The classical Gnedenko theorem establishes that there are three type of possible limit max-stable distributions for maxima of blocks of independent and identically distributed (iid) observations. However, for the types of data to which extreme value models are commonly applied, temporal independence is usually an unrealistic assumption and one could ask about the appropriateness of max-stable models. Luckily, stationary and weekly dependent series follow the same distributional limit laws as those of independent series, although with parameters affected by dependence. For rainfall data, we will play with these results, analyzing *max-stability at work* for rare events estimation and the real impact of "neglecting" iid property.

1 Introduction

When we are dealing with meteorological data there are two situations that matter to differentiate: the case of data concentrated around the average, with no disastrous consequences for the society; on the other hand, the case of data away from the center of the distribution, that can have a very negative impact and which is important to quantify. Typically, one is interested in the analysis of maximal observations and records over time, since these entail negative consequences. The rainfall is a good example of this: the engineering structures associated with extremal precipitation levels, need to be constructed to withstand the extremal behavior of this process; for example, a reservoir must be able to store the amount of rain expected to fall in some specific location.

M.I. Fraga Alves (✉)
DEIO and CEAUL, Faculty of Sciences, University of Lisbon, Lisbon, Portugal
e-mail: mialves@fc.ul.pt

J.-P. Bourguignon et al. (eds.), *Mathematics of Energy and Climate Change*,
CIM Series in Mathematical Sciences 2, DOI 10.1007/978-3-319-16121-1_1

Extreme Value Theory (EVT) is the theory of modeling and measuring events which occur with very small probability, which has proved to be a powerful and useful tool to describe atypical situations that may have a significant impact in many application areas, where knowledge of the behavior of the tail of a distribution is of main interest. The classical result is Gnedenko's theorem [2]. It establishes that there are three types of possible limiting distributions (*max-stable*) for maxima of blocks of observations, which are unified in a single representation—the Generalized Extreme Value (GEV) distribution.

For rainfall data in Barcelos, we will play with *max-stability* and some statistical parametric model approaches to estimate p-return levels associated with $T = 1/p$-year return periods, for p small.

2 Preliminaries

In this section some preliminary concepts are presented. Denote by F the distribution function (DF) underlying the data under study and $F^{\leftarrow}$ its generalized inverse, defined as $F^{\leftarrow}(y) := \inf\{x : F(x) \geq y\}$. Typical design values are:

Definition 1 (T-year Return Level: u_T) A value which is exceeded once in a year with a probability $1/T$

$$u_T = F^{\leftarrow}(1 - 1/T)\ . \tag{1}$$

Definition 2 (u_T-Return Period: T) Average number of years between occurrences of an event of magnitude greater than a predefined high level u_T

$$T = \frac{1}{P[X > u_T]}\ . \tag{2}$$

If in (1) the value $1/T =: p$ is very small, say $p < 1/n$, with n denoting the available sample size, then we are dealing with *high or extreme quantiles* and it is crucial to model rare events.

We cannot simply assume that these atypical values are impossible. Design levels correspond to return periods of 100 years or more and the empirical cumulative distribution function (ECDF) is not enough for making statistical inference!

2.1 Daily Rainfall in Barcelos 1932–2008

Daily Rainfall in Barcelos 1932–2008 The following data is freely available from www.snirh.pt and has also been analyzed in [4, 5], including high quantiles estimation for monthly maxima (Fig. 1). The data analysis here and along the text was done using R package (see [6]).

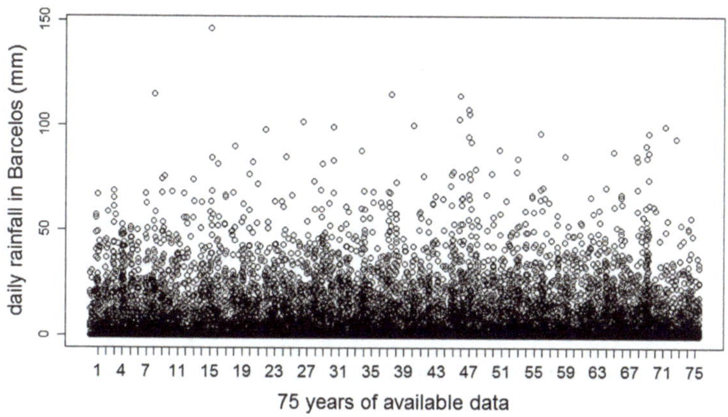

Fig. 1 Daily rainfall in Barcelos 1932–2008

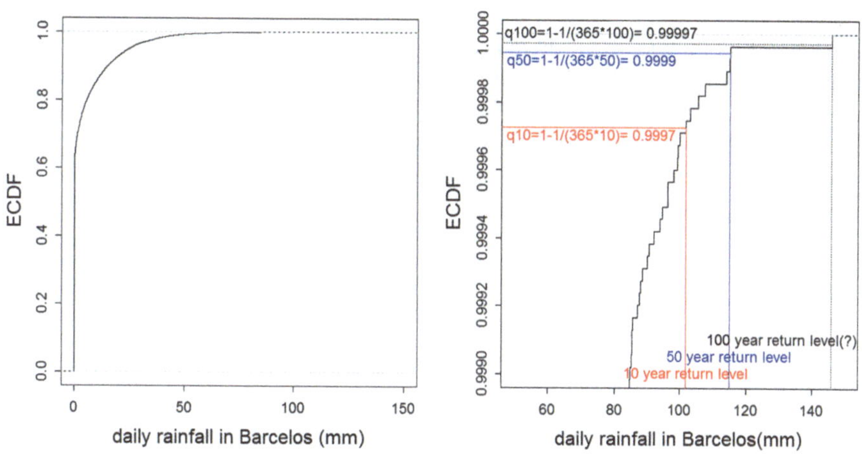

Fig. 2 ECDF and 'zoom' of ECDF for daily rainfall in Barcelos; q_T denotes $q_T := 1 - p_T = 1 - \frac{1}{365 \times T}$

If we make 'zoom' of the ECDF for daily rainfall in Barcelos (see Fig. 2), and aim to estimate the 100-year return level, the best we can do with the ECDF is giving the sample maximum, and the same applies to any T-year return level, with $T > 75$. Consequently, extrapolation is required.

3 "Annual" Maxima Approach or Gumbel Method

EVT provides limit laws for an extrapolation beyond the sample. The classical result is Gnedenko's theorem (see [2]), which establishes that there are three types of possible limiting max-stable distributions for maxima, M_n, of blocks of n iid observations with common DF F, which are unified in a single representation—the GEV distribution

$$G_\gamma(x) = \exp\left\{-[1+\gamma x]_+^{-1/\gamma}\right\}, \qquad \gamma \in \mathbb{R}. \tag{3}$$

[notation: $x_+ := \max(0, x)$]. That is, if there are sequences $a_n > 0$ e b_n, such that $P\left[\frac{M_n - b_n}{a_n} \le x\right] \longrightarrow G(x)$, as $n \to \infty$, for some non-degenerated DF G, then G is of the same type of $G_\gamma(x)$ and we say that F belongs to the max-domain of attraction of G_γ [notation:$F \in \mathscr{D}(G_\gamma)$].

A property associated with the limit distribution is the *max-stability*.

Definition 3 (Max-Stability) A DF G is max-stable if there are real constants $A_k > 0$ e B_k such that

$$G^k(x) = G(A_k x + B_k), \text{ for all } k \ .$$

It is important to note that if there is a limit distribution for the linearly normalized maximum, then that limit distribution must be max-stable. This means that if G corresponds to the GEV DF for some *location/scale* parameters λ/δ, $G(\cdot) \equiv G_\gamma(\cdot;\lambda,\delta)$, then G^k is also a GEV distribution with the same shape γ and for some other *location/scale* associated parameters λ_k/δ_k, $G^k(\cdot) \equiv G_\gamma(\cdot;\lambda_k,\delta_k)$. In other words, taking powers of G results only in a change of *location* and *scale*. In fact, suppose that $\gamma \neq 0$,

$$\begin{aligned} G^k(x) &= \left(\exp\left\{-\left[1+\gamma\left(\frac{x-\lambda}{\delta}\right)\right]^{-1/\gamma}\right\}\right)^k \\ &= \exp\left\{-k\left[1+\gamma\left(\frac{x-\lambda}{\delta}\right)\right]^{-1/\gamma}\right\} \\ &= \exp\left\{-\left[1+\gamma\left(\frac{x-\lambda_k}{\delta_k}\right)\right]^{-1/\gamma}\right\}, \end{aligned}$$

where

$$\lambda_k := \lambda - \frac{\delta}{\gamma}(1-k^\gamma) \quad \text{and} \quad \delta_k := \delta\, k^\gamma.$$

The case Gumbel, $\gamma = 0$, is similar.

Consider the available data—daily rainfall in Barcelos 1932–2008—divided in m blocks, usually years, and pick up the maximum in each block (Fig. 3).

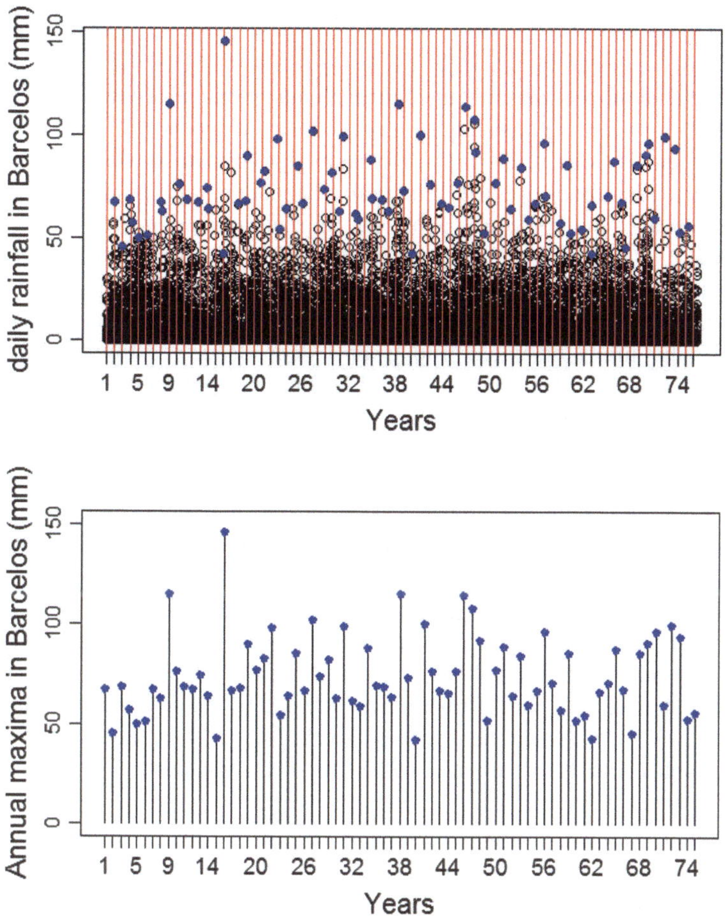

Fig. 3 Blocks of years, daily data and annual maxima (*left*); annual maxima (*right*)

With real λ and positive δ standing for location and scale parameters, fit $G_\gamma(x; \lambda, \delta) := G_\gamma((x - \lambda)/\delta)$ to the annual maximum $Y := \max(X_1, X_2, \cdots, X_n)$, with an available sample of m annual maxima $Y_1, Y_2, \cdots, Y_m$, considered iid, and proceed with γ estimation and also location of location and scale parameters (λ, δ). Afterwards, input $(\hat{\gamma}, \hat{\lambda}, \hat{\delta})$ for rare events estimation, associated to GEV fit for Y in the:

- Return period for level u, $T_u = \dfrac{1}{1 - G_\gamma(u; \lambda, \delta)}$,
- T-year return level, $u \equiv u_T = G_\gamma^{\leftarrow}\left(1 - \frac{1}{T}; \lambda, \delta\right)$.

4 Weak Dependence in Stationary Sequences

Extreme value models were obtained through mathematical arguments that assume an underlying process consisting of a sequence of independent random variables. However, and as Coles refers in [1], for many types of data to which extreme value models are commonly applied, temporal independence is usually an unrealistic assumption and one could ask about the appropriateness of max-stable models. Stationarity, which is a more realistic assumption for many physical processes, corresponds to a series whose variables may be mutually dependent, but whose stochastic properties are homogeneous through time. It is usual to assume a condition that limits the extent of long-range dependence at extreme levels, so that the events $X_i > u$ and $X_j > u$ are approximately independent, provided u is *high enough*, and time points i and j have a large separation. Loosely speaking, extreme events are close to independent at times that are far enough apart. Many stationary series satisfy this property and it is a property that is often plausible for physical processes. For example, knowledge that it rained heavily today might influence the probability of extreme rainfall in one or two days, but not for a specified day in, for instance, three months time (see [1]).

What about the suitability of max-stable models in his case? Provided a series has limited long-range dependence at extreme levels, maxima of stationary series follow the same distributional limit laws as those of independent series. However, the parameters of the limit distribution are affected by the dependence in the series.

Let $X_1, X_2, \cdots$ be a stationary process and $X_1^*, X_2^*, \cdots$ be a sequence of independent variables with the same marginal distribution. Denote $M_n = \max\{X_1, \ldots, X_n\}$ and $M_n^* = \max\{X_1^*, \ldots, X_n^*\}$.

Under suitable regularity conditions, $P\left[\frac{M_n^* - b_n}{a_n} \leq x\right] \longrightarrow G_1(x)$, as $n \to \infty$, for some non-degenerated DF G_1, with real sequences $a_n > 0$ e b_n, if and only if $P\left[\frac{M_n - b_n}{a_n} \leq x\right] \longrightarrow G_2(x)$, where

$$G_2(x) = G_1^{\theta}(x), \quad \text{for a constant } \theta \in (0, 1]\,.$$

The constant θ is designated as *extremal index* and the independence case corresponds to $\theta = 1$. Notice that if G_1 is a GEV distribution, so is G_2, by max-stability. Moreover, if G_1 is GEV with *shape* γ and *location/scale* λ/δ then G_2 is GEV with the same *shape* γ and *location* and *scale* $\lambda_\theta := \lambda - \frac{\delta}{\gamma}(1 - \theta^\gamma)$ and $\delta_\theta := \delta\,\theta^\gamma$.

The Barcelos Rain Case Study In Fig. 4 it is represented the autocorrelation function (ACF) for daily and annual maxima of daily rainfall records, which highlights the absence of a significative dependence for the latter. We should also mention that the tendency is not significative, which was concluded from a preliminary statistical test study.

In Fig. 5 (*right*) it is represented the autocorrelation function (ACF) for very high daily rainfall records, exceeding $u = 42$ mm (*left*), which corresponds to the minimum of annual maxima; this seems to reveal that the events $X_i > u$ and $X_j > u$

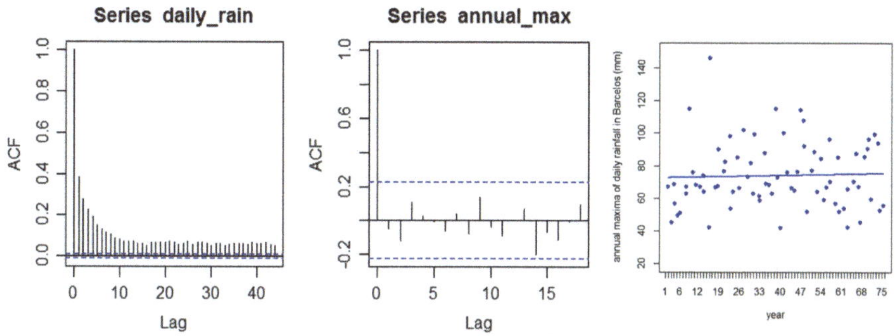

Fig. 4 ACF for daily rainfall (*left*) and for annual maxima (*center*) [R-package]; annual maxima (*right*)

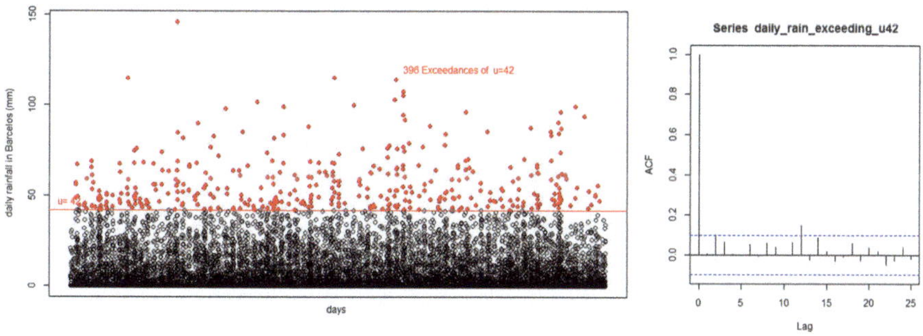

Fig. 5 Daily rainfall exceeding $u = 42$ mm (*left*) and respective ACF (*right*) [R-package]

are approximately independent, provided u is *high enough*, and instants i and j are far apart.

5 TOP Annual Approach: Ten Largest Observations per Year

Consider now the ten largest observations per year (see Fig. 6). The TOP annual approach relies on a convenient parametric model underlying the sample of the r largest observations, picked up for the m years.

Consider the limit joint model for r top order statistics (o.s.), r fixed, with joint limit density function

$$g_{1,\cdots,r}(w_1,\cdots,w_r) := G_\gamma(w_r)\prod_{i=1}^{r}\frac{g_\gamma(w_i)}{G_\gamma(w_i)}, \text{ for } w_1 > \cdots > w_r \,, \tag{4}$$

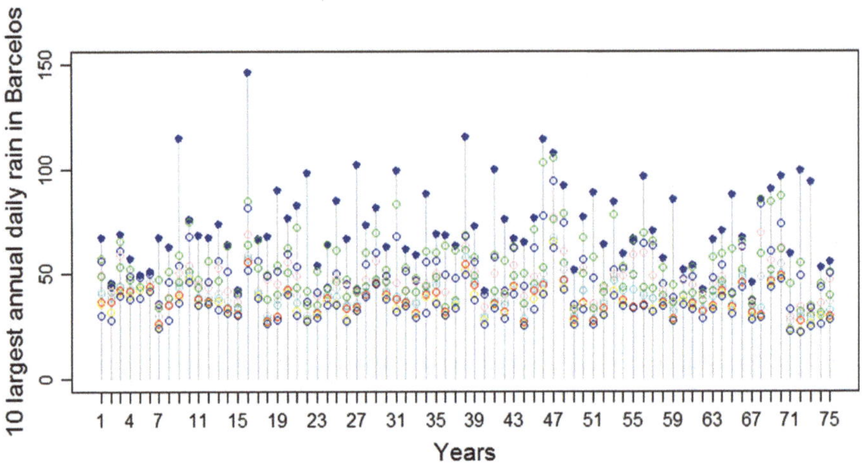

Fig. 6 Blocks of years, daily data and ten top observations per year

Table 1 Parameters estimated by GEV fit to annual maximum, by TO approach, with $r = 1$ (Gumbel method), $r = 5$ and $r = 10$ and 100-year return level estimates

# TO	$\hat{\gamma}$	$\hat{\lambda}$	$\hat{\delta}$	rl 100-year
r = 1	−0.030	65.19	16.00	133.867
r = 5	0.013	66.38	15.60	140.265
r = 10	0.005	66.95	15.04	136.880

[R-library(ismev)]

with $g_\gamma(w) := \frac{\partial G_\gamma}{\partial w}(w)$. In statistical inference for rare events, a possible approach is to model the top observations (TO) available from the sample with that joint structure. More precisely, $F \in \mathscr{D}(G)$ for $a_n > 0$ and b_n if the r-vector

$$\left(\frac{X_{n:n} - b_n}{a_n}, \dots, \frac{X_{n-r+1:n} - b_n}{a_n}\right)$$

has joint limit density function given in (4). In Table 1 the parameters and 100-year return level Maximum Likelihood (ML) estimates are summarized.

6 Return Levels vs. Return Periods: *Empirical and Max-Stability*

Inspired in [3], a graphical representation for the empirical Return Levels vs. Return Periods for both data (days/years), on a compatible scale, is given. It allows dealing with point-in-time and extreme-value distributions at the same time. Consider an empirical graphical representation of (T, u_T),

$$T = \frac{1}{1 - F_Y(u_T)}, \quad \text{with } Y \text{ annual maximum,}$$

with the DF F_Y replaced by its counterpart, ECDF:

- ***black dots***—correspond to $(\hat{T}^Y_{rescaled}, y_{i:n_y})$, with

$$\hat{T}^Y_{rescaled} = \frac{n}{1 - F_{n_y}(y_{i:n_y})},$$

  the rescaled empirical return period, in *days*, where:

  - $F_{n_y}(y_{i:n_y}) = \frac{i}{n_y+1}, \quad i = 1, \ldots, n_y$
    is the ECDF for the annual maxima sample of size n_y, $\{y_i\}_{i=1}^{n_y}$,
  - $n_y = 75$ is the *number of year-periods* available
  - Rescaled according to $n = 365$, the *number of days in one year-period.*

- *grey dots*—correspond to $(\hat{T}_d, x_{j:n_d})$, with X the *daily rainfall*, where

$$\hat{T}_d = \frac{1}{1 - F_{n_d}(x_{j:n_d})},$$

  is the empirical return period, in *days*, where:

  - $F_{n_d}(x_{j:n_d}) = \frac{j}{n_d+1}, \quad j = 1, \ldots, n_d$
    is the ECDF for the daily rainfall sample of size n_d, $\{x_j\}_{j=1}^{n_d}$,
  - $n_d = 27{,}570$ is the *number of days* for the available rainfall data.

Mínguez et al. provide in [3] a similar graphical representation to plot both distributions on a compatible scale of *hours/years*. In the Fig. 7 it is depicted the referred graphical representation, adapted to the present case of scale *days/years*. Note that the true abscissas axis units are days, where the ticks have been rescaled to years.

Consider now a random variable $Y := \max_i X_i$, with $\{X_i\}_{i=1}^n$ an iid sample of X with DF F. Then the DF of Y, F_Y, is identified with $F_Y = F^n$. Supported by EVT, we fit the GEV$(\gamma; \lambda, \delta)$ to Y, which means that the following approximation holds,

$$F^n(x) \approx G_\gamma(x; \lambda, \delta) = \exp\left[-\left(1 + \gamma\frac{x - \lambda}{\delta}\right)_+^{-1/\gamma}\right].$$

On the other hand, by *max-stability* it is possible to approximate the DF of X by a GEV model, $F(x) \approx \left(G_\gamma(x; \lambda, \delta)\right)^{1/n} \equiv G_\gamma(x; \lambda_d, \delta_d)$, with different *location/scale*

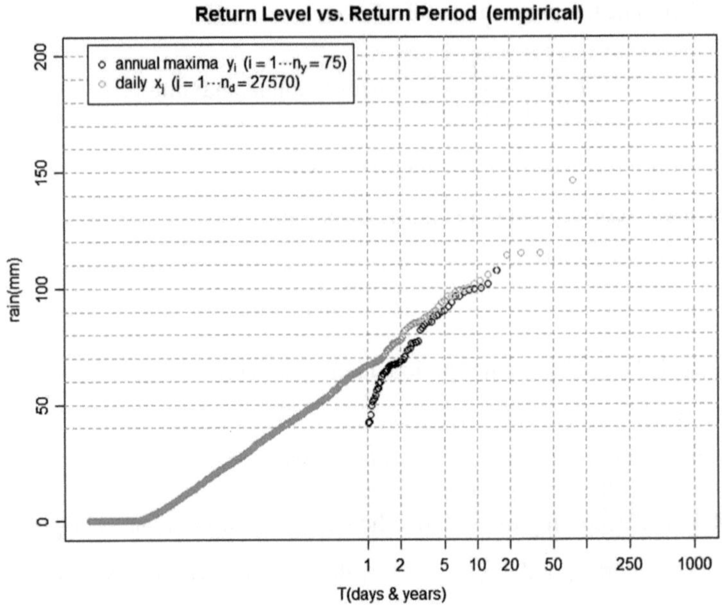

Fig. 7 Scatter plot of $(\hat{T}^{Y}_{rescaled}, y_{i:n_y})$, $n_y = 75$, and $(\hat{T}_d, x_{j:n_d})$, $n_d = 27{,}570$

parameters, (λ_d, δ_d). That is,

$$F(x) \approx \exp\left[-\left(1+\gamma\frac{x-\lambda_d}{\delta_d}\right)_+^{-1/\gamma}\right], \quad \text{with} \quad \begin{cases}\lambda_d := \lambda - \frac{\delta}{\gamma}(1-n^{-\gamma}) \\ \delta_d := \delta n^{-\gamma}\end{cases}. \tag{5}$$

Consequently, the respective return periods will be approximated as

$$T^Y := \frac{1}{1-F_Y(x)} \quad \hookrightarrow \quad T^Y \approx \frac{1}{1-G_\gamma(x;\lambda,\delta)},$$

$$T_d := \frac{1}{1-F(x)} \quad \hookrightarrow \quad T_d \approx \frac{1}{1-G_\gamma(x;\lambda_d,\delta_d)},$$

and the rescaled return period, for compatible scale representation, is

$$T^Y_{rescaled} := \frac{n}{1-F_Y(x)} \quad \hookrightarrow \quad T^Y_{rescaled} \approx \frac{n}{1-G_\gamma(x;\lambda,\delta)}.$$

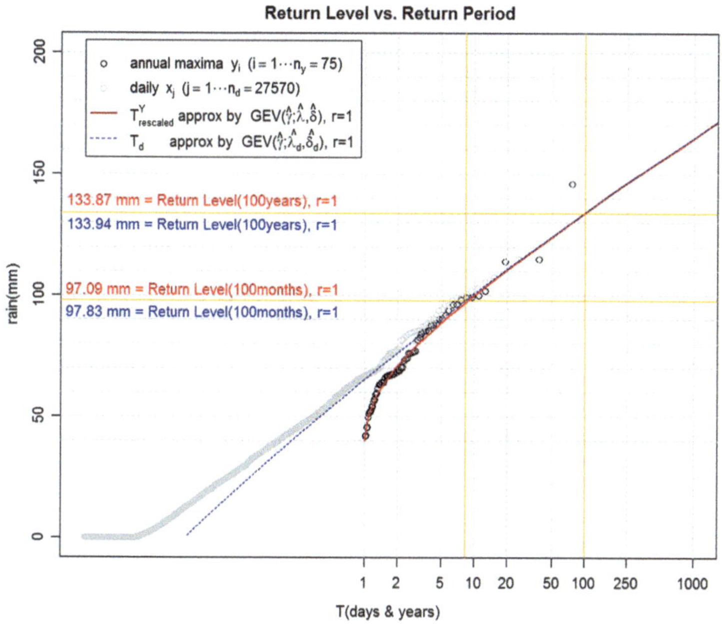

Fig. 8 $r = 1$: approximations for T = 100-years and T = 100-months return levels, supported by *annual-maxima fit* [*above the straight line*] and by *max-stability* [*below straight line*]; the *solid line* corresponds to annual-maxima fit and the *dashed line* to fit by max-stability

In Fig. 8, and for the Gumbel method approach ($r = 1$), approximations are plotted for:

- $T = 100$-*years* return level,

$$u_T \approx G_{\hat{\gamma}}^{\leftarrow}\left(1 - \tfrac{1}{100}; \hat{\lambda}, \hat{\delta}\right) = 133.87 \quad [\textit{annual-maxima fit}]$$

$$u_T \approx G_{\hat{\gamma}}^{\leftarrow}\left(1 - \tfrac{1}{100\times 365}; \hat{\lambda}_d, \hat{\delta}_d\right) = 133.94 \quad [\textit{max-stability}]$$

- $T = 100$-*months* return level,

$$u_T \approx G_{\hat{\gamma}}^{\leftarrow}\left(1 - \tfrac{1}{(100/12)}; \hat{\lambda}, \hat{\delta}\right) = 97.09 \quad [\textit{annual-maxima fit}]$$

$$u_T \approx G_{\hat{\gamma}}^{\leftarrow}\left(1 - \tfrac{1}{100\times 30}; \hat{\lambda}_d, \hat{\delta}_d\right) = 97.83 \quad [\textit{max-stability}]$$

Figure 9 is similar to Fig. 8 for top observations approach, with $r = 5$ and $r = 10$.

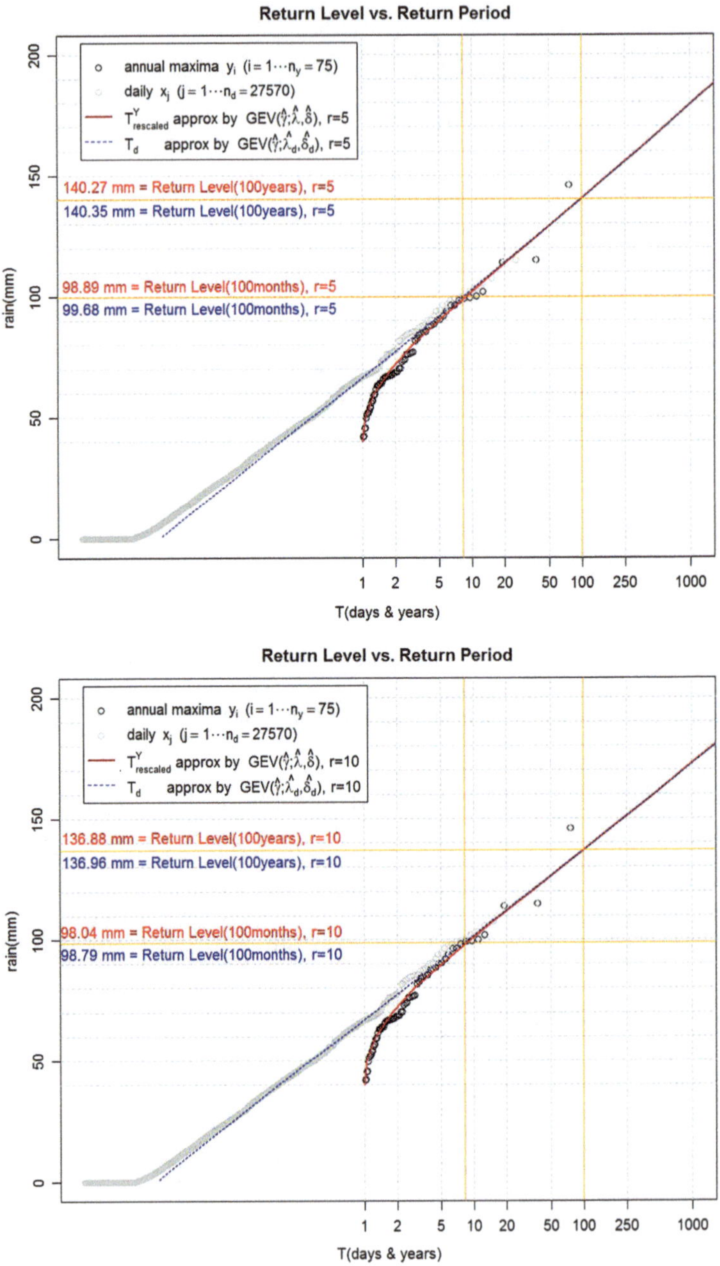

Fig. 9 Approximations for T = 100-years and T = 100-months return levels, supported by *annual-maxima fit* [*above the straight line*] and by *max-stability* [*below straight line*]; the *solid line* corresponds to annual-maxima fit and the *dashed line* to fit by max-stability, by top observations approach with $r = 5$ [*up*] and $r = 10$ [*down*]

It looks like as, for large return levels, the graphics of approximations of T_d and $T^Y_{rescaled}$, respectively, $\tilde{T}_d$ (*dashed line*) and $\tilde{T}^Y_{rescaled}$ (*solid line*) are similar; that is,

$$\frac{1}{1 - G_\gamma(x; \lambda_d, \delta_d)} \approx \frac{n}{1 - G_\gamma(x; \lambda, \delta)}, \qquad \text{as } x \to x^F, \quad x^F := \text{right endpoint of } F.$$

This is easily supported by the following: from max-stability

$$G_\gamma(x; \lambda_d, \delta_d) = \left[G_\gamma(x; \lambda, \delta)\right]^{1/n};$$

consequently,

$$\log G_\gamma(x; \lambda_d, \delta_d) = \frac{1}{n} \log G_\gamma(x; \lambda, \delta)$$

and, as $x \to x^F$,

$$n = \frac{\log G_\gamma(x; \lambda, \delta)}{\log G_\gamma(x; \lambda_d, \delta_d)} = \frac{\log[1 - (1 - G_\gamma(x; \lambda, \delta))]}{\log[1 - (1 - G_\gamma(x; \lambda_d, \delta_d))]} \approx \frac{1 - G_\gamma(x; \lambda, \delta)}{1 - G_\gamma(x; \lambda_d, \delta_d)},$$

where the last approximation comes from $\log(1 - z) \approx z$ for $z \to 0$.

All in all, we conclude that, for large return levels,

$$\frac{1}{1 - G_\gamma(x; \lambda_d, \delta_d)} \approx \frac{n}{1 - G_\gamma(x; \lambda, \delta)},$$

that is,

$$\tilde{T}_d \approx \tilde{T}^Y_{rescaled} \qquad \text{as } x \to x^F.$$

It is also worth mentioning that for small return levels the dashed line of $\tilde{T}_d$ is not far from its empirical counterpart $\hat{T}_d$, the best approximation obtained with $r = 10$ top observations fit.

7 Final Comments

The article deals with the estimation of return levels for daily rainfall data together with the impact of neglecting the usual assumed hypothesis of independence and identical distribution of the observed data. Moreover, although studied by other authors, the Barcelos data set is used here for comparing procedures of analysis in extreme value theory, mainly investigating how the property of max-stability really works in practice. In this respect, it should be emphasized the closeness of the dashed lines of Figs. 8 and 9, built on results of Eq. (5) and arising from the property max-stability, to the grey daily scatter plots of empirical return periods; in a certain sense, and in this example, it seems that *max-stability is at work.*

Acknowledgements Partially supported by PEst-OE/MAT/UI0006/2014 and EXTREMA-FCT/PTDC/MAT/101736/2008.

References

1. Coles, S.: An Introduction to Statistical Modeling of Extreme Values. Springer Series in Statistics. Springer, London (2001)
2. Gnedenko, B.V.: Sur la distribution limite du terme maximum d'une série aléatoire. Ann. Math. **44**, 423–453 (1943)
3. Mínguez, R., Guanche, Y., Méndez, F.J.: Point-in-time and extreme-value probability simulation technique for engineering design. Struct. Saf. **41**, 29–36 (2012)
4. Nascimento, F.F.: Abordagem Bayesiana Não-paramétrica para Análise de Valores Extremos. Ph.D. Thesis. Universidade Federal do Rio de Janeiro (2009)
5. Nascimento, F.F., Gamerman, D., Lopes, H.F.: A semiparametric Bayesian approach to extreme value estimation. Stat. Comput. **22**, 661–675 (2012)
6. R Development Core Team (2011). R: A Language and Environment for Statistical Computing. The R Foundation for Statistical Computing, Vienna. Available online at http://www.R-project.org/ [ISBN: 3-900051-07-0]

Impacts of Vaccination and Behavior Change in the Optimal Intervention Strategy for Controlling the Transmission of Tuberculosis

Temesgen Debas Aweke and Semu Mitiku Kassa

Abstract A dynamical model of TB for two age groups that incorporate vaccination of children at birth, behavior change in adult population, treatment of infectious children and adults is formulated and analyzed. Three types of control measures (vaccination, behavior change and anti-TB treatment strategies) are applied with separate rate for children and adults to analyze the solution of the controlled system by using the concept of optimal control theory. It is indicated that vaccination at birth and treatment for both age groups have impact in reducing the value of the reproduction number ($\mathscr{R}_o$) whereas behavior modification does not have any impact on $\mathscr{R}_0$. Pontryagin's Minimum Principle has been used to characterize the optimal level of controls applied on the model. It is shown that the optimal combination strategy of vaccination, behavior change and treatment for the two age groups can help to reduce the disease epidemic with minimum cost of interventions, in shorter possible time.

1 Introduction

Tuberculosis (TB) is an air born bacterial infectious disease caused mainly by Mycobacterium Tuberculosis which frequently affects the lungs (pulmonary TB) in addition to other organs of the body. The infection of tuberculosis begins when a person inhales infected bacilli that are released from the lungs of an infected person. Two to three weeks after infection the immune system forms tubercles that contains the mycobacteria. Ninety percent of the infections stop here and lay dormant (for an indefinite period of time) possibly never going on to be a detectable active disease[15]. But, some people may develop the disease soon after infection

T.D. Aweke
Department of Mathematics, Mizan-Teppi University
Teppi, Ethiopia
e-mail: temesgen.y2000@gmail.com

S.M. Kassa (✉)
Department of Mathematics, Addis Ababa University
Addis Ababa, Ethiopia
e-mail: semu.mitiku@aau.edu.et

J.-P. Bourguignon et al. (eds.), *Mathematics of Energy and Climate Change*,
CIM Series in Mathematical Sciences 2, DOI 10.1007/978-3-319-16121-1_2

if they have weak immune system (infants and people who have other disease such as HIV/AIDS). If the bacteria do not lay dormant, the bacteria continue to grow until the tubercles invade other portions of the lung and active tuberculosis begins. Mortality rate can be reduced by taking different anti-TB drugs (such as Isoniazid, Rifampicin, Pyrazinamide, Ethambutol and Streptomycin [20]) either in combination or independently after the diagnosis tests such as; microscopic examination of sputum smears, TB skin test, chest x-ray and others. Vaccinating infants and educating susceptible individuals to bring behavior change will help from contracting the disease.

Despite the availability of highly efficacious treatment for decades, TB remains a major global health problem. In 2011, there were an estimated 8.7 million incident cases of TB globally, equivalent to 125 cases per 100,000 population [1]. Due to difficulty of TB diagnosis in children, estimating the burden of TB in children is difficult. Of the 8.7 million incident cases, an estimated 0.5 million were children[1]. The best estimate for the same year is 0.9 million deaths and out of which 64,000 were children [1]. In 2011, there were an estimated 0.22 million incident cases of TB in Ethiopia, which is equivalent to 258 cases per 100,000 population [3, 6]. The best estimate for mortality were 0.15 million among HIV-positive people.

Even if there are big debates concerning efficacy of the vaccine, there is still a vaccine for TB. In line with this, various models have been formulated to assess the impact of vaccination in preventing the spread of TB. Lietman and Blower [13], studied two tuberculosis models to predict the impact of pre-exposure and post-exposure vaccines. Their result stated that pre-exposure vaccine are necessary to prevent a substantial increase in new infections and may be effective in disease eradication compared to post-exposure vaccines. Bekele [2] indicated that vaccination at birth plays a great role in reducing children new incidence cases and the number of deaths caused by TB. He also indicated that waning effect of the vaccine and its effectiveness are the determinant factors in the dynamics. Gaff and Schaefer [8] also analyzed the impacts of vaccination to control the transmission of infectious disease. They formulated an optimal control problem that incorporates vaccination and treatment as an intervention mechanism. They concluded that vaccination is an important mechanism in the presence of treatment to eradicate an epidemic. All these authors didn't analyze the role of individuals behavior change about the general happening of the disease to control an epidemic.

People of developing countries like Ethiopia have a chance to be exposed to infectious diseases such as TB in more crowded areas such as public transportation, prisons and refugee camps. In addition to this, large family members in rural areas in which one of the members is infected with TB bacteria may be considered as risky environment. From such risky environment Mycobacterium tuberculosis (MTB) has a chance to be transmitted from infectious to susceptible people through coughing or sneezing as well as from a contact with sputum of a TB patient. But to minimize the chance of getting TB bacteria the following measures or actions may be taken as control interventions: opening windows of public transportation vehicles while in use; if there is a family member or a friend in a refugee camp who has been coughing for the last two weeks should be advised to go to a health center for diagnosis; if one of their family member is infectious, then separate his/her nutritional materials and use gloves to handle his/her sputum; and wearing mask is also another self-protective mechanism that can help individuals from contracting disease. Most individuals of the population are not well informed or convinced about transmission and control mechanisms of TB. If most individuals get a qualified information about the disease in the above risk environments, then they may protect themselves from infection by applying the above self-protective mechanisms. Some individuals in the population start applying self-protective measures once they have a first hand experience of the disease. Some other individuals try to apply self-initiated measures based on concrete information they got from different sources such as radio and television programs, from written materials and Internet, from expertise and from people that have infection experience. The mechanism from media depends on the quality of the campaign or public effort to bring more impact and expenditure in the dissemination of the information. Since the behavior of individuals may not change easily, preventive measures with behavior change require a huge effort and investment from health sectors, from government and non-governmental organizations in preventing the disease.

Many existing models are based on the assumption that the behavior of individuals to protect themselves against an infectious disease remains constant or unchanged in the course of the outbreak. But in practice, if individuals have got concrete or qualified information about transmission and control methods of the disease, they will start to apply any of the existing self-protective measures. This will help susceptible individuals to reduce the average number of contacts with infectious individuals. This decreases the incidence rate of the disease. Therefore, analyzing the role of behavior modification for controlling the transmission of infectious disease like TB is very important. Kassa and Ouhinou [11] formulated a mathematical model of infectious disease epidemic that incorporates behavior change and treatment. They have shown that their mathematical model can portray the way how the population reacts to an increase in prevalence in the course of an outbreak and how one can plan medical treatment to control disease epidemics. In their analysis, they indicated that behavior modification by society plays an important role in controlling an epidemic, even when some pharmaceutical treatments are being given to the infected ones. If children get BCG vaccine at birth, the probability of getting the disease is lower than non-vaccinated children. Therefore, including these compartments or class of population provide a great significance in the dynamics. In addition to this, some sectors of the population such as children may not be able to learn and hence change their behavior. Therefore, it is necessary to classify the population into children and adults. But these authors didn't include vaccination into their mathematical model to analyze its role and also they didn't classify the population according to their age.

To analyze the role of public health measures and to plan effective control mechanisms for eradication of TB epidemic, it is very important to explore the contribution of vaccination at birth and behavior modification of adults against TB. Therefore, in this paper we incorporate vaccination at birth, behavior modification for susceptible adults as well as treatment of infectious individuals for the two age groups into a dynamical model of TB. Using this model we investigate the dynamics of an epidemic and also apply optimal control theory to propose cost effective public health intervention strategy to control the spread of TB disease.

The paper is organized as follows, in Sect. 2 we describe and formulate the mathematical model consisting of a system of ordinary differential equations that describe the impact of vaccination, behavior change modification when treatment is also considered as a possible intervention mechanism for two age groups. The mathematical analysis of the model is discussed in Sect. 3. Formulation of optimal control problems in the presence of four control parameters is discussed in Sect. 4. In this section the existence of optimal control solutions is also analyzed. Section 5 is about numerical simulation and results. We conclude the paper with a discussion in the last section.

2 Mathematical Models with Vaccination, Behavior Change and Treatment

Since TB dynamics in different age groups vary irrespective of the setting, age consideration may be taken into account when modeling biological systems and diseases such as tuberculosis (TB). Due to this, we classify the total population into children whose age is less than 15 years old and adults whose age is greater than or equal to 15 years old. We represent total birth rate by π, natural rate of mortality for children and adults by μ_c, μ_a respectively whereas d_c, d_a represent the tuberculosis induced death rates of children and adults per capita. Bacillus Calmette-Guerin (BCG) vaccination at birth may not protect the infection 100 % due to the quality of the vaccine and improper usage[2]. The failure of the vaccine may lead children to join latently infected class of children. However, after a certain number of years mostly 10–15 years, the vaccine is assumed to wane and we will have susceptible adults [2]. If we add an educated compartment (E) into age dependent model, it is possible to observe that individuals in educated class are exposed to the infection with a rate smaller than other susceptible individuals. But recruitment rate into educated class varies through time with respect to the lethality of the disease. This recruitment function describes the learning effect of population which can be measured indirectly by observing individuals behavior modification towards exposedness to the disease [10, 11]. The total variable population $N(t)$ of children and adults age groups are subdivided into the following compartments: Vaccinated children (V_c), non-vaccinated Susceptible children (S_c), Latently infected children (L_c), Infectious children (I_c), Treated children (T_c), Susceptible adults (S_a), Educated adults (E), Latently infected adults (L_a), Infectious adults (I_a) and Treated adults (T_a). Vaccinated children become susceptible due to waning at the rate w_c. We are considering that a proportion r of new births per unit of time join the vaccinated children population and the remaining (1-r) proportion per unit of time join the susceptible children class who are not vaccinated. After exposure a proportion q_c shows slow progression to the latent stage while the remaining (1-q_c) proportion will join the infectious class of children. Due to improper usage and quality of the vaccine used, we assume that the vaccine efficacy varies. Thus, if ε measures efficacy of the vaccine then, $(1 - \varepsilon)$ measures the inefficacy of the vaccine in preventing infection with $0 \leq \varepsilon \leq 1$. If a proportion q_a of the infected adults remain in the latent cohort while the remaining (1-q_a) will join the infectious class directly. The behavior function ($e(t)$) is described as a function of the prevalence $p(t)$ of the disease as is used in [10]. At the beginning of the outbreak, people understand very little about the disease and the reaction could be almost inexistent and at high prevalence, susceptible individuals will apply any of the available self protective measures and change their behavior. This implies that $e(p = 0\,\%) = 0$ and $e(p = 100\,\%) = 1$. Therefore,

$$e(p) = \frac{p^n}{p_*^n + p^n} \qquad \text{or equivalently} \qquad e(t) = \frac{(I_c + I_a)^n}{N^n p_*^n + (I_c + I_a)^n},$$

where p_* is the prevalence producing half of the maximum behavioral change value, with $p = \frac{I_c+I_a}{N}$, n is a hill coefficient that portrays the rate of reaction by the population [10]. If we denote by α the mean rate at which susceptible individuals get persuaded and recruited into the educated class per unit of time, αe will give us the actual recruitment rate to the cohort of educated class from the susceptible class. However, every protective measure may not be absolutely effective due to the choice of different measures taken by the population with varying coefficients of effectiveness. If we denote the average effectiveness of all existing self-protective measures for the disease by γ, then $1 - \gamma$ will measure the average failure of self-protective actions. Latently infected children may progress to actively infected children class through endogenous reactivation with rate b_c or reinfection with rate k_c. Similarly latently infected adults may progress to actively infected adult class through reactivation with rate b_a or reinfection with a rate k_a. Recruitment for children from infectious class to treatment class is assumed to be δ and treatment rate for adults or rate of recruitment for adults from infectious class to treatment group is assumed to be σ. After the end of effective treatment it is assumed that, treated children will join the latent class of children at the rate of η and treated adults will join latent class of adult at the rate of τ.

Thus, the dynamics of the TB model can be described by the following deterministic system of nonlinear ODE:

$$\begin{aligned}
\dot{V}_c &= r\pi - (1-\varepsilon)\lambda V_c - A_1 V_c \\
\dot{S}_c &= (1-r)\pi + w_c V_c - (A_2+\lambda) S_c \\
\dot{L}_c &= (1-\varepsilon)\lambda V_c - (A_3 + k_c\lambda) L_c + q_c \lambda S_c + \eta T_c \\
\dot{I}_c &= (1-q_c)\lambda S_c + (b_c + k_c\lambda) L_c - (A_7+\delta) I_c \\
\dot{T}_c &= \delta I_c - (A_2+\eta) T_c \\
\dot{S}_a &= f_1 V_c + f_1 S_c - \alpha e S_a - (\mu_a + \lambda) S_a \\
\dot{E} &= \alpha e S_a - \mu_a E - (1-\gamma)\lambda E \\
\dot{L}_a &= f_1 L_c + q_a \lambda S_a + (1-\gamma)\lambda E - (A_5 + k_a\lambda) L_a + \tau T_a \\
\dot{I}_a &= f_1 I_c + (1-q_a)\lambda S_a + (b_a + k_a\lambda) L_a - (A_8+\sigma) I_a \\
\dot{T}_a &= \sigma I_a + f_1 T_c - (\mu_a + \tau) T_a
\end{aligned}, \qquad (1)$$

where

$$\begin{aligned}
&\dot{x}(t) = \tfrac{dx}{dt}, \\
&A_1 = f_1 + w_c + \mu_c, \quad A_2 = f_1 + \mu_c, \quad A_3 = b_c + f_1 + \mu_c, \\
&A_4 = f_1 + \mu_c + d_c + \delta, \quad A_5 = b_a + \mu_a, \quad A_6 = \mu_a + d_a + \sigma, \\
&A_7 = f_1 + \mu_c + d_c, \quad A_8 = \mu_a + d_a, \quad A_9 = f_1 + \mu_c + \eta
\end{aligned}$$

$$\begin{aligned}
N(t) &= V_c(t) + S_c(t) + L_c(t) + I_c(t) + T_c(t) + S_a(t) + E(t) + L_a(t) + I_a(t) + T_a(t) \\
\dot{N}(t) &= \pi - \mu_c N_c - \mu_a N_a - I_c d_c - I_a d_a
\end{aligned} \qquad (2)$$

To prove boundedness, from (2) the rate of total population can be expressed as:

$$\begin{aligned}
\dot{N} &= \dot{V}_c + \dot{S}_c + \dot{L}_c + \dot{I}_c + \dot{T}_c + \dot{S}_a + \dot{E} + \dot{L}_a + \dot{I}_a + \dot{T}_a \\
\dot{N} &= \pi - \mu_c(V_c + S_c + L_c + I_c + T_c) - \mu_a(S_a + E + L_a + I_a + T_a) \\
&\quad - I_c d_c - I_a d_a \\
\dot{N} &= \pi - \mu_c N_c - \mu_a N_a - (I_c d_c + I_a d_a) \quad \Rightarrow \dot{N} \le \pi - \mu_c N_c - \mu_a N_a \\
\dot{N} &\le \pi - \mu(N_c + N_a) \quad \Rightarrow \dot{N} \le \pi - \mu N \quad \text{where} \quad \mu = \min\{\mu_c, \mu_a\}
\end{aligned},$$

Therefore, $\dot{N} \le \pi - \mu N$. (3)

when we solve this first order linear differential equation, we get

$$N(t) \le \tfrac{\pi}{\mu} + e^{-\mu t}\left(N(0) - \tfrac{\pi}{\mu}\right). \quad \text{Since} \quad e^{-\mu t} \le 1, \quad \text{for} \quad t \ge 0.$$

If $N(0) \le \frac{\pi}{\mu}$, then $N(t) \le \frac{\pi}{\mu}$ for $t \ge 0$.

Thus, the total population is bounded above by $\frac{\pi}{\mu}$. By assuming that infectious classes have the same level of infectivity, the force of infection is given by $\lambda = \frac{c\beta}{N}(I_c + I_a)$. Thus the system in (1) is biologically feasible in the region

$$\Omega = \Big\{(V_c, S_c, L_c, I_c, T_c, S_a, E, L_a, I_a, T_a) \in \mathbb{R}_+^{10} : V_c + S_c + L_c + I_c + T_c + S_a + E + L_a + I_a + T_a \leq \frac{\pi}{\mu}\Big\}.$$

3 Mathematical Analysis

3.1 Equilibrium Points

Solving system (1) simultaneously when the time derivatives are equal to zero gives an expression of the equilibrium points.

- $\dot{V}_c = r\pi - (1-\varepsilon)\lambda V_c - A_1 V_c = 0 \quad \Rightarrow V_c^* = \frac{r\pi}{A_1+(1-\varepsilon)\lambda^*}$
- $\dot{S}_c = (1-r)\pi + w_c V_c - (A_2 + \lambda)S_c = 0$

$\Rightarrow S_c^* = \frac{1}{A_2+\lambda^*}[(1-r)\pi + w_c V_c^*] = \frac{(1-r)\pi(A_1+(1-\varepsilon)\lambda^*)+w_c r\pi}{(A_1+(1-\varepsilon)\lambda^*)(A_2+\lambda^*)}$

- $\dot{L}_c = (1-\varepsilon)\lambda V_c + q_c\lambda S_c - (A_3 + k_c\lambda)L_c + \eta T_c = 0$

$\Rightarrow L_c^* = \frac{1}{A_3+k_c\lambda^*}[(1-\varepsilon)\lambda^* V_c^* + q_c\lambda^* S_c^* + \eta T_c^*]$

- $\dot{I}_c = (1-q_c)\lambda S_c + (b_c + k_c\lambda)L_c - (A_7 + \delta)I_c = 0$

$\Rightarrow I_c^* = \frac{1}{(A_7+\delta)}[(1-q_c)\lambda^* S_c^* + (b_c + k_c\lambda^*)L_c^*]$

- $\dot{T}_c = \delta I_c - (A_2 + \eta)T_c = 0 \quad \Rightarrow T_c^* = \frac{\delta}{(A_2+\eta)}I_c^*$
- $\dot{S}_a = f_1 V_c + f_1 S_c - \alpha e S_a - (\mu_a + \lambda)S_a = 0$

$\Rightarrow S_a^* = \frac{1}{\alpha e+\mu_a+\lambda^*}[f_1 V_c^* + f_1 S_c^*]$ Since $e = \frac{\lambda^n}{\lambda_0^n+\lambda^n}$

then $S_a^* = \dfrac{(f_1 V_c^* + f_1 S_c^*)[(\lambda_0)^n + (\lambda^*)^n]}{\alpha(\lambda^*)^n + (\mu_a + \lambda^*)[(\lambda_0)^n + (\lambda^*)^n]}.$

- $\dot{E} = \alpha e S_a - (\mu_a + (1-\gamma)\lambda)E = 0 \quad \Rightarrow E^* = \frac{1}{\mu_a+(1-\gamma)\lambda^*}\alpha e S_a^*$

$\Rightarrow E^* = \dfrac{\alpha(\lambda^*)^n S_a^*}{(\mu_a + (1-\gamma)\lambda^*)[(\lambda_0)^n + (\lambda^*)^n]}$

- $\dot{I}_a = f_1 I_c + (1-\gamma)\lambda E + (b_a + k_a\lambda)L_a + (1-q_a)\lambda S_a - (A_8 + \sigma)I_a = 0$

$\Rightarrow I_a^* = \frac{1}{(A_8+\sigma)}[f_1 I_c^* + (1-\theta)(1-\gamma)\lambda^* E^* + (1-q_a)\lambda^* S_a^* + (b_a + k_a\lambda^*)L_a^*]$

- $\dot{L}_a = f_1 L_c + (1-\gamma)\lambda E + q_a\lambda S_a - (A_5 + k_a\lambda)L_a + \tau T_a = 0$

$\Rightarrow L_a^* = \frac{1}{A_5+k_a\lambda^*}[f_1 L_c^* + \theta(1-\gamma)\lambda^* E^* + q_a\lambda^* S_a^* + \tau T_a^*]$

- $\dot{T}_a = \sigma I_a + f_1 T_c - (\mu_a + \tau)T_a = 0 \quad \Rightarrow T_a^* = \frac{1}{\mu_a+\tau}[\sigma I_a^* + f_1 T_c^*].$

Let the population in each class at the steady state be denoted by $V_c^*, S_c^*, L_c^*, I_C^*, T_c^*, S_a^*, E^*, L_a^*, I_a^*$ and T_a^*. Then the corresponding force of infection is any of the non-negative roots, $\lambda^* = \frac{c\beta}{N^*}(I_c^* + I_a^*)$ of the above system. Since the mathematical model we have considered have ten compartments and the transitions from one compartment to other compartments are nonlinear, it is difficult to get an explicit expression of the components of the endemic equilibrium point. If we set $\lambda^* = 0$ in the system, we will get the disease free equilibrium point $\mathscr{E}_0 = (V_c^D, S_c^D, L_c^D, I_C^D, T_c^D, S_a^D, E^D, L_a^D, I_a^D, T_a^D)$, where $V_c^D = \frac{1}{A_1} r\pi$, $S_C^D = \frac{1}{A_1 A_2}[(1 - r)\pi A_1 + w_c r\pi]$, $L_c^D = 0$, $I_c^D = 0$, $T_c^D = 0$, $E^D = 0$, $I_a^D = 0$, $L_a^D = 0$, $T_a^D = 0$, $S_a^D = \frac{\pi f_1}{A_1 A_2 \mu_a}[r(A_2 + w_c) + A_1(1 - r)]$.

3.1.1 Reproduction Number

Definition 1 The basic reproduction number, basic reproduction ratio or basic reproductive rate is defined as the average number of secondary infections that occur when one infective is introduced into a completely susceptible host population [14].

We calculate the basic reproduction ratio (number), $\mathscr{R}_0$, using the van den Driessche and Watmough next generation matrix approach from [16] to get

$$\mathscr{R}_0 = \mathscr{R}_{V_c} + \mathscr{R}_{S_c} + \mathscr{R}_{S_a}, \tag{4}$$

where

$$\begin{aligned}
\mathscr{R}_{V_c} &= [\frac{A_5 b_c(f_1 + (A_8 + \sigma)) + f_1 b_a(A_7 + \delta)}{A_3 A_5 (A_7 + \delta)(A_8 + \sigma)}]\frac{(1 - \varepsilon)c\beta}{\mathscr{N}_0} V_C^D, \\
\mathscr{R}_{S_c} &= \{\frac{\begin{array}{c}[b_c A_5(f_1 + (A_8 + \sigma)) + f_1 b_a(A_7 + \delta)]q_c \\ +(A_3 A_5(f_1 + A_8 + \sigma))(1 - q_c)\end{array}}{A_3 A_5 (A_7 + \delta)(A_8 + \sigma)}\}\frac{c\beta}{\mathscr{N}_0} S_c^D \\
\mathscr{R}_{S_a} &= (\frac{b_a q_a + A_5(1 - q_a)}{A_5(A_8 + \sigma)})\frac{c\beta}{\mathscr{N}_0} S_a^D,
\end{aligned} \tag{5}$$

where $\mathscr{N}_0 = V_c^D + S_c^D + L_c^D + I_c^D + T_c^D + S_a^D + E^D + I_a^D + L_a^D + T_a^D$, and $\mathscr{R}_{V_c}, \mathscr{R}_{S_c}, \mathscr{R}_{S_a}$ are the contributions from vaccinated children, susceptible children, and susceptible adults respectively.

3.2 *Stability of Disease Free Equilibrium*

To analyze local stability of disease free equilibrium, let us consider the following points from [16] by writing our system of Eq. (1) as $\dot{x}_i = f_i(x) = \mathscr{F}_i(x) - \mathscr{V}_i(x), i = 1, 2, 3, \ldots, 10$, where $x = (x_1, x_2, x_3, \ldots, x_n)$ with each $x_i \geq 0$, representing the number of individuals in each compartment. For clarity we sort the compartments so that the first m compartments correspond to infected individuals. $\mathcal{V}_i = \mathcal{V}_i^- - \mathcal{V}_i^+$ with the assumption that $\mathcal{F}_i(x)$ is the rate of appearance of new infections in compartment i, $\mathcal{V}_i^+(x)$ is the rate of transfer of individuals into compartment i by all other means, and $\mathcal{V}_i^-(x)$ is the rate of transfer of individuals out of compartment i.

A1. Since each function $\mathcal{F}_i, \mathcal{V}_i^-$, and $\mathcal{V}_i^+$ represents the transfer of individuals, they are all non-negative. This implies that if $x \geq 0$, then $\mathcal{F}_i, \mathcal{V}_i^-, \mathcal{V}_i^+ \geq 0$ for $i = 1, 2, 3, \ldots, 10$.
A2. If a compartment is empty, then there can be no transfer of individuals out of the compartment by death, infection, or any other means. Thus, if $x_i = 0$ then $\mathcal{V}_i^- = 0$.
A3. The incidence of infection for uninfected compartments is zero. Thus, $\mathcal{F}_i = 0$ for $i > m = 4$.
A4. The disease free subspace is invariant. If $x \in X_s$ then $\mathcal{F}_i = 0$ and $\mathcal{V}_i^+ = 0$ for $i = 1, 2, 3, 4$, where X_s is the set of all disease free states *i.e.*, $X_s = \{x \geq 0 | x_i = 0, i = 1, 2, \ldots, m\}$, for our system $m = 4$ since we have four infectious classes.
A5. Disease free equilibrium (DFE) is stable in the absence of new infections. That is, if $\mathcal{F}_i(x)$ is set to be zero, then all eigenvalues of $Df(\mathcal{E}_0)$ have negative real parts. From the calculation for $\mathcal{F}_i(x) = 0$, the eigenvalues of $Df(\mathcal{E}_0)$ are $-A_1, -A_2, -A_3, -A_5, -(A_7 + \delta), -(A_8 + \sigma), -(A_2 + \eta), -\mu_a, -(\mu_a + \tau)$ and $-(\alpha e + \mu_a)$ such that all of them are negative.

Therefore, we have the following theorem,

Theorem 1 *Suppose the disease transmission model is given by (1) where $f(x)$ [right hand side of (1)] satisfy conditions (A1)–(A5). If $\mathcal{E}_0$ is a DFE of the model, then $\mathcal{E}_0$ is locally asymptotically stable if $\mathcal{R}_0 < 1$ and unstable otherwise, where $\mathcal{R}_0$ is the reproduction number defined in (4).*

Proof This is an immediate consequence of Theorem 2 in [16].

4 Formulation of the Control

The possible interventions for TB disease can be categorized as prevention with vaccination, preventive education and treatment to the infected individuals. In this paper we consider these interventions as control parameters.

(a) Vaccination: Increase the rate of vaccinating children at birth. Let the current rate of vaccinating children at birth be r_0 per unit of time for some $r_0 > 0$ to protect children from infection and let also assume that the control function $u_1(t)$ measures the additional rate of recruitment of children for vaccination per unit of time. The cost of vaccinating children becomes expensive as the proportion of non-vaccinated children gets smaller. So, we can add a term $(\frac{V_c}{Nc})^m$ as a coefficient for $u_1^2(t)$ where V_c represents vaccinated children, N_c total population of children and $m > 1$ is any positive constant integer. Numerical investigations suggested to take $m = 10$ for the best fit [8]. Therefore, we took $m = 10$. Then its application in the dynamics is modeled by simply replacing the term r in (1) by $(r_0 + u_1(t))$. Due to limitation of resources $u_1(t)$ is restricted to its maximum vaccination rate $r_{max} > 0$ or r_{max} is the maximum attainable value of u_1 at time t, where $0 \leq r_0 + u_1(t) \leq 1$.

(b) Preventive Education: Preventive mechanisms, here, are self-protective actions an individual may apply due to the information he/she got from different sources. By applying self-initiated protective measures an individual can reduce the risk of contracting the disease. Let the current level of preventive education campaigns by various agents have convinced up to $100 \times (\alpha_0 \times e)\%$ (for some $\alpha_0 > 0$) of population per unit of time to effectively participate in the self protective schemes available to them. If more options of self-protective measures are offered to the population, more individuals may decide to choose and use at least one of the mechanisms. This is an effort made to keep susceptible individuals from getting infection. On the other hand, we can also educate infectious individuals who didn't take part in any of the self-protective actions about the disease to take the medicine properly until the end of the specified time given by the health workers. (This will have an indirect gain to the susceptible once; so we did not include this effect in the model.) Assume that the control function $u_2(t)$ measures the rate at which additional susceptible individuals are convinced to take part in behavior modification. Then its application in the dynamics is modeled by simply replacing the term α in Eq. (1) by $(\alpha_0 + u_2(t))$. However, the cost of the effort in convincing the population for behavior modification becomes expensive as the proportion of the non-convinced susceptible individuals gets smaller [11]. So we can again include the term $(\frac{E}{N_a})^m$ as part of the coefficient for $u_2^2(t)$ in the objective function (8), where N_a represents the total number of population of adults, and E number of adult population in educated class. Because of practicality and economic limitations on the maximum rate behavior modification, we assume that $\alpha_{max} > 0$ is the maximum rate and $0 \leq \alpha_0 + u_1(t) \leq 1$.

(c) Treatment of infected children and adults: Infectious children and adults can be effectively treated within an average treatment period of 6 months [2], provided they take the treatment properly. Assume that the control function $u_3(t)$ measures the rate at which additional infectious children are recruited to treated class at any time t and $u_4(t)$ measures the rate at which additional infectious adults are recruited to treated class at any time t. If the current percentage of treatment per unit of time for children is δ_0 and σ_0 for adults, this control will be seen in the dynamics as $(\delta_0 + u_3(t))I_c(t)$ by replacing $\delta I_c(t)$ and $(\sigma_0 + u_4(t))I_a(t)$ by replacing $\sigma I_a(t)$ in (1). Due to economical and logistic reasons, there are limitations on the maximum rate at which individuals are recruited to get treatment at each time period. Thus, the constant δ_{max} and σ_{max} represent the maximum rate of recruitment for treatment of infected children and adults respectively as well as $0 \leq \delta_0 + u_1(t) \leq 1$ and $0 \leq \sigma_0 + u_1(t) \leq 1$.

When we include the above controls into our model, we will get the following system of equations.

$$
\begin{aligned}
\dot{V}_c &= (r_0 + u_1(t))\pi - (1-\varepsilon)\lambda V_c - A_1 V_c \\
\dot{S}_c &= (1 - r_0 - u_1(t))\pi + w_c V_c - (A_2 + \lambda)S_c \\
\dot{L}_c &= (1-\varepsilon)\lambda V_c - (A_3 + k_c\lambda)L_c + q_c\lambda S_c + \eta T_c \\
\dot{I}_c &= (1-q_c)\lambda S_c + (b_c + k_c\lambda)L_c - A_7 I_c - (\delta_0 + u_3(t))I_c \\
\dot{T}_c &= (\delta_0 + u_3(t))I_c - A_9 T_c \\
\dot{S}_a &= f_1 V_c + f_1 S_c - (\alpha_0 + u_2(t))eS_a - (\mu_a + \lambda)S_a \\
\dot{E} &= (\alpha_0 + u_2(t))eS_a - \mu_a E - (1-\gamma)\lambda E \\
\dot{L}_a &= f_1 L_c + q_a\lambda S_a + (1-\gamma)\lambda E - (A_5 + k_a\lambda)L_a + \tau T_a \\
\dot{I}_a &= f_1 I_c + (1-q_a)\lambda S_a + (b_a + k_a\lambda)L_a - A_8 I_a - (\sigma_0 + u_4(t))I_a \\
\dot{T}_a &= (\sigma_0 + u_4(t))I_a + f_1 T_c - (\mu_a + \tau)T_a,
\end{aligned}
\tag{6}
$$

where $\lambda = \frac{c\beta}{N}(I_c + I_a), -r_0 \leq u_1(t) \leq 1 - r_0, -\alpha_0 \leq u_2(t) \leq 1 - \alpha_0, -\delta_0 \leq u_3(t) \leq 1 - \delta_0, -\sigma_0 \leq u_4(t) \leq 1 - \sigma_0$ for all $t \in [0, t_f]$.

4.1 Reproduction Number with Controls

By using the same technique as in Sect. 3.1.1, it is possible to calculate the basic reproduction number from Eq. (6) in the presence of controls using the next generation matrix. Hence,

$$\mathscr{R}_0(u) = \mathscr{R}_{vc}(u) + \mathscr{R}_{sc}(u) + \mathscr{R}_{sa}(u), \tag{7}$$

where

$$
\begin{aligned}
\mathscr{R}_{V_c}(u) &= [\frac{b_a f_1(\delta_0 + u_3 + A_7) + b_c A_5(f_1 + A_8 - (\sigma_0 + u_4))}{A_3A_5A_8(A_7 + \delta_0 + u_3)}]\frac{(1-\varepsilon)c\beta}{\mathscr{N}_0}V_c^D, \\
\mathscr{R}_{S_c}(u) &= \{\frac{[b_a f_1(A_7 + \delta_0 + u_3) + b_c A_5(f_1 + A_8 + (\sigma_0 + u_4))]q_c}{A_3A_5A_8(A_7 + \delta_0 + u_3)} \\
&\quad + \frac{A_3A_5(f_1 + A_8 - (\sigma_0 + u_4))(1-q_c)}{A_3A_5A_8(A_7 + \delta_0 + u_3)}\}\frac{c\beta}{\mathscr{N}_0}S_c^D, \\
\mathscr{R}_{S_a}(u) &= (\frac{b_a q_a + A_5(1-q_a)}{A_5A_8})\frac{c\beta}{\mathscr{N}_0}S_a^D,
\end{aligned}
$$

which shows that vaccination and treatment have visible impact on the value of $\mathscr{R}_o$.

With (6) and given initial population size of each compartment, *our main goal is to find or propose the best strategy in terms of either in combination or independent efforts of vaccination, education and treatment that will minimize the costs of interventions*. If we know the initial value population size and the control trajectory, i.e., the values of $\mathbf{u}(t)$ over the whole time interval $0 < t < T$, then we can integrate (6) to get the state trajectory over the same time interval. We want to choose the control trajectory so that the state and control trajectories minimize the objective function:

$$J(u_1, u_2, u_3, u_4) = \int_0^{t_f} [C_1 I_c(t) + C_2 I_a(t) + \frac{B_1}{2}(\frac{V_c}{N_c})^m u_1^2(t) + \frac{B_2}{2}(\frac{E}{N_a})^m u_2^2(t) + \frac{B_3}{2} u_3^2(t) + \frac{B_4}{2} u_4^2(t)]\, dt \tag{8}$$

where the constants C_1, C_2 and $B_i, i = 1, 2, 3, 4$ can be considered as values that indicate the importance of one type of intervention over the other. $C_1 I_c$ and $C_2 I_a$ represent the number of infectious children and adults respectively, whereas the terms $\frac{B_1}{2}(\frac{V_c}{N_c})^m u_1^2$, $\frac{B_2}{2}(\frac{E}{N_a})^m u_2^2$, $\frac{B_3}{2} u_3^2$, and $\frac{B_4}{2} u_4^2$ represent the costs of vaccine, education and treatment for children and adults respectively. Since the implementation of any public health intervention has increasing costs when a higher fraction of the population is reached, we take a non-linear cost function like the quadratic. So, we seek to find optimal controls $u_1^*, u_2^*, u_3^*, u_4^*$ such that

$$J(u_1^*, u_2^*, u_3^*, u_4^*) = \min_U J(u_1, u_2, u_3, u_4), \tag{9}$$

where $U = \{(u_1(t), u_2(t), u_3(t), u_3(t), u_4(t)) | u_1(t), u_2(t), u_3(t), u_4(t)\}$ is the set of Lebesgue integrable functions, with $u_1(t) \in [-r_0, 1 - r_0]$, $u_2(t) \in [-\alpha_0, 1 - \alpha_0]$, $u_3(t) \in [-\delta_0, 1 - \delta_0]$, $u_4(t) \in [-\sigma_0, 1 - \sigma_0]$

4.1.1 Existence and Characterization of Optimal Control Solution

The first task will be to examine conditions that can assure the existence of a solution to our optimal control problem.

Theorem 2 (Existence of Optimal Control Solution) *There exists an optimal control $u_1^*(t)$, $u_2^*(t)$, $u_3^*(t)$, $u_4^*(t)$ and corresponding solutions V_c^*, S_c^*, L_c^*, I_c^*, T_c^*, S_a^*, E^*, L_a^*, I_a^*, T_a^* to the state initial value problem (6) and (9) that minimizes $J(u_1, u_2, u_3, u_4)$ over U.*

Proof The non trivial requirements on the set of admissible controls U and on the set of endpoint conditions are verified from Fleming and Rishel's Theorem [7].

A. The set of all solutions to system (6) with corresponding control functions in U is non-empty.
B. The state system can be written as a linear function of the control variables with coefficients dependent on time and the state variables.
C. The integrand $\mathscr{L}$ in (8) from the objective function with $\mathscr{L} = C_1 I_c(t) + C_2 I_a(t) + \frac{B_1}{2}(\frac{V_c}{N_c})^m u_1^2(t) + \frac{B_2}{2}(\frac{E}{N_a})^m u_2^2(t) + \frac{B_3}{2} u_3^2(t) + \frac{B_4}{2} u_4^2(t)$ is convex on U and

additionally it satisfies $\mathscr{L}(\mathbf{x},\mathbf{u},t) \geq \delta_1 \mid (u_1,u_2,u_3,u_4) \mid^\beta -\delta_2$, where $\delta_1 > 0$ and $\beta > 1$.

In order to establish condition A, we refer to Picard-Lindelöf's theorem from [5, 9]. If the solutions to the state equations are bounded and if the state equations are Lipschitz in the state variables, then there is a unique solution corresponding to every admissible control $\mathbf{u}$. From (3) it is indicated that the total population is bounded from below by a positive spermium N_0 and bounded above by $\frac{\pi}{\mu}$ as well as each of the state variables are bounded. With the bounds established above, it follows that the state system is continuous and bounded. It is equally direct to show the boundedness of the partial derivatives with respect to the state variables in the state system, which establishes that the system is Lipschitz with respect to the state variables (see [4]). This completes the proof that condition A holds.

Condition B is verified by observing the linear dependence of the state equations on controls u_1, u_2, u_3 and u_4. Finally, to verify condition C, since linear combination of convex functions are also convex the integrand $\mathscr{L}(\mathbf{x},\mathbf{u},t)$ is convex on U. To prove the boundedness of $\mathscr{L}$ we note that by the definition of U, we have

$$B_4u_4^2 \leq B_4. \quad \text{Since } u_4 \in [0,1], \quad \frac{B_4}{2}u_4^2 \leq \frac{B_4}{2} \quad \Rightarrow \frac{B_4}{2}u_4^2 - \frac{B_4}{2} \leq 0.$$

$$\begin{aligned}\mathscr{L}(\mathbf{x},\mathbf{u},t) &= C_1I_c(t) + C_2I_a(t) + \frac{B_1}{2}u_1^2(t)(\frac{V_c}{N_c})^m + \frac{B_2}{2}(\frac{E}{N_a})^m u_2^2(t) + \frac{B_3}{2}u_3^2(t)\\ &\quad + \frac{B_4}{2}u_4^2\\ &\geq \frac{B_1}{2}u_1^2(t)(\frac{V_c}{N_c})^m + \frac{B_2}{2}(\frac{E}{N_a})^m u_2^2(t) + \frac{B_3}{2}u_3^2(t) + \frac{B_4}{2}u_4^2(t) - \frac{B_4}{2}\end{aligned}$$

$$\Rightarrow \quad \mathscr{L}(\mathbf{x},\mathbf{u},t) \geq \min\big(\frac{B_1}{2}(\frac{V_c}{N_c})^m, \frac{B_2}{2}(\frac{E}{N_a})^m, \frac{B_3}{2}, \frac{B_4}{2}\big)(u_1^2+u_2^2+u_3^2+u_4^2) - \frac{B_4}{2}.$$

$$\Rightarrow \quad \mathscr{L}(\mathbf{x},\mathbf{u},t) \geq \min\big(\frac{B_1}{2}(\frac{V_c}{N_c})^m, \frac{B_2}{2}(\frac{E}{N_a})^m, \frac{B_3}{2}, \frac{B_4}{2}\big)|(u_1,u_2,u_3,u_4)|^2 - \frac{B_4}{2}.$$

Therefore $\mathscr{L}(\mathbf{x},\mathbf{u},t) \geq \delta_1|(u_1,u_2,u_3,u_4)|^\beta - \delta_2$

where $\delta_1 = \min\big(\frac{B_1}{2}(\frac{V_c}{N_c})^m, \frac{B_2}{2}(\frac{E}{N_a})^m, \frac{B_3}{2}, \frac{B_4}{2}\big), \quad \delta_2 = \frac{B_4}{2}$ and $\beta = 2$.□

The necessary conditions arise from Pontryagin's minimum principle (PMP). To apply this principle we convert (6)–(9) into a problem of minimizing a hamiltonian, H with respect to u_1, u_2, u_3, u_4. Then the hamiltonian is given by

$$\begin{aligned}H(\mathbf{x},\mathbf{u},h,t) &= C_1I_c(t) + C_2I_a(t) + \tfrac{B_1}{2}u_1^2(t)(\tfrac{V_c}{N_c})^m + \tfrac{B_2}{2}(\tfrac{E}{N_a})^m u_2^2(t)\tfrac{B_3}{2}u_3^2(t)\\ &\quad + \tfrac{B_4}{2}u_4^2(t) + \textstyle\sum_{i=1}^{10} h_i f_i\end{aligned}$$

$$\begin{aligned}
\Rightarrow \quad H(\mathbf{x},\mathbf{u},h,t) = {} & [C_1 I_c(t) + C_2 I_a(t) + \tfrac{B_1}{2}u_1^2(t)(\tfrac{V_c}{N_c})^m + \tfrac{B_2}{2}(\tfrac{E}{N_a})^m u_2^2(t) \\
& + \tfrac{B_3}{2}u_3^2(t) + \tfrac{B_4}{2}u_4^2(t)] \\
& + h_1[(r_0 + u_1(t))\pi - (1-\varepsilon)\lambda V_c - A_1 V_c] \\
& + h_2[(1 - r_0 - u_1(t))\pi + w_c V_c - (A_2 + \lambda)S_c] \\
& + h_3[(1-\varepsilon)\lambda V_c - (A_3 + k_c\lambda)L_c + q_c\lambda S_c + \eta T_c] \\
& + h_4[(1-q_c)\lambda S_c + (b_c + k_c\lambda)L_c - A_7 I_c - (\delta_0 + u_3(t))I_c] \\
& + h_5[(\delta_0 + u_3(t))I_c - A_9 T_c] \\
& + h_6[f_1 V_c + f_1 S_c - (\alpha_0 + u_2(t))eS_a - (\mu_a + \lambda)S_a] \\
& + h_7[(\alpha_0 + u_2(t))eS_a - \mu_a E - (1-\gamma)\lambda)E] \\
& + h_8[f_1 L_c + q_a\lambda S_a + (1 - \alpha_0 - u_2(t))E - (A_5 + k_a\lambda)L_a + \tau T_a] \\
& + h_9[f_1 I_c + (1-q_a)\lambda S_a + (b_a + k_a\lambda)L_a - A_8 I_a - (\sigma_0 + u_4(t))I_a] \\
& + h_{10}[(\sigma_0 + u_4(t))I_a + f_1 T_c - (\mu_a + \tau)T_a]
\end{aligned}, \tag{10}$$

where each f_i is the right hand side of the differential equation of the ith state variable of (6), $\mathbf{x} = (V_c, S_c, L_c, I_c, T_c, S_a, E, L_a, I_a, T_a)$, $\mathbf{u} = (u_1, u_2, u_3, u_4)$, $\mathbf{h} = (h_1, h_2, h_3, h_4, h_5, h_6, h_7, h_8, h_9, h_{10})$. If $(u_1^*, u_2^*, u_3^*, u_4^*)$ is an optimal control yet to be determined, then from Pontryagin's Minimum Principle we have:

(a) The minimum conditions in the interior of the control region:

$$\begin{aligned}
& \tfrac{\partial H}{\partial u_i} = 0, i = 1,2,3,4 \\
& \bullet\ \tfrac{\partial H}{\partial u_1} = 0 \Rightarrow B_1(\tfrac{V_c}{N_c})^m u_1(t) + h_1\pi - h_2\pi = 0 \\
& \text{Therefore, } u_1(t) = \tfrac{\pi}{B_1}(\tfrac{V_c}{N_c})^{-m}(h_2 - h_1) \\
& \bullet\ \tfrac{\partial H}{\partial u_2} = 0 \Rightarrow B_2(\tfrac{E}{N})^m u_2(t) - h_6 eS_a + h_7 eS_a - h_8 E = 0 \\
& \text{Therefore, } u_2(t) = \tfrac{1}{B_2}(\tfrac{E}{N_a})^{-m}[h_8 E + (h_6 - h_7)eS_a] \\
& \bullet\ \tfrac{\partial H}{\partial u_3} = 0 \Rightarrow B_3 u_3(t) - h_4 I_c + h_5 I_c = 0 \\
& \text{Therefore, } u_3(t) = \tfrac{1}{B_3}(h_4 - h_5)I_c \\
& \bullet\ \tfrac{\partial H}{\partial u_4} = 0 \Rightarrow B_4 u_4(t) - h_9 I_a + h_{10} I_a = 0 \\
& \text{Therefore, } u_4(t) = \tfrac{1}{B_4}(h_9 - h_{10})I_a
\end{aligned} \tag{11}$$

(b) The transversality conditions: $h_i(t_f) = 0, i = 1,2,3,\ldots,10$. Moreover, from the conditions that $u_1(t) \in [-r_0, 1-r_0]$, $u_2(t) \in [-\alpha_0, 1-\alpha_0]$, $u_3(t) \in [-\delta_0, 1-\delta_0]$, $u_4(t) \in [-\sigma_0, 1-\sigma_0]$ for all $t \in [0, t_f]$ we arrive at:

$$\begin{aligned}
u_1^* &= \min\left\{1 - r_0, \max\left\{-r_0, \tfrac{\pi}{B_1}(\tfrac{V_c}{N_c})^{-m}(h_2 - h_1)\right\}\right\} \\
u_2^* &= \min\left\{1 - \alpha_0, \max\left\{-\alpha_0, \tfrac{1}{B_2}(\tfrac{E}{N})^{-m}[h_8 E + (h_6 - h_7)eS_a]\right\}\right\} \\
u_3^* &= \min\left\{1 - \delta_0, \max\left\{-\delta_0, \tfrac{1}{B_3}(h_4 - h_5)I_c\right\}\right\} \\
u_4^* &= \min\left\{1 - \sigma_0, \max\left\{-\sigma_0, \tfrac{1}{B_4}(h_9 - h_{10})I_a\right\}\right\}
\end{aligned} \tag{12}$$

(c) The adjoint equation:

- $$\begin{aligned}\dot{h}_1(t) &= -\frac{\partial H}{\partial V_c}\\ &= -\frac{B_1}{2}\frac{m}{N_c^2}(N_c - V_c)(\frac{V_c}{N_c})^{m-1}u_1^2 + \frac{B_2}{2}\frac{m}{E}(\frac{E}{N_a})^{m+1}u_2^2 + A_1h_1 - h_2w_c - f_1h_6\\ &+c\beta(1-\varepsilon)\frac{(I_c+I_a)}{N^2}(N - V_c)(h_1 - h_3) + c\beta k_c\frac{(I_c+I_a)}{N^2}(h_4 - h_3)L_c\\ &+c\beta\frac{(I_c+I_a)}{N^2}(q_ch_3 + (1-q_c)h_4 - h_2)S_c + c\beta\frac{(I_c+I_a)}{N^2}(q_ah_8 + (1-q_a)h_9 - h_6)S_a\\ &+c\beta k_a\frac{(I_c+I_a)}{N^2}(h_9 - h_8)L_a + np_*^nN^{n-1}(\alpha_0 + u_2)\frac{(I_c+I_a)^n}{(p_*^nN^n+(I_c+I_a)^n)^2}(h_7 - h_6)S_a\\ &-c\beta(1-\gamma)\frac{(I_c+I_a)}{N^2}Eh_7\end{aligned}$$

- $$\begin{aligned}\dot{h}_2(t) &= -\frac{\partial H}{\partial S_c}\\ &= \frac{B_1}{2}\frac{m}{V_c}(\frac{V_c}{N_c})^{m+1}u_1^2 + \frac{B_2}{2}\frac{m}{E}(\frac{E}{N_a})^{m+1}u_2^2 + c\beta(1-\varepsilon)\frac{(I_c+I_a)}{N^2}V_c(h_3 - h_1) + A_2h_2\\ &+c\beta\frac{(I_c+I_a)}{N^2}(N - S_c)(h_2 + q_ch_3 - (1-q_c)h_4) + c\beta k_c\frac{(I_c+I_a)}{N^2}(h_4 - h_3)L_c\\ &-c\beta(1-\gamma)\frac{(I_c+I_a)}{N^2}Eh_7 - np_*^nN^{n-1}(\alpha_0 + u_2)\frac{(I_c+I_a)^n}{(p_*^nN^n+(I_c+I_a)^n)^2}(h_7 - h_6)S_a\\ &+c\beta\frac{(I_c+I_a)}{N^2}(q_ah_8 + (1-q_a)h_9 - h_6)S_a + c\beta k_a\frac{(I_c+I_a)}{N^2}(h_9 - h_8)L_a - f_1h_6\end{aligned}$$

- $$\begin{aligned}\dot{h}_3(t) &= -\frac{\partial H}{\partial L_c}\\ &= \frac{B_1}{2}\frac{m}{V_c}(\frac{V_c}{N_c})^{m+1}u_1^2 + \frac{B_2}{2}\frac{m}{E}(\frac{E}{N_a})^{m+1}u_2^2 + c\beta(1-\varepsilon)\frac{(I_c+I_a)}{N^2}V_c(h_3 - h_1)\\ &+c\beta k_c\frac{(I_c+I_a)}{N^2}(h_4 - h_3)(N - L_c) + c\beta\frac{(I_c+I_a)}{N^2}(q_ch_3 + (1-q_c)h_4 - h_2)S_c\\ &+c\beta\frac{(I_c+I_a)}{N^2}((1-q_a)h_9 + q_ah_8 - h_6)S_a + c\beta k_a\frac{(I_c+I_a)}{N^2}(h_9 - h_8)L_a\\ &-c\beta(1-\gamma)\frac{(I_c+I_a)}{N^2}h_7E - f_1h_8 - h_4b_c + A_3h_3\\ &+np_*^nN^{n-1}(\alpha_0 + u_2)\frac{(I_c+I_a)^n}{(p_*^nN^n+(I_c+I_a)^n)^2}(h_7 - h_6)S_a\end{aligned}$$

- $$\begin{aligned}\dot{h}_4(t) &= -\frac{\partial H}{\partial I_c}\\ &= \frac{B_1}{2}\frac{m}{V_c}(\frac{V_c}{N_c})^{m+1}u_1^2 + \frac{B_2}{2}\frac{m}{E}(\frac{E}{N_a})^{m+1}u_2^2 + c\beta(1-\varepsilon)\frac{(N-(I_c+I_a))}{N^2}V_c(h_1 - h_3) - C_1\\ &+c\beta\frac{(N-(I_c+I_a))}{N^2}(h_2 - q_ch_3 - (1-q_c)h_4)S_c + c\beta k_c\frac{(N-(I_c+I_a))}{N^2}(h_3 - h_4)L_c\\ &-c\beta\frac{(N-(I_c+I_a))}{N^2}(q_ah_8 + (1-q_a)h_9 - h_6)S_a - c\beta k_a\frac{(N-(I_c+I_a))}{N^2}(h_8 - h_9)L_a\\ &+(\delta_0 + u_3)(h_4 - h_5) + c\beta(1-\gamma)\frac{(N-(I_c+I_a))}{N^2}h_7E + A_7h_4 - f_1h_9\\ &-\frac{np_*^nN^{n-1}(\alpha_0+u_2)(I_c+I_a)^{n-1}(N-(I_c+I_a))}{(p_*^nN^n+(I_c+I_a)^n)^2}(h_7 - h_6)S_a\end{aligned}$$

- $$\begin{aligned}\dot{h}_5(t) &= -\frac{\partial H}{\partial T_c}\\ &= \frac{B_1}{2}\frac{m}{V_c}(\frac{V_c}{N_c})^{m+1}u_1^2 + \frac{B_2}{2}\frac{m}{E}(\frac{E}{N_a})^{m+1}u_2^2 + c\beta(1-\varepsilon)\frac{(I_c+I_a)}{N^2}V_c(h_3 - h_1) + A_9h_5\\ &+c\beta k_c\frac{(I_c+I_a)}{N^2}(h_4 - h_3)L_c + c\beta\frac{(I_c+I_a)}{N^2}(q_ch_3 + (1-q_c)h_4 - h_2)S_c - f_1h_{10}\\ &+c\beta\frac{(I_c+I_a)}{N^2}(q_ah_8 + (1-q_a)h_9 - h_6)S_a + c\beta k_a\frac{(I_c+I_a)}{N^2}(h_9 - h_8)L_a - \eta h_3\\ &-c\beta(1-\gamma)\frac{(I_c+I_a)}{N^2}h_7E + np_*^nN^{n-1}(\alpha_0 + u_2)\frac{(I_c+I_a)^n}{(p_*^nN^n+(I_c+I_a)^n)^2}(h_7 - h_6)S_a\end{aligned}$$

- $$\begin{aligned}\dot{h}_6(t) &= -\frac{\partial H}{\partial Sa}\\ &= \frac{B_2}{2}\frac{m}{E}(\frac{E}{N_a})^{m+1}u_2^2 + c\beta(1-\varepsilon)\frac{(I_c+I_a)}{N^2}V_c(h_3 - h_1) + \mu_ah_6\\ &+c\beta k_c\frac{(I_c+I_a)}{N^2}(h_4 - h_3)L_c + c\beta\frac{(I_c+I_a)}{N^2}(q_ch_3 + (1-q_c)h_4 - h_2)S_c - f_1h_{10}\\ &+c\beta\frac{(I_c+I_a)}{N^2}(h_6 - q_ah_8 - (1-q_a)h_9)(N - S_a) + c\beta k_a\frac{(I_c+I_a)}{N^2}(h_9 - h_8)L_a\\ &-\frac{(\alpha_0+u_2)(I_c+I_a)^n(p_*^nN^n+(I_c+I_a)^n-np_*^nN^{n-1})}{(p_*^nN^n+(I_c+I_a)^n)^2}(h_7 - h_6)S_a\\ &-c\beta(1-\gamma)\frac{(I_c+I_a)}{N^2}h_7E\end{aligned}$$

- $$\begin{aligned}\dot{h}_7(t) &= -\frac{\partial H}{\partial E}\\ &= \frac{B_2}{2}\frac{m}{E}\left(\frac{E}{N_a}\right)^{m+1}u_2^2 + \mu_a h_7 - (1-\alpha_0-u_2)h_8\\ &+c\beta(1-\varepsilon)\frac{(I_c+I_a)}{N^2}V_c(h_3-h_1) + c\beta\frac{(I_c+I_a)}{N^2}((1-q_c)h_4+q_c h_3-h_2)S_c\\ &+c\beta k_c\frac{(I_c+I_a)}{N^2}(h_4-h_3)L_c + c\beta\frac{(I_c+I_a)}{N^2}(q_a h_8+(1-q_a)h_9-h_6)S_a\\ &+c\beta k_a\frac{(I_c+I_a)}{N^2}(h_9-h_8)L_a + np_*^n N^{n-1}(\alpha_0+u_2)\frac{(I_c+I_a)^n}{(p_*^n N^n+(I_c+I_a)^n)^2}(h_7-h_6)S_a\\ &+c\beta(1-\gamma)\frac{(I_c+I_a)}{N^2}(N-E)h_7\end{aligned}$$

- $$\begin{aligned}\dot{h}_8(t) &= -\frac{\partial H}{\partial L_a}\\ &= \frac{B_2}{2}\frac{m}{E}\left(\frac{E}{N_a}\right)^{m+1}u_2^2 + c\beta(1-\varepsilon)\frac{(I_c+I_a)}{N^2}V_c(h_3-h_1) + A_5 h_8\\ &+c\beta\frac{(I_c+I_a)}{N^2}S_c(q_c h_3+(1-q_c)h_4-h_2) + c\beta k_c\frac{(I_c+I_a)}{N^2}(h_4-h_3)L_c - b_a h_9\\ &-c\beta(1-\gamma)\frac{(I_c+I_a)}{N^2}Eh_7 + np_*^n N^{n-1}(\alpha_0+u_2)\frac{(I_c+I_a)^n}{(p_*^n N^n+(I_c+I_a)^n)^2}(h_7-h_6)S_a\\ &+c\beta\frac{(I_c+I_a)}{N^2}(q_a h_8+(1-q_a)h_9-h_6)S_a + c\beta k_a\frac{(I_c+I_a)}{N^2}(h_9-h_8)(N-L_a)\end{aligned}$$
- $$\begin{aligned}\dot{h}_9(t) &= -\frac{\partial H}{\partial I_a}\\ &= \frac{B_2}{2}\frac{m}{E}\left(\frac{E}{N_a}\right)^{m+1}u_2^2 + c\beta(1-\varepsilon)\frac{(N-(I_c+I_a))}{N^2}V_c(h_1-h_3)\\ &+c\beta\frac{(N-(I_c+I_a))}{N^2}(h_2-q_c h_3-(1-q_c)h_4)S_c - c\beta k_c\frac{(N-(I_c+I_a))}{N^2}(h_4-h_3)L_c\\ &+c\beta\frac{(N-(I_c+I_a))}{N^2}(h_6-q_a h_8-(1-q_a)h_9)S_a + c\beta k_a\frac{(N-(I_c+I_a))}{N^2}(h_8-h_9)L_a\\ &+(\sigma_0+u_4)(h_9-h_{10}) - c\beta(1-\gamma)\frac{(N-(I_c+I_a))}{N^2}h_7 E - C_2 + A_8 h_9\\ &-\frac{np_*^n N^{n-1}(\alpha_0+u_2)(I_c+I_a)^{n-1}(N-(I_c+I_a))}{(p_*^n N^n+(I_c+I_a)^n)^2}(h_7-h_6)S_a\end{aligned}$$

- $$\begin{aligned}\dot{h}_{10}(t) &= -\frac{\partial H}{\partial T_a}\\ &= -\frac{B_2}{2}\frac{m}{E}\left(\frac{E}{N_a}\right)^{m+1}u_2^2 + c\beta(1-\varepsilon)\frac{(I_c+I_a)}{N^2}V_c(h_3-h_1) - \tau h_8\\ &+c\beta\frac{(I_c+I_a)}{N^2}(q_c h_3+(1-q_c)h_4-h_2)S_c + c\beta k_c\frac{(I_c+I_a)}{N^2}(h_4-h_3)L_c + \mu_a h_{10}\\ &-c\beta(1-\gamma)\frac{(I_c+I_a)}{N^2}Eh_7 - np_*^n N^{n-1}(\alpha_0+u_2)\frac{(I_c+I_a)^n}{(p_*^n N^n+(I_c+I_a)^n)^2}(h_7+h_6)S_a\\ &+c\beta\frac{(I_c+I_a)}{N^2}(q_a h_8+(1-q_a)h_9-h_6)S_a + c\beta k_a\frac{(I_c+I_a)}{N^2}(h_9-h_8)L_a + \tau h_{10}\end{aligned}$$

5 Numerical Simulation and Results

We simulate the result by using fourth order Runge-Kutta method. The process begins with an initial guess on the control variable. Then, the state equations are simultaneously solved forward in time and the adjoint equations are solved backward in time. The control is updated by inserting the new values of state and adjoint variables into its characterization, and the process is repeated until convergence occurs (Fig. 1).

The initial conditions for the state variables are estimated as follows. According to the World Bank [18] estimation of the population size of Ethiopia for 2010 and 2011, the percentage of children is 41 %. From a total 84,734,000 population in 2011, there were a total of 34,740,940 children and 49,993,060 adult population. In each age group, we assume that initially 80 % are susceptible and 20 % are infected. We subtract the number of infectious children and adults from these 20 % to get the number of latently infected individuals in each age class. We took total birth to be equal to the average of 9 years births and obtained 2,747,945 per year. With regard

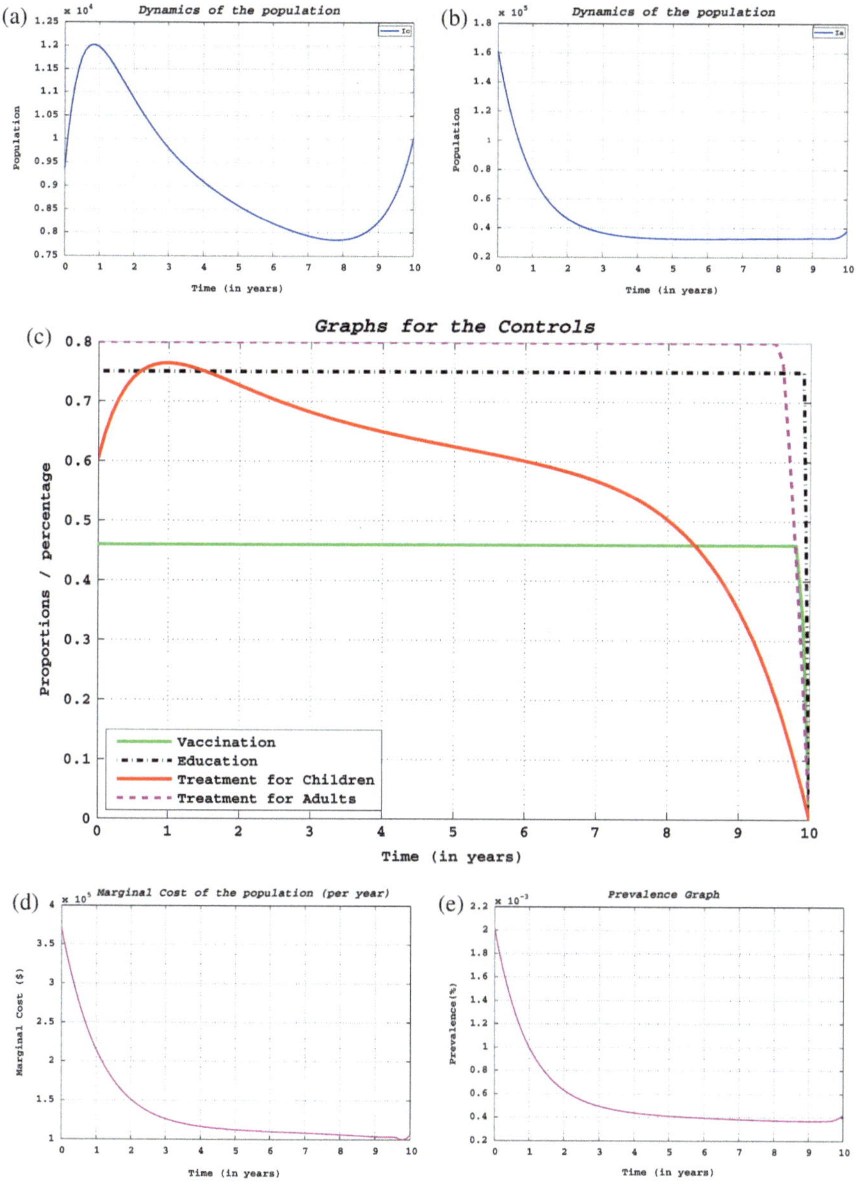

Fig. 1 $C_1 = 3$, $C_2 = 2$, $B_1 = 8 \times 10^5$, $B_2 = 4 \times 10^5$, $B_3 = 6 \times 10^4$, $B_4 = 3 \times 10^4$. (**a**) The graph of the number of infectious children when all controls applied. (**b**) The graph of the number of infectious adults when all controls applied. (**c**) The graph of controls. (**d**) The graph of marginal cost of interventions. (**e**) The graph of prevalence

Table 1 Value and source of parameters

Parameters	Symbol	Value	Source
Current rate of vaccinating children at birth	r_0	0.54	[17]
Efficacy of the vaccine	ε	0.8	[12]
Effective number of contacts	c	7	Assumed
Transmission probability	β	0.12	Assumed
Endogenous reactivation rates for children	b_c	0.001	[2]
Endogenous reactivation rates for adults	b_a	0.003	[2]
Reinfection rates for children	k_c	0.015	[2]
Reinfection rates for adults	k_a	0.02	[2]
Natural mortality rate for children	μ_c	0.0133	[19]
Natural mortality rate for adults	μ_a	0.01768	[2]
Tuberculosis induced death rates of children	d_c	0.06	Assumed
Tuberculosis induced death rates of adults	d_a	0.03	Assumed
Proportion of slow rout to the latent stage in children	q_c	0.8	Assumed
Proportion of slow rout to the latent stage in adults	q_a	0.9	[2]
Cure rate for children	η	0.75	Assumed
Cure rate for adults	τ	0.85	Assumed
Treatment rate for actively infected children	δ_0	0.2	Assumed
Treatment rate for actively infected adults	σ_0	0.06	Assumed
The rate at which the BCG vaccine wanes	w_c	0.6667	[2]
Per capita ageing functions	f_1	0.6667	[2]

to vaccination at birth, the vaccination coverage of Ethiopia in 2011 were 54 % per year as indicated in [17]. This implies that the total number of vaccinated children in 2011 were 1,418,447. We assumed that 20 % of the susceptible adult population are convinced or well informed about the transmission, prevention and treatment of the disease. Since most of Ethiopian people live in areas which are far away from hospitals or health centers, it is not possible to vaccinate every child at birth. Therefore, we assumed that maximum attainable percentage of vaccinating children to be 80 % per year. Similarly we expect to convince up to 75 % of susceptible individuals per year to change their behavior about the disease as well as the maximum rate of recruitment for treatment is 90 % per year (other parameter values are given in Table 1).

The result can be summarized by observing the role of various interventions from the graph of prevalence (Fig. 2a) and from the graph of cost (Fig. 2b). In the simulation, we first used all controls (u_1, u_2, u_3 and u_4) to optimize the objective function **J**, in the second phase we set all controls to zero and optimized the objective function, in the third phase we set the controls u_2 and u_3 to zero and optimize the objective function over the controls u_1 and u_4, then we used u_2 and u_3 only by setting u_1 and u_4 to zero in order to optimize the objective function, finally we set the controls u_1 and u_3 to zero in order to optimize the objective function over the controls u_2 and u_4. When we apply all interventions simultaneously, we got the lowest prevalence (see Fig. 2a) with minimum cost of intervention (see

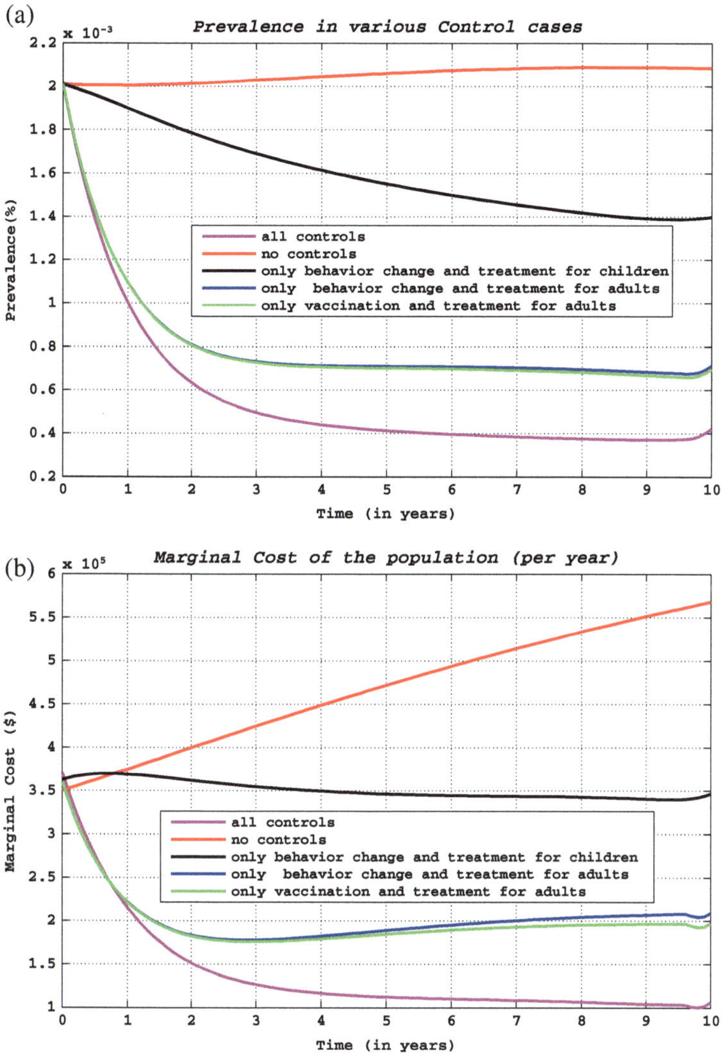

Fig. 2 (**a**) The graph of the prevalence in various cases of the controls; (**b**) the graph of the marginal cost of the interventions (per year) in various cases of controls

Fig. 2b). On other hand, the values of the prevalence and the marginal cost increase if we optimize the objective function in the absence of any intervention. We can also observe that only implementation of treatment for children and self-protective measures in the absence of vaccination and treatment for adults are not sufficient to decrease the prevalence like implementation of all controls (see Figs. 2a, b and 3). In the strategy where all the controls are being used, it is optimal to apply all existing resources to each of the control measures at the beginning. Since our aim

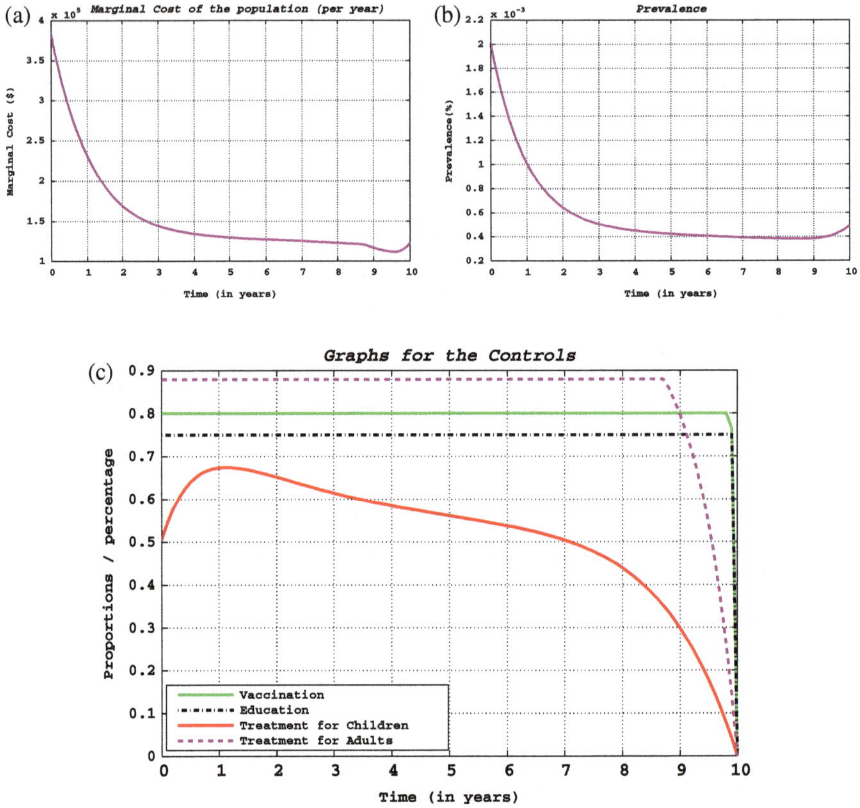

Fig. 3 $B_1 = 4\times10^5, B_2 = 6\times10^5, B_3 = 8\times10^4$ and $B_4 = 6\times10^4$. When we decrease the weight of vaccination by half and increase the weight of other interventions, the graph of marginal cost of interventions is greater than 1×10^5 dollar per year. Similarly the graph of prevalence decrease until it approaches to 0.04 %. When we observe the graph of controls, efforts of recruiting children for treatment starts to decrease before other efforts start to decrease and never increase beyond 70 %. Unlike the graph of efforts for vaccination in Fig. 1c, the graph for efforts of vaccinating children goes to its maximum rate. (**a**) Marginal cost of interventions; (**b**) prevalence; (**c**) controls

is to minimize the cost function, the graph of some state variables may rise-up. For example, during minimizing the objective function the number of infectious children increases for a moment and then it decreases down (see Fig. 1a).

When we minimize the cost of vaccination by half, and when we increase other costs, the result from Fig. 4b indicates that there will be no significant difference on the graph of prevalence but as we can observe from the graph of controls (see Fig. 4d) efforts of treatment for children never increase beyond 70 % and efforts of vaccination starts to fall starting from fourth year of intervention period.

Generally, to get the best result the effort of vaccinating children at birth and educating the population on existing self-preventive mechanisms should be given more emphasis at the next level in priority. With all the controls employed simultaneously

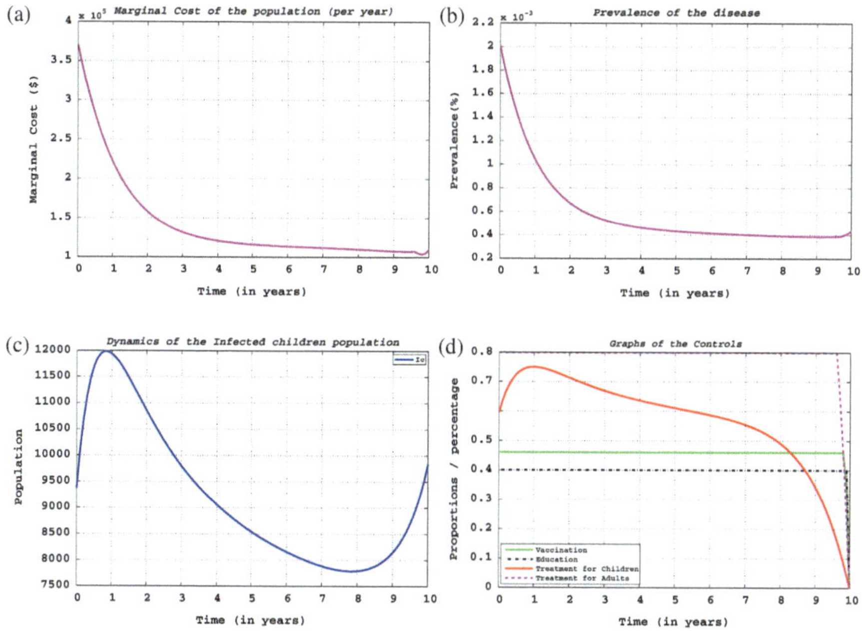

Fig. 4 In the absence of disease induced deaths of children and adults by keeping other parameters. When we ignore disease induced deaths, efforts of convincing adults to take part in behavior modification decreases its rate to 0.4. In addition to these it requires huge effort of treating children. (**a**) Marginal cost of interventions; (**b**) prevalence; (**c**) infectious children; (**d**) controls

in an optimal way, the prevalence can possibly drop to less than 0.04 % within 10 years of intervention period. In the absence of interventions/controls the prevalence increases (Figs. 5 and 6).

6 Discussion and Conclusion

In this paper, we analyzed a dynamical model of tuberculosis (TB) classified into two age groups that takes vaccination at birth and behavior modification of the population into account. Since organized data about burden of the disease for children and adults is not available, we used data from WHO global report about Ethiopia. Even, in the WHO report specific data about children is missing. Therefore, we have estimated and used some assumptions in taking values for some of the parameters for numerical purposes. Numerical simulations of the resulting optimality system showed that, vaccination as well as behavior modification by the society have a great impact in controlling the epidemic. In the presence of all interventions, the prevalence could decrease below 0.04 % within 10 years (see Fig. 2a). In the absence of any intervention, the prevalence increases. It was

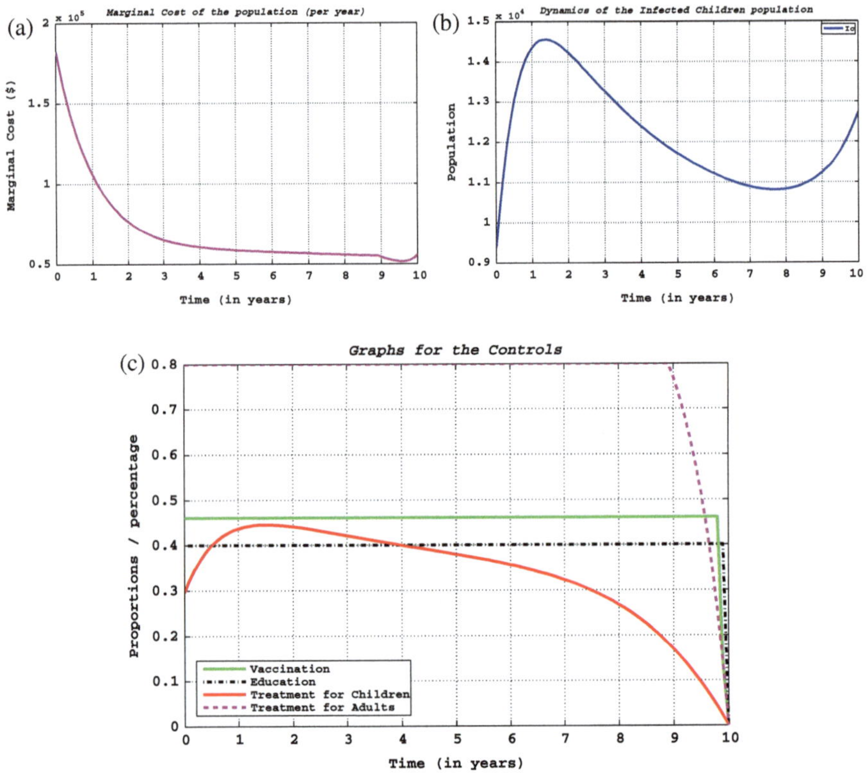

Fig. 5 $C_1 = 1$, $C_2 = 1$, when we put the weight for infectious children and adults equal to unity, the marginal cost of interventions decreases and reaches to 0.5×10^5 dollars per year. Efforts of vaccinating children, convincing adults for behavior modification and treatment of children doesn't go beyond the rate 0.5. But the number of infectious children starts to increase up after small decrease. (**a**) Marginal cost of interventions; (**b**) infectious children; (**c**) controls

also indicated from Fig. 2b that, inclusion of all controls resulted in minimum cost of intervention. Therefore, increase the effort of vaccinating children at birth and educate the adult society about the disease to bring behavior change in addition to treating the infected once with anti-TB drugs will help to decrease the prevalence significantly. Optimal control theory is used to explain dominance of the vaccination and behavioral change interventions as compared to treatment. The optimality system also proposes the cost effective way of controlling the disease when vaccination, behavior change as well as treatment of children and treatment of adults are being implemented on the population at the same time. Based on the result obtained from numerical simulation, we recommend the following points: to eradicate the disease from the country, TB patients should take medicines prescribed by their doctors properly until the end of the treatment period, because failure of the treatment may lead to an increase in the prevalence; since primary source of infection for children is from close contacts with adults, educating the society to

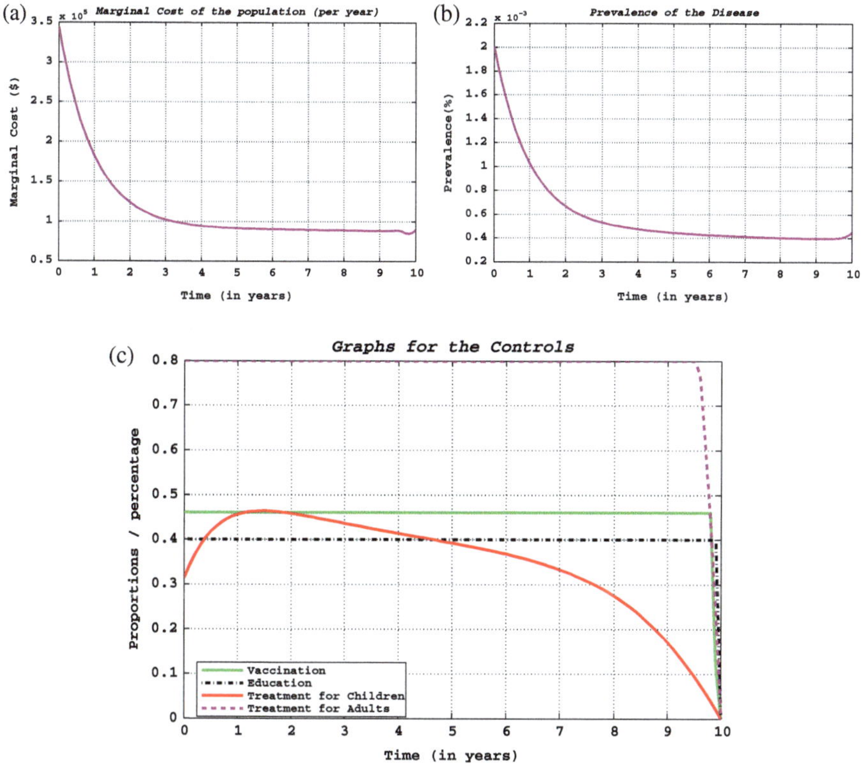

Fig. 6 $C_1 = 1$, $C_2 = 2$, when we double the weight for infectious adults, the marginal cost of intervention increases as compared to graph of the cost in Fig. 5a. But there is no significant difference on the graph of other values. (**a**) Marginal cost of interventions; (**b**) prevalence; (**c**) controls

modify their behavior is very important and this also helps children not to get the infection from adults. In general, if health policy makers consider to use all these interventions optimally, it is possible to decrease the prevalence of the disease in a resource limited areas.

Acknowledgements This work is in part supported by the Swedish International Science Program (ISP) at the Department of Mathematics, Addis Ababa University.

References

1. Barbu V., Precupanu T.: Convexity and Optimization in Banach Spaces, 4th edn. Springer, Dordrecht (2010)
2. Bekele B.T.: Modeling tuberculosis dynamics in children and adults in the presence of vaccination. M.Sc. thesis, Stellenbosch University (2010)

3. Central Statistical Agency Ethiopia, and ICF International: Ethiopia Demographic and Health Survey 2011. Addis Ababa, Ethiopia and Calverton, Maryland, USA: Central Statistical Agency and ICF International (2012)
4. Coddington, E.A.: An Introduction to Ordinary Differential Equations. Prentice-Hall, Englewood Cliffs (1961)
5. Coddington, E.A., Levinson, N.: Theory of Ordinary Differential Equations. McGraw Hill, New York (1955)
6. Federal Democratic Republic of Ethiopia Ministry of Health: Health and Health Related Indicators (2011)
7. Fleming, W.H., Rishel R.W.: Deterministic and Stochastic Optimal Control Applications of Mathematics, vol. 1. Springer, New York (1975)
8. Gaff, H., Schaefer, E.: Optimal control applied to vaccination and treatment strategies for various epidemiological models. Math. Biosci. Eng. **6**(3), 469492 (2009)
9. Grass, D., Caulkins, J.P., Feichtinger, G., Tragler, G., Behrens, D.A.: Optimal Control of Nonlinear Processes, with Applications in Drugs, Corruption, and Terror. Springer, Berlin (2008)
10. Kassa, S.M., Ouhinou, A.: Epidemiological models with prevalence dependent endogenous self-protection measure. Math. Biosci. **229**, 41–49 (2011)
11. Kassa, S.M., Ouhinou, A.: The impact of self-protective measures in the optimal interventions for controlling infectious diseases of human population. J. Math. Biol. (2014). doi:10.1007/s00285-014-0761-3
12. Kernodle, D.: Decrease in the effectiveness of Bacille Calmette-Guerin vaccine against pulmonary tuberculosis. Clin. Infect. Dis. **51**(2), 177–184 (2010)
13. Lietman, T., Blower, S.: Potential impact of tuberculosis vaccine as epidemic control agents. Clin. Infect. Dis. **30**(3), 316–322 (2000)
14. Ma, S., Xia, Y.: Mathematical Understanding of Infectious Disease Dynamics, vol. 16. World Scientific, Singapore (2009)
15. Silva, C.J., Torres, D.F.M.: Optimal control strategies for tuberculosis treatment: a case study in Angola. Center for Research and Development in Mathematics and Applications, Department of Mathematics, Portugal; Numer. Algebra Control Optim. **2**(3), 601–617 (2012)
16. Van den Driessche, P., Watmough, J.: Reproduction numbers and sub-threshold endemic equilibria for compartmental models of disease transmission. Math. Biosci. **180**, 29–48 (2002)
17. World Health organization: Reported estimates of BCG coverage. http://apps.who.int/immunization_monitoring/en/globalsummary/timeseries/tscoveragebcg.htm (2013). Accessed 20 Apr 2013
18. World Bank: http://www.worldbank.org/indicator/SP.DYN.CBRT.IN (2013). Accssed on 19 Apr 2013
19. Trading Economics: http://www.tradingeconomics.com/sub-saharan-africa/death-rate-crude-per-1-000-people-wb-data.html (2013). Accessed 16 Aug 2013
20. Index Mundi: http://www.indexmundi.com/g/g.aspx?c=et&v=30 (2013). Accssed 21 April 2013

Modeling of Extremal Earthquakes

Margarida Brito, Laura Cavalcante, and Ana Cristina Moreira Freitas

Abstract Natural hazards, such as big earthquakes, affect the lives of thousands of people at all levels. Extreme-value analysis is an area of statistical analysis particularly concerned with the systematic study of extremes, providing an useful insight to fields where extreme values are probable to occur. The characterization of the extreme seismic activity is a fundamental basis for risk investigation and safety evaluation. Here we study large earthquakes in the scope of the Extreme Value Theory. We focus on the tails of the seismic moment distributions and we propose to estimate relevant parameters, like the tail index and high order quantiles using the geometric-type estimators.

In this work we combine two approaches, namely an exploratory oriented analysis and an inferential study. The validity of the assumptions required are verified, and both geometric-type and Hill estimators are applied for the tail index and quantile estimation. A comparison between the estimators is performed, and their application to the considered problem is illustrated and discussed in the corresponding context.

1 Introduction

Earthquakes are a worldwide and ever present menace, threatening to occur at any second. A severe earthquake is one of the most frightening and destructive phenomena of nature. Experiencing an earthquake is a terrible experience, the lived moments are reported as full of panic, terror, and death. For survivors, the terrible

M. Brito (✉) • L. Cavalcante
Faculdade de Ciências, Universidade do Porto, Porto, Portugal
e-mail: mabrito@fc.up.pt; laucavalcante@fc.up.pt

A.C.M. Freitas
Faculdade de Economia, Universidade do Porto, Porto, Portugal
e-mail: amoreira@fep.up.pt

J.-P. Bourguignon et al. (eds.), *Mathematics of Energy and Climate Change*,
CIM Series in Mathematical Sciences 2, DOI 10.1007/978-3-319-16121-1_3

images remain in their memory and become part of their daily lives, as well as the constant fear of the possibility of the next big earthquake which may take lives and separate families forever. It is estimated that there are about one million earthquakes per year. However, the vast majority occurs in the midst of oceans or in sparsely populated regions, and they pass relatively unnoticed by the population. There are annually about 20 earthquakes that cause significant damage. On average, only one catastrophic earthquake occurs every year, and a highly catastrophic one every 5 years.

Since the underlying phenomena responsible for the occurrence of an earthquake are still very far from being completely understood, it is rather important to collect as much data as possible and categorize it in order to be able to provide some insight on how to diminish their negative impacts, in particular, in what concerns the reduction of number of deaths and economic losses. This is an important challenge requiring a large multidisciplinary effort. In this work, we perform a statistical analysis taking into account specific features of big earthquakes. When we are dealing with extreme events, the classical statistical models are inappropriate for the statistical modeling of earthquake size. Hence, we are particularly interested in the study of the tail distribution of the data.

The Extreme Value Theory (EVT) is one field of statistics that has been devised to study these extreme events using only a limited amount of data (see e.g. [1], and references therein). In the study of earthquakes, the EVT is a relevant tool, providing important information, such as the estimation of the probability of occurring a large earthquake over a long period of time or high quantiles (see e.g. [22]).

In the present work we consider the seismic activity in Philippines and Vanuatu Islands. The data sets are taken from the Harvard Seismic Catalog and the tail behavior of the distributions of large earthquakes seismic moments is characterized using EVT techniques. In order to apply these methods, a preliminary data analysis is performed to investigate the validity of the usual underlying assumptions. The geometric-type and the Hill estimator, as well as its bias corrected versions, are considered for the estimation of the tail index and are employed for the quantile estimation. A comparison between the estimators is carried out and their performance is discussed carefully.

All the analysis is supported by graphical tools that show, in a clear way, the features of the data that are regarded as most relevant to the study being addressed.

The paper is organized as follows. Some important concepts and results about EVT and earthquakes are briefly presented in Sect. 2. The investigation, in order to verify the validity of the usual assumptions and the analysis of the seismic moments, are performed in Sect. 3. Some final comments about the study, including an interpretation of the results in terms of the frequencies of seismic moment exceedances, are provided in Sect. 4.

2 Essential Notions of EVT and Earthquakes

2.1 Extreme Value Theory

The Extreme Value Theory is a powerful and fairly robust framework to study the tail behavior of a distribution, since it encompasses a set of probabilistic results that allow characterizing and modeling the extreme values behavior. In this way, the EVT is very useful in making statistical inferences about rare events in several areas of knowledge (e.g. meteorology, hydrology, insurance, environment, etc.), and its use may enable the implementation of appropriate prevention procedures.

More concretely, through this theory, extreme values may be modeled using the limiting distribution of the maxima of the random variables or of its excesses over a threshold. Thus, the statistical basis for applications of EVT is constituted by the following two main limit theorems.

Theorem 1 (Fisher-Tippett-Gnedenko Theorem) *Let $X_1, X_2, \ldots, X_n$ be independent and identically distributed (i.i.d.) random variables (r.v.) with distribution function (d.f.) F and $M_n = \max(X_1, X_2, \ldots, X_n)$ denote the maximum of the n observations. If a sequence of real numbers $a_n > 0$ and b_n exists such that*

$$\lim_{n\to\infty} P\left(\frac{M_n - b_n}{a_n} \leq x\right) = \lim_{n\to\infty} F^n\left(a_n x + b_n\right) = G\left(x\right),$$

then if G is a non degenerate d.f., it belongs to one of the following types

$$\text{Type I (Gumbel)}: \quad \Lambda\left(x\right) = \exp\{-\exp\left(-x\right)\},\ x \in \mathbb{R};$$

$$\text{Type II (Fréchet)}: \ \Phi_\alpha\left(x\right) = \begin{cases} 0, & x \leq 0, \\ \exp\left(-x^{-1/\gamma}\right), & x > 0; \end{cases}$$

$$\text{Type III (Weibull)}: \Psi_\alpha\left(x\right) = \begin{cases} \exp\{-\left(-x\right)^{1/\gamma}\}, & x > 0, \\ 1, & x \geq 0; \end{cases}$$

for all continuity points of G.

If a d.f. F satisfies the conditions of the theorem, it is said that F belongs to the domain of attraction of G $\left(F \in DA(G)\right)$.

These three types of distributions may be combined into the single d.f.

$$G_\gamma\left(x\right) = \begin{cases} \exp\left(-\left(1+\gamma x\right)^{-1/\gamma}\right), & \text{for } 1+\gamma x > 0,\ \gamma \neq 0, \\ \exp\left(-\exp\left(-x\right)\right), & \text{for } x \in \mathbb{R},\ \gamma = 0, \end{cases}$$

where γ is the shape parameter, known as tail index, determining the weight of the right tail of the underlying d.f. F. This distribution is known as the Generalized Extreme Value (GEV) distribution.

Theorem 2 (Pickands-Balkema-de Haan Theorem) *Let $X_1, X_2, \ldots, X_n$ be a sample of n i.i.d. r.v. with d.f. F, x^F the right endpoint of F and $F_{X-u|X>u}(x) = P\{X - u \le x \mid X > u\}$ the excess d.f. over a (high) threshold u. Then,*

$$F \in DA(G_\gamma) \text{ iff } \lim_{u \to x^F} \sup_{0 \le x < x^F - u} \left| F_{X-u|X>u}(x) - H_{\gamma,\sigma_u}(x) \right| = 0,$$

where $H_{\gamma,\sigma_u}(x)$ represents the Generalised Pareto Distribution, given by:

$$H_{\gamma,\sigma_u}(x) = \begin{cases} 1 - \left(1 + \gamma \frac{x-u}{\sigma_u}\right)^{-1/\gamma}, & \text{for } 1 + \gamma \frac{x-u}{\sigma_u} > 0,\ \gamma \neq 0, \\ 1 - \exp(-\frac{x-u}{\sigma_u}), & \text{for } x \ge u,\ \gamma = 0, \end{cases}$$

where γ, u, $\sigma_u > 0$ are the shape, location, and scale parameter depending on threshold u, respectively.

Similarly with GEV, using another parameterization, the GPD is separated into three families depending on the value of the shape parameter:

- Type I (Exponential): $H(x) = 1 - \exp(-x)$, if $\gamma = 0$,
- Type II (Pareto): $H(x) = 1 - x^{-1/\gamma}$, if $\gamma > 0$,
- Type III (Beta): $H(x) = 1 - (-x)^{-1/\gamma}$, if $\gamma < 0$.

These two theorems state that, under their conditions, the limit distribution of the normalized maximum is the GEV distribution, and that the limit of the excess d.f. is the GPD. Hence, they are fundamental to make possible real-world applications.

In order to perform a correct inference about extreme events from the accessible data, it is necessary to properly select the extreme observations following some criterion. There are two primary methods to define such extreme observations which arise from the two main results of the classical EVT: the Block Maxima method, also known as Gumbel's approach, and the Peaks Over Threshold method.

The Block Maxima (BM) method consists in dividing the data in equal sized blocks, taking the maximum observation in each block and studying its asymptotic distribution. In the Peaks Over Threshold (POT) method one considers a certain high threshold and then studies the asymptotic distribution of the excesses over this high threshold.

Accordingly, as with the data set under study, one must be aware to consider both methods' disadvantages when applying them. One major drawback of the BM method is that only one observation in a block is used, resulting in a final sample of small size. On other hand, this method is more robust with respect to eventual dependence between the observations.

Since our interest is centered in the frequencies of exceedances of certain critical values, here we adopt the POT approach that picks up all relevant high observations and seems to make better use of the available information.

In modeling the extreme value distribution, the main issue to be solved is the parameter estimation. The shape parameter γ is of great interest in the analysis of the tails, since it characterizes the behavior of extremes. This parameter indicates the heaviness of the tail distribution, the tail being heavier for larger values γ. It also plays a crucial role in the estimation of other extreme events' parameters, namely in high quantiles estimation. In practice, the tail index is associated to the frequency with which extreme events occur and the high order quantiles are levels that are exceeded with a small probability. The adequate estimation of these quantities is the most important problem.

We assume that $X_1, X_2, \ldots, X_n$ is a sample of i.i.d. r.v. with d.f. F and denote by $X_{(1,n)} \leq X_{(2,n)} \leq \cdots \leq X_{(n,n)}$ the corresponding order statistics (o.s.). The estimation of γ is based on the k top o.s., where $k = k_n$ is an intermediate sequence of positive integers ($1 \leq k < n$), that is,

$$k \to \infty, \qquad \frac{k}{n} \to 0 \quad \text{as} \quad n \to \infty. \tag{1}$$

Several estimators have been proposed for the estimation of γ (see e.g. [6, 10, 18, 20]). Here we consider the following estimator for $\gamma > 0$, the geometric-type (GT) estimator

$$\widehat{GT}\,(k) = \sqrt{\frac{M_n^{(2)} - \left[M_n^{(1)}\right]^2}{\frac{1}{k}\sum_{i=1}^{k}\log^2(n/i) - \left(\frac{1}{k}\sum_{i=1}^{k}\log(n/i)\right)^2}} \tag{2}$$

where

$$M_n^{(j)}\,(k) = \frac{1}{k}\sum_{i=1}^{k}\left(\log X_{(n-i+1,n)} - \log X_{(n-k,n)}\right)^j. \tag{3}$$

We also consider the commonly used Hill estimator (see [18]) defined by

$$\hat{H}\,(k) = \frac{1}{k}\sum_{i=1}^{k} \log X_{(n-i+1,n)} - \log X_{(n-k,n)}. \tag{4}$$

The asymptotic properties of these aforementioned estimators were investigated and, under certain conditions, they share some common desirable properties, such as consistency and asymptotic normality (cf. [2, 9, 17]).

The problem of estimating high order quantiles has received increased attention as a useful tool in data modeling, which has been performed in a wide variety of problems in many different scientific areas. This field addresses interesting questions such as the size of some extreme event that will only occur with a given small probability, or the expected time until the realization of an extreme event.

The classical quantile estimator was proposed by [23],

$$\hat{\chi}^{W}_{1-p} = X_{(n-k,n)} \left(\frac{k}{np} \right)^{\hat{\gamma}},$$

where $\hat{\gamma}$ is a consistent estimator of γ.

Using general quantile techniques and the POT methodology, the well known POT estimator for high quantiles above the threshold $X_{(n-k,n)}$ arises naturally and is given by

$$\hat{\chi}^{P}_{1-p} = \frac{\left(\frac{k}{np}\right)^{\hat{\gamma}} - 1}{\hat{\gamma}} \cdot X_{(n-k,n)} M_n^{(1)} + X_{(n-k,n)}, \qquad p < \frac{k}{n}, \tag{5}$$

where $\hat{\gamma}$, $X_{(n-k,n)} M_n^{(1)}$ and $u = X_{(n-k,n)}$ are, respectively, suitable estimators of the shape, scale and location parameters of the Generalized Pareto Distribution.

In the present work both the $\widehat{GT}(k)$ and $\hat{H}(k)$ are used to estimate γ. The high quantiles are estimated considering (5) and using $\widehat{GT}(k)$ and $\hat{H}(k)$ as estimators of γ. The asymptotic behavior of these quantile estimators was studied and their asymptotic normality was proved (cf. [3, 8, 10]).

The problem of reducing the bias of these tail index estimators was addressed in [3], where were proposed the following two asymptotic equivalent geometric-type bias corrected estimators

$$\overline{\widehat{GT}}(k) = \widehat{GT}(k) \left(1 - \frac{\beta \left(\frac{n}{k}\right)^{\rho}}{(1-\rho)^2} \right),$$

and

$$\overline{\overline{\widehat{GT}}}(k) = \widehat{GT}(k)\, exp \left\{ -\frac{\beta}{(1-\rho)^2} \left(\frac{n}{k}\right)^{\rho} \right\}.$$

Hill bias corrected estimators may be found in [4], namely

$$\overline{\hat{H}}(k) = \hat{H}(k) \left(1 - \frac{\beta \left(\frac{n}{k}\right)^{\rho}}{1-\rho} \right)$$

and

$$\overline{\overline{\hat{H}}}(k) = \hat{H}(k)\, exp \left\{ -\frac{\beta}{1-\rho} \left(\frac{n}{k}\right)^{\rho} \right\},$$

where ρ and β are the shape and scale parameters.

Here, in order to get bias corrected high quantiles estimators, we also consider the form (5), based on the above bias corrected estimators.

The accurate estimation of the tail index is very important, also because of its great influence on the estimation of other relevant parameters of rare events, such as the right endpoint of the underlying d.f. F. Since the impact of its influence can be considerable, the appropriate estimation of γ is fundamental in obtaining a suitable quantile estimator with a good performance.

2.2 Earthquakes

In general, everything in nature tends to an equilibrium. Due to the thermodynamic equilibrium, the constituents of the Earth's interior are in constant motion. Boosted by this movement, which causes friction with its bottom, the tectonic plates move and interchange slowly, thereby contributing to the constant evolution of the terrestrial relief.

The earthquakes mainly arise due to forces, within the earth's crust, tending to displace one mass of rock relative to another. Each time the plates interact with each other, a large amount of energy is accumulated in its rocks. When its elasticity limit is reached, they will fracture and instantly release all the energy that had been accumulated during the elastic deformation. That causes vibrations, called seismic waves, which travel outwards in all directions from the fault and give rise to violent motions at the earth's surface, unleashing an earthquake.

Therefore, earthquakes are natural shocks that occur as a result of this sudden release of huge amounts of the energy that has been slowly-accumulated over many years. If the earthquake is large enough, the seismic waves are recorded on seismographs around the world, and can cause the ground to quake strongly.

Earthquakes do not occur at random, but are distributed according to a well-defined pattern. About 90 % of earthquake activity is associated with plate-boundary processes, so the global seismicity patterns reveals a strong correlation between plate boundaries and the presence of intercontinental fault zones, indicating that earthquakes often occur at tectonic plate boundaries. We can say, without committing a gross error, that the alignments of earthquakes indicate the boundaries of tectonic plates.

After the initial fracture, a number of secondary ruptures, corresponding to the progressive adjustment of fractured rocks, may occur, causing successive lower intensity earthquakes called aftershocks. If these vibrations occur at the sea floor, they can produce a long and smooth waving that in shallow water becomes authentic water columns known as tidal waves or tsunamis.

Therefore, earthquakes represent one of the most energetic and rapid manifestations of the planet's internal dynamics.

The scientific analysis of earthquakes requires means of measurement, and the size of an earthquake has been measured in several ways. The early methods used a kind of numerical scale based on a synthesis of observed effects, called the *intensity* scales. Some attempts to relate intensity to the amplitude of ground motion led to a quantity called *magnitude*, based on the records of ground amplitudes, normalized for their variation with regard to the distance from the earthquake epicenter. However, the known magnitudes present a saturation point which does not allow for a correct estimation of the true earthquake size for larger earthquakes, underestimating it. Moreover, it turns out that larger earthquakes, which have larger rupture surfaces, systematically radiate more long-period energy. Nowadays, the measurement that is adopted preferably for scientific studies is the *seismic moment* of the displaced ground (see e.g. [7, 19]). This measurement avoids the saturation problem, since it does not have an intrinsic upper bound, and describes the size of an earthquake as an essential combination of physical quantities.

The seismic moment, M, provides more accurate measures of the energy released from an earthquake, taking into account the rock properties, such as its rigidity, μ, the area of the fault plane that actually moves, A, and the amount of movement on the fault, D, combining these three factors in the following form

$$M = \mu AD.$$

Because many people do not really know the meaning of this measure, and given that the magnitude scale has been used for a very long time, the need to convert it into some kind of magnitude scale came about. These factors have resulted in the definition of a new magnitude scale, the moment magnitude, m_w, based on the seismic moment

$$m_w = \frac{2}{3}\left(\log M - 16.1\right), \tag{6}$$

where M is in units of dyne-cm.

The seismic moment, based on classical mechanics, provides, in this way, a uniform scale of earthquake size, and is considered the most consistent measure for accurate quantification of the energy released from an earthquake.

3 Extreme Value Modeling of Earthquake Data

In this section, we analyze the tail behavior of the distribution of the seismic moments, following the POT approach. We begin by describing the data considered for this study. We perform an exploratory data analysis, where we discuss which type of distribution may model the large seismic moments as well as the properties of stationarity and independence of the data. Then we proceed to the estimation of the tail parameters of the seismic moment distribution.

3.1 Description of the Earthquake Data

We consider the earthquake data obtained from the Harvard Seismic Catalog, available at the Global Centroid-Moment-Tensor (CMT) web page (cf. e.g. [11, 12, 14]). Here, we restrict the area of study to earthquakes occurring within the Philippines and Vanuatu Islands, and the analysis was performed in a similar way for both regions. In particular, we extract and analyze the information on their seismic moments covering the period 01.01.1976–31.12.2010. The original data sets contain 1255 events for Philippines Islands, and 1012 events for Vanuatu Islands. However, in order to apply the POT method we selected an adequate and large enough level $u = 10^{24}$ dyne-cm, that corresponds to a moment magnitude $m_w \approx 5.27$, the same value considered in related works such as in [21]. The observations under this threshold were removed. Since we detect a failure in the data acquisition of the Vanuatu Islands until 01-01-1980, we shall consider only the Vanuatu Islands data subsequent to this date. So, the final data sets, on which the following analysis is based, consider 821 cases for Philippines Islands and 647 cases for Vanuatu Islands. We did not exclude aftershocks because, besides excluding a great fraction of the range of seismic moments considered, the removal would introduce a bias in the parameters estimation (cf. e.g. [21]). Since the considered region has a lot of deep earthquakes, they were not excluded as well. In Fig. 1 the seismic moments of Philippines and Vanuatu Islands over the above mentioned period are plotted.

3.2 Preliminary Data Analysis

Before considering the problem of estimating the tail parameter γ, it is important to discuss if the Pareto-type model provides a plausible fit to the seismic moment

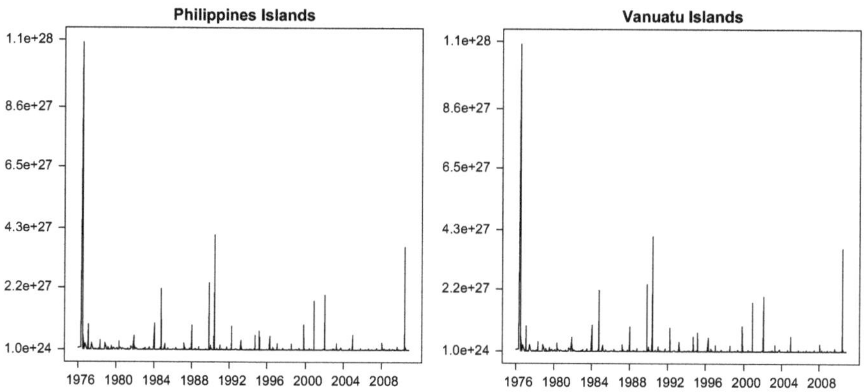

Fig. 1 Seismic moments of Philippines (*left*) and Vanuatu (*right*) Islands

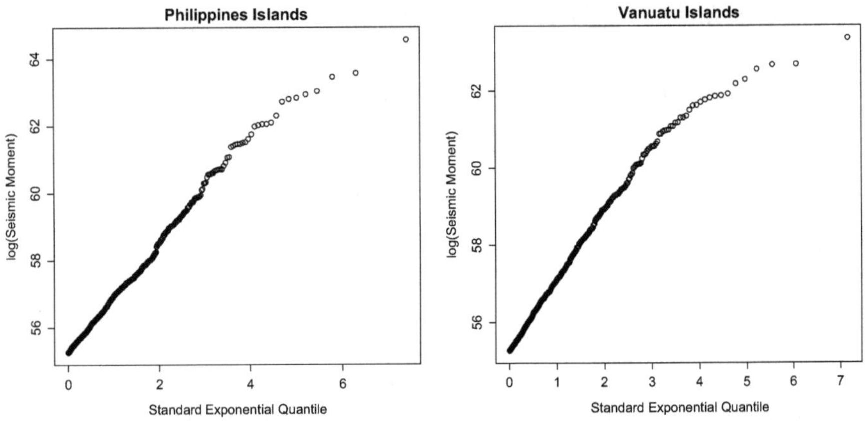

Fig. 2 Pareto QQ plot for Philippines (*left*) and Vanuatu (*right*) Islands seismic moment data

distributions of the data under study. This can be achieved graphically through quantile-quantile (QQ) plots, which constitute a very informative and powerful tool to graphically evaluate how close two distributions are from each other.

Usually, as in this case, the most convenient comparison is between the empirical quantiles and the quantiles of the assumed parametric distribution. If the sample data and the reference distribution are derived from populations with a common distribution, the QQ plot should have a linear form.

Since we believe our data is heavy tailed, we present the Pareto QQ plots of our data sets in Fig. 2.

In the case $Y \stackrel{D}{=} \log X$, where X and Y are Pareto and Exponential distributed r.v., respectively, then the usual Pareto QQ plots are Exponential QQ plots of the log-transformed data.

In the resulting scatterplot, a linear pattern is evident, which is indicative of the good agreement between observed values and the values predicted by the model. If we analyze the behavior of the QQ plots, we may remark that, with the exception of the extreme upper points, which are based on a small number of extreme values, the plots are approximately linear. Hence, the visual impressions based on the Pareto QQ plots suggest that the Vanuatu and Philippines Islands earthquake data sets do seem to exhibit heavy tails ($\gamma > 0$).

We analyse the stationarity of the data under study. More precisely, in the line of the study of Corral [5], we investigate if the mean value defined for any property of the earthquake occurrence process is approximately the same for different time windows. We plot the normalized cumulative number of earthquakes versus time.

The linear behavior that we can observe in Fig. 3 indicates that the mean seismic rate is approximately constant, and so, the data may be considered homogeneous in time.

For the application of the EVT we must analyse the independence of the data.

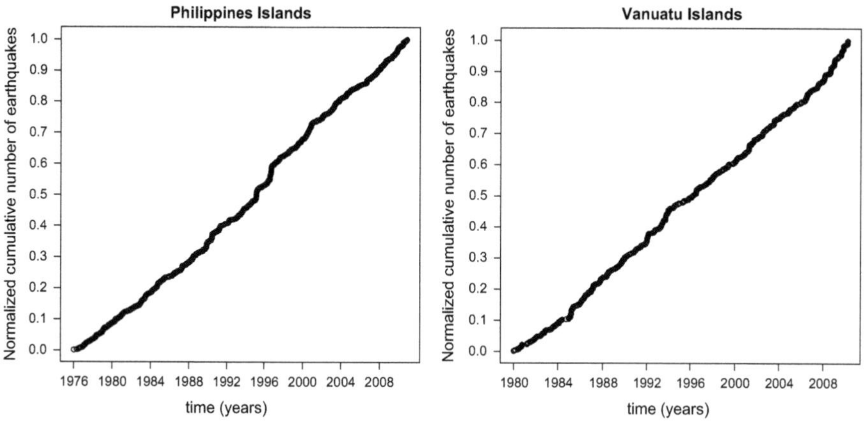

Fig. 3 Cumulative number of earthquakes normalized by the total number in the period considered as a function of time, for seismicity of Philippines (*left*) and Vanuatu (*right*) Islands with $M \geq 10^{24}$

In our case, the goal is to investigate the existence of dependence between consecutive seismic moments, i.e, verify how the seismic moment of one event, M_{i-1}, influences the seismic moment of the next, M_i. For that, let us consider the conditional probability density determined by

$$\frac{P\left(\eta \leq M_i < \eta + \Delta_\eta \mid M_{i-1} \geq M'_c\right)}{\Delta_\eta},$$

where M'_c is the threshold considered on the previous magnitude when this condition is imposed. Here we denote the initial threshold, u, as M_c, and the condition $M \geq M_c$ is always satisfied (see e.g. [5]).

The conditional probability density of a seismic moment is then defined as the probability of the seismic moments are within a small interval of values, divided by the length of the small interval, Δ_η, tending to zero, considering only the cases in which the seismic moment of the immediately previous event is bigger than a threshold M'_c.

If the seismic moment M_i is independent of M_{i-1}, then, as it is well known, the conditional distribution of M_i given that $M_{i-1} \geq M'_c$, $M'_c \geq M_c$, is identical to the unconditional distribution of M_i. Note that the case $M_c = M'_c$ gives the unconditional distribution of the considered data.

We observe in Fig. 4 that, in general, the different empirical densities, using different thresholds M'_c, share the same properties, which suggest the independence of seismic moments M_i with regards to their history. The small oscillations between the densities may be caused by the errors associated to the finite sample and the eventual dependence is apparently too weak to lead to major differences in the distributions.

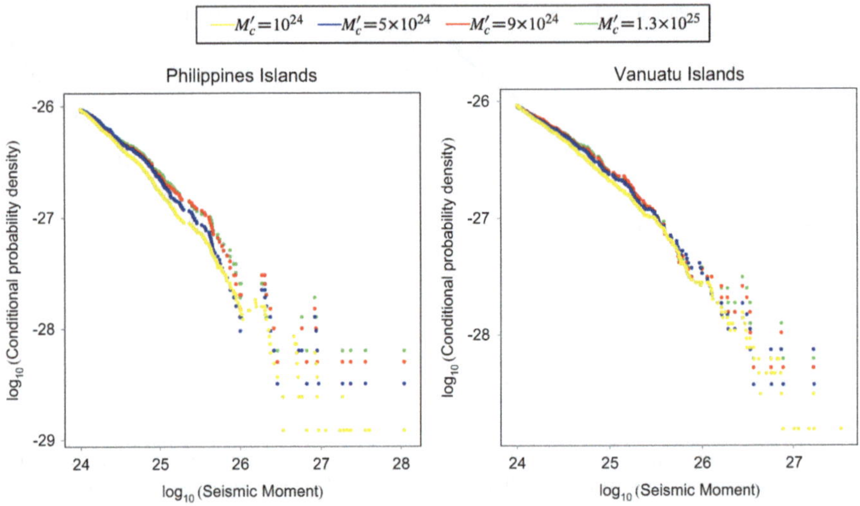

Fig. 4 Conditional probability densities of earthquake seismic moments, for seismicity of Philippines (*left*) and Vanuatu (*right*) Islands, evaluated using different thresholds M_c' and with a constant $M_c = 10^{24}$ ($\Delta_\eta = 10^{25}$)

3.3 Estimation of Tail Parameters

In this section we formalize our main objective of investigating the extremal behavior of large earthquakes and how the proposed estimators behave with this type of data.

Then, we discuss the estimation of the tail parameters through the POT approach. The GT and the Hill estimators are considered for the estimation of the tail index and are employed on POT estimator for the quantile estimation.

Some graphical plots illustrate the tail parameters of large earthquake data, as a function of k.

From the presented bias corrected estimators, we can easily note that the bias dominant components are dependent on second order parameters, shape ρ and scale β. To illustrate the behavior of the corrected estimators we consider the suitable estimators of the parameter ρ proposed by [13]

$$\hat{\rho}_n^{(\tau)}(k) = -\left|\frac{3\left(T_n^{(\tau)}(k) - 1\right)}{T_n^{(\tau)}(k) - 3}\right|, \tag{7}$$

where

$$T_n^{(\tau)}(k) = \begin{cases} \dfrac{\left(M_n^{(1)}(k)\right)^{\tau} - \left(M_n^{(2)}(k)/2\right)^{\tau/2}}{\left(M_n^{(2)}(k)/2\right)^{\tau/2} - \left(M_n^{(3)}(k)/6\right)^{\tau/3}}, & \text{if } \tau > 0 \\ \dfrac{\log\left(M_n^{(1)}(k)\right) - \frac{1}{2}\log\left(M_n^{(2)}(k)/2\right)}{\frac{1}{2}\log\left(M_n^{(2)}(k)/2\right) - \frac{1}{3}\log\left(M_n^{(3)}(k)/6\right)}, & \text{if } \tau = 0, \end{cases}$$

with M_n^j as in (3), and the β estimator obtained in [15]

$$\hat{\beta}_{\hat{\rho}}(k) = \left(\frac{k}{n}\right)^{\hat{\rho}} \frac{\left(\frac{1}{k}\sum_{i=1}^{k}\left(\frac{i}{k}\right)^{-\hat{\rho}}\right)\frac{1}{k}\sum_{i=1}^{k}U_i - \frac{1}{k}\sum_{i=1}^{k}\left(\frac{i}{k}\right)^{-\hat{\rho}}U_i}{\left(\frac{1}{k}\sum_{i=1}^{k}\left(\frac{i}{k}\right)^{-\hat{\rho}}\right)\frac{1}{k}\sum_{i=1}^{k}\left(\frac{i}{k}\right)^{-\hat{\rho}}U_i - \frac{1}{k}\sum_{i=1}^{k}\left(\frac{i}{k}\right)^{-2\hat{\rho}}U_i}, \tag{8}$$

where

$$U_i = i\left(\log\frac{X_{(n-i+1,n)}}{X_{(n-i,n)}}\right),$$

with $1 \le i \le k < n$.

It is known that the external estimation of ρ and β at a larger k value than the one used for γ-estimation has clear advantages, allowing bias reduction without increasing the asymptotic variance (see e.g. [4]). In line with other studies, and among some suggestions (see e.g. [16]), the level that seemed most appropriate to consider in illustrations is

$$k_h = \lfloor n^{1-\epsilon} \rfloor, \text{ for some } \epsilon > 0 \text{ small}, \tag{9}$$

where $\lfloor x \rfloor$ denotes the integer part of x.

We remark that the class of estimators of ρ presented above, and consequently also the β estimators, is dependent on a tuning parameter $\tau \ge 0$. Then, firstly we need to choose the tuning parameter τ, in which we will support the estimation of the second order parameters ρ and β.

For this use, we consider in (9), $\epsilon = 0.005$ and $\epsilon = 0.001$, i.e, we use the following k_h levels:

$$k_{h1} = \lfloor n^{0.995} \rfloor \quad \text{and} \quad k_{h2} = \lfloor n^{0.999} \rfloor. \tag{10}$$

As usual, the means whereby we do this choice, passes by portraying the sample paths of $\hat{\rho}_\tau(k)$ in (7) for the values $\tau \in \{0, 0.5, 1\}$, as functions of k, in order to analyze the variations that it causes in their behavior, and use the following

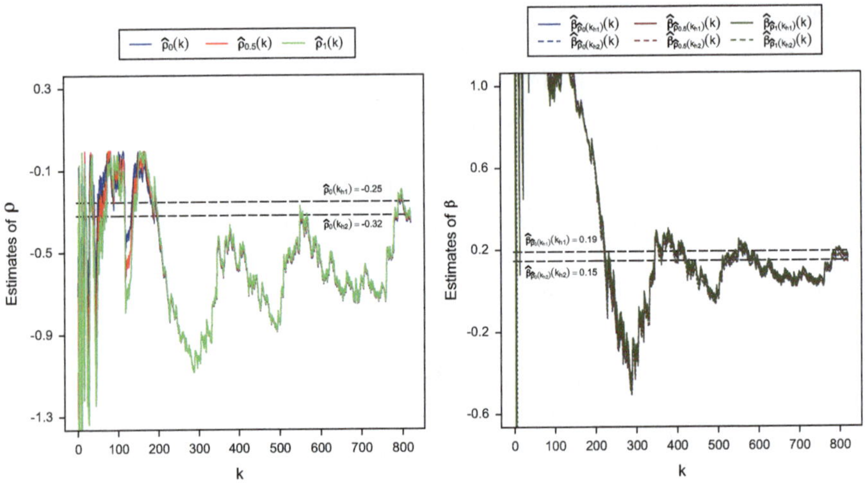

Fig. 5 Estimates of the second order parameters ρ (*left*) and β (*right*) for seismicity of Philippines Islands

algorithm as a stability criterion for large values of k:

1. Consider $\hat{\rho}_\tau(k)$, $\tau \in \{0, 0.5, 1\}$, for the integer values $k \in (\lfloor n^{0.995} \rfloor, \lfloor n^{0.999} \rfloor)$ and compute their median, denoted by χ_τ;
2. Choose the *tuning parameter* $\tau^* = \arg\min_\tau \sum_k (\hat{\rho}_\tau(k) - \chi_\tau)^2$;
3. Compute the ρ estimates $\hat{\rho}_{\tau^*}(k_{h1})$ and $\hat{\rho}_{\tau^*}(k_{h2})$, and the β estimates $\hat{\beta}_{\rho_{\tau^*}(k_{h1})}(k_{h1})$ and $\hat{\beta}_{\rho_{\tau^*}(k_{h2})}(k_{h2})$, with k_{h1} and k_{h2} given by (10).

The Figs. 5 and 6 show the sample paths of the second order parameter estimators, $\hat{\rho}$ and $\hat{\beta}$, based on the Philippines and Vanuatu seismic moment observations, respectively.

We can see that the sample paths of $\hat{\rho}$, for the three different values of τ, have very similar behavior. It is however apparent that the behavior of $\hat{\rho}$ is slightly better when considering $\tau = 0$, especially for data concerning the Vanuatu Islands. Since in both cases the algorithm described above also points to the choice of $\tau = 0$, we choose this value of τ to estimate ρ.

Thus, for Philippines Islands, we have $k_{h1} = \lfloor 821^{0.995} \rfloor = 793$ and $k_{h2} = \lfloor 821^{0.999} \rfloor = 815$, that is, the corresponding estimates of ρ are $\hat{\rho}_0(793) \approx -0.25$ and $\hat{\rho}_0(815) \approx -0.32$ and the corresponding estimates of β are $\hat{\beta}_{\hat{\rho}_0(793)}(793) \approx 0.19$ and $\hat{\beta}_{\hat{\rho}_0(815)}(815) \approx 0.15$, represented both graphically through straight lines. Doing the same procedure to Vanuatu Islands, we have $k_{h1} = \lfloor 647^{0.995} \rfloor = 626$ and $k_{h2} = \lfloor 647^{0.999} \rfloor = 642$, that is, the corresponding estimates of ρ are $\hat{\rho}_0(626) \approx -0.20$ and $\hat{\rho}_0(642) \approx -0.25$ and the corresponding estimates of β are $\hat{\beta}_{\hat{\rho}_0(626)}(626) \approx 0.51$ and $\hat{\beta}_{\hat{\rho}_0(642)}(642) \approx 0.44$.

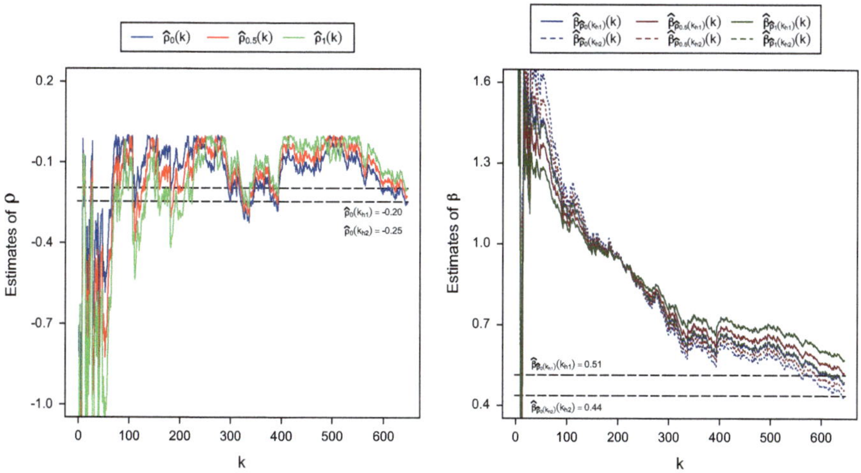

Fig. 6 Estimates of the second order parameters ρ (*left*) and β (*right*) for seismicity of Vanuatu Islands

Since from the $\hat{\beta}$ sample paths, there are no readily apparent significant differences between the use of k_{h1} or k_{h2}, and due to the fact that the tail index estimation is more affected by the ρ fluctuations than the β ones, we use both levels in the rest of the study.

Moreover, here we also present a possible optimal level k_0 of top observations to consider when the geometric-type estimator is used to estimate γ, through the minimization of the asymptotic mean square error ($AMSE$) of the geometric-type estimator. Considering the following distributional representation of the geometric-type estimator (see [3, Theorem 2.2]).

$$\widehat{GT}\,(k) \overset{D}{=} \gamma + \frac{\gamma}{2\sqrt{k}}Q_n - \frac{\gamma}{\sqrt{k}}P_n + \frac{A\left(\frac{n}{k}\right)}{(1-\rho)^2} + o_p\left(A\left(\frac{n}{k}\right)\right) + O_p\left(\frac{\log^2 k}{k}\right),$$

we get what we need to calculate the $AMSE(\widehat{GT})$ and provide for their minimization

$$\frac{\partial}{\partial k}\left[AMSE\left(\widehat{GT}\right)\right] = 0 \Longleftrightarrow \frac{\partial}{\partial k}\left[V\left(\widehat{GT}\right) + \left(Bias\left(\widehat{GT}\right)\right)^2\right] = 0$$

$$\Longleftrightarrow \frac{\partial}{\partial k}\left[\frac{2\gamma^2}{k} + \left(\frac{\gamma\beta}{(1-\rho)^2}\right)^2\left(\frac{n}{k}\right)^{2\rho}\right] = 0.$$

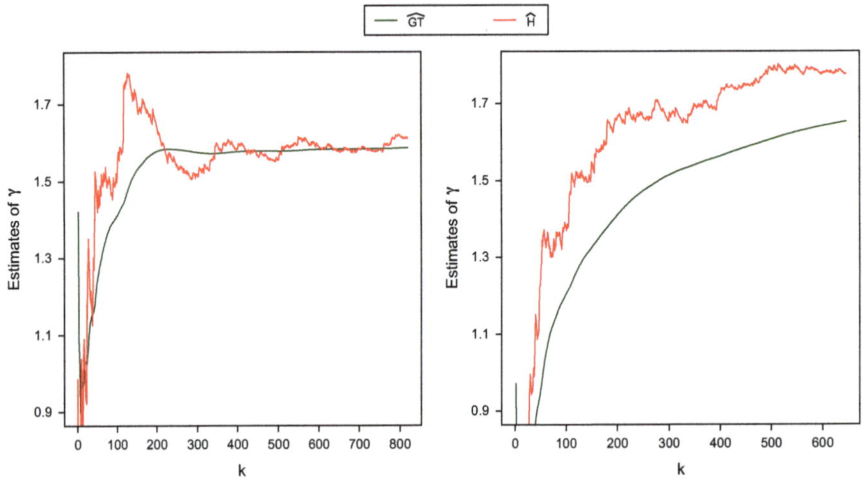

Fig. 7 Plot for the GT estimator, $\widehat{GT}$, and for the Hill estimator, $\hat{H}$, of γ, for seismicity of Philippines (*left*) and Vanuatu (*right*) Islands

Solving the equation in order to k and denoting the result as $k_0^{\widehat{GT}}$, we obtain

$$k_0^{\widehat{GT}} = \left[\frac{(1-\rho)^2}{-2\rho\beta^2}\right]^{1/(1-2\rho)} n^{-2\rho/(1-2\rho)}.$$

Although this is not the optimal value for the bias corrected estimators, the value of the tail index and quantiles calculated with the geometric-type estimator at the $k_0^{\widehat{GT}}$ level is represented in some illustrations for comparison.

As a first step, we estimate the tail index, γ, using the GT and Hill's estimators.

Concerning the shape parameter γ, Fig. 7 displays the estimated values of the GT and Hill estimators, as a function of k, for Philippines and Vanuatu Islands data. As can be observed, for Philippines Islands data both estimators stabilize around the same value of γ, which is 1.6, with identical scatter plots for moderate and high values of k, although it is worth to give emphasis to the smoothness that the geometric-type estimator displays.

For the Vanuatu Islands data, though not so explicit as to the Philippines data, the behavior of GT tends to stabilize around the value of 1.64 as k increases. The same is true for the Hill estimator around the value of 1.78, although in a slightly more erratic way.

The GT estimator presents the best performance specially for Philippines Islands data, displaying almost a straight line around 1.58 for k-values larger than 300.

In Fig. 8 it is possible to compare the behavior of the GT estimator with its corrected versions, $\overline{\widehat{GT}}$ and $\overline{\overline{\widehat{GT}}}$. We note that the corrected estimators maintain the good behavior; that is, they have less variation in the initial values of k, and

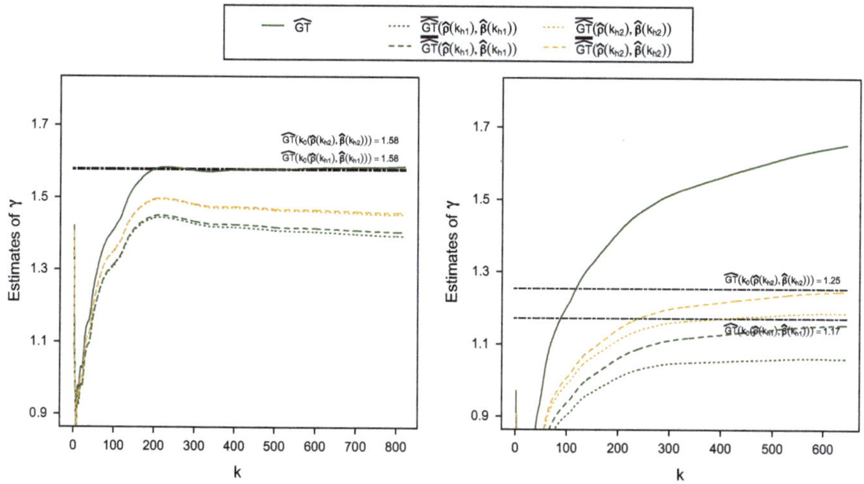

Fig. 8 Plot for the GT estimator, $\widehat{GT}$, and for the corresponding GT bias corrected estimators, $\overline{\widehat{GT}}$ and $\overline{\overline{\widehat{GT}}}$, of γ, for seismicity of Philippines (*left*) and Vanuatu (*right*) Islands

stabilize at slightly lower values than the uncorrected estimator. Depending on the unknown value of the tail index parameter that we seek, this type of behavior seems to be indicative of a better performance of the corrected estimators. Particularly for Vanuatu Islands data, this improvement seems to be evident since the corrected estimators begin to stabilize sooner than the non corrected ones, showing a very satisfactory behavior, to the right from the initial values of k.

In order to make the comparison between the bias corrected GT estimators and the Hill ones, we draw the sample paths of one against the other.

We might see from Fig. 9 that the estimates provided by the corrected Hill estimators are around the same values of the estimates given by the corrected GT estimators. However, it is quite clear that the Hill estimators hold a rather irregular behavior compared to the GT estimators, especially for smaller values of k.

It is suggestive that the value of γ that best describes the seismic moment of the Philippines Islands is a little below 1.5, and that of the Vanuatu Islands is slightly above 1.

As in most of the applications, the main interest lays not on the tail index but in the quantiles of the extreme distributions, which are more stable and robust. Now we analyze the sample paths of the quantiles estimators. We estimate the values of POT high quantiles estimator, in (5), based on the GT and Hill estimators, as a function of k, for Philippines and Vanuatu Islands data, considering the percentile 99 %. Each tail index estimator leads to a different estimation of large quantiles, which is also dependent on k. The straight dashed line represents the estimate of the empirical 99 % quantile. When more than one straight line is present, the empirical quantile is represented by the inferior one.

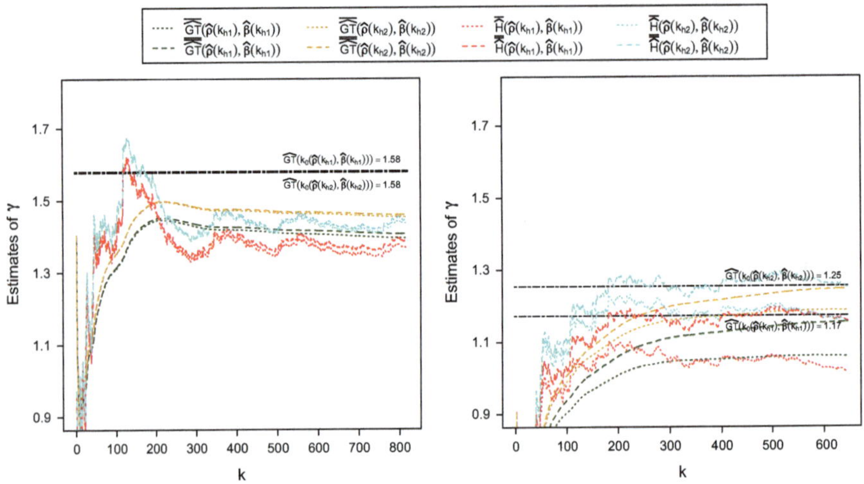

Fig. 9 Plot for the GT bias corrected estimators, $\overline{\widehat{GT}}$ and $\overline{\overline{\widehat{GT}}}$, and for the Hill ones, $\overline{\widehat{H}}$ and $\overline{\overline{\widehat{H}}}$, of γ, for seismicity of Philippines (*left*) and Vanuatu (*right*) Islands

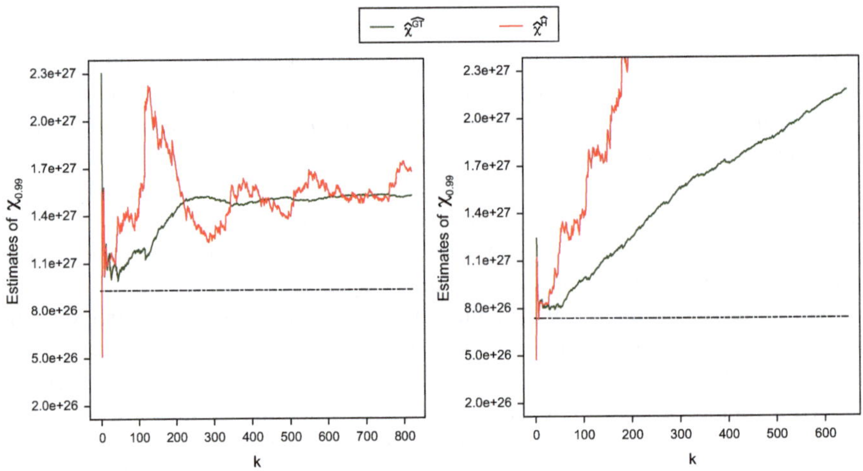

Fig. 10 Plot for the 99-quantile estimators based on the GT estimator, $\hat{\chi}^{\widehat{GT}}$, and on the Hill estimator, $\hat{\chi}^{\hat{H}}$, of $\chi_{0.99}$, for seismicity of Philippines (*left*) and Vanuatu (*right*) Islands (empirical quantiles $\chi_{0.99} = 9.29 \times 10^{26}$ and $\chi_{0.99} = 7.37 \times 10^{26}$, for Philippines and Vanuatu Islands, respectively)

We might see from Fig. 10 that, for the Philippines Islands, both estimates do not present values close to the empirical quantile. For values of k larger than 300, the estimates tend to stabilize, and it is apparent that this stabilization process is significantly more regular for the GT based quantiles estimator. The uneven performance that the Hill quantile plot shows make it extremely hard to decide upon

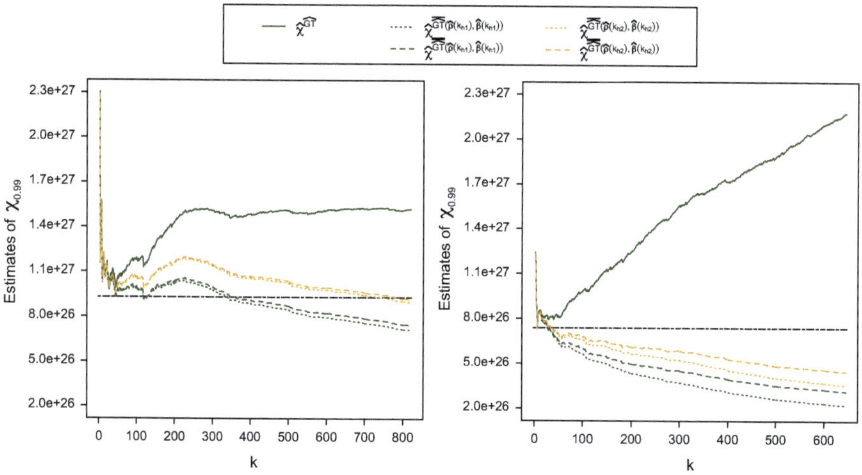

Fig. 11 Plot for the 99-quantile estimators based on the GT estimator, $\widehat{\chi}^{GT}$, and on the corresponding geometric-type bias corrected estimators, $\overline{\widehat{\chi}}^{GT}$ and $\overline{\overline{\widehat{\chi}}}^{GT}$, of $\chi_{0.99}$, for seismicity of Philippines (*left*) and Vanuatu (*right*) Islands (empirical quantiles $\chi_{0.99} = 9.29 \times 10^{26}$ and $\chi_{0.99} = 7.37 \times 10^{26}$, for Philippines and Vanuatu Islands, respectively)

a specific value for k. For the Vanuatu Islands the behavior of both estimators is not the best, but the Hill based quantiles estimator presents a much more irregular behavior.

Now comparing the GT based quantiles estimator with its corrected versions, we can observe in Fig. 11 that the improvement caused by the correction is quite remarkable. It is also worth noting that considering the k_{h2} level to estimate the second order parameters, the performance seems to be a little better. Also in Fig. 11, and for the Philippines Islands data, it can be seen that the quantile value calculated using the geometric-type estimator at its optimal levels $\widehat{k_0^{GT}}$, represented by the superior straight lines, almost coincides with the value of the quantiles estimator based on the geometric-type estimation for k-values larger than 200, which highlights the fairly stable behavior of this quantiles estimator in this range of values.

In Fig. 12, we can observe that the bias corrected Hill quantiles estimators present estimate values very similar to the ones presented by the bias corrected GT quantiles estimators. Although the corrected Hill quantiles estimators, using the k_{h2} level to compute the second order parameters, appear to have values more close to the empirical quantile than the corresponding corrected GT quantiles estimators, in case of Philippines Islands only for k-values greater that 300, their erratic and much less stable behavior may be a factor of considerable disadvantage.

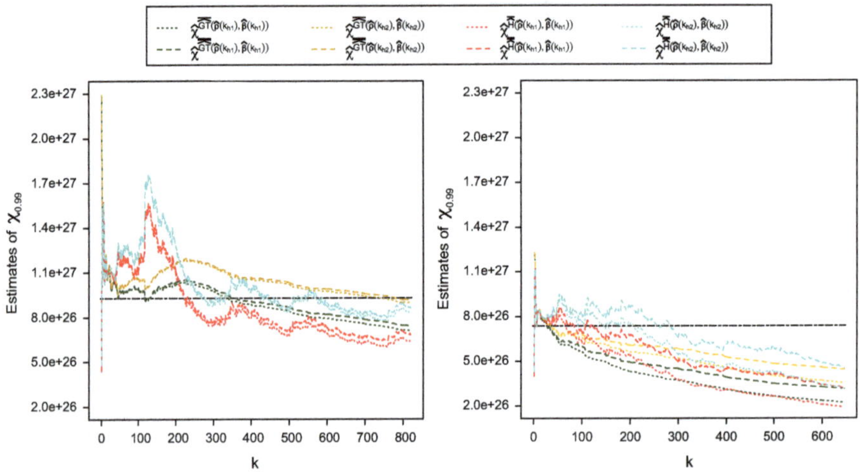

Fig. 12 Plot for the 99-quantile estimators based on the geometric-type bias corrected estimators, $\hat{\overline{\chi}}^{GT}$ and $\hat{\overline{\overline{\chi}}}^{GT}$, and on the Hill bias corrected estimators, $\hat{\chi}^{\overline{H}}$ and $\hat{\chi}^{\overline{\overline{H}}}$, of $\chi_{0.99}$, for seismicity of Philippines (*left*) and Vanuatu (*right*) Islands (empirical quantiles $\chi_{0.99} = 9.29 \times 10^{26}$ and $\chi_{0.99} = 7.37 \times 10^{26}$, for Philippines and Vanuatu Islands, respectively)

4 Final Considerations

In this study we consider the seismic moments of the Philippines and Vanuatu Islands larger than the level 10^{24} recorded during 35 years. We begin by analyzing the data in order to investigate the presence of heavy tails, the stationarity and the independence of the observations. In this way, we verify that the exceedances can be modeled by heavy tailed distributions. We use the geometric-type estimator and its bias corrected versions for estimating the tail index and high quantiles. For the sake of comparison we also consider the corresponding Hill estimators.

The geometric-type estimator shows a better performance when compared to the Hill estimator, namely it is worth emphasizing the contrast between the smoothed behavior of the geometric-type estimator and the irregular behavior exhibited by the Hill estimator.

It is well known that the considerable bias that appears in several estimators reveals a difficult problem that goes well beyond the application. In order to deal with this problem we also study and apply corrected versions of the geometric-type estimator. As expected, its performance is improved. We may emphasize that in some situations the Hill's bias corrected estimators present an erratic and less stable behavior. This is a real disadvantage for example in choosing a specific value for k.

In general, it is possible to conclude that the smoother behavior is a common quality shared by the estimates obtained for the GT tail index estimators, as by GT-based quantiles estimates, which show a very small variability, reflecting the more regular behavior of the GT estimators.

Regarding the case of Philippines Islands, and when considering the geometric-type estimator, we obtain an estimate for the seismic moment 0.99-quantile of 1.51×10^{27}. In a more practical way, we may say that it is expected that one out of a hundred earthquakes has a seismic moment larger than 1.51×10^{27}. Since in average there are 23.43 earthquakes per year, we may say that an earthquake exceeding a seismic moment of 1.51×10^{27} is expected to happen in Philippines Islands once in every 4.35 years. Moreover, we may also conclude that the probability of occurring an earthquake with seismic moment larger than 1.51×10^{27} next year is approximately $1 - 0.99^{23.43}$, that is, 21%.

As one knows, the performance of the estimators depends on the distribution of the data, and there is not an uniformly agreed best estimator. Nevertheless, from results of practical example conducted here, one could say that, for this type of data, the GT estimator turns out to be the best choice for tail index estimator, and the POT estimator when used for high quantiles.

On the whole, the application of the EVT to the problem under study seems quite promising since it provides reasonable estimates of the tails of the seismic moment distribution.

Acknowledgements ACMF is partially supported by FCT grant SFRH/BPD/66174/2009 and LC is supported by FCT grant SFRH/BD/60642/2009. All three authors are supported by FCT project PTDC/MAT/120346/2010. Research funded by the European Regional Development Fund through the programme COMPETE and by the Portuguese Government through the FCT—Fundação para a Ciência e a Tecnologia under the project PEst—C/MAT/UI0144/2013. The authors also thank the referees for their comments.

References

1. Beirlant, J., Goegebeur, Y., Segers, J., Teugels, J.: Statistics of Extremes: Theory and Applications. Wiley, Chichester (2004)
2. Brito, M., Freitas, A.C.M.: Limiting behaviour of a geometric estimator for tail indices. Insur. Math. Econ. **33**, 221–226 (2003)
3. Brito, M., Cavalcante, L., Freitas, A.C.M.: Bias corrected geometric-type estimators. Preprint CMUP 2014-6 (2014)
4. Caeiro, F., Gomes, M.I., Pestana, D.: Direct reduction of bias of the classical Hill estimator. Revstat **3**, 113–136 (2005)
5. Corral, A.: Dependence of earthquake recurrence times and independence of magnitudes on seismicity history. Tectonophysics **424**, 177–193 (2006)
6. Csörgő, S., Deheuvels, P., Mason, D.M.: Kernel estimates of the tail index of a distribution. Ann. Stat. **13**, 1050–1077 (1985)
7. Day, R.W.: Geotechnical Earthquake Engineering McGraw-Hill, New york (2002)
8. de Haan, L., Rootzén, H.: On the estimation of high quantiles. J. Stat. Plann. Inference **35**, 1–13 (1993)
9. Deheuvels, P., Haeusler, E., Mason, D.M.: Almost sure convergence of the Hill estimator. Math. Proc. Camb. Philos. Soc. **104**, 371–381 (1988)
10. Dekkers, A.L.M., Einmahl, J.H.J., de Haan, L.: A moment estimator for the index of an extreme-value distribution. Ann. Stat. **17**, 1833–1855 (1989)

11. Dziewonski, A.M., Chou, T.-A., Woodhouse, J.H.: Determination of earthquake source parameters from waveform data for studies of global and regional seismicity. J. Geophys. Res. **86**, 2825–2852 (1981)
12. Ekström, G., Nettles, M., Dziewonski, A.M.: The global CMT project 2004–2010: centroid-moment tensors for 13,017 earthquakes. Phys. Earth Planet. Inter. **200–201**, 1–9 (2012)
13. Fraga Alves, M.I., Gomes, M.I., de Haan, L.: A new class of semi-parametric estimators of the second order parameter. Port. Math. **60**, 193–213 (2003)
14. Global CMT Catalogue: Available from http://www.globalcmt.org/ (2013). Last accessed Aug 2013
15. Gomes, M.I., Martins, M.J.: "Asymptotically unbiased" estimators of the tail index based on external estimation of the second order parameter. Extremes **5**, 5–31 (2002)
16. Gomes, M.I, Martins, M.J., Neves, M.: Improving second order reduced bias extreme value index estimation. Revstat **5**, 177–207 (2007)
17. Haeusler, E., Teugels, J.L.: On asymptotic normality of Hill's estimator for the exponent of regular variation. Ann. Stat. **13**, 743–756 (1985)
18. Hill, B.M.: A simple approach to inference about the tail of a distribution. Ann. Stat. **3**, 1163–1174 (1975)
19. Howell, B.F. Jr.: An Introduction to Seismological Research. Cambridge University Press, Cambridge (1990)
20. Pickands, J.: Statistical inference using extreme order statistics. Ann. Stat. **3**, 119–13 (1975)
21. Pisarenko, V.F., Sornette, D.: Characterization of the frequency of extreme events by Generalised Pareto Distribution. Pure Appl. Geophys. **160**, 2343–2364 (2003)
22. Pisarenko, V.F., Sornette, D., Rodkin, M.V.: Distribution of maximum earthquake magnitudes in future time intervals, application to the seismicity of Japan (1923–2007). Earth Planets Space **62**, 567–578 (2010)
23. Weissman, I.: Estimation of parameters and large quantiles based on the k largest observations. J. Am. Stat. **73**, 812–815 (1978)

Detonation Wave Solutions and Linear Stability in a Four Component Gas with Bimolecular Chemical Reaction

F. Carvalho, A.W. Silva, and A.J. Soares

Abstract We consider a four component gas undergoing a bimolecular chemical reaction of type $A_1 + A_2 \rightleftharpoons A_3 + A_4$, described by the Boltzmann equation (BE) for chemically reactive mixtures. We adopt hard-spheres elastic cross sections and modified line-of-centers reactive cross sections depending on both the activation energy and geometry of the reactive collisions. Then we consider the hydrodynamic limit specified by the reactive Euler equations, in an earlier stage of the chemical reaction, when the gas is far from equilibrium (slow chemical reaction). In particular, the rate of the chemical reaction obtained in this limit shows an explicit dependence on the reaction heat and on the activation energy. Starting from this kinetic setting, we study the dynamics of planar detonation waves for the considered reactive gas and characterize the structure of the steady detonation solution. Then, the problem of the hydrodynamic linear stability of the detonation solution is treated, investigating the response of the steady solution to small rear boundary perturbations. A numerical shooting technique is used to determine the unstable modes in a pertinent parametric space for the considered problem. Numerical simulations are performed for the Hydrogen-Oxygen system and some representative results are presented, regarding the steady detonation wave solution and linear stability.

F. Carvalho (✉)
Instituto Politécnico de Viana do Castelo, Centro de Matemática, Universidade do Minho, Braga, Portugal
e-mail: filipecarvalho@esce.ipvc.pt

A.J. Soares
Centro de Matemática, Universidade do Minho, Braga, Portugal
e-mail: ajsoares@math.uminho.pt

A.W. Silva
Instituto Federal do Paraná, Curitiba, Brazil
e-mail: adriano.silva@ifpr.edu.br

J.-P. Bourguignon et al. (eds.), *Mathematics of Energy and Climate Change*, CIM Series in Mathematical Sciences 2, DOI 10.1007/978-3-319-16121-1_4

1 Introduction

In a recent paper, Carvalho and Soares [1] consider a model for a two component reacting gas mixture in the framework of the Boltzmann equation and develop a detailed analysis of the dynamics and linear stability of steady detonation wave. This analysis refers to a theoretical detonating mixture of constituents A and B undergoing a generic reversible reaction of type $A+A \rightleftharpoons B+B$. The mathematical treatment of the linear stability of steady detonation waves developed in [1] is rather satisfactory and the numerical technique proposed there can be viewed as an efficient procedure to study the stability problem. Paper [1] also includes an extensive investigation of the stability problem and the results obtained numerically show a rather good qualitative agreement with other results known in literature. However, paper [1] does not explore the study of concrete detonation examples neither the validation of the proposed numerical procedure with respect to the available experimental data. This can be an interesting improvement of the results presented in paper [1].

On the other hand, there exists an increasing interest in detonation physics, from both the experimental and numerical point of view, due to the related engineering applications, and safety and military issues. Experimental observations and numerical studies [2–5] indicate that the detonation, especially in gases, tends to be unstable. Therefore, the stability analysis of detonation waves remains an interesting topic that has been quite investigated in recent years due to the computational advances.

Motivated by all these aspects, in the present paper we apply the numerical procedure proposed in paper [1] to a different chemically reactive system and investigate the stability of detonation waves in a concrete explosive Hydrogen-Oxygen system. This kinetic formulation refers to a four component gas and adopts a more realistic model of reactive cross sections which modifies the standard line-of-centers model by introducing the dependence on the geometry of the reactive collision. Then we study the dynamics and hydrodynamic linear stability of steady detonation wave solutions, described by the reactive Euler equations obtained in the hydrodynamic limit proper of the initial stage of the chemical reaction.

Numerical simulations are performed for the Hydrogen-Oxygen system and some representative results are presented, regarding the steady detonation wave solution and linear stability.

The present paper constitutes the first part of a work in progress and the numerical results presented here are still limited. We intend to develop a more detailed numerical analysis of the stability problem and, at the same time, to compare our results with other numerical and experimental results available in literature. We expect that such comparisons can be used to reinforce the validity of the numerical procedure presented in paper [1] and consolidate the robustness of the kinetic model proposed in paper [6].

2 The Reactive System Modelling

The model adopted in this paper for the reacting gaseous mixture is the one proposed in paper [6] for a four component mixture of constituents A_1, A_2, A_3 and A_4 undergoing the reversible bimolecular chemical reaction

$$A_1 + A_2 \rightleftharpoons A_3 + A_4. \tag{1}$$

Here we include the principal features of the model, with more emphasis on those aspects necessary for our analysis. For a detailed description of the model, see paper [6] and also paper [7] for the foundational aspects of the theory.

2.1 Modelling Aspects

The constituents of the gas have molecular masses m_1, m_2, m_3 and m_4, molecular diameters d_1, d_2, d_3 and d_4, and binding energies E_1, E_2, E_3 and E_4, respectively. The heat of the chemical reaction is specified by the balance of the binding energies as $Q_R = E_3 + E_4 - E_1 - E_2$. Molecular masses are such that $m_1 + m_2 = m_3 + m_4$, as prescribed by the chemical law. The molecules collide among themselves through binary elastic scattering, and reactive encounters according to the chemical law (1). For elastic scattering, the differential cross sections $\sigma_{\alpha\beta}$ are assumed to correspond to a hard-sphere potential,

$$\sigma_{\alpha\beta} = d_{\alpha\beta}^2, \qquad \text{with} \quad d_{\alpha\beta} = \frac{1}{2}(d_\alpha + d_\beta). \tag{2}$$

For reactive encounters, the differential cross sections are assumed with activation energy and dependent on the geometry of the collision, given by

$$\sigma_{12}^\star = \begin{cases} 0 & \text{for} \quad \gamma_{12} < \varepsilon_f^\star, \\ \mathbf{s}_f d_{12}^2 \left[1 - \frac{2\varepsilon_f}{\mu_{12}(g_{12}\cdot\mathbf{k}_{12})^2}\right] & \text{for} \quad \gamma_{12} \geq \varepsilon_f^\star, \end{cases} \tag{3}$$

$$\sigma_{34}^\star = \begin{cases} 0 & \text{for} \quad \gamma_{34} < \varepsilon_r^\star, \\ \mathbf{s}_r d_{34}^2 \left[1 - \frac{2\varepsilon_r}{\mu_{34}(g_{34}\cdot\mathbf{k}_{34})^2}\right] & \text{for} \quad \gamma_{34} \geq \varepsilon_r^\star, \end{cases} \tag{4}$$

where μ_{12} and μ_{34} are reduced masses, γ_{12} and γ_{34} relative translational energies in the direction of the line joining the centers of the colliding molecules, ε_f and ε_r forward and reverse activation energies, $\mathbf{s}_f$ and $\mathbf{s}_r$ are the corresponding steric factors, $\mathbf{k}_{12}$ and $\mathbf{k}_{34}$ unit collision vectors joining the centers of the two colliding molecules pointing from the center of the A_2 and A_4-particle to the center of A_1 and A_3-particle. Moreover, g_{12} is the pre-collisional asymptotic relative velocity of the constituents A_1 and A_2, and g_{34} is the pre-collisional asymptotic relative velocity of

A_3 and A_4. These kinetic parameters are given by

$$\mu_{12} = \frac{m_1 m_2}{m_1 + m_2}, \qquad \mu_{34} = \frac{m_3 m_4}{m_3 + m_4},$$

$$\gamma_{12} = \frac{\mu_{12}\,(g_{12}\cdot\mathbf{k}_{12})^2}{2kT}, \quad \gamma_{34} = \frac{\mu_{34}\,(g_{34}\cdot\mathbf{k}_{34})^2}{2kT},$$

$$\varepsilon_f^\star = \frac{\varepsilon_f}{kT}, \quad \varepsilon_r^\star = \frac{\varepsilon_r}{kT}, \quad Q_R^\star = \frac{Q_R}{kT} \quad \text{with} \quad \varepsilon_r^\star \equiv \varepsilon_f^\star - Q_R^\star,$$

where k represents the Boltzmann constant and T the mixture temperature. Definitions (3) and (4) mean that a reactive collision occurs only when the relative translational energy in the direction of the line joining the centers of the molecules is larger than the activation energy.

Assuming that relativistic and quantum effects are absent, elastic collisions obey the classical laws of mechanics. Therefore, elastic collisions between A_α and A_β molecules, with asymptotic pre-collisional velocities $\mathbf{c}_\alpha$ and $\mathbf{c}_\beta$ and asymptotic post-collisional velocities $\mathbf{c}'_\alpha$ and $\mathbf{c}'_\beta$, respect the following conservation laws of linear momentum and total energy,

$$m_\alpha \mathbf{c}_\alpha + m_\beta \mathbf{c}_\beta = m_\alpha \mathbf{c}'_\alpha + m_\beta \mathbf{c}'_\beta, \tag{5}$$

$$\frac{1}{2} m_\alpha \mathbf{c}_\alpha^2 + \frac{1}{2} m_\beta \mathbf{c}_\beta^2 = \frac{1}{2} m_\alpha \mathbf{c}_\alpha'^2 + \frac{1}{2} m_\beta \mathbf{c}_\beta'^2. \tag{6}$$

Furthermore, reactive collisions respect the following conservation laws of linear momentum and total energy (kinetic plus chemical link energy)

$$m_1 \mathbf{c}_1 + m_2 \mathbf{c}_2 = m_3 \mathbf{c}_3 + m_4 \mathbf{c}_4, \tag{7}$$

$$\frac{1}{2} m_1 \mathbf{c}_1^2 + \frac{1}{2} m_2 \mathbf{c}_2^2 = \frac{1}{2} m_3 \mathbf{c}_3^2 + \frac{1}{2} m_4 \mathbf{c}_4^2 + Q_R. \tag{8}$$

2.2 The Model Equations

The state of a reacting gaseous mixture in the phase space (spanned by the positions $\mathbf{x}$ and velocities $\mathbf{c}_\alpha$) is characterized, at the mesoscopic level, by the set of distribution functions $f_\alpha \equiv f(\mathbf{x}, \mathbf{c}_\alpha, t)$, with $\alpha = 1, \ldots, 4$, in such a way that the number of molecules of the contituent A_α in the volume element $d\mathbf{x}d\mathbf{c}_\alpha$ around the position $\mathbf{x}$ and velocity $\mathbf{c}_\alpha$, at time t, is given by $f_\alpha d\mathbf{x}d\mathbf{c}_\alpha$.

The reactive Boltzmann equation that describes the phase space evolution of the distribution functions f_α, if we consider no external forces and neglect internal degrees of freedom, is given by

$$\frac{\partial f_\alpha}{\partial t} + \sum_{i=1}^{3} c_i^\alpha \frac{\partial f_\alpha}{\partial x_i} = \mathcal{Q}_\alpha^E + \mathcal{Q}_\alpha^R. \tag{9}$$

Above, $\mathcal{Q}_\alpha^E$ and $\mathcal{Q}_\alpha^R$ represent the elastic and the reactive collision terms, respectively, and might be defined as follows

$$\mathcal{Q}_\alpha^E = \sum_{\beta=1}^{4} \mathcal{Q}_{\alpha\beta}^E, \quad \text{with} \quad \mathcal{Q}_{\alpha\beta}^E = \int \Big(f'_\alpha f'_\beta - f_\alpha f_\beta\Big) d_{\alpha\beta}^2 \big(\mathbf{g}_{\beta\alpha} \cdot \mathbf{k}_{\beta\alpha}\big) d\mathbf{k}_{\beta\alpha} d\mathbf{c}_\beta, \tag{10}$$

$$\mathcal{Q}_{1(2)}^R = \int \left[f_3 f_4 \left(\frac{m_1 m_2}{m_3 m_4}\right)^3 - f_1 f_2 \right] \sigma_{12}^\star \left(g_{12} \cdot \mathbf{k}_{12}\right) d\mathbf{k}_{12} d\mathbf{c}_{2(1)}, \tag{11}$$

$$\mathcal{Q}_{3(4)}^R = \int \left[f_1 f_2 \left(\frac{m_3 m_4}{m_1 m_2}\right)^3 - f_3 f_4 \right] \sigma_{34}^\star \left(g_{34} \cdot \mathbf{k}_{34}\right) d\mathbf{k}_{34} d\mathbf{c}_{4(3)}. \tag{12}$$

The elastic terms $\mathcal{Q}_\alpha^E$ incorporate the mixture effects whereas the reactive terms $\mathcal{Q}_\alpha^R$ include all other effects associated to the chemical reaction, in particular a redistribution of mass and transfer of energy.

2.3 *The Consistency of the Model*

Some properties are very important in order to show the mathematical and physical consistency of the model. One of these properties states that elastic collisions do not modify the number of molecules of each constituent. This result is ensured by the following statement about the elastic terms defined in (10),

$$\int_{\mathbb{R}^3} \mathcal{Q}_\alpha^E \, d\mathbf{c}_\alpha = 0, \qquad \alpha = 1, 2, 3, 4. \tag{13}$$

On the other hand, the reactive encounters imply that the variation of the number of molecules of constituents A_1 and A_2 is the same and, at the same time, it is opposite to the variation of the number of molecules of constituents A_3 and A_4. This result is stated by the following property on the reactive terms defined in (11) and (12),

$$\int_{\mathbb{R}^3} \mathcal{Q}_1^R \, d\mathbf{c}_1 = \int_{\mathbb{R}^3} \mathcal{Q}_2^R \, d\mathbf{c}_2 = -\int_{\mathbb{R}^3} \mathcal{Q}_3^R \, d\mathbf{c}_3 = -\int_{\mathbb{R}^3} \mathcal{Q}_4^R \, d\mathbf{c}_4. \tag{14}$$

There are some known physical collisional invariants, that is, macroscopic quantities that do not chance during an elastic collision or reactive encounter. As a consequence, a good and consistent model must reflect this situation. From the mathematical point of view, a function $\psi = (\psi_1, \psi_2, \psi_3, \psi_4)$ is a collisional invariant for the considered model (9)–(12) if the following conditions hold

$$\sum_{\alpha=1}^{4} \int_{\mathbb{R}^3} \psi_\alpha \left(\mathcal{Q}_\alpha^E + \mathcal{Q}_\alpha^R\right) d\mathbf{c}_\alpha = 0. \tag{15}$$

The present modelling ensures the conservation of the partial number densities of certain pairs of constituents, namely one reactant and one product. This is a consequence of properties (13) and (14) and reproduces the correct balance of chemical exchange rates. The corresponding collision invariants can be chosen as suitable functions $\psi = (\psi_1, \psi_2, \psi_3, \psi_4)$, defined by $(\psi_1, \psi_2, \psi_3, \psi_4) = (1, -1, 0, 0)$, $(\psi_1, \psi_2, \psi_3, \psi_4) = (0, 1, 1, 0)$, $(\psi_1, \psi_2, \psi_3, \psi_4) = (0, 1, 0, 1)$. Moreover, the molecular conservation laws (5)–(8) imply the conservation of the linear momentum components and total energy of the mixture. The corresponding collision invariants can be assumed as functions $\psi = (\psi_1, \psi_2, \psi_3, \psi_4)$ such that $\psi_\alpha = m_\alpha c_1^\alpha$, $\psi_\alpha = m_\alpha c_2^\alpha$, $\psi_\alpha = m_\alpha c_3^\alpha$, for the linear momentum components, and $\psi_\alpha = E_\alpha + \frac{1}{2}\mathbf{c}_\alpha^2 m_\alpha$ for the total energy. In the present model, the set of all collisional invariants constitute a 7-dimensional linear space.

The consistency of the model based on the properties stated in this subsection allow to derive the macroscopic picture of the kinetic modelling in terms of certain macroscopic variables and balance equations specifying the evolution of such variables. In particular, the macroscopic variables are defined as average quantities taken over the distribution functions f_α by integrating with respect to the velocities $\mathbf{c}_\alpha$ (see paper [6] for the definitions). The balance equations, in turn, are obtained as a set of seven conservation equations associated to the collisional invariants introduced above, together with the rate equation of the model specifying the evolution of the progress variable of the chemical reaction.

For sake of brevity, these equations are omitted here in their general formulation. For the analysis developed in the present paper, it is enough to consider the one-dimensional version of these equations, formulated in its hydrodynamic limit of Euler level, and this will be the main subject of Sect. 3.

2.4 Thermodynamical Equilibrium

The reactive mixture is in thermodynamical equilibrium when the elastic and reactive collisional terms are such that

$$\mathscr{Q}_\alpha^E + \mathscr{Q}_\alpha^R = 0, \qquad \alpha = 1, \ldots, 4. \tag{16}$$

In particular, for the present model, condition (16) implies the vanishing of the elastic collisional terms, that is

$$\mathscr{Q}_\alpha^E = 0, \qquad \alpha = 1, \ldots, 4, \tag{17}$$

and therefore condition (17) defines a state known in literature as a state of mechanical equilibrium. When all constituents are at the same temperature T, the mixture reaches a state of mechanical equilibrium if and only if the distribution functions f_α are Maxwellians, defined by

$$f_\alpha^M = n_\alpha \left(\frac{m_\alpha}{2\pi kT}\right)^{\frac{3}{2}} \exp\left[-\frac{m_\alpha(\mathbf{c}_\alpha - \mathbf{v})^2}{2kT}\right], \quad \alpha = 1, \dots, 4, \tag{18}$$

where n_α is the number density of constituent A_α, and $\mathbf{v}$ is the mean velocity of the whole mixture, (see paper [6] for the definitions). The above Maxwellians (18) do not ensure, in general, the vanishing of the reactive collisional operators and thus do not define a state of thermodynamical equilibrium for the reactive mixture. The only distribution function that ensures the thermodynamical equilibrium is the thermodynamical Maxwellian distribution given by

$$M_\alpha = n_\alpha^{\text{eq}} \left(\frac{m_\alpha}{2\pi kT}\right)^{\frac{3}{2}} \exp\left[-\frac{m_\alpha(\mathbf{c}_\alpha - \mathbf{v})^2}{2kT}\right], \quad \alpha = 1, \dots, 4, \tag{19}$$

where n_α^{eq}, for $\alpha = 1, \dots, 4$, represent number densities constrained to the law of mass action for the considered model, namely

$$\ln\left[\frac{n_1^{\text{eq}} n_2^{\text{eq}}}{n_3^{\text{eq}} n_4^{\text{eq}}} \left(\frac{\mu_{34}}{\mu_{12}}\right)^{\frac{3}{2}}\right] = Q_R^\star, \tag{20}$$

which represents the chemical equilibrium condition of the model. Distribution functions (19) define the unique equilibrium solutions of Eq. (9).

3 The Reactive Euler Equations in the Hydrodynamic Limit

The reactive Euler equations of the model can be derived from the Boltzmann equations (9), when an approximate solution of Eq. (9) has been obtained for a prescribed chemical regime.

3.1 Approximate Solution of the Boltzmann Equation

In the present study, we assume that the chemical reaction is in its initial stage corresponding to consider a slow reaction for which the gas mixture is far from chemical equilibrium. In this regime, the elastic collisions are more frequent than reactive encounters. Using the Chapman-Enskog methodology, which is rather common in kinetic theory [8], it is possible to obtain an approximate solution of Eq. (9) consistent with the prescribed chemical regime.

In paper [6], starting from the appropriate scaling of the Eq. (9), the Chapman-Enskog methodology has been combined with second-order Sonine expansions of the distribution functions f_α around the Maxwellians f_α^M defined in Eq. (18), and the authors obtained the following approximate solution,

$$f_\alpha = f_\alpha^M \left[1 + a_\alpha \left(\frac{15}{8} - \frac{5m_\alpha (c_\alpha - v)^2}{4kT} + \frac{m_\alpha^2 (c_\alpha - v)^4}{8k^2T^2} \right) \right], \quad \alpha = 1, 2, 3, 4, \tag{21}$$

where the coefficients a_α are determined by the following equations

$$-\frac{(1 - e^{-\mathscr{A}^\star}) \mathbf{s}_f d_{12}^2 e^{-\varepsilon_f^\star} \left[1 - 4\varepsilon_f^\star + 3\varepsilon_f^\star E(\varepsilon_f^\star) e^{\varepsilon_f^\star} \right] \mu_{12}^{3/2} n_1 n_2}{8m_\alpha^2} \tag{22}$$

$$= \sum_{\beta=1}^{4} \frac{n_\alpha n_\beta \sqrt{\mu_{\alpha\beta}}}{(m_\alpha + m_\beta)^2} d_{\alpha\beta}^2 \left\{ \frac{10m_\alpha^2 + 8m_\alpha m_\beta + 13m_\beta^2}{m_\alpha + m_\beta} a_\alpha - 15\mu_{\alpha\beta} a_\beta \right\}$$

for $\alpha = 1, 2,$

$$\left\{ 1 - 4\varepsilon_f^\star + 3\varepsilon_f^\star E(\varepsilon_f^\star) e^{\varepsilon_f^\star} - 4Q_R^\star \left[1 + Q_R^\star - \varepsilon_f^\star E(\varepsilon_f^\star) e^{\varepsilon_f^\star} (3 - Q_R^\star) \right] \right\} \tag{23}$$

$$\times \frac{(1 - e^{-\mathscr{A}^\star}) \mathbf{s}_f d_{12}^2 e^{-\varepsilon_f^\star} \mu_{34}^5 n_1 n_2}{8m_\alpha^2 \mu_{12}^{7/2}}$$

$$= \sum_{\beta=1}^{4} \frac{n_\alpha n_\beta \sqrt{\mu_{\alpha\beta}}}{(m_\alpha + m_\beta)^2} d_{\alpha\beta}^2 \left\{ \frac{10m_\alpha^2 + 8m_\alpha m_\beta + 13m_\beta^2}{m_\alpha + m_\beta} a_\alpha - 15\mu_{\alpha\beta} a_\beta \right\}$$

for $\alpha = 3, 4.$

Above, $\mathscr{A}^\star$ is the affinity of the chemical reaction [8], and $E(\varepsilon_f^\star)$ represents the exponential integral $E(\varepsilon_f^\star) = \int_{\varepsilon_f^\star}^{+\infty} \frac{e^{-y} dy}{y}$. In the present paper, we omit the details of the methodology, they are given in paper [6].

The approximate solution (21) with coefficients specified by expressions (22) and (23) includes the non-equilibrium effects induced by the chemical reaction.

3.2 Reactive Euler Equations

The reactive Euler equations of the model are obtained from the balance equations, here omitted for sake of brevity, when all macroscopic quantities are expressed in terms of the approximate solution (21)–(23). They have the form

$$\frac{\partial}{\partial t} n_2 + \sum_{i=1}^{3} \frac{\partial}{\partial x_i} (n_2 v_i) = \tau_2, \tag{24}$$

$$\frac{\partial}{\partial t}(n_1 - n_2) + \sum_{i=1}^{3} \frac{\partial}{\partial x_i}((n_1 - n_2)v_i) = 0, \tag{25}$$

$$\frac{\partial}{\partial t}(n_2 + n_3) + \sum_{i=1}^{3} \frac{\partial}{\partial x_i}((n_2 + n_3)v_i) = 0, \tag{26}$$

$$\frac{\partial}{\partial t}(n_2 + n_4) + \sum_{i=1}^{3} \frac{\partial}{\partial x_i}((n_2 + n_4)v_i) = 0, \tag{27}$$

$$\sum_{i=1}^{3} \frac{\partial}{\partial t}(\varrho v_i) + \sum_{i=1}^{3}\sum_{j=1}^{3} \frac{\partial}{\partial x_j}(p\delta_{ij} + \varrho v_i v_j) = 0, \tag{28}$$

$$\frac{\partial}{\partial t}\left[\frac{3}{2}nkT + \sum_{\alpha=1}^{4} n_\alpha E_\alpha + \frac{1}{2}\varrho v^2\right] + \sum_{i=1}^{3} \frac{\partial}{\partial x_i}\left[\sum_{j=1}^{3} p\delta_{ij}v_j + \left(\frac{3}{2}nkT + \sum_{\alpha=1}^{4} n_\alpha E_\alpha + \frac{1}{2}\varrho v^2\right) v_i\right] = 0, \tag{29}$$

where v_i, n, ρ and p are spatial components of the mean velocity, number density, mass density and pressure of the mixture. The production term τ_2 in Eq. (24) is the reaction rate which specifies the progress of the chemical reaction, given by

$$\tau_2 = n_3 n_4 \tau_r - n_1 n_2 \tau_f, \tag{30}$$

where τ_f and τ_r are forward and backward reaction rates given by

$$\tau_f = \tau^{(0)} \left\{1 - \frac{1 - 4\varepsilon_f^\star + 3\varepsilon_f^\star E(\varepsilon_f^\star)e^{\varepsilon_f^\star}}{1 - \varepsilon_f^\star E(\varepsilon_f^\star)e^{\varepsilon_f^\star}} \frac{a_1 m_2^2 + a_2 m_1^2}{8(m_1 + m_2)^2}\right\}, \tag{31}$$

$$\tau_r = \left(\frac{m_1 m_2}{m_3 m_4}\right)^{\frac{3}{2}} e^{Q_R^\star} \tau^{(0)} \left\{1 - \frac{a_3 m_4^2 + a_4 m_3^2}{8(m_3 + m_4)^2} \times \frac{1 - 4\left(\varepsilon_f^\star + Q_R^\star + Q_R^{\star 2}\right) + \varepsilon_f^\star E(\varepsilon_f^\star)e^{\varepsilon_f^\star}\left(3 + 12Q_R^\star + 4Q_R^{\star 2}\right)}{1 - \varepsilon_f^\star E(\varepsilon_f^\star)e^{\varepsilon_f^\star}}\right\}, \tag{32}$$

with $\tau^{(0)}$ being the forward reaction rate coefficient given by

$$\tau^{(0)} = \sqrt{\frac{8\pi kT}{\mu_{12}}} \mathsf{s}_f^2 d_{12}^2 e^{-\varepsilon_f^\star} \left(1 - \varepsilon_f^\star E(\varepsilon_f^\star)e^{\varepsilon_f^\star}\right). \tag{33}$$

Equations (24)–(29) are the reactive Euler equations of the considered model in the adopted chemical regime. They define a closed system and constitute the governing equations of the model. Formally, such equations are similar to the corresponding ones obtained from a phenomenological theory in fluid dynamics. The interesting feature of such equations is that the reaction rate τ_2 has been constructed from a kinetic approach and then has an explicit representation completely justified by the microscopic kinetic model.

4 Steady Detonation Wave Solutions

In this section we use the model of Sects. 2 and 3 to study the problem of the propagation of steady detonation waves in an explosive quaternary mixture, following the qualitative description of the Zeldovich, von Neumann and Doering (ZND) theory [2, 3].

4.1 The ZND Model of Detonation

The well known ZND theory proposes a very simple physical model of detonation with finite chemical reaction zone. The ZND configuration of the detonation solution is represented in Fig. 1. The solution consists of a strong planar, non-reactive, shock front propagating with constant velocity D, greater or equal to its minimum allowed value which is called the Chapman-Jouguet velocity, towards a quiescent gas mixture ahead of the wave. The shock front compresses the mixture, renders the pressure to very high values so that the ignition process takes place. An exothermic chemical reaction initiates and takes place in the finite reaction zone

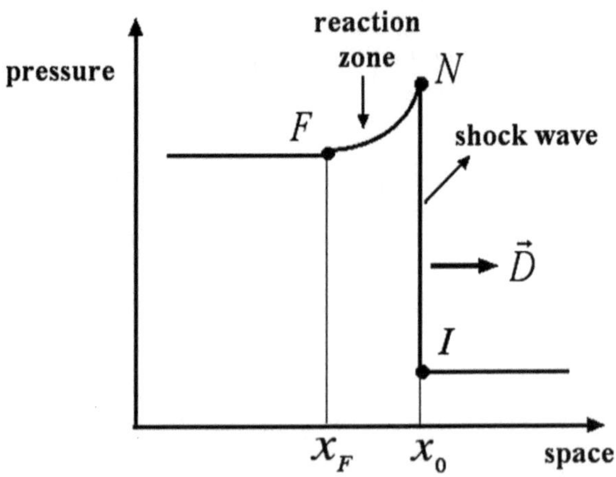

Fig. 1 ZND profile of a one dimensional steady detonation wave

following the shock wave until the equilibrium is reached. The initial state of the quiescent mixture, ahead of the shock wave, is denoted by I. The von Neumann state, just ahead of the shock, represents the state with very high pressure where the chemical reaction initiates. The chemical reaction proceeds in the reaction zone of finite length attached to the shock front, until the equilibrium final state denoted by F. The entire ZND configuration is steady with respect to the shock wave front.

The ZND theory gives a simplified but recognized description of a steady detonation wave. It is commonly used in literature as the first step in understanding and explaining the complex dynamics of real detonations in gases.

4.2 The Mathematical Approach

From the mathematical point of view, the ZND detonation solution is described by the reactive Euler equations (24)–(29), formulated in the one dimensional form and referred to the steady normalized frame attached to the shock wave. The governing equations for the detonation problem become

$$\frac{d}{dx}\Big[(v-D)\,n_2\Big] = Dt_c\tau_2, \tag{34}$$

$$\frac{d}{dx}\Big[(v-D)\,(n_1-n_2)\Big] = 0, \tag{35}$$

$$\frac{d}{dx}\Big[(v-D)\,(n_2+n_3)\Big] = 0, \tag{36}$$

$$\frac{d}{dx}\Big[(v-D)\,(n_2+n_4)\Big] = 0, \tag{37}$$

$$\frac{d}{dx}\Big[(v-D)\,\varrho v + nkT\Big] = 0, \tag{38}$$

$$\frac{d}{dx}\left[(v-D)\left(\frac{3}{2}nkT + \frac{\varrho v^2}{2} + \sum_{\alpha=1}^{4} E_\alpha n_\alpha\right) + nkTv\right] = 0, \tag{39}$$

where D is the constant shock wave velocity, $x_s = \frac{x-Dt}{Dt_c}$ the normalized steady variable, $t_c = \frac{1}{4n^+d_{12}^2}\sqrt{\frac{M}{\pi kT^+}}$ a characteristic time, where $M = m_1 + m_2$ is the total mass of the reactants and the superscript $+$ refers to the initial state I. For sake of simplicity, the steady variable x_s will still be denoted by the plain symbol x.

The steady detonation wave problem is solved in two different steps. In the first step, we solve the shock problem to characterize the von Neumann state. This is a pure algebraic problem associated to the Rankine-Hugoniot (RH) jump conditions, and no chemistry is involved. In the second step, we characterize all states within the reaction zone. This is a differential problem associated to the rate law of the chemical reaction and the chemistry plays a relevant role.

Von Neumann State Since no chemistry is involved in the shock problem, the rate equation (34) becomes of conservative type. The integration of the resulting system (34)–(39) between the initial state I and the von Neumann state N leads to the RH jump conditions in the form

$$n_\alpha (v - D) = -n_\alpha^+ D, \quad \alpha = 1, 2, 3, 4, \tag{40}$$

$$\varrho v (v - D) + nkT = n^+ kT^+, \tag{41}$$

$$\left(\frac{3}{2} nkT + \frac{\varrho v^2}{2} + \sum_{\alpha=1}^{4} E_\alpha n_\alpha\right)(v - D) + nkTv \tag{42}$$

$$= -\left(\frac{3}{2} n^+ kT^+ + \sum_{\alpha=1}^{4} E_\alpha n_\alpha^+\right) D.$$

For each value of the shock velocity D, the RH conditions (40)–(42) characterize the von Neumann state $(n_1, n_2, n_3, n_4, v, T)$ behind the shock wave, when the initial state $(n_1^+, n_2^+, n_3^+, n_4^+, 0, T^+)$ is assigned.

States in the Reaction Zone The intermediate states within the reaction zone describe sequential states of the chemical process and are characterized by integrating the rate equation (34) with initial conditions at the von Neumann state. Using the RH conditions (40) for $\alpha = 1, 3, 4$ together with (41) and (42), we can write the rate equation in the form

$$\frac{d}{dx} n_2 = \frac{D t_c \tau_2}{v - D + n_2 \frac{dv}{dn_2}}. \tag{43}$$

For each value of the shock velocity D, and starting from the von Neumann state characterized in the previous step, Eq. (43) together with RH conditions (40) for $\alpha = 1, 3, 4$ as well as (41) and (42) completely characterizes all states in the reaction zone. In particular, the final state of chemical equilibrium is obtained when the reaction rate τ_2 vanishes.

The mathematical approach and the solution procedure just described allow to obtain the reaction zone profiles for pressure, mean velocity, temperature, mass density and also the calculation of the wave thickness and other relevant properties in the detonation mechanism.

4.3 Numerical Results for Detonation Waves in the H_2-O_2 System

In this section we perform some numerical simulations for one dimensional steady detonation waves propagating in the hydrogen-oxygen mixture. We are particularly interested in the elementary chemical reaction

$$OH + H_2 \rightleftharpoons H + H_2O \tag{44}$$

that is involved in the realistic multi-step detonation mechanism of the hydrogen-oxygen mixture.

The initial state of the fresh quiescent mixture and the reference input data for the reaction heat Q_R and forward activation energy ε_f are chosen as follows

$$\begin{aligned} & n_{OH} = 0.1\,\text{mol/l}, \quad n_{H_2} = 0.2\,\text{mol/l}, \quad n_H = 0.03\,\text{mol/l}, \quad n_{H_2O} = 0.02\,\text{mol/l}, \\ & v = 0\,\text{ms}^{-1}, \quad T = 298.15\,\text{K}, \\ & Q_R = -63.3\,\text{kJ/mol}, \quad \varepsilon_f = 13.8\,\text{kJ/mol}. \end{aligned} \tag{45}$$

Since $Q_R < 0$, the forward chemical reaction is exothermic. Our representation of the detonation wave structure is determined using the mathematical modelling described in Sects. 4.1 and 4.2. We obtain some detonation profiles for different values of the detonation wave velocity, namely

$$D = 3120\,\text{ms}^{-1}, \quad D = 3130\,\text{ms}^{-1}, \quad D = 4400\,\text{ms}^{-1}, \quad D = 4500\,\text{ms}^{-1}. \tag{46}$$

Figures 2, 3 and 4 show the reaction zone profiles for pressure, temperature and mean velocity, when $D = 3120\,\text{ms}^{-1}$ and $D = 3130\,\text{ms}^{-1}$. Figures 5, 6 and 7 show the corresponding profiles when $D = 4400\,\text{ms}^{-1}$ and $D = 4500\,\text{ms}^{-1}$. These

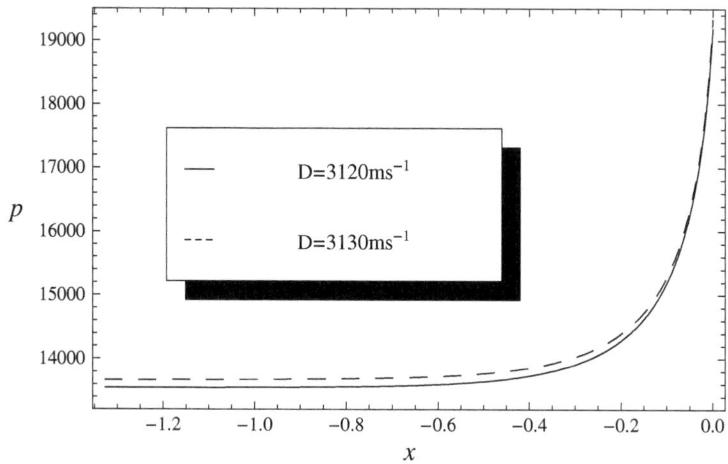

Fig. 2 Pressure profile in the reaction zone for two different wave velocities, $D = 3120\,\text{ms}^{-1}$ (*solid line*) and $D = 3130\,\text{ms}^{-1}$ (*dashed line*)

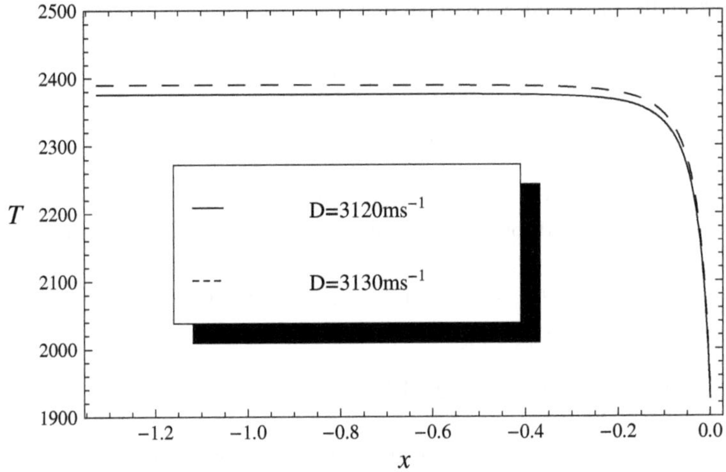

Fig. 3 Temperature profile in the reaction zone for two different wave velocities, $D = 3120\,\text{ms}^{-1}$ (*solid line*) and $D = 3130\,\text{ms}^{-1}$ (*dashed line*)

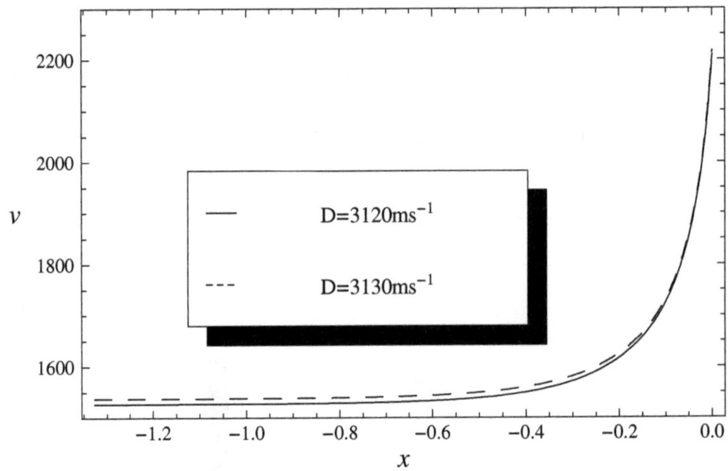

Fig. 4 Mean velocity profile in the reaction zone for two different wave velocities, $D = 3120\,\text{ms}^{-1}$ (*solid line*) and $D = 3130\,\text{ms}^{-1}$ (*dashed line*)

figures reproduce the typical ZND configuration for the diagrammed macroscopic variables. In particular, the pictures reveal that, as expected, the width of the reaction zone, that is the wave thickness, decreases with increasing values of the detonation velocity D.

To be more precise, the wave thickness is the spatial distance from the shock front to the equilibrium final state, reached when the reaction rate τ_2 vanishes and

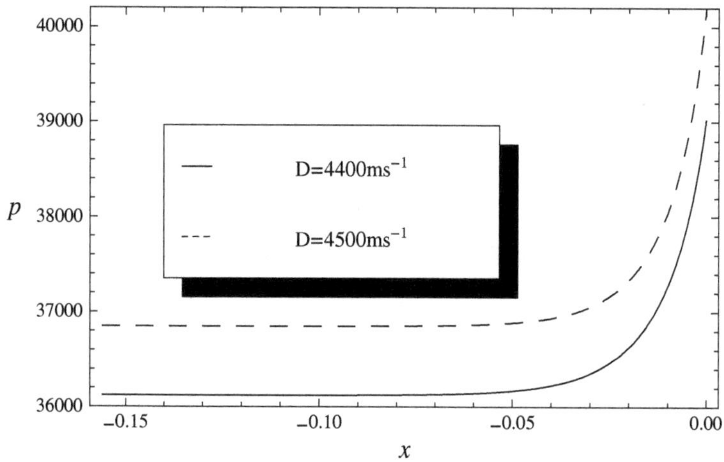

Fig. 5 Pressure profile in the reaction zone for two different wave velocities, $D = 4400\,\mathrm{ms}^{-1}$ (*solid line*) and $D = 4500\,\mathrm{ms}^{-1}$ (*dashed line*)

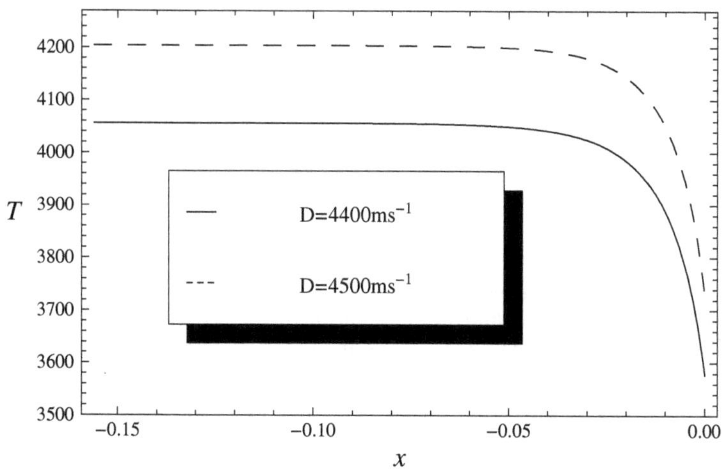

Fig. 6 Temperature profile in the reaction zone for two different wave velocities, $D = 4400\,\mathrm{ms}^{-1}$ (*solid line*) and $D = 4500\,\mathrm{ms}^{-1}$ (*dashed line*)

n_2 becomes constant, see Eq. (43). Thus, considering that the mixture reaches the chemical equilibrium when $\frac{d}{dx}n_2 < 10^{-6}$, the wave thickness can be determined for the detonation velocities defined in (46). The results are given in Table 1.

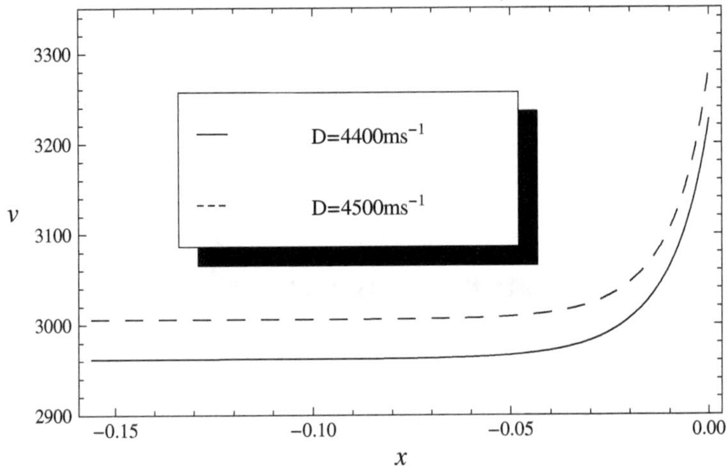

Fig. 7 Mean velocity profile in the reaction zone for two different wave velocities, $D = 4400\,\text{ms}^{-1}$ (*solid line*) and $D = 4500\,\text{ms}^{-1}$ (*dashed line*)

Table 1 Wave thickness for different values of the detonation velocity D

D	$3120\,\text{ms}^{-1}$	$3130\,\text{ms}^{-1}$	$4400\,\text{ms}^{-1}$	$4500\,\text{ms}^{-1}$
Wave thickness	1.340	1.327	0.156	0.142

5 Linear Stability Analysis

Experimental and numerical studies show that detonations tend to be structurally unstable, particularly in gases, see [2, 3]. The reaction zone is extremely sensitive to small perturbations and the detonation wave typically exhibits oscillating instabilities, which become more pronounced when the shock front propagates with velocity close to its minimum value. The evolution of such instabilities and a systematic analysis about the unstable modes, neutral stability boundaries and growth rates of the instabilities can be of crucial importance in the interpretation of the complex detonation mechanism. From the mathematical point of view, this stability analysis can be developed using a normal-mode linear approach of the steady planar detonation solution. This linear approach is valid when one investigates the effects induced by small perturbations and assumes that the steady structure of the detonation wave is not significantly modified.

In this section we formulate the linear stability problem for the steady detonation solution characterized in Sect. 4, then we describe the numerical technique used in the simulations and, finally, we present some results about the detonation instability.

5.1 Stability Problem

The problem is formulated assuming that a small perturbation is instantaneously assigned at the rear boundary and a distortion on the shock wave position occurs. As a result, the shock distortion affects the steady character of the reaction zone and the objective is to investigate the dynamics of the perturbations induced on the macroscopic variables representing the steady detonation wave in the reaction zone.

First, the one-dimensional closed governing Eqs. (34)–(39) are transformed to the coordinate frame attached to the perturbed wave. A new wave coordinate is introduced, x, which measures the distance from the perturbed shock,

$$x = x^{\ell} - \psi(t), \qquad \text{with} \qquad \psi(t) = Dt + \tilde{\psi}(t), \tag{47}$$

where x^{ℓ} is the laboratory coordinate, $\psi(t)$ the position of the perturbed wave in the laboratory frame, and $\tilde{\psi}(t)$ the spatial displacement of the perturbed shock with respect to its unperturbed position. The shock position in the new frame is $x = 0$ and the shock velocity is $D(t) = D + \tilde{\psi}'(t)$. Then, to describe the oscillatory behaviour of the instabilities, we perform a normal mode expansion of the steady state variables and perturbed shock position, in the form

$$z(x,t) = z^*(x) + e^{at}\,\overline{z}(x) \qquad \text{and} \qquad \psi(t) = \overline{\psi}\,e^{at}, \qquad \text{with} \qquad a, \overline{\psi} \in \mathbb{C}, \tag{48}$$

where we have used the vectorial notation for the state fields, $z = [n_1\ n_2\ n_3\ n_4\ v\ p]^T$. Here, $z^*(x)$ represents the state vector of the steady solution, $\overline{z}(x)$ the vector of the spatial disturbances of the steady state fields, $\overline{\psi}$ the disturbance amplitude parameter, and a is a perturbation parameter such that $\operatorname{Re} a$ and $\operatorname{Im} a$ are the perturbation growth rate and frequency, respectively.

We linearize Eqs. (34)–(39) by means of expansions (48) and normalize the state variables with respect to the complex amplitude parameter $\overline{\psi}$. For sake of simplicity, we keep the original notation $\overline{z}$ for the normalized variables. The resulting equations constitute the stability equations of the present problem and have the form

$$Da\overline{n}_2 + \left(v^* - D\right)\frac{d}{dx}\overline{n}_2 + (\overline{v} - a)\frac{d}{dx}n_2^* + \overline{n}_2\frac{d}{dx}v^* + n_2^*\frac{d}{dx}\overline{v} = \overline{\tau}_2, \tag{49}$$

$$Da(\overline{n}_1 - \overline{n}_2) + \left(v^* - D\right)\frac{d}{dx}(\overline{n}_1 - \overline{n}_2) + (\overline{v} - a)\frac{d}{dx}(n_1^* - n_2^*) + (\overline{n}_1 - \overline{n}_2)\frac{d}{dx}v^* + (n_1^* - n_2^*)\frac{d}{dx}\overline{v} = 0, \tag{50}$$

$$Da(\overline{n}_2 + \overline{n}_3) + \left(v^* - D\right)\frac{d}{dx}(\overline{n}_2 + \overline{n}_3) + (\overline{v} - a)\frac{d}{dx}(n_2^* + n_3^*) + (\overline{n}_2 + \overline{n}_3)\frac{d}{dx}v^* + (n_2^* + n_3^*)\frac{d}{dx}\overline{v} = 0, \tag{51}$$

$$Da(\overline{n}_2+\overline{n}_4)+\left(v^*-D\right)\frac{d}{dx}(\overline{n}_2+\overline{n}_4)+(\overline{v}-a)\frac{d}{dx}(n_2^*+n_4^*) \qquad (52)$$

$$+(\overline{n}_2+\overline{n}_4)\frac{d}{dx}v^*+(n_2^*+n_4^*)\frac{d}{dx}\overline{v}=0,$$

$$\varrho^*Dt_ca\overline{v}+\frac{d}{dx}\overline{p}+\varrho^*(\overline{v}-a)\frac{d}{dx}v^*+(\overline{\varrho}\frac{d}{dx}v^*+\varrho^*\frac{d}{dx}\overline{v})\left(v^*-D\right)=0, \qquad (53)$$

$$Dt_ca\overline{p}+\frac{5}{3}\left(p^*\frac{d}{dx}\overline{v}+\overline{p}\frac{d}{dx}v^*\right)+\left(v^*-D\right)\frac{d}{dx}\overline{p}+(\overline{v}-a)\frac{d}{dx}p^*=\frac{2Q_R^\star\overline{\tau}_2}{3}, \qquad (54)$$

where $\overline{\tau}_2$ is the linearized representation of the reaction rate, given by

$$\overline{\tau}_2=(n_3^*\overline{n}_4+n_4^*\overline{n}_3)\tau_r^*+(n_1^*\overline{n}_2+n_2^*\overline{n}_1)\tau_f^*. \qquad (55)$$

We also linearize the Rankine-Hugoniot conditions (40)–(42) using the normal mode expansions (48), obtaining

$$\overline{n}_\alpha(0)=\frac{\left(n_\alpha^*-n_\alpha^+\right)a-n_\alpha^*\overline{v}(0)}{v^*-D}, \qquad \alpha=1,2,3,4, \qquad (56)$$

$$\overline{v}(0)=\frac{3\varrho^+v^{*2}+\frac{3}{2}\left(p^*-p^+\right)-\frac{3}{2}D\varrho^+v^*+\sum_{\alpha=1}^4E_\alpha n_\alpha}{-\varrho^*\left(v^*-D\right)^2+\frac{5}{2}p^*}\,a, \qquad (57)$$

$$\overline{p}(0)=-\varrho^+av^*-\left(v^*-D\right)\varrho^*\overline{v}(0). \qquad (58)$$

From the linearization procedure, we obtain twelve real first-order homogeneous ODE's, Eqs. (49)–(54), to be considered in the reaction zone from $x=0$ at the von Neumann state to $x=x_F$ at the equilibrium final state, with twelve real initial conditions given by Eqs. (56)–(58). The equations involve twelve unknowns specified by the real and imaginary parts of $\overline{n}_1,\overline{n}_2,\overline{n}_3,\overline{n}_4,\overline{v},\overline{p}$. Since the equations involve the complex perturbation parameter a, the ODE system is not closed, and an additional closure condition is needed. We assign the following boundary condition, at $x=x_F$, initially proposed and justified by Buckmaster and Ludford in [9],

$$\overline{v}(x_F)+a=\frac{-1}{\gamma\varrho_{eq}^*c_{eq}^*}\,\overline{p}(x_F), \qquad (59)$$

where γ is the ratio of specific heats, and c_{eq}^* and ϱ_{eq}^* are the isentropic sound speed and mixture mass density at $x=x_F$.

The stability problem just formulated will be numerically solved as described in the next subsection.

5.2 Numerical Technique

The numerical technique used in this paper to treat the stability problem is based on an iterative shooting algorithm proposed in [1]. The algorithm combines the numerical approach developed by Lee and Stewart in paper [10] with the original ideas advanced by Erpenbeck in paper [11].

Broadly speaking, the technique consists in choosing, first, a trial value for the perturbation parameter a, then solving the ODE's (49)–(54) with initial conditions (56)–(58) and, finally, verifying if the solution previously obtained satisfies the boundary condition (59). If the boundary condition is satisfied, the corresponding solution of (49)–(54) and (56)–(58) represents a stability solution, that is a solution of the stability problem.

After finding a solution of the stability problem, for a trial value of a, the last step is straightforward. In fact, it only requires to determine if the solution of the stability problem produces a stable or an unstable mode of propagation, and this is as follows: If $\mathrm{Re}\, a > 0$, then the parameter a results in an unstable mode; if $\mathrm{Re}\, a < 0$, then it results in a stable mode.

The conclusion of the stability analysis is the following. The steady detonation solution is stable when all solutions of the stability problem result in stable modes of propagation. Conversely, it is unstable when at least one solution of the stability problem results in an unstable mode.

The main numerical difficulty of the stability analysis is to find solutions of the stability problem. In fact, an arbitrary trial value of a does not satisfy, in general, the boundary condition (59). This difficulty has been solved by Carvalho and Soares in paper [1], using a numerical technique that combines ideas and methodologies of previous works, as mentioned above.

Accordingly, we introduce the residual function $\mathscr{H}$ in a fixed domain $\mathscr{R} \subset \mathbb{C}$,

$$\mathscr{H}(a) = \overline{v}(x_F) + a + \frac{1}{\gamma \varrho_{eq}^* c_{eq}^*}\, \overline{p}(x_F)\,, \quad a \in \mathscr{R} \subset \mathbb{C}, \tag{60}$$

and notice that the zeros of $\mathscr{H}$ satisfy the boundary condition (59). Thus, resorting to the argument principle, first used by Erpenbeck in [11], and taking into account that the function $\mathscr{H}$ has no poles in $\mathscr{R}$, we count the number Z of zeros of $\mathscr{H}$ by means of the expression

$$Z = \frac{1}{2\pi i} \int_k^\ell \frac{\mathscr{H}'(\zeta(t))}{\mathscr{H}(\zeta(t))}\, \| \zeta'(t) \|\, dt, \tag{61}$$

where $\zeta : [k, \ell] \to \mathbb{C}$ is a path smooth by parts, describing the contour of $\mathscr{R}$ in the positive sense.

Starting from these ideas, the numerical procedure used to solve the stability problem consists in nine steps described below.

1. Choose the domain $\mathscr{R}$ in the complex plane where we intend to look for eigenvalues of the stability problem.
2. Define a path ζ describing the contour of $\mathscr{R}$ in the positive sense.
3. Select a great number of trial values a_j in the contour of $\mathscr{R}$.
4. Introduce further trial values b_j defined by $b_j = a_j + 10^{-6}$.
5. Solve the stability governing Eqs. (49)–(54) with initial conditions (56)–(58) for each trial value a_j and b_j for the perturbation parameter a.
6. Evaluate the residual function $\mathscr{H}$ at each point a_j and b_j.
7. Estimate the derivative $\mathscr{H}'(a_j)$ by the quotient $(\mathscr{H}(b_j) - \mathscr{H}(a_j))/(b_j - a_j)$.
8. Estimate the mean value μ of the function $\frac{\mathscr{H}'(\zeta(t))}{\mathscr{H}(\zeta(t))} \parallel \zeta'(t) \parallel$ using a suitable sample and a 99 % confidence interval.
9. Count the number of zeros of the residual function $\mathscr{H}$, approximating the expression on the right hand-side of Eq. (61) by $Z = \frac{1}{2\pi i}(k - \ell)\mu$.

The details of the numerical procedure and a rather complete discussion on the technique can be found in [1].

5.3 *Results and Discussion*

Using the numerical procedure described in the previous subsection, we were able to obtain some results on the stability behaviour of the steady detonation solution described in Sect. 4. Considering the reference input data indicated in (45) and the values of the detonation wave velocity D referred in (46), we obtain some estimations for the number of unstable modes in the region $\mathscr{R}$ defined by

$$0.001 < \mathrm{Re}\, a < 0.1 \qquad \text{and} \qquad 0.001 < \mathrm{Im}\, a < 0.1\,.$$

The estimations are presented in Table 2 and represent very preliminary results on the stability behaviour of the steady detonation solution described in Sect. 4.

Table 2 Estimations for the unstable modes in the region $\mathscr{R}$, for fixed values of the reaction heat Q_R^* and activation energy ϵ_f^*, and for different values of the detonation velocity D

$D\,(\mathrm{ms}^{-1})$	Number of unstable modes
3120	12 to 334
3130	158 to 684
4400	0
4500	0

The determination of instability solutions is a very complex task and a time consuming problem. In Table 2 we present some estimations for the number of instability modes obtained for different values of the detonation velocity. These results are still rough approximations and should be improved. In fact, it is known that the number of instability modes grows as the detonation velocity approaches its minimum value, see for example [2, 10]. The results presented in Table 2 show that unstable modes exist for lower values of the detonation velocity, as it is expected from the literature. However, since the confidence intervals obtained for $D = 3120\,\mathrm{ms}^{-1}$ and $D = 3130\,\mathrm{ms}^{-1}$ show a significant overlap, we are not able to compare the number of instability modes for these two detonation velocities.

The results obtained in the present paper are still very limited and should be improved. This will be addressed in a work in progress, where more accurate simulations will be conducted and further detailed results will be included. Among several interesting topics, we intend to investigate the following two issues: the limit detonation velocity characterizing stable solutions for other detonating of Hydrogen-Oxygen mixtures defined in terms of different constituent concentrations; the relation between the detonation wave velocity and the number of instability modes. Moreover other simulations will be performed oriented to compare our results with others available in literature, obtained from numerical studies and experimental works.

Acknowledgements This paper was partially supported by the Research Centre of Mathematics of the University of Minho through Portuguese funds from the "Fundação para a Ciência e a Tecnologia", through the Project PEstOE/MAT/UI0013/2014. The author Silva, A.W. acknowledges the support by the Instituto Federal do Paraná—Campus of Curitiba, Brazil.

References

1. Carvalho, F., Soares, A.J.: On the dynamics and linear stability of one-dimensional steady detonation waves. J. Phys. A Math. Theor. **45**, 255501, 1–23 (2012)
2. Fickett, W., Davis, W.C.: Detonation, Theory and Experiment. University of California Press, Berkeley (1979)
3. Lee, J.H.S.: The Detonation Phenomenon. Cambridge University Press, Cambridge (2008)
4. Oran, E.S., Boris, J.P.: Numerical Simulation of Reactive Flow. Cambridge University Press, New York (2001)
5. Oran, E.S., Weber Jr, J.W., Stefaniw, E.I., Lefebvre, M.H., Anderson Jr, J.D.: A numerical study of a two-dimensional H2-O2-Ar detonation using a detailed chemical reaction model. Combust. Flame **113**, 147–163 (1998)
6. Kremer, G.M., Silva, T.G.: Analysis of the reaction rate coefficients for slow bimolecular chemical reactions. Braz. J. Phys. **42**, 400–409 (2012)
7. Kremer, G.M., Silva, A.W., Alves, G.M.: On inelastic reactive collisions in kinetic theory of chemically reacting gas mixtures. Phys. A **389**, 2708–2718 (2010)
8. Kremer, G.M.: Introduction to the Boltzmann Equation and Transport Processes in Gases. Springer, Berlin (2010)

9. Buckmaster, J.D., Ludford G.S.S.: The effect of structure on the stability of detonations I. Role of the induction zone. In: Twenty-First Symposuim (International) on Combustion, The Combustion Institute, pp. 1669–1676. Elsevier, Amsterdam (1986)
10. Lee, H.I., Stewart, D.S.: Calculation of linear detonation stability: one dimensional instability of plane detonation. J. Fluid Mech. **216**, 103–132 (1990)
11. Erpenbeck, J.J.: Stability of steady-state equilibrium detonations. Phys. Fluids **5**, 604–614 (1962)

Mathematical Aspects of Coagulation-Fragmentation Equations

F.P. da Costa

Abstract We give an overview of the mathematical literature on the coagulation-like equations, from an analytic deterministic perspective. In Sect. 1 we present the coagulation type equations more commonly encountered in the scientific and mathematical literature and provide a brief historical overview of relevant works. In Sect. 2 we present results about existence and uniqueness of solutions in some of those systems, namely the discrete Smoluchowski and coagulation-fragmentation: we start by a brief description of the function spaces, and then review the results on existence of solutions with a brief description of the main ideas of the proofs. This part closes with the consideration of uniqueness results. In Sects. 3 and 4 we are concerned with several aspects of the solutions behaviour. We pay special attention to the long time convergence to equilibria, self-similar behaviour, and density conservation or lack thereof.

1 Introduction: Some Processes and Models

Coagulation (coalescence, agglomeration, aggregation) and fragmentation phenomena are ubiquitous in many scientific disciplines, such as: Physical [41, 73, 83], Astronomical [194], Chemical [227], Atmospheric [188, 196], Biological [6, 174], Environmental [96, 151], as well as in several technological processes [24, 86, 97].

Their quantitative modelling can be achieved by several mathematical approaches, such as those using stochastic processes, computer simulations, or by the mathematical or numerical analysis of certain types of differential equations, generally called coagulation-fragmentation equations. We will centre most of our attention in a class of these equations, the discrete coagulation-fragmentation equations, but will also refer to the so called continuous case. Our goal is to review the most important Mathematical Analysis results about existence, uniqueness and

F.P. da Costa (✉)
Departamento de Ciências e Tecnologia, Universidade Aberta, Rua da Escola Politécnica 141-147, 1269-001 Lisboa, Portugal

Centro de Análise Matemática, Geometria e Sistemas Dinâmicos, Departamento de Matemática, Instituto Superior Técnico, Universidade de Lisboa, Av. Rovisco Pais, 1049-001 Lisboa, Portugal
e-mail: fcosta@uab.pt; fcosta@math.tecnico.ulisboa.pt

J.-P. Bourguignon et al. (eds.), *Mathematics of Energy and Climate Change*, CIM Series in Mathematical Sciences 2, DOI 10.1007/978-3-319-16121-1_5

several properties of the solutions, with special attention to dynamical aspects, in a way that is accessible to anyone with a background in differential equations but with no previous contact with these equations. To this end, we shall try to present the results, the ideas of the proofs, and some of their details, in the most reader friendly way we can, often detailing simpler situations more deeply and just glossing over more technically demanding proofs, or even just referring the result and calling the reader's attention to the original articles. This, we think, will help the reader to gain a feel for the subject without getting too much bogged down in the technicalities, at the same time that will show him/her directions for a deeper study of the issues.

In order to provide an overview as broad as possible of the field, in this introductory part, in addition to those systems that will be the focus of out attention later on, we will present a number of coagulation-fragmentation equations that have been studied in the mathematical literature, although we will not enter into much detail, and, in several cases, will not refer to them again afterwards.

The mathematical literature on this type of equations has had a huge growth in the last two decades, so much so that a comprehensive review of the field is no longer possible in a work such as this. And this is not only true about the mathematical literature, but even more so about the mathematical modelling literature, as well as the extremely rich and variegated contributions coming from Physics and other scientific and technological areas. This scientific literature is also a seemingly inexhaustible fountain of interesting and difficult mathematical problems, and so every mathematician must spend as much (if not more) time being acquainted with it than he/she will spend with the mathematical literature. This feature is also reflect in the list of references of the present work.

Although the Mathematical Analysis of the coagulation-fragmentation equations is a relatively recent research area, there exists already a number of reviews that are useful to whoever wants to obtain an overview of the problems, methods, and existing results. Two of the most recent reviews [133, 219], have important overlaps with topics we deal with in this work. Another recent review, largely (but not exclusively) centered on the study of coagulation-fragmentation-diffusion systems using duality arguments and entropy and entropy-dissipation methods is [68]. A very interesting introduction to some of the classical models in this field, including a discussion of the physical ideas involved, is [43]. Another interesting reference, although somewhat outdated, is [74]. Drake [72] was the first review article about mathematical modelling of these problems and although very few of its content is mathematically rigorous it is still an interesting source for the literature before the 1970s. Finally, although somewhat outside the scope of our work, it is important to point out the review article of Aldous [2] that seems to had a tremendous importance in calling the attention of probabilists to this area of research which resulted in the contemporaneous explosion of works using a stochastic approach to the coagulation and fragmentation processes (see, for instance, [21] and references therein). Similarly to what happens with the studies using a stochastic approach, those on numerical analysis (see, e.g., [88, 118, 119, 141, 142, 189, 226]) are outside the scope of this work; a natural consequence of space limitations and also of my lack of expertise in those fields.

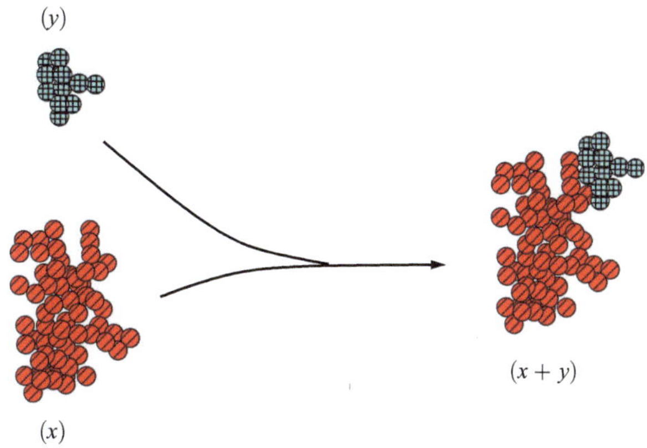

Fig. 1 Scheme of the coagulation process of an x-cluster and a y-cluster

1.1 Coagulation and Fragmentation Processes

By coagulation (or coalescence, or agglomeration, or aggregation) one means a class of phenomena by means of which there is an increase in the size (or mass) of particles through their collision with other similar, smaller, particles. In the overwhelming majority of cases simultaneous collision of more than two particles are extremely rare and are not considered.

In Fig. 1 we present a schematic coagulation event between a particle of size (or mass) x, called x-cluster, and another of size (or mass) y, y-cluster. As will be pointed out below there are modelling situation in which particle sizes vary continuously (in $\mathbf{R}^+$) and others for which the size is assumed discrete, indexed in $\mathbf{N}^+$.

The reciprocal process of spontaneous fragmentation is, as the name implies, that by which a given particle breaks up and originates two, or more than two, smaller particles, and is schematically presented in Fig. 2.

A different type of fragmentation that is sometimes considered in the literature is the collision induced fragmentation, also known as non-linear fragmentation: we can consider this process as made of two consecutive steps: a coagulation step forming an extremely unstable cluster that (in the time scale of the full process) instantaneously breaks into two or more smaller aggregates, as schematically illustrated in Fig. 3.

The differential equations of coagulation-fragmentation type are one of the attempts at the mathematical modelling of the processes schematically represented in Figs. 1, 2, and 3. All these equations can be considered as equations of structured population dynamics, as they model the dynamics of systems of particles with some kind of internal structure (due to size, mass, or some other characteristic). Indeed, there are cases where standard equations of biological population dynamics pop up in the study of coagulation systems, although not in a direct and obvious way (e.g.: see [53, 60]), but the fact remains that, due to the special structure of the coagulation-fragmentation, the general methods of structured population dynamics (as in, e.g., [221]) are usually not relevant (however, see Sect. 1.10.4). A much more important

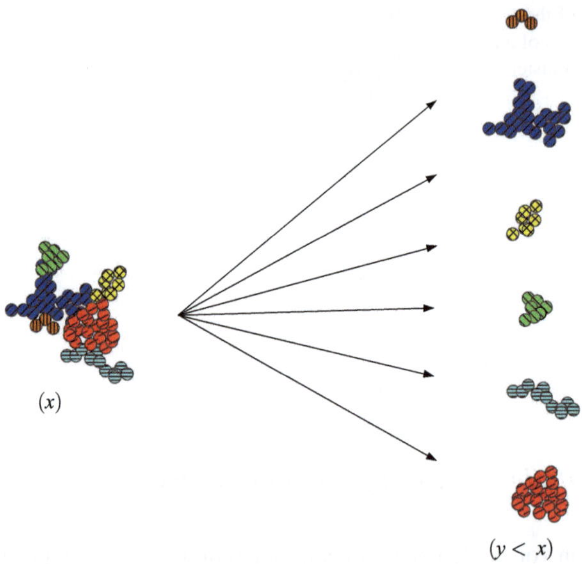

Fig. 2 Scheme of the spontaneous (or linear) multiple fragmentation process of an x-cluster into several smaller y-clusters

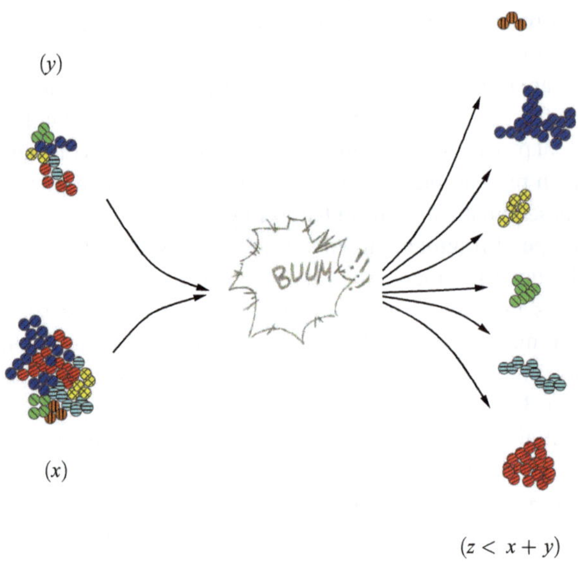

Fig. 3 Scheme of the collisional (or non-linear) fragmentation process

connection, as far as mathematical methods are concerned, is with kinetic theory, as is made plain in some of the most recent studies on the long time behaviour of solutions and self-similarity we shall refer to below.

In the following parts of the present section we briefly present some of the coagulation and fragmentation models that have been more widely studied from the mathematical point of view.

1.2 Smoluchowski's Coagulation Equations

The coagulation differential equation was originally proposed in 1916 by the physicist von Smoluchowski as a model for the kinetics of colloid formation [204, 205], and in spite of the fact that it is at present one of the best studied, there are still has a number of important open mathematical problems about it.

Let us represent the coagulation process of Fig. 1 in the following notation, usual in chemical kinetics:

$$(x) + (y) \overset{a(x,y)}{\longrightarrow} (x + y) \ ,$$

where $a(x, y)$ is the rate of the coagulation reaction among an x-cluster and a y-cluster, usually called the coagulation coefficient or coagulation kernel. Often these coefficients depend only on the mass of the clusters, but there are cases where it is important to consider dependencies on some other characteristic (cf. cases in Sects. 1.9 and 1.10) or upon time [212, 223]. The only general mathematical property imposed by all physical situations is the symmetry and non-negativity of the coagulation kernel: $a(x, y) = a(y, x) \geq 0$.

When the cluster masses assume only discrete values, multiples of a smaller quantity considered as unity (the mass of the 1-cluster, or *monomer*), the usual notation for $a(x, y)$ is $a_{x,y}$, and the usual letters to denote cluster sizes are $i, j, k, \ldots$, instead of $x, y, \ldots$.

In Table 1 we collected some coagulation kernels occurring in the literature (see references cited in [49, 72, 133]).

Let us start by considering the case of discrete masses, which was also the one considered by Smoluchowski. Assuming the system is spatially homogeneous, we represent the concentration (or density) at time t of the j-cluster by $c_j = c_j(t)$, and denote by $c = (c_j)$ the vector of concentrations of all clusters. Assuming valid the mass action law of chemical kinetics the rate of chance of c_j is given by the differential equation

$$\dot{c}_j = Q_c(c)(j) \tag{1}$$

where $\dot{c}_j$ denotes the time derivative c_j and $Q_c(c)(j)$ is the mathematical function that represents the coagulation (hence the subscript c) reaction terms affecting the j

Table 1 Some coagulation kernels $a(x, y)$ occurring in the literature

$a(x, y)$	Comments
1	Approximately Brownian coagulation
	Linear chain polymerization
$x + y$	Polymerization of branched chains of ARB_{f-1} type ($f \gg 1$)
	Limit case of gravitation coagulation
$x^{-2/3} + y^{-2/3}$	Diffusional growth of supported metal crystalites
xy	Polymerization of branched chains of RA_f type ($f \gg 1$)
$x^\alpha y^\beta + x^\beta y^\alpha$	A general case including e.g. Golovin, Stockmayer, etc
$(x^{1/3} + y^{1/3})(x^{-1/3} + y^{-1/3})$	Brownian coagulation (continuum regimen)
$(x^{1/3} + y^{1/3})^2(x^{-1} + y^{-1})^{1/2}$	Brownian coagulation (free molecular regimen)
$(x^{1/3} + y^{1/3})^3$	Tangential coagulation (linear velocity profiles)
$(x^{1/3} + y^{1/3})^7$	Tangential coagulation (non-linear velocity profiles)
$(x^{1/3} + y^{1/3})^2 \lvert x^{1/3} - y^{1/3}\rvert$	Gravitational deposition (particles bigger than $\sim 50\,\mu$m)
$(x^\alpha + y^\alpha)^\beta \lvert x^\gamma - y^\gamma\rvert$	Ballistic coagulation ($\alpha, \beta, \gamma \geq 0, \alpha\beta + \gamma \leq 1$)

component of the concentration vector c. There are two contributions to this reaction term:

1. the creation of j-clusters due to the reactions of smaller clusters with appropriate masses, $(j - k) + (k) \to (j)$, with $k = 1, \ldots, j - 1$, and $j \geq 2$, to which corresponds the term

$$Q_1(c)(j) := \frac{1}{2}\sum_{k=1}^{j-1} a_{j-k,k}c_{j-k}c_k\,, \tag{2}$$

and defining $Q_1(c)(1) := 0$;
2. the destruction of j-clusters due to the coagulation reactions of a j-cluster and any other present in the system, $(j) + (k) \to (j+k)$, with $k = 1, 2, \ldots$. Not imposing an *a priori* upper bound on the size of the clusters, to this process corresponds the term

$$Q_2(c)(j) := c_j \sum_{k=1}^{\infty} a_{j,k}c_k\,. \tag{3}$$

Hence, the right-hand side of (1) is

$$\begin{aligned} Q_c(c)(j) &:= Q_1(c)(j) - Q_2(c)(j) \\ &= \frac{1}{2}\sum_{k=1}^{j-1} a_{j-k,k}c_{j-k}c_k - c_j\sum_{k=1}^{\infty} a_{j,k}c_k\,, \quad j \in \mathbf{N}. \end{aligned} \tag{4}$$

The discrete Smoluchowski system is the system of a countable number of ordinary differential equations (1) with the right-hand side given by (4).

In several cases it is preferable to consider the version of Smoluchowski equations for which the cluster masses can be any positive real number. This continuous version, first considered by Müller in 1928 [72, 164], can be written as the following integro-differential equation

$$\partial_t c(t,x) = Q_c(c)(t,x) \tag{5}$$

with $c(t,x)$ the concentration (or density) of x-clusters at time t, and

$$\begin{aligned} Q_c(c)(t,x) &:= Q_1(c)(t,x) - Q_2(c)(t,x) \\ &:= \frac{1}{2}\int_0^x a(x-y,y)c(t,x-y)c(t,y)dy - c(t,x)\int_0^\infty a(x,y)c(t,y)dy. \end{aligned} \tag{6}$$

The first mathematical works about (1) seem to have been the papers by McLeod [154, 155]. On the continuous version (5) the first mathematical papers are those by Morgenstern [163] and Melzak [158] (this last one also including fragmentation, see Sect. 1.5). In the last two decades there has been a huge progress in our understanding of several questions about existence, uniqueness, regularity, and asymptotic behaviour of solutions to these equations, and part of these results will be presented below.

1.3 Oort-Hulst-Safronov Coagulation Equations

Another coagulation equation that has received some attention is the Oort-Hulst-Safronov equation [65, 177], [194, Chapter 8], that was first proposed to model astronomy phenomena. This equation has also the general form (5) but differs in the way $Q_1(c)(t,x)$ and $Q_2(c)(t,x)$ are defined:

1. the x-clusters creation rate depend on a kind of mean value size of the clusters, namely:

$$Q_1(c)(t,x) := -\partial_x\left(c(t,x)\int_0^x ya(x,y)c(t,y)dy\right); \tag{7}$$

2. the destruction of x-clusters due to the coagulation with other existing clusters only occurs through the reaction with higher masses, i.e., through a kind of sedimentation of smaller clusters onto larger ones:

$$Q_2(c)(t,x) := c(t,x)\int_x^\infty a(x,y)c(t,y)dy\,. \tag{8}$$

As a consequence, some solution properties are distinct from the corresponding ones in the Smoluchowski's system, the most notable of them is the finite speed of propagation of the support of solutions [75], which contrasts strongly to what happens in Smoluchowski's (see, for instance, [46] for the discrete case). Notwithstanding these differences in behaviour, both these equations are related and can be seen as limit cases of one-parameter families of cluster equations [75, 120].

1.4 Fragmentation Equations

The first references to fragmentation processes took place in the context of chemical studies on polymer degradation (see, e.g., [197]). The first reference to the mathematical modelling of the spontaneous fragmentation process seems to have been done, using probabilistic methods, by Kolmogorov [112], who also suggested and supervised the later study [89] by Filippov. The first non-probabilistic mathematical reference to these processes is included in Melzak work [158] about the continuous coagulation-fragmentation system. For the discrete version the first reference seems to be the paper [206] by Spouge. As the spontaneous fragmentation process can be modelled by a linear equation (see below), the modern approach to these problems is intimately connected with tools and methods from Linear Functional Analysis, and an excellent introduction to them can be seen in [7, Chapters 8 and 9].

In this section we will consider the spontaneous fragmentation case; other processes, such as collisional fragmentation or volumetric dispersion, which are also related with coagulation processes will be presented later on in Sect. 1.7.

We can represent the fragmentation process of Fig. 2 by[1]

$$(x) \xrightarrow{B(x)} (y_1) + (y_2) + \dots ,$$

where $B(x)$ is the rate of fragmentation of x-clusters. Let $\psi(x,y)$ be the average number of y-clusters produced by the fragmentation of an x-cluster. In the case of discrete masses, denoted by $i, j, k, \dots$, we use the traditional notation B_j and $\psi_{j,k}$, instead of $B(j)$ and $\psi(j,k)$.

Mass conservation in each simple fragmentation reaction implies that the total mass of daughter particles must be equal to the mass of the original particle, namely

$$\int_0^x y\psi(x,y)dy = x \qquad \text{or} \qquad \sum_{k=1}^{j-1} k\psi_{j,k} = j, \tag{9}$$

for the continuous and for the discrete cases, respectively.

[1]The notation is not very good since it suggests there can be at most a countable number of daughter particles (y_k): in fact, there is no *a priori* reason preventing the distribution to be continuous.

For definiteness, let us consider the discrete case. Assuming the mass action law the rate of change of c_j due to spontaneous fragmentation processes is given by the differential equation

$$\dot{c}_j = Q_f(c)(j)\,, \tag{10}$$

where $Q_f(c)(j)$ encodes the fragmentation reaction contributions (hence the subscript f) to the evolution of the j-cluster concentration, that can be expressed by

$$Q_f(c)(j) := -Q_3(c)(j) + Q_4(c)(j), \quad j \in \mathbf{N}, \tag{11}$$

where

1. the destruction of j-clusters due to the fragmentation $(j) \to (k) + \ldots$, is given by

$$Q_3(c)(j) := B_j c_j \quad \text{and} \quad Q_3(c)(1) := 0\,; \tag{12}$$

2. the creation of j-clusters due to fragmentation of bigger clusters is modelled by

$$Q_4(c)(j) := \sum_{k=1}^{\infty} B_{j+k} \psi_{j+k,j} c_{j+k}\,. \tag{13}$$

A frequent assumption, valid, for instance, in the degradation of polymers [227], is that of binary fragmentation, i.e., each fragmenting cluster produces only two daughter particles, and thus, by the symmetry of the physical process, $\psi_{i,j} = \psi_{i,i-j}$. Hence, (9) implies $\sum_{k=1}^{j-1} \psi_{j,k} = 2$, which has the obvious interpretation that the average number of daughter particles in each fragmentation is equal to two.[2]

In this case, denoting by $b_{j,k}$ the rate constant for the binary fragmentation reaction $(j + k) \to (j) + (k)$, i.e., $b_{j,k} := B_{j+k}\psi_{j+k,k}$, we conclude that $B_j = \frac{1}{2}\sum_{k=1}^{j-1} b_{j-k,k}$ and thus the right-hand side of the binary fragmentation reaction is

$$Q_f(c)(j) := -\frac{1}{2}\sum_{k=1}^{j-1} b_{j-k,k} c_j + \sum_{k=1}^{\infty} b_{j,k} c_{j+k}. \tag{14}$$

As with the case of Smoluchowski's equations, the continuous versions of the fragmentation equations consist in integro-differential equations obtained formally by substituting the sums by integrals.

The fragmentation mechanism is mathematically encoded in the functions $B(x)$, $\psi(x,y)$ and $b(x,y)$ and the only general property these functions must obey, on physical grounds, is non-negativity. Furthermore, the binary fragmentation coefficients must be symmetric: $b(x,y) = b(y,x)$.

[2] As it should, in a binary fragmentation…

In the mathematical literature it is common to assume growth conditions such as $b(x,y) \leq (x+y)^{\gamma}$, or $b(x,y) \leq x^{\gamma} + y^{\gamma}$, or $b(x,y) \geq (x+y)^{\gamma}$, etc., or other conditions, such as the *strong fragmentation* [34, 45], and the *weak fragmentation* [29, 30, 36, 48]. In models of some specific phenomena particular fragmentation kernels need to be considered (e.g.: see [103]).

An assumption that is particularly important from the physical viewpoint, corresponding to the occurrence of microscopic reversibility, is called the detailed balance condition. This presupposes the simultaneous existence of coagulation and fragmentation processes (cf. next section) and, informally, it says that it must exist an equilibrium (i.e., time independent) solution to each of the individual reactions

$$(j) + (k) \rightleftharpoons (j+k).$$

The detailed balance condition is the following: there exists a positive sequence (M_j), with $M_1 = 1$, such that

$$a_{j,k}M_jM_k = b_{j,k}M_{j+k}. \tag{15}$$

The sequence (M_j) is physically interpreted as the system's partition function [13, 36].

1.5 Coagulation-Fragmentation Equations

The coagulation-fragmentation equations are the system that describes phenomena where coagulation and fragmentation processes are simultaneously present. As such, it has the form

$$\dot{c}_j(t) = Q_c(c(t))(j) + Q_f(c(t))(j)\,, \tag{16}$$

or

$$\partial_t c(t,x) = Q_c(c)(t,x) + Q_f(c)(t,x)\,, \tag{17}$$

in the discrete and in the continuous case, respectively.

Possibly the first explicit reference to this system in the literature is in [23] treating phenomena of polymerization and de-polymerization in chemistry. The authors consider the discrete version of the equations, binary fragmentation, and reaction kernels independent of the cluster sizes, $a_{j,k} \equiv a$, $b_{j,k} \equiv b$.

The first mathematical study about the existence of solutions is Melzak's 1957 paper cited in the last section [158], that considers the continuous system with bounded kernels. The extension to unbounded kernels started more than three decades latter in Stewart's papers [210, 211]. For the discrete system the first existence result was published by Spouge [206] in 1984, valid for bounded fragmentation coefficients. More general results were obtained by Ball and Carr [12], by Laurençot [126], by the author [45], among many others. Analogous existence results where obtained for similar equations modelling an Ising spin system with Glauber dynamics by Kreer [116].

Long time behaviour of solutions to coagulation-fragmentation systems is a rather difficult problem, not yet completely understood. The first significant contribution was by Aizenman and Bak in 1979, for the continuous system with constant coefficients [1]. In the last couple of decades a number of important papers have greatly advanced our understanding of the dynamic behaviour of solutions of both the discrete and the continuous coagulation-fragmentation equations. Some of these contributions will be analysed below.

The vast majority of the mathematical studies have considered binary fragmentation, but some other types of fragmentation processes have also been considered [84–86, 139, 215] and we shall briefly refer to them in Sect. 1.7.

1.6 Becker-Döring Equations

The original Becker-Döring model was proposed in 1935 in the context of nucleation studies [17] (formation of liquid droplets in a supersaturated vapour) in which the concentrations of large clusters are so small that one assumes the only relevant reactions are those of coagulation between a cluster and a monomeric particle, and the fragmentation of a cluster by shedding off a single monomeric particle at a time, as schematically illustrated in Fig. 4.

Even at very low densities it is not physically reasonable to expect the Becker-Döring to be a good approximation [182]. However, the rich (and difficult) mathematics of the Becker-Döring system, together with the fact that some of its properties are believed to also hold in more general systems whose mathematical

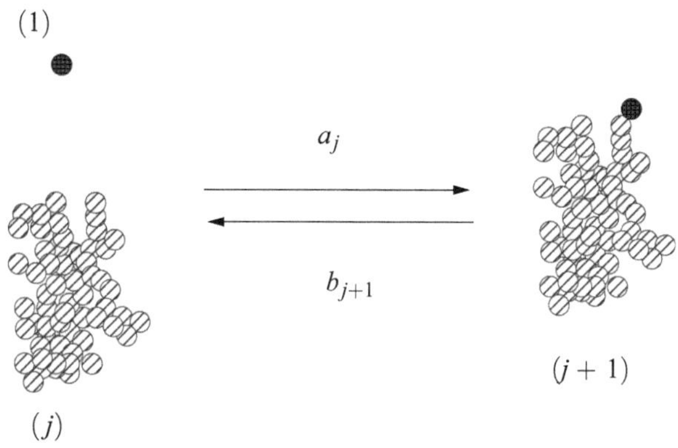

Fig. 4 Scheme of the Becker-Döring processes

study is considerably more complex, have turned the Becker-Döring system into a paradigmatic model in coagulation-fragmentation studies whose importance and contribution to the understanding of the issues involved can hardly be overstated [202]. Even from the physical and mathematical modelling points of view, Becker-Döring like systems continue to this day to be proposed and studied [73, 107].

In the original version of the Becker-Döring system the monomer concentration was assumed to be time independent. It was Burton [28] and Penrose and Lebowitz [182] who first considered the current version of the equations, in which the mass of the system is formally constant and thus the monomer concentration has to change with time, and by this turning them into a non-linear system of the differential equations that is a particular case of the coagulation-fragmentation system with the rate kernels satisfying the restriction[3]

$$a_{j,k} = b_{j,k} = 0 \quad \text{if } j \wedge k > 1. \tag{18}$$

Due to historical reasons, the notation used in Becker-Döring system is slightly different from the one that would be obtained by substituting (18) into (4),(14) and the result into (16). For $j > 1$ let us define $a_j := a_{j,1}$ and $b_{j+1} := b_{j+1,1}$. let $a_1 = \frac{1}{2}a_{1,1}$, $b_2 = \frac{1}{2}b_{2,1}$, and remember that the rate coefficients $a_{j,k}$ e $b_{j,k}$ are invariant under permutation of the subscripts. Thus, the Becker-Döring system is usually written as

$$\begin{cases} \dot{c}_1 = -J_1(c) - \sum_{j=1}^{\infty} J_j(c), \\ \\ \dot{c}_j = J_{j-1}(c) - J_j(c), \quad j \geq 2, \end{cases} \tag{19}$$

where $J_j(c) := a_j c_1 c_j - b_{j+1} c_{j+1}$.

1.7 Equations with Non-linear Fragmentation

As pointed out in Sect. 1.1 there is a fragmentation mechanism that is quiet different from the spontaneous fragmentation considered in Sects. 1.4–1.6, and that mathematically originates non-linear contributions to the equations. There are some situations in astrophysics and atmospheric sciences where this non-linear

[3] In this work we shall use the notation $x \wedge y = \min\{x, y\}$ and $x \vee y = \max\{x, y\}$ and analogously for the comparison of more than two numbers.

fragmentation model has been used [194, 208, 209] and also in the mathematics literature there has been some interest (see, for example [42, 76, 213, 222]).

Considering the discrete case, assume that the collision between a k-cluster and a $(j-k)$-cluster can give rise to a j-cluster with probability $w_{j-k,k}$, or, with probability $1 - w_{j-k,k}$, to a variable number of daughter particles with total mass equal to j. Observe that, in contradistinction to spontaneous fragmentation, in this collisional fragmentation process some of the daughter particles can be *larger* than any of the original clusters.

As usual, the equations are of the type

$$\dot{c}_i = Q_d(c)(i) \tag{20}$$

where the reaction term has the following additive contributions:

1. formation of i-clusters by coagulation of smaller clusters of appropriate size, say $i-j$ and j, with probability $w_{i-j,j}$, to which corresponds the term
$$\frac{1}{2}\sum_{j=1}^{i-1} w_{i-j,j}a_{i-j,j}c_{i-j}c_j\,; \tag{21}$$
2. destruction of i-clusters by their collision with any other cluster, independently of the final result be a coagulation or a fragmentation, which corresponds to a contribution $Q_2(c)(i)$ given by (3);
3. the formation of an i-cluster as the result of a collision followed by instantaneous fragmentation, with probability complementary to the one above, whose term reads as
$$\frac{1}{2}\sum_{j=i+1}^{\infty}\sum_{k=1}^{j-1} \Psi^i_{j-k,k}(1 - w_{j-k,k})a_{j-k,k}c_{j-k}c_k, \tag{22}$$
where $\Psi^i_{j-k,k}$ gives the distribution of size i fragments produced by collisional fragmentation of $j-k$ and k-clusters. Thus, this function is analogous to the function ψ in the spontaneous fragmentation models (cf. Sect. 1.4). Observe that Ψ has to satisfy the identity $\Psi^i_{j,k} = \Psi^i_{k,j}$, and, considering that each reaction conserves mass, also
$$\sum_{i=1}^{j+k-1} i\,\Psi^i_{j,k} = j + k.$$

Hence, the general coagulation-fragmentation system with collisional fragmentation (20) is

$$\dot{c}_i = \frac{1}{2}\sum_{j=1}^{i-1} w_{i-j,j}a_{i-j,j}c_{i-j}c_j - c_i\sum_{j=1}^{\infty} a_{i,j}c_j +$$
$$+\frac{1}{2}\sum_{j=i+1}^{\infty}\sum_{k=1}^{j-1} \Psi^i_{j-k,k}(1-w_{j-k,k})a_{j-k,k}c_{j-k}c_k.$$

Naturally, to the right-hand side of this system one can add the spontaneous fragmentation term $Q_f(c)(i)$ given by (11).

The specification of the functions $\Psi^i_{j,k}$, $w_{j,k}$ and $a_{j,k}$ allows the modelling of particular cases of interest such as, for example, that considered in [209] where collisional fragmentation always produces only monomers.

The first mathematical study of these equations is due to Laurençot and Wrzosek [139]. An analogous (continuous) system, with the imposition of a maximum cluster size was proposed by Fasano and co-workers in the context of liquid-liquid dispersions in chemical engineering [84–86] (see also [215]). Another similar system was considered in studies of polymerization reactions with catalysed fragmentation [111].

1.8 Diffusive Coagulation-Fragmentation Equations

The previous approaches to coagulation-fragmentation processes assumed spatially homogeneous systems and so the cluster densities are independent of the spacial location. However, the spacial dependence of the cluster densities, and in particular the consideration of diffusive effects, has been recognized important is several situations [22, 71, 178].

The discrete version of these systems can be written as

$$\dot{c}_j = \nabla_z(d_j\nabla_z c_j) + Q_c(c)(j) + Q_f(c)(j)\,, \ \text{in } \Omega\times\mathbf{R}^+\subseteq\mathbf{R}^n\times\mathbf{R}^+ \tag{23}$$

where the diffusion coefficients $d_j = d_j(z,c)$ are non-negative functions, and adequate conditions are imposed to $c_j = c_j(z,t)$ on the boundary $\partial\Omega$, or to their decay at infinity.

The first mathematical study about these systems seems to be by Slemrod [199], but the first general existence and uniqueness result (without fragmentation, $Q_f(c)(j) \equiv 0$) is [19] by Bénilan and Wrzosek. In the last decade a growing number of papers have been published about systems like (23), or its continuous analogue, dealing with existence, uniqueness, and behaviour of solutions. In this context the contributions by Amann, Laurençot, Mischler, Wrzosek, among others, are extremely important (see, for example, [3–5, 68, 79, 121, 129, 132, 137, 138, 224, 225]). In the present chapter we shall not further consider these works.

1.9 Equations with Kinetic and Transport Terms

Another type of space dependence in these cluster equations is the introduction of transport terms, first considered in meteorological studies, in particular in models of cloud formation [20, 143]. The goal is to model the convection of clusters due to a given velocity field. We shall exemplify with the case studied in [74, Chapters 10 and 11]. Let $z \in \Omega \subseteq \mathbf{R}^3$ denote the space variable, $v(z,t,x) \in \mathbf{R}^3$ the velocity of the x-cluster at time t and position z, and let $r(z,t,x) \in \mathbf{R}$ be the rate of change of the concentration $c(z,t,x)$ by condensation or evaporation of droplets of size x in the space-time location (z,t). The equation, first proposed by Levin and Sedunov, and by Berry, is

$$\partial_t c(z,t,x) + \partial_x(r(z,t,x)c(z,t,x)) + \nabla_z(v(z,t,x)c(z,t,x)) = Q_c(c)(t,x), \qquad (24)$$

where $Q_c(c)$ is defined by (6), with the concentrations also dependent of the z variable, but the coagulation kernel only dependent on the masses.

In this model the velocity field is an "exterior" field where the clusters are embedded. In particular, the coagulation reactions are not influenced by the velocity field v. A different possibility is to consider the coagulation process depending on the local velocity, that is, considering the velocity field not as some field carrying the clusters, but essential as the field that describes the local velocity of each cluster. This more detailed model, analogous to the viewpoint used in kinetic theory, was first considered in the context of discrete velocity models by Slemrod and co-workers [201, 203], and, more recently, in the general case, among others by Escobedo et al. in [81] where they proved results on global existence of weak solutions and their convergence as $t \to \infty$.

Let $c(z,t,x,p)$ be the concentration of x-clusters with linear momentum p, and located at z at time t. The system studied in [81] is

$$\partial_t c + v \cdot \nabla_z c = Q_c(c), \qquad (25)$$

where $v = p/x$, and the coagulation term $Q_c = Q_1 - Q_2$ is defined by

$$Q_1(c) := \frac{1}{2}\int_{\mathbf{R}^3}\int_0^x a(y', y-y')c(\cdot,\cdot,y')c(\cdot,\cdot,y-y')dx'dp'$$

$$Q_2(c) := \int_{\mathbf{R}^3}\int_0^x a(y,y')c(\cdot,\cdot,y)c(\cdot,\cdot,y')dx'dp'$$

where $y := (x,p) \in \mathbf{R}^+ \times \mathbf{R}^3$, etc.

A similar model, without the space dependence, was considered before by Baranger in [15] and by Roquejoffre and Villedieu in [193]. Still another model was studied by Fournier and Mischler [94, 95], in which, although there is also no spacial dependence, there are, additionally to the binary collisions resulting in coagulating events, other binary elastic collisions (modelled by Boltzmann collision operator) and inelastic collisions (modelled by a granular collision operator).

Naturally, the analysis of this type of equations makes use of methods closely related to those used in studies of Boltzmann's and related kinetic equations. In this chapter we will not consider these works further.

1.10 Other Models

Other models have been considered in the literature. We shall now describe some of them.

1.10.1 Multi-Index Cluster Models

In the systems of previous sections clusters were characterized by a single "internal" quantity, their mass or volume, and, in some cases, by some "external" ones, such as the spacial position or velocity. However, for some applications one needs to characterize the existing clusters by additional variables identifying relevant physical quantities.

One obvious case is in co-polymerization reactions when there are two monomeric species, A and B say, and it is important to keep track of the way a given cluster is made, not only by the total number of monomeric particles, but accounting for how many of each monomeric species a given cluster is formed. Thus, in the simplest situation a cluster has to be described by a vector subscript (i_A, i_B) informing that the cluster is made by i_A units of the monomeric species A and by i_B units of B. This approach was used in kinetic studies of micelles and vesicle formation (cf. e.g. [64]) where the following two component Becker-Döring like system was proposed:

$$\frac{d}{dt}c_{i,j} = J^A_{i-1,j}(c) - J^A_{i,j}(c) + J^B_{i,j-1}(c) - J^B_{i,j}(c), \quad i,j \in \mathbf{N}^+ \setminus \{1\},$$

with the terms of microscopic balance for monomer A given by $J^A_{i,j}(c) := a_{i,j}c_{1,0}c_{i,j} - b_{i+1,j}c_{i+1,j}$ and those for B by $J^B_{i,j}(c) := \alpha_{i,j}c_{0,1}c_{i,j} - \beta_{i,j+1}c_{i,j+1}$, with the meaning of the symbols analogous to that in the usual Becker-Döring equation (19).

A similar case occurs when the clusters are made of two phases of the same substance and one needs to keep track of the quantities of each of them. An example is presented in [186], where it is considered that each particle has a continuously varying mass x, of which $\alpha \leq x$ is the mass of one of the phases (ice or liquid

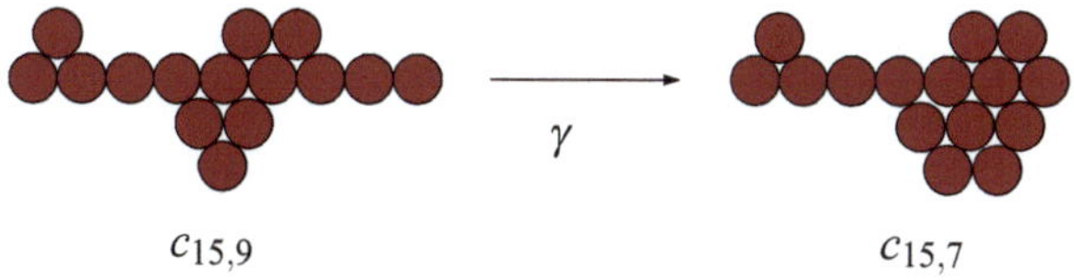

Fig. 5 An example of an internal geometric rearrangement "reaction" of a $(15, 9)$-cluster to a $(15, 7)$-cluster, at a rate $\gamma c_{15,9}$

water). Using a notation analogous to that in Sect. 1.2, the coagulation operator correspondent to (6) is now

$$
\begin{aligned}
Q_c(c)(x,\alpha) &= \frac{1}{2}\int_0^x \int_0^\alpha a(x-y,\alpha-\beta;y,\beta)c(\cdot,x-y,\alpha-\beta)c(\cdot,y,\beta)dyd\beta - \\
&= -c(\cdot,x,\alpha)\int_0^\infty \int_0^\infty a(x,\beta;y,\beta)c(\cdot,y,\beta)dyd\beta.
\end{aligned}
$$

Other model requiring a multi-index is considered in [220], where each cluster is characterized by its mass j and also by another subscript $k \leq j$ describing its shape in the sense that it reflects its diameter. In this case a cluster can not only be subject to coagulation and fragmentation reactions, but also to internal rearrangement "reactions" that are a mere change in its geometry, as Fig. 5 attempts to exemplify.

A further model of this type was used in the study of surface capping in cell-antibody interactions [37, 63]. In this case a j-cluster can be represented by a graph for which each of the j nodes stands for a monomeric unit in the cluster. All nodes are potentially of maximum degree three but not all of them have this valency at a given particular time. The $k < j$ nodes with degree one (the leaves of the graph) are particularly important in this model and so the clusters are characterized by the pair (j, k) and their dynamic was studied using an adapted version of the Becker-Döring system.

Finally, another case with an associated graph (in this case a tree) was studied in [56], motivated by the study of self-organized criticality in [99]. The model consists in a coagulation system for the evolution of clusters described by a pair (p, q), where p is the "order" and q its mass, and where the reactions are schematically represented by

$$(i,j) + (k,m) \to (\vee(i,k,(i \wedge k) + 1), j + m).$$

This means that the mass satisfies the usual additivity, and the order satisfy the Horton-Strahler rules. Each cluster is represented by an edge of a tree and a reaction between two clusters corresponds to the respective edges concurring in a node (cf. Fig. 6). In [56] a coagulation system for the time evolution of the concentrations

Fig. 6 Illustration of the Horton-Strahler rules in the orders of the edges of a tree

$c_{i,j}$ of the (i,j)-clusters is studied, as well as the evolution of some mesoscopic quantities, like the total number of clusters of a given order.

1.10.2 Annihilation Models

We now briefly consider a class of systems in which clusters are also made up of two monomer species but merit a reference outside Sect. 1.10.1 because part of the physical processes involved are significantly different from the usual coagulation in that cluster can annihilate each other. The two-species coagulation-annihilation system describe the time evolution of the concentration of clusters of two different particle species (A and B, say) in which the A-particle clusters [resp., B-particle clusters] undergo coagulation between themselves, symbolically

$$A_\mu + A_\lambda \xrightarrow{K_a} A_{\mu+\lambda} \quad \text{[resp., } B_\mu + B_\lambda \xrightarrow{K_b} B_{\mu+\lambda}\text{]},$$

but when an A-particle cluster and a B-particle cluster come together, they annihilate each other, and in the simplest such model the annihilation is complete, i.e., for all μ and λ,

$$A_\mu + B_\lambda \xrightarrow{L} \emptyset,$$

where $\emptyset$ represents a physically inert species. In the physics literature these processes have been approached through a variety of techniques. Using a mass action approach as in coagulation studies, one of the first works seems to be Ben-Naim and Krapivsky [18] where, for the case of discrete cluster sizes, it is assumed that reactions rates are independent of the cluster sizes and all have the same value, $K_a(\mu,\lambda) = K_b(\mu,\lambda) = L(\mu,\lambda) = 2$. In that work, the authors investigated the time evolution of the system and the existence, or non-existence, of a universal similarity behaviour of the solutions. More recently, Laurençot and van Roessel [135] considered these same issues in the case of continuous cluster sizes with reaction rates still independent of the cluster sizes but with the coagulation rates $K_a(\mu,\lambda) = K_b(\mu,\lambda) = k$ possibly different from the annihilation kernel $L(\mu,\lambda) = L$, a case that had already been considered by Krapivsky in [115] for the discrete case with $k = 2$. Still within the context of rate coefficients independent of cluster sizes, in [60] da Costa et al. extended [135] by considering the possibility of the coagulation rates of A-clusters and of B-clusters to be different from each other, i.e., $K_a(\mu,\lambda) = K_a$, $K_b(\mu,\lambda) = K_b$, and $L(\mu,\lambda) = L$, where K_a, K_b, L are positive constants, otherwise unrestricted. A slightly more general process has also been proposed consisting in an incomplete annihilation between A and B-particles (see, e.g., [115]), which means that the A and B species still annihilate each other but only to the extent corresponding to their respective sizes, which is represented schematically by

$$A_\mu + B_\lambda \longrightarrow \begin{cases} A_{\mu-\lambda} & \text{if } \mu > \lambda \\ \emptyset & \text{if } \mu = \lambda \\ B_{\lambda-\mu} & \text{if } \mu < \lambda \end{cases} .$$

Coagulation-annihilation with incomplete annihilation are hard to analyse mathematically and the method used for the complete annihilation case, based on Laplace transform techniques, do not seem to work. This difficulty has led to the consideration of some toy models that still show some interesting behaviour. One such model was first introduced by Redner et al. in [191] and a similar but more general one was introduced in [108]. These toy models retain the incomplete annihilation process but get rid of both, the two different monomeric species, and the coagulation reactions. The process is schematically represented by

$$A_j + A_k \overset{a_{j,k}}{\longrightarrow} A_{|j-k|},$$

where $A_0 := \emptyset$. Still assuming that there is no destruction of mass in each individual reaction, it now makes more sense to think of j as the size of the cluster "active part", being the difference between $j + k$ and $|j - k|$ the size of the resulting cluster that has become inactive after the reaction. One illustration of this is in Fig. 7.

The dynamics of this cluster system is governed by the following equations, called the RBK cluster system in [59],

$$\dot{c}_j = \sum_{k=1}^{\infty} a_{j+k,k}c_{j+k}c_k - \sum_{k=1}^{\infty} a_{j,k}c_jc_k, \qquad j = 1, 2, \ldots, \tag{26}$$

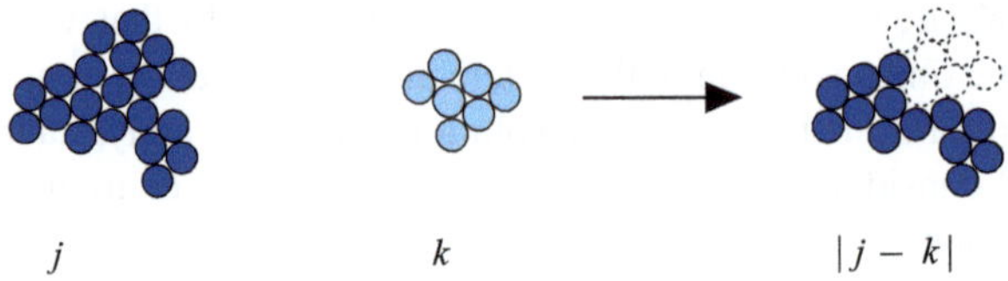

Fig. 7 Schematic reaction in the RBK coagulation-annihilation model

whose mathematical study started only recently [59, 61]. It is worth noticing the similarities, and also the differences between the RBK system (26) and Smoluchowski's coagulation equations.

1.10.3 Coagulation of Intervals in the Real Line

The models considered in this section are toy models for maturation and ageing processes in physical systems far from equilibrium and have been considered extensively in the Physics literature. Below we shall concentrate exclusively on mathematical works. It is also interesting to remark that some of these models can be seen as a kind of "dual" processes to the RBK system with mono-disperse initial data [108].

The first of these models was studied in 1992 by Carr and Pego [39]. Their motivation was the studies of metastability in solutions to reaction-diffusion systems of Chafee-Infante type $u_t = \varepsilon^2 u_{xx} + u - u^3$ in the bounded interval $(0, 1)$ with homogeneous Neumann conditions at the boundary, and very small diffusion coefficient ε. This rather interesting behaviour had been discovered and studied by the same authors and by Fusco and Jack Hale in a series of remarkable papers (cf. [38, 98] for an introduction to those results). It consists in the fact that a typical solution rapidly approaches functions that, in spite of not being equilibria (and being far from one) are practically time independent for an extraordinarily large interval of time (of the order $e^{1/\varepsilon}$). The graph of these functions are essentially constant but for what happens in the neighbourhood of a finite number N of points of $(0, 1)$, where very sharp transitions take place. When, due to the extra slow dynamics, two of these transition layers finally come close to one another, the dynamics has a markedly increase in its speed in such a way that the transition layers suddenly collapse and disappear, after which the dynamics returns to its exponentially slow pace.

The mean field coagulation like model [39, 100] for this behaviour was first derived by [165] and is the following: consider $N \gg 1$ points arbitrarily chosen in the interval $(0, 1)$ (these points represent the location of the transition layers in the solution to the reaction-diffusion equation); assume the following process with discrete time: in each time unit look for the shortest interval in the partition of $(0, 1)$ defined by the N chosen points and eliminate the two points that are its boundary, thus producing the fusion of that interval with its nearest neighbours (this

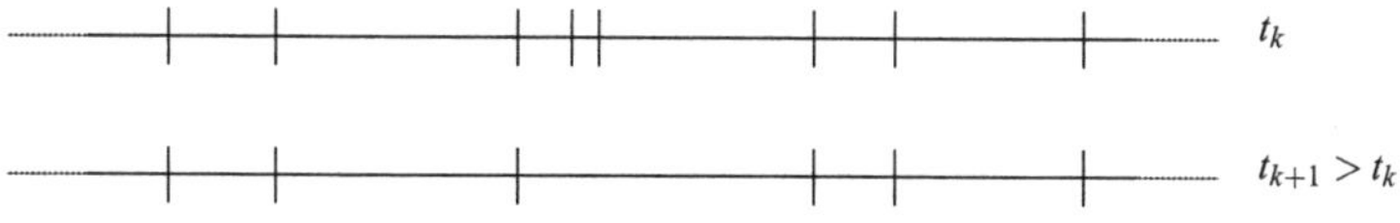

Fig. 8 A schematic interval's coagulation process

corresponds to the quick collapse of the transition layers) and having as a result the reduction of the number of points to $N-2$ in the next time unit (cf. Fig. 8). Denoting by $f(x,t)$ the density, at time t, of the distribution of the number of intervals by unit length, the total number of intervals by unit length is

$$N(t) = \int_0^\infty f(x,t)dx.$$

Let $\mathscr{L}(t)$ be the smallest interval at time t. The time evolution of f due to the process described above has the following two contributions:

1. formation of an interval with length x by coalescence of an interval of length $\mathscr{L}(t)$ with two intervals of lengths y and $x - y - \mathscr{L}(t)$,
2. disappearance of an interval of length x by coalescence with any other interval.

The differential equation for f is

$$\partial_t f(x,t) = \frac{f(\mathscr{L}(t),t)\dot{\mathscr{L}}}{N^2(t)} \left[\int_0^\infty f(y,t)f(x-y-\mathscr{L}(t),t)dy - 2f(x,t)N(t) \right], \tag{27}$$

where $f(x,t) = 0$ if $x < \mathscr{L}(t)$. In [39] the time scale was chosen so that the expected number of coagulation events per unit time is $f(\mathscr{L}(t),t)\frac{d\mathscr{L}}{dt} = 1$. In [100] the time was parametrized by the smallest interval size, i.e., $\mathscr{L}(t) = t$, and the system was written for the probability density $\rho_t(x) := f(x,t)/N(t)$, instead of f.

A more general model, also studied by Carr and Pego [40], is a generalization of previous models by Derrida et al. [67] and by Pesz and Rodgers [185]. The difference relative to the previous model is that now, in each unit of time, the smallest interval is divided in α^{-1} parts according to a probability density $d\nu(\alpha)$ and these parts are randomly redistributed by the remaining intervals in the partition.

Let $X(t)$ be the size of the smallest interval at time t, $\varphi(x,t)$ be the expected number of intervals with length larger than or equal to x at time t normalize by the initial number of intervals, and let $N(t)$ be the normalized total number of intervals at time t. Then, the dynamics is determined by the equation

$$\partial_t \varphi(x,t) = -\frac{\dot{N}(t)}{N(t)} \int_0^\infty \Big(\varphi(x - \alpha X(t), t) - \varphi(x,t) \Big) \alpha^{-1} d\nu(\alpha)$$

subject to

$$N(t) = \varphi(x,t), \quad \text{for } -\infty < x \leq X(t).$$

Another model was proposed and rigorously studied by Menon et al. [162]. The corresponding process is the following: at each time step choose an integer $k \geq 1$ at random with probability p_k, and merge the smallest interval with k randomly chosen intervals.

With $\rho_t^{\star k}$ denoting the k-fold self convolution of ρ_t, and the remaining variables as above, the dynamics is described by the equation

$$\partial_t f(x,t) = f(\mathscr{L}(t),t)\dot{\mathscr{L}} \sum_{k=1}^{\infty} p_k\Big(\rho_t^{\star k}(x-\mathscr{L}) - k\rho_t(x)\Big), \quad \text{with } x > \mathscr{L}(t).$$

In [162] the time scale was taken as $t = N(t)^{-1}$, and the analysis was based on the method of Gallay and Mielke [100].

1.10.4 Proliferation Models in Population Dynamics

The mathematical studies about proliferation processes in biological populations, being them of individuals, cells, or biochemical molecules, have resulted in an appreciable diversity of differential equation used as models [184].

One of these equations, to model the time evolution of a cell population undergoing mitosis, by which a cell of size x is broken into two of sizes $x/2$ at a rate $B(x)$, is the following [184, Chapter 4]

$$\partial_t n(x,t) + \partial_x n(x,t) = -B(x)n(x,t) + 4B(2x)n(2x,t),$$

where $n(x,t)$ is the density of cells of size x at time t. It is possible to generalize this process by assuming that a cell can be broken into α equal daughter cells with sizes x/α [55]. On the other hand, if the fragmentation process allows the two daughter cells to have distinct sizes, the differential equation for the density $n(x,t)$ has the typical form of a fragmentation equation with mass transport [184] (cf. Sects. 1.4 and 1.9 above):

$$\partial_t n(x,t) + \partial_x n(x,t) = -B(x)n(x,t) + \int_x^{\infty} b(x,y)n(y,t)dy.$$

In order to model more specific situations, the mathematical models can become correspondingly more complex. As an example that has recently received some attention, we can point to models of growth and proliferation of prions (i.e., of proteins with transmissible pathological conformations) responsible for the Bovine Spongiform Encephalopathy ("Mad Cow Disease") [102, 187] and several mathematical models have already been object of a rigorous analysis [136, 198, 216]. According to the contemporary biological understanding, there are two basic prion forms, a normal, non-infectious, monomeric one (denoted by PrP^{C} in the literature) and an infectious polymeric form (PrP^{Sc}) formed by the polymerization of the monomeric form. Above a certain critical size n, the PrP^{Sc} seems to be have the strong tendency to rapidly bond with the monomers. The PrP^{Sc} has also break up into polymers below the critical size that are quickly degraded into PrP^{C} monomers. Denoting by $y_0(t)$ the PrP^{C} concentration and by y_i the concentration of PrP^{Sc} polymeric chains made up of i monomers, the differential equation model is the following [102, 187]

$$\begin{aligned}\dot{y}_0 &= \lambda - dy_0 - y_0\sum_{i=n}^{\infty}\beta_i y_i + 2\sum_{j=1}^{n-1}\sum_{i=n+j}^{\infty} jb_i y_i + 2\sum_{j=1}^{n-1}\sum_{i=n}^{n+j-1} ib_i y_i\\ \dot{y}_i &= \beta_{i-1}y_0 y_{i-1} - \beta_i y_0 y_i - a_i y_i - (i-1)b_i y_i + 2\sum_{j=i+1}^{\infty} b_j y_j,\end{aligned}$$

where $a_i, b_i, \beta_i, \lambda$ e n are positive constants. Continuous mass versions of these equations were also considered in the literature [102, 136, 198, 216].

1.11 *Other Problems About Coagulation and Fragmentation Models: Relation with Particle Models*

To finish this introductory part, we will refer to a different type of mathematical studies of coagulation and fragmentation equations. So far, the mathematical works that we have referred to were those that, starting with a given differential equation, have as goal the study of (some) properties of its solutions (say: existence, uniqueness, regularity, mass conservation, long time behaviour, self-similarity). In the following sections of this chapter this is also the theme we will be interested in, but in the present section we consider another important class of problems that have attracted some attention: starting from more fundamental non-equilibrium Statistical Physics assumptions in terms of stochastic processes, to obtain the coagulation equations as some kind of thermodynamic limit of these processes. Here we will just present, in a brief way adapted from [192], the kind of approach used, and direct the interested reader to the works of, among others, Guiaş [105], Norris [175], Großkinsky and co-workers [104], Kolokoltsov [114], Rezakhanlou [192], and Fournier and co-workers [66, 92].

In the microscopic model one initially considers a collection of $N \gg 1$ particles randomly distributed in points $x_i \in \mathbf{R}^d$, with $d \geq 2$ and $i \in I = \{1, 2, \ldots, N\}$. Each particle have an integer mass $m_i \in \mathbf{N}^+$ and is animated with a Brownian motion with diffusion constant $2d(m_i)$. When two particles of masses m_i and m_j are at a distance from one another equal to $\|x_i - x_j\| = \varepsilon > 0$ they can coagulate to made a particle of mass $m_i + m_j$, randomly located in any of the positions x_i or x_j, with probability dependent of the masses of the original particles[4]; Due to this coagulation process the number of particles in the system diminishes with time and so the indexing set is time dependent , $I_{q(t)} \subset I$. One assumes that the dynamics of this particle system $q(t) := \{(x_i(t), m_i(t)) | i \in I_{q(t)}\}$ is a Markov process with infinitesimal generator $\mathscr{L} = \mathscr{A}_{\text{dif}} + \mathscr{A}_{\text{c}}^{\varepsilon}$, where $\mathscr{A}_{\text{dif}}$ is the contribution of the Brownian motion between collisions, and $\mathscr{A}_{\text{c}}^{\varepsilon}$ is the coagulation term. For this microscopic process one defines the empirical measure

$$g_n(dx, t) = \frac{1}{K_\varepsilon} \sum_i \delta_{x_i(t)}(dx) \mathbf{1}(m_i(t) = n).$$

Being $K_\varepsilon \overset{\varepsilon \to 0}{\longrightarrow} \infty$ an appropriate scaling factor and Z a constant (the total macroscopic density), one can prove that in the thermodynamic limit, i.e., when $N \to \infty$ keeping $N/K_\varepsilon = Z$, the measure $g_n(dx, t)$ converges to a measure $c_n(x, t)dx$ in the following sense

$$\lim_{N\to\infty} \mathbf{E}_N \left| \int_{\mathbf{R}^d} J(x, t)(g_n(dx, t) - c_n(x, t)dx) \right| = 0,$$

for all test functions J bounded and continuous in $\mathbf{R}^d \times [0, \infty)$. The density $c_n(x, t)$ of the limit measure solves the coagulation equation with diffusive terms (23).

2 Existence and Uniqueness of Solutions to Discrete Coagulation-Fragmentation Systems

In this section we shall review results about existence and uniqueness of solutions to coagulation-fragmentation systems, with special emphasis to the discrete case. Note, however, that most of the results in one case have equivalent in the other, and a rigorous relation between the two can be established [131]. We start with the special case of Smoluchowski's coagulation equation because of its importance, historically and conceptually. Most of the section will be devoted to existence results. Uniqueness will be treated in the last part.

We start by briefly presenting the most relevant spaces needed afterwards.

[4]In other versions of this coagulation process of stochastic particles it is assumed the resulting particle is located at the centre of mass $\frac{x_i m_i + x_j m_j}{m_i + m_j}$ [176], in still others coagulation can happen within a whole interval of distances between the particles and not only at the distance ε [104]

2.1 Finite Density Spaces

With the notation introduced in Sect. 1.2, let $c_j(t)$ be the concentration of j-clusters at time t and, without loss of generality, assume the mass of a j-cluster is j. Thus, the quantity $\rho(t) := \sum_{j=1}^{\infty} jc_j(t)$ can be interpreted as the total density of the system (total mass, assuming the volume is constant) and it is reasonable to impose that solutions to (16) must have finite density, which means that, for all $t \geq 0$, the solution must be an element of the Banach space $X_1 \subset \ell^1$ of finite density solutions defined by

$$X_1 := \left\{ c = (c_j) \in \mathbf{R}^{\mathbf{N}^+} : \|c\|_1 := \sum_{j=1}^{\infty} j|c_j| < \infty \right\}. \tag{28}$$

In many situations it is important to consider other Banach spaces, namely

$$X_\alpha := \left\{ c = (c_j) \in \mathbf{R}^{\mathbf{N}^+} : \|c\|_\alpha := \sum_{j=1}^{\infty} j^\alpha |c_j| < \infty \right\}, \quad \alpha \geq 0. \tag{29}$$

Some of these spaces have also physical meaning, for example, the norm in $X_0 = \ell^1$ is a quantity proportional to the total number of clusters. Due to the physical meaning associated to the coagulation and fragmentation equations we will consider only non-negative solutions, i.e., those remaining in the non-negative cone of the relevant space X_α,

$$X_\alpha^+ := \left\{ c \in X_\alpha : c_j \geq 0, \forall j \right\}. \tag{30}$$

It is not hard to prove [44, Theorem 1.2.1] that the spaces X_α with the norms $\|\cdot\|_\alpha$ constitute a compact and normal scale of Banach spaces, which means that, for all $\beta > \alpha \geq 0$, $X_\beta \subset X_\alpha$ with the inclusions being continuous, dense, and compact, and for all $c \in X_\beta$ it holds that $\|c\|_\alpha \leq \|c\|_\beta$, and the following interpolation inequality is also valid

$$\forall 0 \leq \alpha < \beta < \gamma, \ \forall c \in X_\gamma, \ \ \|c\|_\beta^{\gamma-\alpha} \leq \|c\|_\alpha^{\gamma-\beta} \|c\|_\gamma^{\beta-\alpha}.$$

This scale is also regular, meaning that the norm of the dual spaces X'_α is a logarithmically convex function of the parameter α, but this result in not needed in what follows.

For the continuous version of the coagulation-fragmentation equations (17), where the cluster masses are in $\mathbf{R}^+ = (0, \infty)$, one defines the relevant spaces in an analogous way, but with the difference that the need to control what happens to very small clusters, and the non-existence of an inclusion relation in the $L^p(\mathbf{R}^+)$

spaces similar to what exists in the ℓ^p, leads to the following finite density space

$$Y_1 := L^1(\mathbf{R}^+, (1+y)dy) = L^1(\mathbf{R}^+, dy) \cap L^1(\mathbf{R}^+, ydy),$$

where dy is the Lebesgue measure on $\mathbf{R}$. The norm in this space is

$$\|\cdot\|_{Y_1} := \|\cdot\|_{L^1(\mathbf{R}^+,dy)} + \|\cdot\|_{L^1(\mathbf{R}^+,ydy)}.$$

2.2 *Discrete Smoluchowski Equations*

Let us consider the Cauchy problem for Smoluchowski's coagulation system (1)–(4),

$$\begin{aligned} \dot{c}_j &= \frac{1}{2}\sum_{k=1}^{j-1} a_{j-k,k}c_{j-k}c_k - c_j\sum_{k=1}^{\infty} a_{j,k}c_k, \\ c_j(0) &= c_{j0}, \end{aligned} \tag{31}$$

The approach to questions of existence to (31) that have been most fruitful so far consists in its approximation by finite n-dimensional truncations for which one can prove that their solutions $c^n(t)$ approach, in an adequate sense, a function $c(t)$ which can be proved to be a solution of the infinite dimensional system (31). This approach was used from the very first mathematical works, in the coagulation system by McLeod [154], in the coagulation-fragmentation systems by Spouge [206] and in the Becker-Döring, by Ball et al. [13].

A different approach that has been occasionally used in continuous coagulation-fragmentation systems consists in the use of fixed point theorems and operator semi-group theory, techniques that were pioneered by Melzak [158] and by Aizenman and Bak [1]. The approach using semi-group theory has been very successful in the study of (linear) fragmentation systems (cf., for example, [14, 153]).

In the present chapter we will only use the approach based on truncation. There are essentially two finite n-dimensional truncations used in the literature: the *n-maximal truncation* and the *n-minimal truncation,* in the designation introduced in [49]. The first one corresponds to the following system of n ordinary differential equations for the phase space vector $(c_1, c_2, \ldots, c_n)$:

$$\dot{c}_j = \frac{1}{2}\sum_{k=1}^{j-1} a_{j-k,k}c_{j-k}c_k - c_j\sum_{k=1}^{n-j} a_{j,k}c_jc_k, \quad j \in \{1, \ldots, n\}. \tag{32}$$

The second corresponds to the system

$$\dot{c}_j = \frac{1}{2}\sum_{k=1}^{j-1} a_{j-k,k}c_{j-k}c_k - \sum_{k=1}^{n} a_{j,k}c_jc_k, \quad j \in \{1,\dots,n\} \tag{33}$$

for the same phase space vector. A $2n$-dimensional truncation analogous to the n-minimal truncation, for which the $2n$-dimensional vector is $(c_1, c_2, \dots, c_{2n})$, is the following [123]:

$$\begin{aligned} \dot{c}_j &= \frac{1}{2}\sum_{k=1}^{j-1} a_{j-k,k}c_{j-k}c_k - \sum_{k=1}^{n} a_{j,k}c_jc_k, \; j \in \{1,\dots,n\} \\ \dot{c}_j &= \frac{1}{2}\sum_{k=j-n}^{n} a_{j-k,k}c_{j-k}c_k, \qquad j \in \{n+1,\dots,2n\}. \end{aligned} \tag{34}$$

The starting point of the analysis consists in considering an appropriate and rigorous version of the formal identity, which is a weak version of the coagulation equation:

$$\sum_{j=1}^{\infty} g_jc_j(t) - \sum_{j=1}^{\infty} g_jc_j(\tau) = \frac{1}{2}\int_\tau^t \sum_{j=1}^{\infty}\sum_{k=1}^{\infty}(g_{j+k} - g_j - g_k)a_{j,k}c_j(s)c_k(s)ds, \tag{35}$$

where one assumes that $0 \le \tau \le t$, and (g_j) is a non-negative test sequence.

From this equality (or for a rigorous version of it) one extracts the *a priori* estimates needed for the proofs. For instance, from (35) one can infer that the only *a priori* estimate expected to hold in X_1^+ is the boundedness of density (corresponding to the test sequence $g_j = j$), and also that the stronger the growth rate of the coefficients, the harder to get the estimates one needs and the more likely it is to expect the need of some control via assumptions on the higher moments $M_p(c) := \sum_{j=1}^{\infty} j^p c_j(t)$, $p > 1$. (Note that for non-negative sequences $M_p(c) = \|c\|_p$.)

Before proceeding it is necessary to be precise about what is meant by a solution [12, 123]:

Definition 1 Let $T \in (0, +\infty]$ and $c_0 = \big(c_{j0}\big) \in [0, +\infty)^{\mathbb{N}^+}$. A solution $c = (c_j)$ of (31) in $[0, T)$ is a sequence of non-negative continuous functions satisfying $\forall j \ge 1$ and $\forall t \in (0, T)$,

(i) $c_j \in C([0, T))$

(ii) $\displaystyle\sum_{j=1}^{\infty} a_{j,k}c_j \in L^1(0, t)$

(iii) $\displaystyle c_j(t) = c_{j0} + \int_0^t \Big(\frac{1}{2}\sum_{k=1}^{j-1} a_{j-k,k}c_{j-k}(s)c_k(s) - \sum_{k=1}^{\infty} a_{j,k}c_j(s)c_k(s)\Big)ds$

A problem that immediately arises is to know if solutions with non-negative [resp. positive] initial data remain non-negative [resp. positive] for all later times. This problem was first studied in the Becker-Döring case [13], and afterwards in the coagulation-fragmentation in [34, 36]. For the Smoluchowski coagulation system the following result was proved in [46]:

Theorem 1 ([46]) *Let $a_{j,k} > 0$ for all j, k. Take any $c_0 \in X_1^+$ and let c be a solution of (31) in $[0, T) \subset [0, +\infty)$. For each $t \in [0, T)$, let $\mathscr{J}(t)$ be the set of subscripts j for which $c_j(t) > 0$. Then $\mathscr{J}(t) \equiv \mathscr{J}$ is independent of t and is given by $\mathscr{J} = span_{\mathbf{N}_0}(\mathscr{J}(0)) := \left\{ j = \sum_i n_i p_i : p_i \in \mathscr{J}(0), n_i \in \mathbf{N}_0, \max_i n_i > 0 \right\}$.*

It is easy to conclude from this result that if, for some subscript p, we have $c_p(0) > 0$ then it is always true that $c_p(t) > 0$ for all $t > 0$. On the other hand, the proof of the result implies that if $c_p(0) = 0$ then, either $c_p(t) = 0, \forall t > 0$ (if $p \notin \mathrm{span}_{\mathbf{N}_0}(\mathscr{J}(0))$), or it will be always positive (if $p \in \mathrm{span}_{\mathbf{N}_0}(\mathscr{J}(0))$.)

The proof of the theorem uses in a fundamental way the following way to write Smoluchowski's equation:

$$c_j(t)E_j(t) = c_\tau E_j(\tau) + \int_\tau^t E_j(s)R_j(s)ds \tag{36}$$

where

$$E_j(t) := \exp\left(\sum_{k=1}^{\infty} a_{j,k}c_k(t)\right), \qquad R_j(t) := \frac{1}{2}\sum_{k=1}^{j-1} a_{j-k,k}c_{j-k}(t)c_k(t),$$

and $R_1(t) \equiv 0$.

These positivity results are also relevant for the coagulation-fragmentation systems, for which this method (with the obvious modification in the definitions of E_j and R_j) was first used in [34, 36], where it was also used to prove that, for all $t > 0$, all components j of $(c_j(t))$ are strictly positive (i.e., $\mathscr{J} \equiv \mathbf{N}$) provided that, for all natural numbers k, the coefficients $a_{1,k}$ and $b_{1,k}$ are positive.

Naturally, the existence results depend on the hypothesis on the coagulation coefficients $a_{j,k}$. It is interesting to observe that not only the growth but also the structure of the coefficients is crucial for the existence of solution, as it is clear in the following results. Let us start by considering the following coagulation coefficients, that will be called *multiplicative type coefficients*:

(H1) There exists non-negative sequences (r_j) and $(\alpha_{j,k})$ such that

$$a_{j,k} = r_j r_k + \alpha_{j,k}, \tag{37}$$

and one of the two conditions is satisfied:

$$\lim_{j\to\infty} \frac{r_j}{j} = 0, \quad \lim_{j\to\infty} \frac{\alpha_{j,k}}{j} = 0, \quad \forall k \geq 1, \tag{38}$$

or

$$\inf_{j\geq 1}\frac{r_j}{j}=R>0,\quad \alpha_{j,k}\leq Kr_jr_k,\quad \forall j,k\geq 1, \tag{39}$$

for non-negative constants R and K.

For this kind of coefficients we have the following result due to Laurençot [123], and Leyvraz and Tschudi [148],

Theorem 2 *Assuming (H1) and being $c_0\in X_1^+$, there exists at least one solution c of (31) in $[0,+\infty)$ such that, for all $t\in[0,+\infty)$, it holds true that $c(t)\in X_1^+$ and $\|c(t)\|_1\leq\|c_0\|_1$.*

Sketch of proof The basic idea of the proof is, by taking limits as $n\to\infty$ in the sequence of solutions of n-truncated systems, to obtain a non-negative continuous function and to prove that this function is, in fact, a solution to the Cauchy problem for Smoluchowski's equation (31).

Being a bit more precise, assume the coagulation coefficients satisfy (38). Then, the existence of a function $c=(c_j)$ that is limit of the sequence of solution (c_j^N) of the truncated systems [for instance, using (32)] is a consequence of the Ascoli-Arzela theorem, due to the compact inclusion of X_1 in $X_{(r)}:=\Big\{c=(c_j)\in\mathbf{R}^{\mathbf{N}^+}:\|c\|_{(r)}<\infty\Big\}$, where $\|c\|_{(r)}:=\sum_{j=1}^{\infty}r_j|c_j|$, and the equiboundedness and uniform equicontinuity of the sequence of solutions to the truncated systems. Another way to prove the existence of a limit as $N\to\infty$ of the sequence of truncated solutions (c_j^N) is due to Ball and Carr [12] and consists in the application of Helly's theorem [113, pp. 370–371] to an equibounded sequence of uniformly bounded variation functions built from the solutions (c_j^N).

By (38), the *a priori* uniform boundedness of the density of (c_j^N) is sufficiently strong to conclude the following estimate (uniform in N_k)

$$\sum_{i=M}^{N_k}r_ic_i^{N_k}\leq\sup_{i\geq M}\frac{r_i}{i}\sum_{i=M}^{N_k}ic_i^{N_k}\leq\|c_0\|_1\sup_{i\geq M}\frac{r_i}{i}, \tag{40}$$

which allows us to control the infinite sum in the right-hand side of (32) and to obtain the pointwise limit

$$\lim_{k\to\infty}\left|\sum_{j=1}^{N_k}a_{i,j}c_j^{N_k}-\sum_{j=1}^{\infty}a_{i,j}c_j\right|=0. \tag{41}$$

By the dominated convergence theorem, implies that we can take limits as $N\to\infty$ and prove that c is a solution of (31).

If the coagulation coefficients satisfy (39) instead of (38) the problem is harder and the argument has to be modified. The difficulty of this case arises from the fact that the uniform bound on the density of (c_j^N) is not strong enough to control the terms in the second sum of the right-hand side of the truncated system. A way to overcome this problem, due to Laurençot [123], uses truncation (34).

The first part of the proof consists in getting the existence of a function c that is limit of the solutions (c_j^N) of the truncated systems. This is a consequence of the compact injection of $H^1(0,T)$ in $C(0,T)$. We start by pointing out that the version of (35) for the solutions (c_j^N) of (34) is

$$\sum_{j=1}^{2N} g_j c_j^N(t) - \sum_{j=1}^{2N} g_j c_j^N(\tau) = \frac{1}{2}\int_\tau^t \sum_{j=1}^{N}\sum_{k=1}^{N}(g_{j+k} - g_j - g_k)a_{j,k}c_j^N(s)c_k^N(s)ds, \tag{42}$$

and the needed estimates are concluded by exploiting this g-moment propagation equation. Choosing in (42) $g_j = j\mathbf{1}_{\{1,\dots,N\}}$, $g_j = 1$, and $g_j = j^{1/2}\mathbf{1}_{\{1,\dots,N\}}$, one gets

$$\sum_{j=1}^{N} jc_j^N(t) \le \sum_{j=1}^{N} jc_j^N(\tau) \le \sum_{j=1}^{N} jc_{j0} \tag{43}$$

$$\sum_{j=1}^{2N} c_j^N(t) + \frac{1}{2}\int_0^t \Big|\sum_{j=1}^{N} r_j c_j^N(s)\Big|^2 ds \le \sum_{j=1}^{2N} c_{j0} \tag{44}$$

$$\int_\tau^t \Big|\sum_{j=M}^{N} r_j c_j^N(s)\Big|^2 ds \le 4\Big(\sum_{j=1}^{N} j^{1/2}c_j^N(\tau)\Big)M^{-1/2}. \tag{45}$$

These *a priori* estimates imply that, for every $T \in (0,+\infty)$ and $N \ge j$, $c_j^N(t) \le \|c_0\|_0$ in $[0,T]$ and

$$\left\|\frac{dc_j^N}{dt}\right\|_{L^2(0,T)} \le T^{1/2}\left(\sum_{i=1}^{j-1} a_{i,j-i}\right)\|c_0\|_0^2 + \sqrt{2}(1+K)r_j\|c_0\|_0^{3/2}. \tag{46}$$

Hence, (c_j^N) is bounded in $H^1(0,T)$ and thus is relatively compact in $C(0,T)$. Using a diagonalization argument, one can conclude the existence of a subsequence $(c_j^{N_k})$ converging in $C(0,T)$ to some function $c = (c_j)$ as $N_k \to \infty$.

This convergence, together with the estimates (43) and (45) imply the version of (41) for the present case,

$$\lim_{k\to\infty}\left\|\sum_{j=1}^{N_k} a_{i,j}c_j^{N_k} - \sum_{j=1}^{\infty} a_{i,j}c_j\right\|_{L^2(0,T)} = 0, \tag{47}$$

which is the main ingredient to pass to the limit in the ith equation of (34), thus concluding the proof that c is a solution of (31). ■

Let us now consider *additive* coagulation coefficients, i.e., those satisfying the condition

(H2) There exist non-negative sequences (r_j) and $(\alpha_{j,k})$ such that

$$a_{j,k} = r_j + r_k + \alpha_{j,k}, \tag{48}$$

and also $0 \leq \alpha_{j,k} \leq K(j+k)$, for some constant $K \geq 0$.

If the sequences (r_j) and $\alpha_{j,k}$ are sublinear and satisfy (38) the argument of Leyvraz and Tschudy presented above can be adapted to obtain an existence proof also in this case. When $r_j \leq$ (const.)j it is necessary to modify those arguments: it is still possible to use Helly's theorem to prove the convergence of a subsequence of the sequence of solutions to truncated systems but the remaining proof needs to be changed using an identity like (35) for the evolution of the partial sums $\sum_{j=m}^{N} g_j c_j^N(t)$. This approach, due to Ball and Carr [12], is also applicable to the coagulation-fragmentation system, and so we leave a more detailed presentation to the next section. The result that is proved is the following:

Theorem 3 *Let $K > 0$ be a constant and assume $a_{j,k} \leq K(j+k)$. Let $c_0 \in X_1^+$. Then, there exists at least one solution c of (31) in $[0, +\infty)$ such that $c(t) \in X_1^+$ e $\|c(t)\|_1 \leq \|c_0\|_1$, for all $t \in [0, +\infty)$.*

An important distinction between systems with multiplicative and additive coefficients is that, in the last case, there are no solutions to the Cauchy problems (31) when the coefficients grow superlinearly, which certainly contrasts with what happen in the multiplicative case, as seen in Theorem 2.

This somewhat surprising non-existence result is a consequence of the following two theorems, to which we shall return in Sect. 4 on density conservation:

Theorem 4 ([12]) *Assume (H2) and let $c_0 \in X_1^+$. Then, for every $T > 0$, all solution c of (31) in $[0, T)$ conserve density, $\|c(t)\|_1 = \|c_0\|_1$.*

Theorem 5 ([35, 69]) *Let $T \in (0, +\infty]$, and let $C_L, C_U > 0$ and $\beta \geq \alpha > 1$ be constants such that $C_L(j^\alpha + k^\alpha) \leq a_{j,k} \leq C_U(jk)^\beta$. Let $c_0 \neq 0$ be an arbitrary element of X_1^+. Then, there are no solutions c of (31) in $[0, T)$ that conserve density in $[0, \tau)$, $\forall \tau \leq T$.*

Clearly, one can multiply the number of existence results indefinitely by making assumptions about the coefficients different from (H1) or (H2), but for these studies to be of any relevance it is necessary that the assumptions are either inordinately general, or of special interest for the applications. The case of the Becker-Döring type coefficients falls into this last class (cf. Sect. 1.6):

$$a_{j,k} = 0 \quad \text{if } j \wedge k > N,$$

where $N \geq 1$ is a fixed constant. The classic case corresponds to $N = 1$ [182] and is the only important one for the applications (cf., e.g., [16]). This means that the only non-zero coefficients are $a_{j,1}$ $(= a_{1,j})$ and the coagulation system is sometimes called the "addition model" (cf. [124]). The Cauchy problem (31) for these models with

$c_0 \in X_1^+$ has density conserving solutions in every interval $[0, T)$ when $a_{j,1} \sim \mathcal{O}(j)$ [13], and, by arguments similar to those used with assumption (H2), do not have any solution in whatever non degenerate time interval if $a_{j,1}$ is superlinear (satisfying somewhat technical but not very restrictive conditions) [124]. We shall return to these addition systems later in the chapter.

2.3 Discrete Coagulation-Fragmentation Equations

Let us now turn our attention to the problem of existence of solutions to the initial value problem for discrete coagulation-fragmentation equations (16), that we now write as follows:

$$\begin{aligned} \dot{c}_j &= \frac{1}{2}\sum_{k=1}^{j-1} W_{j-k,k}(c) - \sum_{k=1}^{\infty} W_{j,k}(c), \\ c_j(0) &= c_{j0}, \end{aligned} \tag{49}$$

where $W_{j,k}(c) := a_{j,k}c_jc_k - b_{j,k}c_{j+k}$.

The approximation of these systems by finite dimensional truncations works as in the previous section. The systems are analogous to those then considered, (32)–(34), the main difference being the substitution of $a_{p,q}c_pc_q$ by $W_{p,q}(c)$. With this minor change, and with the additional condition $\sum_{j=1}^{\infty} b_{j,k}c_j \in L^1(0,t)$, we obtain a definition of solution for coagulation-fragmentation systems of the same type as Definition 1.

As pointed out in Sect. 1.5, the first mathematical work on these equations was due to Spouge [206], who considered sublinear coagulation coefficients $a_{j,k} \leq r_jr_k$ with $r_j \sim \mathcal{O}(j)$ as $j \to \infty$, and somewhat restrictive conditions on the fragmentation coefficients.

More recent results, valid for much more general coefficients were proved by Ball and Carr [12], da Costa [45], Laurençot [126], and others, and it is to these that we will now turn our attention. The fundamental technique of these works is, as in the coagulation system considered in the previous section, the exploitation of the evolution of appropriate g-moments of the solutions (c^N) of the N-truncated systems as a way to obtain *a priori* estimates useful for taking limits as $N \to \infty$.

The version of (35) for solutions (c_j^N) of the N-truncated maximal system that is useful in this study is

$$\begin{aligned} \sum_{j=m}^{N} g_jc_j^N(t) - \sum_{j=m}^{N} g_jc_j^N(\tau) = \int_\tau^t \Bigg(& \frac{1}{2}\sum_{T^1_{m,N}} (g_{j+k} - g_j - g_k)W_{j,k}(c) + \\ & + \frac{1}{2}\sum_{T^2_{m,N}} g_{j+k}W_{j,k}(c) + \sum_{T^3_{m,N}} (g_{j+k} - g_k)W_{j,k}(c) \Bigg) ds, \end{aligned} \tag{50}$$

where $T^p_{m,N}$ are the following subsets of $\mathbf{N} \times \mathbf{N}$: $T^1_{m,N} := \{j, k \geq m, j + k \leq N\}$, $T^2_{m,N} := \{j, k \leq m-1, m \geq j+k \leq N\}$, and $T^3_{m,N} := \{j \leq m-1, k \geq m, j+k \leq N\}$, with the sum defined to be zero if the corresponding set is empty.

Let us start with the case where the coagulation coefficients are of the type $a_{j,k} \leq K(j + k)$, for some positive constant K. This condition includes the case $a_{j,k} \leq$ (const.)$(jk)^{1/2}$, but not other important cases such as $a_{j,k} \leq$ (const.)jk, that will be considered afterwards.

Theorem 6 ([12, 126]) *Let $a_{j,k} \leq K(j + k)$, where K is an arbitrary positive constant. Let c_0 be any element of X_1^+. Then, there exists at least one solution c of (31) defined on $[0, +\infty)$ and satisfying $\|c(t)\|_1 = \|c_0\|_1$.*

The original proof of this theorem is due to Ball and Carr [12, Theorems 2.4 and 2.5]. In what follows we present a simpler version by Laurençot [126] that is based on the adaptation and generalization of a classical result of de la Vallée Poussin [190, Theorem I.1.2-2] which, *grosso modo,* guarantees that every integrable function has an higher integrability property (cf. Lemma 1 below). It is this additional integrability that allows the deduction of an *a priori* estimate to pass to the limit $N \to \infty$ in the sequence (c^N)

A noteworthy aspect of these proofs is that no assumptions are made on the binary fragmentation coefficients (apart from the general ones of positivity and symmetry). The result of [126] is even applicable to equations with multiple fragmentation (10)–(11), but here we will particularize for the case of binary fragmentation (49) .

Sketch of proof To get a function c as limit of solutions (c^N) to the truncated systems we proceed as in the proof of Theorem 2, applying Helly's theorem to an auxiliary sequence [12]. The fundamental problem is to prove that the limit function is a solution to the Cauchy problem. It is on this problem that we will centre our attention.

Let $\mathscr{K}_1$ be the subset of $C^1([0, +\infty)) \cap W^{2,\infty}_{\mathrm{loc}}(0, +\infty)$ whose elements are non-negative convex functions U such that $U(0) = 0$, $U'(0) \geq 0$, and U' is concave. Let $\mathscr{K}_{1,\infty} \subset \mathscr{K}_1$ be the set of those functions that, additionally, satisfy

$$\lim_{x\to+\infty} U'(x) = \lim_{x\to+\infty} \frac{U(x)}{x} = +\infty. \tag{51}$$

The following lemma is an extension of a result of de la Vallée Poussin that is useful in what follows:

Lemma 1 ([122, 140]) *Let $(\Omega, \mathscr{B}, \mu)$ be a measure space, and let $w \in L^1(\Omega, \mathscr{B}, \mu)$. Then, there exists a function $U \in \mathscr{K}_{1,\infty}$ such that $U(|w|) \in L^1(\Omega, \mathscr{B}, \mu)$.*

In applying this lemma to our case $\Omega = \mathbf{N}$, $\mathscr{B} = 2^{\mathbf{N}}$, and, for $I \in \mathscr{B}$, define $\mu(I) := \sum_{i\in I} c_{0j}$, where $c_0 \in X^+$ is the initial condition of the Cauchy problem (31). Since $c_0 \in X^+$ we have $(x \mapsto x) \in L^1(\Omega, \mathscr{B}, \mu)$ and, by Lemma 1, there exists a function $U_0 \in \mathscr{K}_{1,\infty}$ such that $(x \mapsto U_0(x)) \in L^1(\Omega, \mathscr{B}, \mu)$, and so

$$\sum_{i=1}^{\infty} U_0(i) c_{0i} < \infty. \tag{52}$$

Observe that, in the sense of (51), U_0 grows faster at infinity than the identity and thus (52) provides a stronger decay of the initial condition c_0. As with the coagulation equations, the equation for the evolution of the U-moments of solutions (c^N) to the truncated systems is essential to obtain the needed estimates. For that we need to know that, for every $U \in \mathscr{K}_1$, there exists a positive constant m_U such that $(i+j)(U(i+j) - U(i) - U(j)) \leq m_U(iU(j) + jU(i))$, for all $i, j \in \mathbf{N}$. Using this inequality in (50) with $m = 1$ e $g = U_0$ we obtain, for every $0 \leq t \leq T < +\infty$,

$$\sum_{j=1}^{N} U_0(j) c_j^N(t) \leq C(T), \tag{53}$$

$$0 \leq \int_0^T \sum_{i=1}^{N-1} i \sum_{j=i+1}^{N} \Big(\frac{U_0(j)}{j} - \frac{U_0(i)}{i} \Big) b_{i,j-i} c_j^N(s) ds \leq C(T), \tag{54}$$

where by $C(T)$ we denote constants depending on T, and also of K, c_0 and U_0.

The same estimates are valid if in (53) we sum only up to $i \leq M$ and in (54) only up to $i \leq M-1$ and $j \leq M$, with $M \leq N-1$. Taking limits, first $N \to +\infty$, and then $M \to \infty$, we conclude that

$$\sum_{j=1}^{\infty} U_0(j) c_j(t) \leq C(T), \tag{55}$$

$$0 \leq \int_0^T \sum_{i=1}^{\infty} i \sum_{j=i+1}^{\infty} \Big(\frac{U_0(j)}{j} - \frac{U_0(i)}{i} \Big) b_{i,j-i} c_j(s) ds \leq C(T), \tag{56}$$

and from this it follows that $\sum_{j=1}^{\infty} a_{i,j} c_j \in L^1(0,T)$, and $\sum_{j=i+1}^{\infty} b_{i,j-i} c_j \in L^1(0,T)$. From (53) we deduce the following estimate, similar to (40),

$$\sum_{j=M}^{N-i} a_{i,j} c_j^N \leq 2iK \sup_{j \geq M} \frac{j}{U_0(j)} \sum_{j=M}^{N-i} U_0(j) c_i^N \leq C(i,T) \sup_{j \geq M} \frac{j}{U_0(j)}, \tag{57}$$

which, together with the analogous one obtained from (55) and with Lebesgue dominated convergence theorem, allow us to control the tails of the series corresponding to the coagulation terms and get

$$\lim_{N \to \infty} \left\| \sum_{j=1}^{N-i} a_{i,j} c_i^N c_j^N - \sum_{j=1}^{\infty} a_{i,j} c_i c_j \right\|_{L^1(0,T)} = 0. \tag{58}$$

The corresponding limit for the fragmentation terms, namely

$$\lim_{N\to\infty}\left\|\sum_{j=i+1}^{N} b_{i,j-i}c_j^N - \sum_{j=i+1}^{\infty} b_{i,j-i}c_j\right\|_{L^1(0,T)} = 0, \tag{59}$$

results from (54), (56) and the dominated convergence theorem.

This concludes the proof that the function c obtained as the limit of the truncations (c^N) when $N \to \infty$, is a solution of (49). To prove that the norm of c is equal to the norm of the initial condition we again use (53) and (55) in order to write, with $N \geq M-1 \geq 2$ arbitrary,

$$\begin{aligned}\Big|\|c(t)\|_1 - \|c_0\|_1\Big| &\leq \sum_{j=1}^{M-1} j|c_j^N(t) - c_j(t)| + \sum_{j=N+1}^{\infty} jc_{0j} + \sum_{j=M}^{N} jc_j^N(t) + \sum_{j=M}^{\infty} jc_j(t) \\ &\leq \sum_{j=1}^{M-1} j|c_j^N(t) - c_j(t)| + \sum_{j=N+1}^{\infty} jc_{0j} + 2C(T)\sup_{j\geq M}\frac{j}{U_0(j)}, \end{aligned} \tag{60}$$

which, by the arbitrariness of M and N, implies that $\|c(t)\|_1 = \|c_0\|_1$. ■

If the coagulation coefficients do not satisfy the bound $a_{j,k} \leq K(j+k)$, but only the weaker condition $a_{j,k} \leq K(jk)^\alpha$, with $\alpha \in [0,1]$, there are also several existence theorems for which it is also necessary to impose, in addition to the growth condition on the coefficients, some conditions on their structure, as well as restrictions upon the fragmentation coefficients. As we pointed out in page 114 a first result of this type, by Spouge [206], is the following (written with the hypothesis of binary fragmentation).

Theorem 7 ([206]) *Let $a_{j,k} \leq \mathcal{O}(j)\mathcal{O}(k)$, as $j,k \to +\infty$ where K is an arbitrary positive constant, and let $b_{j,k}$ satisfy $\sum_{k=1}^{j-1} b_{j,k} \leq Q$ and $b_{j,k} \leq \mathcal{O}(k)$ when $k \to +\infty$, for j fixed, where $Q > 0$ is a constant. Let $c_0 \neq 0$ be an arbitrary sequence in X_1^+. Then, there exists at least one solution c of (49) defined in $[0, +\infty)$.*

The proof of this result, like the one of Theorem 2, uses Helly's and Ascoli-Arzela theorems in order to obtain a solution of the Cauchy problem (49) by taking the limit $N \to \infty$ in the sequence of solutions to truncated systems [206].

Another existence result, obtained in [45] with a so called *strong fragmentation condition* on the fragmentation coefficients, is the following:

Theorem 8 ([45]) *Let $a_{j,k} \leq K_a(jk)^\alpha$, with constants $K_a > 0$ and $\alpha \leq 1$. Let $b_{j,k}$ be such that $\sum_{j=1}^{\lfloor\frac{r-1}{2}\rfloor} j^\mu b_{j,r-j} \geq K_f(\mu)r^{\gamma+\mu}$, where μ, γ and $K_f(\mu)$ are non-negative constants, and $\gamma > \alpha$. Take any element $c_0 \in X_1^+$. Then, there exists at least one solution c of (49) defined on $[0, +\infty)$. The solutions of (49) obtained as limits of solutions of maximally truncated systems are unique and satisfy $\|c(t)\|_1 = \|c_0\|_1$, $\forall t \geq 0$.*

Observe that the strong fragmentation condition used in this theorem is satisfied by fragmentation coefficients of the type

$$b_{j,k} \sim (j+k)^{\beta} \quad \text{or} \quad b_{j,k} \sim (jk)^{\beta}, \quad \text{with} \quad \beta > -1. \tag{61}$$

Sketch of proof The basic ingredient of the proof is the regularizing effect the strong fragmentation condition has on some higher moments, a fact that allows us to obtain the needed *a priori* estimates. This regularization result consists in the local integrability of moments $\|c(\cdot)\|_{1+\gamma-\varepsilon}$, $\forall\varepsilon$, of functions c that are obtained as weak-$*$ limits of sequences of solutions to truncated systems. This idea had already been used by Carr [34] in the study of the asymptotic behaviour of solutions when the coagulation coefficients satisfy the conditions of Theorem 6. The main difference between the tools used in [34] and in [45] is that the differential inequality for the evolution of higher moments of solutions c^N to truncated systems is now

$$\frac{d}{dt}\|c^N\|_{\mu} \le \mathscr{C}_0 + \mathscr{C}_1\|c^N\|_{\mu}^{\alpha_1} - \mathscr{C}_2\|c^N\|_{\mu}^{\alpha_2}, \tag{62}$$

where $\mu \ge 1+\alpha$, $\alpha_1 = 1+\frac{2\alpha-1}{\mu-1}$, $\alpha_2 = 1+\frac{\gamma}{\mu-1}$ and $\mathscr{C}_j$ positive constants dependent only on α, γ, μ and $\|c_0\|$. In [34], the inequality akin to (62) has the right-hand side of the Bernoulli equation and thus can be explicitly solved by a standard change of variable. In the case of (62) the analysis is less direct but one can prove that $\|c^N\|_{\mu}$ satisfies the inequality

$$\|c^N\|_{\mu} \le [(\nu-1)At]^{-\frac{1}{\nu-1}}, \tag{63}$$

for every constant $\nu \in (1, \alpha_2)$, and for constants $A = A(\nu, \alpha_1, \alpha_2, \mathscr{C}_0, \mathscr{C}_1, \mathscr{C}_2) > 0$ appropriately chosen. By taking the limit $N \to \infty$ in (63) we obtain the local integrability of the μ-moments ($\mu < 1+\gamma$) of functions obtained as weak-$*$ limits of solutions c^N. It is this local integrability of $(1+\gamma-\varepsilon)$-moments that is the *a priori* estimate which, together with the dominated convergence theorem and Fatou's lemma, allows taking limits in the truncated equations, both in the coagulation and in the fragmentation terms. ■

To finish this section it is worth observing that the results hereby presented do not cover all possible conditions on the coefficients or definitions of solution. In particular, note that if the fragmentation coefficients decay more rapidly that what is determined by the estimates (61) the above results are not applicable to the case $a_{j,k} \sim \mathscr{O}(jk)$ [Theorem 7, due to Spouge, requires $a_{j,k} \sim o(j)o(k)$]. The existence of solution in a case close to this critical growth case, where the coefficients have the structure $a_{j,k} = j^{\alpha}k + k^{\alpha}j$, with $\alpha \in (0,1)$, was obtained in [78] for continuous coagulation-fragmentation systems, as a consequence of estimates proved for the study of the gelation problem. The analysis presented in that paper can also be applied to discrete equations and will be analysed later in Sect. 4.

2.4 On the Uniqueness of Solutions

As in the case of existence studies, the results about uniqueness have been obtained under several different assumptions about the rate coefficients. The approach used by these studies consists, essentially, in assuming the existence of two distinct solutions c and d to the Cauchy problem (31) or (49), and then proving that some moment of the function $|x| := |c-d|$ satisfies a differential inequality which implies $x \equiv 0$ (cf. e.g. [12, 13, 45, 126]).

In order to get the needed differential inequalities one needs to control the evolution of certain moments, which requires the imposition of restrictions, either on the class of solutions under consideration, or on the coefficients, that are usually more stringent than those required in order to prove existence. As an illustration we present the case, studied in [12], where all coagulation coefficients are bounded, which already contains the main ingredients used in more general cases:

Theorem 9 ([12]) *Let $a_{j,k} \leq K$ where $K > 0$ is a constant. Take as initial condition any $c_0 \in X_1^+$. Then, there exists one and only one solution c to (49) defined in $[0, +\infty)$ and satisfying $\|c(t)\|_1 = \|c_0\|_1$ for all $t \geq 0$.*

Sketch of proof Assuming there are two solutions of the initial value problem (49), c and d, define $x := c - d$ and consider the function $\psi_1(t) := \|x(t)\|_1$. The version of (35) with $g_j = j\mathbf{1}(j \leq n)$ and c substituted by $|x|$ gives

$$\sum_{j=1}^{n} j|x_j(t)| = \int_0^t \Big(U_n(s) + V_n(s)\Big)ds, \tag{64}$$

where

$$U_n := \frac{1}{2}\sum_{T_{1,n}^1}(f_{j+k} - f_j - f_k)(W_{j,k}(c) - W_{j,k}(d)), \quad V_n := -\sum_{T_{1,n}^4} f_j(W_{j,k}(c) - W_{j,k}(d)),$$

with $f_j := j\ \mathrm{sgn}(x_j)$, $T_{1,n}^4 := \{1 \leq j \leq n, j+k \geq n+1\}$ and $T_{1,n}^1$ was previously defined in page 115. Noting that $W_{j,k}(c) - W_{j,k}(d) = (c_j x_k + d_k x_j)a_{j,k} - b_{j,k}x_{j+k}$, and using $a_{j,k} \leq K$, we get $\sum_{T_{1,n}^1}(f_{j+k} - f_j - f_k)(c_j x_k + d_k x_j)a_{j,k} \leq$ const.ψ, and thus, because $-(f_{j+k} - f_j - f_k)x_{j+k} \leq -((j+k) - j - k)|x_{j+k}| = 0$, we have $\int_0^t U_n(s)ds \leq (\text{const.})\psi(t)$. The limit $\int_0^t V_n(s)ds \to 0$ as $n \to \infty$ is obtained using the condition on the coefficients and the hypothesis that c and d are density conserving, which is natural since these last conditions are equivalent to

$$\sum_{j=1}^{n} jc_j(t) - \sum_{j=1}^{n} jc_{j0} = -\int_0^t \sum_{T_{1,n}^4} jW_{j,k}(c(s))ds \overset{n\to\infty}{\longrightarrow} 0, \tag{65}$$

and similarly for d.

With these estimates we can write

$$\psi_1(t) \leq (\text{const.}) \int_0^t \psi_1(s) ds, \tag{66}$$

and hence, by Gronwall's inequality, $\psi_1 \equiv 0$, implying uniqueness of density conserving solutions. ■

Observe that Theorem 9 imposes a very strong boundedness condition on the coagulation coefficients but none on the fragmentation ones (apart from the basic ones of non-negativity and symmetry used to prove existence). Observe also that the theorem establishes uniqueness just in the family of density conserving solutions.

This type of restrictions occur also in other cases. In [45] it is proved that, under the hypothesis of Theorem 8, density conserving solutions are unique. The proof uses the same ideas as presented before but the estimates for U_n and V_n are now obtained using the following integrability result $\|c(\cdot)\|_{1+\alpha} \in L^1(0,t)$, $\forall t < \infty$, of solutions c to (49). This additional regularity, similar to what was used in the proof of Theorem 8, has to be established for all solutions of (49), not only for those obtained by taking limits of solutions to truncated problems, and this is achieved by a kind of step-by-step argument first used by Carr in [34]. The estimates finally result in the following inequality, similar to (66),

$$\psi_1(t) \leq \text{const.} \int_0^t \varphi(s)\psi_1(s) ds, \tag{67}$$

where $\varphi(s) = K_a\|c(s)\|_{1+\alpha} + K_a\|d(s)\|_{1+\alpha}$.

Another uniqueness result, similar to the one in Theorem 9, is proved in [126] and complements the existence result whose proof was presented in Theorem 6. Under the condition $a_{j,k} \leq A_j + A_k$, where $A_i \leq K_a i$, it is shown, by a proof like the one above, that uniqueness holds in the class of density conserving solutions satisfying the integrability condition $\sum_{j=1}^{\infty} jA_jc_j \in L^1(0,t)$, for each $t < \infty$. There is a natural problem that immediately comes to mind, which is the existence of solutions with this additional regularity, or, better still, to know what are the additional conditions (if any) that need to be imposed at $t = 0$ that ensure this extra regularity at later times. The answer to this problem was given by Laurençot in [126] and generalizes previous similar results by Carr and da Costa [36]. Before presenting the result we need to introduce the following notation: we say that a function U is an element of $\mathscr{K}_2$ if it is non-negative, convex, belongs to $C^2([0,+\infty))$, satisfies $U(0) = U'(0) = 0$, its derivative is a convex function and there exists a positive constant K_U such that $U'(2x) \leq K_U U'(x)$ for all $x \geq 0$. (The functions $x \mapsto x^m$ are in $\mathscr{K}_2$ if $m \geq 2$.)

Proposition 1 ([126]) *Let $a_{j,k} \leq K(j+k)$, where $K > 0$ is a constant. Let $c_0 \in X_1^+$ be such that there exists $U \in \mathscr{K}_1 \cup \mathscr{K}_2$ with $\sum_{j=1}^{\infty} U(j)c_{j0} < \infty$. Then, there exists at* *least one solution c of (31) defined in* $[0, +\infty)$, *satisfying* $\|c(t)\|_1 = \|c_0\|_1$ *and, for each* $t < \infty$,

$$\sup_{s\in[0,t]} \sum_{j=1}^{\infty} U(j)c_j(s) < \infty.$$

This type of results, usually called "propagation of moments", are very useful for the study of the long time behaviour of solutions and we shall return to them in the next section.

Imposing growth restrictions on the kinetic coefficients it is possible to prove uniqueness without further regularity restrictions on the initial data. An example, due to Ball and Carr [12], is the following:

Theorem 10 ([12]) *Let* $K > 0$ *and* $\alpha \in \left[0, \frac{1}{2}\right]$ *be constants such that, for all natural numbers* j, k, n_0, *it holds* $a_{j,k} \leq K(jk)^{\alpha}$, $\sum_{j=1}^{\lfloor (k+1)/2 \rfloor} j^{1-\alpha} b_{k-j,j} \leq Kk^{1-\alpha}$ *and* $\sum_{j=n_0}^{\lfloor (r+1)/2 \rfloor} j^1 b_{r-j,j} \leq Kr$, *for* $r \geq 2n_0$. *Let* $c_0 \in X_1^+$ *be arbitrary. Then, there exists only one solution c of (49) defined on* $[0, T)$.

Note that, due to $(jk)^{\alpha} \leq (jk)^{1/2} \leq \frac{1}{2}(j + k)$, Theorem 6 can be applied to this case and this means that, under these conditions, solutions to (49) are unique and conserve density. However, note that the present result is not a uniqueness theorem in the class of density conserving solutions (as in the result of [46] cited above) but in the universe of all solutions to the Cauchy problem (49) in the sense of Definition 1. The proof of this theorem uses the method presented above for Theorem 9. The only relevant difference is that now is more convenient to get estimates on $\psi_{1-\alpha}(t) := \|x(t)\|_{1-\alpha}$ instead of $\psi_1(t) = \|x(t)\|_1$. The final result, from which uniqueness easily follows, is the inequality (66) with ψ_1 substituted by $\psi_{1-\alpha}$.

A natural question at this point is to know to what extent the cases not covered by these uniqueness theorems correspond to real cases of non-uniqueness. Clearly, the complete elucidation of this problem means a complete characterization of uniqueness, something not yet achieved at present. However, there are examples of non-uniqueness that seem to provide evidence that this problem is not simple. To conclude we present one of these examples, due to Ball and Carr [12]. Note that it is an example about the linear fragmentation system.

Example 1 ([12]) Let $a_{j,k} = 0$ and $b_{j,k} = 1$. Then, $c_j(t) = e^{-(j-1)t/2}\left(1 - e^{-t/2}\right)^2$ is a solution of (49) with initial condition $c_0 \equiv 0$.

Observe that the assumptions in Example 1 are included in those considered in Theorem 9. Hence, we know that density conserving solutions are unique. As a solution of (49) with $c_0 \equiv 0$ is the identically zero solution, the above example implies that we have non-uniqueness. Due to the linearity of the system, we can obtain analogous non-uniqueness results for other initial conditions. For further discussion on these issues see [7].

3 Long-Time Behaviour of Solutions to Coagulation-Fragmentation Systems

We shall now review some of the most important mathematical aspects of the long time behaviour of solutions. The large number of results in the literature forces us to make some choices about the results we will cover. We will keep the approach at the level used in the previous section so as to give the reader not only a guide to the literature but also to the ideas and some details of the proofs of the existing results.

3.1 *Convergence to Equilibria and Phase Transitions*

In this section we consider results about the convergence to equilibria of solutions to the discrete coagulation-fragmentation systems.

Since, as was pointed out in the Introduction, these equations can be seen as a mathematical model of an isolated chemical system, it is natural to expect solutions to converge to some equilibrium as $t \to +\infty$. In fact, in the usual chemical kinetics models this is exactly what normally happens. We shall see that in the infinite dimensional coagulation-fragmentation systems the asymptotic behaviour is much more interesting, even surprising, and a behaviour that is physically interpreted as a dynamical phase transition can take place under appropriate conditions.

3.1.1 Strong Fragmentation Systems

Let us start with the *strong fragmentation* case. This case was studied by Carr [34] and by Fournier and Mischler [93] and these conditions on the fragmentation coefficients were also considered by da Costa in [45] for the study of existence of solutions discussed above (cf. Theorem 8). The technique used in [34, 93] is based on the fact, pointed out in the discussion of the proof of Theorem 8, that this assumption on the fragmentation coefficients imply the boundedness of some higher moments, which implies the solution is pre-compact in X_1^+ for the norm topology. The existence of a Lyapunov function and the application of LaSalle's invariance principle were the tools that allowed Carr to prove the convergence of the solution to a unique equilibrium. Estimates based on the regularity of higher order moments were essential to obtain the exponential convergence to equilibria by Fournier and Mischler, valid for sufficiently small initial data. The result of [34] is the following:

Theorem 11 ([34]) *Let $K, K_f > 0$, $\alpha \in [0, 1]$ and $\gamma > \alpha$ be constants such that, for all natural numbers j, k, the following holds $a_{j,k} \leq K(j^\alpha + k^\alpha)$, $a_{1,k}, b_{1,k} > 0$ and $\sum_{j=1}^{\lfloor \frac{r-1}{2} \rfloor} j^\mu b_{j,r-j} \geq K_f(\mu) r^{\gamma+\mu}$. Assume that the detailed balance condition (15) is satisfied and that, for some $q \geq 1$, the partition function (M_j) satisfies*

$\liminf_{j\to\infty} M_j^{1/j^q} > 0$. *Let* $\rho \geq 0$ *be arbitrary. Then, there exists a time-independent solution* c^ρ *of the coagulation-fragmentation equations, with density* ρ, *such that, for all initial data* $c_0 \in X_1^+$ *with density* $\|c_0\| = \rho$, *the unique solution* $c(\cdot)$ *of (49) with constant density satisfies* $\|c(t) - c^\rho\|_m \overset{t\to\infty}{\longrightarrow} 0$, *for all* $m \geq 1$.

Sketch of proof With these assumptions on the coefficients the moments of solutions c^N to the truncated systems satisfy the following differential inequality, analogous to (62),

$$\frac{d}{dt}\|c^N\|_\mu \leq \mathscr{C}_0\|c^N\|_\mu - \mathscr{C}_1\|c^N\|_\mu^{\alpha_2}, \tag{68}$$

where $\mu > 1$, $\alpha_2 = 1 + \frac{\gamma}{m-1}$, $\mathscr{C}_0$, and $\mathscr{C}_1$ are positive constants. The standard change of variables used to solve the Bernoulli ordinary differential equations, $\|c^N\|_\mu \mapsto u := \|c^N\|_\mu^{1-\alpha_2}$, can be used to solve explicitly this differential inequality and taking $N \to \infty$ we conclude the solutions of (49) that are obtained as limits of truncated system satisfy

$$\|c\|_\mu \leq A\left(1 - e^{-Bt}\right)^{-\frac{\mu-1}{\gamma}}, \tag{69}$$

where $\mu > 1$, A and B are positive constants.

The strong fragmentation condition also implies that the $(1 + \gamma - \varepsilon)$-moments of every density conserving solution to (49) are integrable, as stated above (cf. page 120), and this implies the uniqueness of density conserving solutions. This fact means that there exists a semi-group of operators $T(\cdot)$ defined by $T(\cdot)c_0 := c$, where c is the unique density conserving solutions of (49). The inequality (69) implies that, for each $\mu > 1$ and $\tau > 0$, $\cup_{t\geq\tau}T(t)c_0$ is a bounded set of X_μ, and so, by the compact inclusion among the spaces X_α (cf. page 107) it is a pre-compact subset of X_1^+. Hence, for each initial condition $c_0 \in X_1^+$, the solution $T(t)c_0$ has a non-empty invariant ω-limit set $\omega(c_0) \subset X_\mu$, for all $\mu \geq 1$. What remains to be proved it that $\#\omega(c_0) = 1$ and its single element is an equilibrium, i.e., a time independent solution with density $\rho = \|c_0\|$, and that this equilibrium is independent of the initial condition, provided its density is ρ. It is at this point that the detailed balance condition is used, and the existence of a Lyapunov functions plays a central role.

For every $d_1 \geq 0$ define the sequence $d = (d_j)$ by

$$d_j := M_j\,(d_1)^j\,. \tag{70}$$

Clearly, the detailed balance condition implies that

$$W_{j,k}(d) = a_{j,k}d_jd_k - b_{j,k}d_{j+k} = a_{j,k}M_jM_k\,(d_1)^{j+k} - b_{j,k}M_{j+k}\,(d_1)^{j+k} = 0,$$

and so d is a stationary solution of (49) with $c_0 = d$ if and only if $d \in X_1^+$. The positivity is obvious from (70), but the fact that d has finite density requires a little more care. In order to study the density of $d = (d_j) = \left(M_j(d_1)^j\right)$ it is natural to consider the function $z \mapsto F(z) : [0, +\infty) \to [0, +\infty]$ defined by

$$F(z) := \sum_{j=1}^{\infty} jM_j z^j. \tag{71}$$

Let $z_s \in [0, +\infty]$ be the convergence radius of this series, and let $\rho_s := \sup_{z\in[0,z_s)} F(z)$. Clearly, there are three distinct cases for z_s: if $z_0 = 0$ then $\rho_s = 0$ and the only equilibrium solution is the zero solution; if $z_s = +\infty$ then $\rho_s = +\infty$ and for each $\rho \geq 0$ there exists a unique equilibrium (70) with density $\|d\|_1 = F(d_1)$; finally, if $z_s \in (0, +\infty)$, it can happen either $\rho_s = +\infty$ or $\rho_s < +\infty$, and in this last case there are no equilibria with densities $\rho > \rho_s$. This case will be important in the next section but under the strong fragmentation conditions that we are currently considering it is possible to prove that $\rho_s = +\infty$, and so, for every $\rho > 0$ there exists an equilibrium given by (70) with density ρ [34]. We will denote this equilibrium by c^ρ.

The discovery of Lyapunov functions for coagulation-fragmentation systems, related with the free energy, or with the entropy, of the physical system, was first made by Aizenman and Bak [1] for continuous systems with constant rate coefficients. For more general systems, the existence of a Lyapunov function seems to have been first identified, at a formal level, by Buhagiar in the Becker-Döring system (cf. ref. cit. [13]) and was first used in a mathematically rigorous way by Ball et al. in [13]. Our presentation will follow this paper closely, although the boundedness of higher moments due to our present strong fragmentation condition greatly simplifies it.

Let $c \in X_1^+$ and consider the function

$$V(c) := \sum_{j=1}^{\infty} c_j \left(\log \frac{c_j}{M_j} - 1\right), \tag{72}$$

where the term in the sum is defined to be zero if the corresponding c_j is zero. The continuity and minimization properties of this functional were established in [13] and will be presented next: let $V(c) = G(c) - F_m(c)$, with

$$G(c) = \sum_{j=1}^{\infty} c_j(\log c_j - 1), \quad F_m(c) = \sum_{j=1}^{\infty} j^m c_j \log M_j^{1/j^m}. \tag{73}$$

It is not hard to prove that G is finite and sequentially weak-$*$ continuous in X_1^+. As the radius of convergence of the series (71) is positive, $\limsup_{j\to\infty} M_j^{1/j} < \infty$, and we conclude that V is bounded below in

$$X_{1,\rho}^+ := \left\{c \in X_1^+ : \|c\|_1 = \rho\right\}, \tag{74}$$

c^ρ is the only minimizer of V in $X^+_{1,\rho}$, and every minimizing sequence $\left(c^{(j)}\right)$ of V in $X^+_{1,\rho}$ converges to c^ρ strongly in X_1. If $\liminf_{j\to\infty} M_j^{1/j^q} > 0$, for some $q \geq 1$, then V is bounded above in $X_m \cap X^+_{1,\rho}$ and is continuous in this set if $m \geq q$.

For solutions (c^n) of the Cauchy problem for the maximally truncated coagulation-fragmentation system, the following holds:

$$V(c^n(t)) + \int_\tau^t D_n(c^n(s))ds = V(c^n(\tau)) \tag{75}$$

where

$$D_N(c^n) := \frac{1}{2} \sum_{j+k\leq N} H_{j,k}(c^n) \tag{76}$$

and

$$\begin{aligned} H_{j,k}(c) &:= (a_{j,k}c_jc_k - b_{j,k}c_{j+k})(\log(M_{j+k}c_jc_k) - \log(M_jM_kc_{j+k})) \\ &= (a_{j,k}c_jc_k - b_{j,k}c_{j+k})(\log(a_{j,k}c_jc_k) - \log(b_{j,k}c_{j+k})) \ \geq \ 0, \end{aligned} \tag{77}$$

where the second equality comes the detailed balance condition (15), and the positivity from the fact that solutions have all their components positive (cf. page 110) and $\forall x, y > 0$, $(x-y)(\log x - \log y) \geq 0$.

Since for every $m \geq 1$ we have $c^n \to c$ strongly in X^+_m, by the continuity of V we conclude that $V(c^n(t)) \to V(c(t))$, for every $t \geq \tau > 0$. Fixing a positive integer N we have $D_n(c^n) \geq D_N(c^n)$ for $n \geq N$, and thus,

$$\liminf_{n\to\infty} \int_\tau^t D_n(c^n(s))ds \geq \int_\tau^t D_N(c(s))ds$$

which, letting $N \to \infty$, gives

$$V(c(t)) + \int_\tau^t D(c(s))ds \leq V(c(\tau)), \tag{78}$$

with

$$D(c) := \frac{1}{2} \sum_{j,k\geq 1} H_{j,k}(c). \tag{79}$$

This concludes the proof that V is a Lyapunov function for (49).

With these ingredients is now easy to prove the existence of one, and only one, equilibrium c^ρ with density ρ [which has necessarily the form (70)], and to get the characterization of $\omega(c_0)$: since this set must consist of solutions along which the

Lyapunov function is constant, this implies, by (78) and density conservation, that $\omega(c_0) = \{c^\rho\}$, with $\rho = \|c_0\|$, as we wanted to prove. ■

As the main ingredient of the above proof is, as already pointed out, the finiteness of higher order moments, Theorem 11 can be adapted, without further difficulties, to the conditions considered in [45], with the condition on the coagulation coefficients changed to $a_{j,k} \leq K(jk)^\alpha$, with $\alpha \leq 1$ [50].

To finish this section it is interesting to observe that the detailed balance condition is *not* a necessary condition to get convergence to equilibria. In fact, in [93], Fournier and Mischler proved that, if $a_{j,k} \leq K(jk)^\alpha$ and $L(j+k)^\gamma \leq b_{j,k} \leq K_f(j+k)^s$, with $\alpha \in [0,1]$, $\gamma > -2(1-\alpha)$ and $s, \gamma \in (-1,\infty)$, then, for all $\rho = \|c_0\|$ sufficiently small,[5] the solution $T(t)c_0$ satisfies

$$\|T(t)c_0 - \hat{c}\|_2 \leq Ke^{-\kappa t}, \quad \forall t \geq 1, \tag{80}$$

where the constants $K, \kappa > 0$ depend only on α, γ, K_c, L, and ρ, and $\hat{c}$ is the only equilibrium of the system with density ρ. Note that this result establishes an exponential rate of convergence to equilibria.

The proof of this result is also based on the finiteness of higher order moments, which comes from the lower bound on the fragmentation coefficients. More specifically, under the stated conditions one proves the following contraction property: there exists a T^* such that, for all $t \geq T^*$, and all solutions c and d of the coagulation-fragmentation system with initial data c_0 and d_0, respectively, both with density ρ, the following inequality holds

$$\frac{d}{dt}\|c(t) - d(t)\|_2 \leq -\kappa\|c(t) - d(t)\|_2. \tag{81}$$

This differential inequality is obtained from

$$\frac{d}{dt}\|c - d\|_2 \leq \left(2K\|c + d\|_3 - \frac{L}{16}\right)\|c - d\|_2,$$

and from an estimate on the third moment of $c + d$ proving that one can estimate its value uniformly in time by a quantity smaller than the absolute value of the negative term, provided the density ρ of the initial condition (and hence of the solutions at later times, since they conserve density) is sufficiently small. It is likely that this

---

[5] The precise technical condition used in [94], possibly not necessary, is that ρ satisfies the inequality

$$128\frac{K_c\rho}{L} + 2\left(\frac{32K_c\rho}{L}\right)^{2+\frac{1+2\alpha}{\gamma+2(1-\alpha)}} < 1.$$

restriction on the initial density can be improved, but so far this nice result is the best one available on rates of convergence with strong fragmentation conditions.

3.1.2 Weak-Fragmentation Systems

When fragmentation is weak (in a sense to be made precise soon) an extraordinarily interesting phenomenon occurs which is physically interpreted as corresponding to the existence of a dynamic phase transition in the system being modelled by (49). The phenomenon is the following: there exists a critical density $\rho_s \in (0, \infty)$ such that

(i) if the initial condition c_0 has density $\rho > \rho_s$, then the solution c to (49) converges weak-$*$, but not strongly, to the only equilibrium c^{ρ_s} with density ρ_s (supercritical case),
(ii) if the initial condition c_0 has density $\rho \leq \rho_s$, then the solution c to (49) converges strongly in X_1^+ to the unique equilibrium c^{ρ} with density ρ (subcritical case).

Note that in case (i) the density of the ω-limit solution, c^{ρ_s}, is strictly smaller than the density of the solution to (49) in every time instant $t < \infty$, whereas in case (ii) the density is also conserved in the limit.

Before going through a brief history of this result and analysing its proof, it is interesting to attend to a possible phase transition interpretation of this behaviour.

If we consider that each component c_j of the solution $c = (c_j)$ represents the concentration of a microscopic j-cluster in a certain physical state, a gas say, and that ρ is the vapour density, the quantity ρ_s can be interpreted as the saturation density of the system. Thus, if the system is in a supersaturated state (i.e., a state with $\rho > \rho_s$,) there is no vapour equilibrium state with that density and the system will evolve to an equilibrium with density exactly equal to the saturated density. The excess density $\rho - \rho_s$ disappears from the gaseous system via condensation, which corresponds to the formation of another physical phase not modelled by none of the variables c_j but, heuristically, corresponding to a cluster incommensurably bigger than j, for every j.

If the system is in a saturated or in a sub-saturated state, with density $\rho \leq \rho_s$, then its evolution proceeds to the unique equilibrium with that density, the density being conserved along the process and in the limit state.

The behaviour described above was first observed in the context of the Becker-Döring equations by Ball et al. in [13]. The (weak-$*$) convergence to an equilibrium is proved using a Lyapunov function, as in the case of strong fragmentation described earlier (cf. page 124), however in the present case the finiteness of higher order moments is not valid, fact that increases the difficulty of the proof of orbit pre-compacity and the identification of the limit density in the subcritical case. In order to obtain these results, in [13] there was the need to impose the following extra decay condition on the initial data (typically, an exponential decay [13, Eq.(5.10)]):

$$\sum_{j=1}^{\infty} \frac{c_{0j}}{M_j z_s^j} < \infty, \quad \text{where } z_s \text{ is the only solution of } F(z_s) = \rho_s, \tag{82}$$

which provided the seeked for control on the tail of the solution $(c_j(t))$. This restriction was later eliminated in [11] by noting that, for the variables $x_n := \sum_{j=n}^{\infty} jc_j$, it is possible to construct a supersolution independently of the decay behaviour of the initial condition x_0, and this implies pre-compacity of the orbit in X_1^+ and thus strong convergence in X_1, which has as consequence that the limit equilibrium has the same density of the solution at finite times.

The extension of this result to coagulation-fragmentation systems (49) more general than the Becker-Döring was marred with several difficulties and was only truly achieved two decades later, with the work of Cañizo [30]. A first attempt was made by Carr and da Costa in [36] where, in order to prove strong convergence in the subcritical case, the following generalized Becker-Döring assumption was used:

$$a_{j,k} = b_{j,k} = 0 \quad \text{if} \quad j \wedge k > N, \tag{83}$$

where N is a fixed positive integer. In the classic Becker-Döring case $N = 1$ [cf. (18)]. With this assumption, with the restriction (82) on the initial data, and with some technical assumptions on the kinetic coefficients, like the ones used in [13], it was possible to rigorously prove the behaviour described in (i) and (ii) above. A subsequent attempt to overcome the restrictions about the regularity of the initial condition was made by da Costa in [48] but was only partially successful: the result obtained, inspired in the method of [11], only allows to draw conclusions for initial data in X_1^+ that, although has no extra decay requirement, needs to have its density ρ bounded above by a bound like $\rho_N \sim \mathcal{O}(N^{-1})$ when $N \to \infty$. This certainly seems to point to the fact that the method is not only insufficient to deal with the generalized Becker-Döring, but it really is totally inadequate for the general coagulation-fragmentation (that corresponds, formally, to take $N \to \infty$).

Notwithstanding these failures, the idea to construct supersolutions, introduced in [11], was a good one and could finally be carefully exploited by Cañizo in [29] to get rid of the restriction on the initial density imposed in [48]. With the assumptions on the coefficients used in previous works [36, 48], Cañizo used an argument similar to the one of Ball and Carr and proved [29, Proposition 3.3] that, if $z < z_s$, $\lambda \in (1, \frac{z_s}{z})$, and if (λ_j) is a decreasing sequence such that

$$\frac{\lambda_{j-1} - \lambda_j}{\lambda_j - \lambda_{j+1}} < \lambda,$$

then, when the initial condition (in the x_n variables introduced above) satisfy $x_n(0) \leq \lambda_n$ for all n, then there exist positive constants C and n_0 such that $x_n(t) \leq C\lambda_n$ for all $n \geq n_0$ and $t > 0$. The proof of this pre-compacity result is based on the following differential inequality for $H_j(\cdot) := (x_j(\cdot) - C\lambda_j)^+$, where $u^+ = u \vee 0$,

$$\frac{d}{dt}\sum_{j=n_0}^{\infty} H_j \leq \text{(const.)}\sum_{j=n_0}^{\infty} H_j,$$

and on the application of Gronwall's lemma.

In the remaining of this section we shall present a more general result, also due to Cañizo [30], that overcomes the need to impose (83) and proves the phase transition behaviour for the general coagulation-fragmentation system and, in a sense, completes the work started by Ball, Carr and Penrose in 1986 with the Becker-Döring system. Cañizo work [30] imposes an additional decay to the initial data $c_0 \in X_1^+$ but it is typically the existence of a moment of order smaller than two, and not an exponential decay, as in [13, 36]. This very mild restriction is more than compensated by the fact that the result is valid for the general coagulation-fragmentation equations, and not only to the restrictive Becker-Döring versions. At present it is not clear if this restriction is essential. The hypotheses considered in [30] are the following:

(H3) There exists constants $K > 0$, $\gamma \in \mathbf{R}$ and $\lambda \in [0, 1)$ such that

$$a_{j,k},\, b_{j,k} \leq K\left(j^\lambda + k^\lambda\right) \tag{84}$$

$$\sum_{j=1}^{i-1} b_{j,i-j} \leq K i^\gamma, \quad \forall i \geq 1. \tag{85}$$

(H4) There exists a positive sequence (M_j) satisfying (15).

(H5) $\lim M_j^{1/j} = z_s^{-1} \in (0, \infty)$ and $\rho_s := F(z_s) \in (0, \infty]$, where F is given by (71).

(H6) The sequence $(M_j z_s^j)$ is monotone decreasing.

(H7) There exists a constant $K_1 > 0$ such that

$$a_{j,1} \geq K_1 j^\lambda, \quad \forall j \geq 1 \tag{86}$$

(H8) The initial data are $c_0 \in X_\mu^+$, with $\mu := \max\{2 - \lambda, 1 + \lambda, 1 + \gamma\}$.

The main result, extending to coagulation-fragmentation the result of [13], is the following:

Theorem 12 ([30]) *Assume (H3)–(H8). Let c be a solution of (49) with (constant) density $\rho = \|c\|_1 = \|c_0\|_1$. The following holds:*

(i) If $\rho > \rho_s$, then $c(t) \overset{*}{\rightharpoonup} \left(M_j z_s^j\right)$ when $t \to +\infty$.

(ii) If $\rho \leq \rho_s$, then $c(t) \to c^{eq}$ strongly in X_1, when $t \to +\infty$, where c^{eq} is the only equilibrium solution with density ρ.

Sketch of proof The general strategy to prove this result was already used in [13] and in [29, 36] and consists of proving that, if a solution converges weak-$*$ to an equilibrium with density strictly smaller than the critical one ρ_s, then the convergence is actually strong in the norm topology of X_1 and thus the limit density is equal to the initial one.

That all solutions of (49) converge weak-$*$ to equilibria had already been proved in [36]: assuming (H3)–(H8), solutions c to (49) satisfy $c(t) \overset{*}{\rightharpoonup} c^\rho$ when $t \to \infty$, for some $\rho \leq \min\{\|c_0\|, \rho_s\}$, where c^ρ is the only equilibrium solution with density ρ. As in the case of strong fragmentation presented earlier, the proof of this result of weak-$*$ convergence to equilibria is based on the existence of a Lyapunov function. The proof in [30] is based on the Lyapunov function, on the control of the density by the $(2-\lambda)$-moment, and in an estimate that implies that the growth of this moment is at most linear in t.

Cañizo also proved a result on the rate of convergence to equilibria using a slightly different Lyapunov function introduced by Jabin and Niethammer in the study of the rate of convergence to equilibria in Becker-Döring systems [109]: for $c \in X_1^+$, let V be defined by (72) and, for $z \in (0, z_s]$, define the *energy of c relative to the equilibrium* $\left(M_j z^j\right)$ by the expression

$$\mathcal{V}_z(c) := V(c) - (\log z)\sum_{j=1}^{\infty} jc_j + \sum_{j=1}^{\infty} M_j z^j. \tag{87}$$

Observe that, if $\rho_s < \infty$ and if we choose z so that $c^{\text{eq}} = \left(M_j z^j\right)$ satisfies $\|c^{\text{eq}}\| = \|c\|$, then, it is easy to conclude that $\mathcal{V}_z(c) = V(c) - V(c^{\text{eq}})$, which justifies the name given to $\mathcal{V}_z(c)$.

The fundamental inequality used in the proof of Theorem 12, relating the density of a positive sequence $c = (c_j)$ with its $(2-\lambda)$-moment, is

$$\|c\| - \sum_{j=1}^{\infty} jM_j c_1^j \leq C\sqrt{D}\sqrt{\|c\|_{2-\lambda}}, \tag{88}$$

where C is a positive constant and $D := D(c)$ is given by

$$D(c) := \sum_{j=1}^{\infty} a_j M_j \left(\frac{c_1 c_j}{M_j} - \frac{c_{j+1}}{M_{j+1}}\right)\left(\log\frac{c_1 c_j}{M_j} - \log\frac{c_{j+1}}{M_{j+1}}\right), \tag{89}$$

where $a_j = a_{j,1}$ if $j \geq 2$, and $a_1 = \frac{1}{2}a_{1,1}$. The inequality is valid under the assumptions (H3)–(H7) and provided $c_1 \in (0, z_s)$ and $c \in X_{2-\lambda}$. The function D is called the *Becker-Döring free energy dissipation rate* in [30], which is a natural designation since it was proved in [13] that the time evolution of the Lyapunov function V along solutions $c(t)$ of the Becker-Döring system satisfies

$$V(c(t)) = V(c(0)) - \int_0^t D(c(s))ds.$$

Observe that, as pointed out before, $\forall x, y > 0$, $(x-y)(\log x - \log y) \geq 0$, and so the function $D(c)$ is non-negative.

Its should be noted that (88) is a purely algebraic relation valid for certain sequences c and has nothing to do with these sequences being solutions of some differential equation. It is merely a consequence of the following estimate about the tail of the series $\sum_j jM_j c_1^j$, which is valid under the same assumptions,

$$\sum_{i=j+1}^{\infty} iM_i c_1^i \leq jM_{j+1} c_1^{j+1}.$$

Under the hypotheses of Theorem 12, taking a solution $c = c(t)$ of (49) with initial condition $c_0 \in X_{2-\lambda}$, and using the approximation of c by solutions to the truncated systems, we can prove that $\|c(t)\|_{2-\lambda}$ satisfies the differential inequality $\frac{d}{dt}\|c(t)\|_{2-\lambda} \leq$ (const.)ρ^2, and thus, for some constant C independent of t, it holds that

$$\|c(t)\|_{2-\lambda} \leq C(1+t) \tag{90}$$

As stated above the idea of the proof consists in showing that if a solution with initial density ρ_0 converges weak-$*$ to an equilibrium with density $\rho < \rho_s$, then it converges strongly in X_1 and thus $\rho = \rho_0$.

Let us assume that $c(t) \overset{*}{\rightharpoonup} c^{\text{eq}}$ when $t \to +\infty$, where $c^{\text{eq}} = \left(M_j z^j\right)$ and $z < z_s$. Thus $c_1(t) \to z < z_s$ and $c_1(t) \leq \frac{z+z_s}{2} < z_s$, for all times $t > t_0$, where t_0 is sufficiently large. Using (88) and (90) we know that $\rho - \rho_1(t) \leq C\sqrt{D}\sqrt{1+t}$, for $t \geq t_0$, where $\rho_1(t) = \left\|\left(M_j c_1(t)^j\right)\right\|_1$, and C is a constant. By continuity of (71) in the interior of its interval of convergence, $\rho_1(t) \to \rho_z := \|c^{\text{eq}}\|$, as $t \to \infty$.

Now, either $\rho - \rho_1(t) > 0$ after some t_1, or there exists a sequence $t_n \to \infty$ such that $\rho - \rho_1(t_n) \leq 0$. Let us start by the first possibility: if $\rho - \rho_1(t) > 0$ for all $t > t_1$ the previous inequality gives the estimate $D(c(t)) \geq \frac{(\rho-\rho_1(t))^2}{1+t}C^{-2}$ and the evolution of V along solutions satisfies

$$V(t) = V(t_1) - \int_{t_1}^t D_{CF}(c(s))ds \tag{91}$$

$$\leq V(t_1) - \int_{t_1}^t D(c(s))ds \tag{92}$$

$$\leq V(t_1) - C^{-2}\int_{t_1}^t \frac{(\rho - \rho_1(s))^2}{1+s}ds, \tag{93}$$

where D_{CF} is the *coagulation-fragmentation free energy dissipation rate* defined by

$$D_{CF}(c) := \frac{1}{2}\sum_{i,j=1}^{\infty} a_{i,j}M_iM_j\left(\frac{c_ic_j}{M_iM_j} - \frac{c_{i+j}}{M_{i+j}}\right)\left(\log\frac{c_ic_j}{M_iM_j} - \log\frac{c_{i+j}}{M_{i+j}}\right). \tag{94}$$

It is worth calling the reader attention to the fact that, although the formal derivation of the evolution equation (91) is trivial, it rigorous proof is far from being simple [36, Theorem 5.2]. The inequality (92) is due to the obvious fact that $D_{CF}(c) \geq D(c) > 0$. The Lyapunov function V is bounded from below along solutions; using this result the integral in the right-hand side of (93) has to be bounded from above and, since $\rho_1(t) \to \rho_z$, we conclude that $\rho_z = \rho$. But then, since $c(t) \overset{*}{\rightharpoonup} c^{\mathrm{eq}}$ and $\|c(t)\| = \rho = \rho_z = \|c^{\mathrm{eq}}\|$, [13, Lemma 3.3] implies that $c(t) \to c^{\mathrm{eq}}$ strongly in X_1.

It remains to consider the possibility of existence of a sequence $t_n \to \infty$ such that $\rho - \rho_1(t_n) \leq 0$. In this case we would have $\rho \leq \rho_1(t_n) \to \rho_z$ and thus $\rho \leq \rho_z$. By the lower semicontinuity of the norm of X_1 with respect to weak-$*$ convergence, we have $\rho_z \leq \rho$ and thus $\rho_z = \rho$. This concludes the proof. ■

Once the long-time limit of solutions in Theorem 12 has been proved, a natural problem to consider is to clarify the way the limit equilibrium solution is approached as $t \to \infty$. Recall that in the strong fragmentation case solutions converge to the limit equilibrium exponentially fast, at least for sufficiently small initial data (cf. page 126). Note that, for that type of coefficients, the critical density is infinite and so all solutions are subcritical.

Under weak fragmentation conditions the long-time behaviour of solutions is expected to be richer, and the cases of supercritical and subcritical densities are thought to exhibit distinct behaviours. However, at present, rigorous results about these aspects are restricted to the Becker-Döring case, and we shall briefly review them next.

For the Becker-Döring equations with subcritical initial density a recent paper by Cañizo and Lods [31] improved previous results by Jabin and Niethammer [109] and prove that, under appropriate assumptions that include exponentially decaying initial conditions, subcritical solutions converge exponentially fast to the limit equilibrium, and give an estimate for the convergence rate. The precise statement of the result requires the introduction of some hypotheses:

(H9) On the coagulation coefficients:

$$a_j = \mathcal{O}(j) \quad \text{as } j \to \infty, \qquad \lim_{j\to\infty}\frac{a_{j+1}}{a_j} = 1, \qquad \inf_j a_j > 0.$$

(H10) On the fragmentation coefficients: $b_j = \mathcal{O}(j)$ as $j \to \infty$.

(H11) On the partition function: $\lim_{j\to\infty}\frac{M_j}{M_{j+1}} =: z_s \in (0, +\infty)$

Under these conditions the following was proved:

Theorem 13 ([31]) *Assume (H9)–(H11). Let c be a solution of (19) with initial condition* $M := \sum_{j=1}^{\infty} e^{\nu j} c_j(0) < +\infty$, *for some* $\nu > 0$, *and let* $z > 0$ *be the monomer density of the corresponding limit equilibrium* $c^{eq} = \left(M_j z^j\right)$. *Then, there exists* $\bar{\nu} \in (0, \nu)$ *and* $\lambda_\star > 0$, *such that, for every* $\eta \in (0, \bar{\nu})$, *there is* $C > 0$, *depending only on* ρ, η, M, *and* $\sum_{j=1}^{\infty} e^{\eta j} M_j z^j$, *such that the following holds for all* $t \geq 0$,

$$\sum_{j=1}^{\infty} e^{\eta j} \left| c_j(t) - M_j z^j \right| \leq C e^{-\lambda_\star t},$$

where, if $\lim_{j\to\infty} a_j = +\infty$, *we can take* $\lambda_\star^{-1} = \sup_k \left(\sum_{j=k+1}^{\infty} M_j z^j\right)\left(\sum_{j=1}^{k} \frac{1}{a_j M_j z^j}\right)$.

The proof of this theorem is rather lengthy and involved and we will not delve with it here: the interested reader should consult the original paper [31]. We just point out that the main tool is an appropriate linearisation of the Becker-Döring equations around the equilibrium c^{eq} and the proof of an appropriate spectral gap in suitable sequence spaces.

Let us now briefly consider the case of supercritical solutions to the Becker-Döring system. As stated in Theorem 12(i), solutions (all of them conserve initial density ρ in finite times) converge to the limit equilibrium with critical density $\rho_s < \rho$. That, in general, this convergence can be a complicated dynamical process has been shown by Penrose in [180], where he proved that, for certain nonequilibrium initial conditions, each component of the solution of (19) remain exponentially close to the initial condition for a time that is exponentially long in $(\rho - \rho_s)^{-1}$, after which it converges to the critical equilibrium. This is a metastability behaviour that, in some sense, agrees with the physical fact that nucleation processes in supersaturated mixtures are exceedingly slow processes.

The occurrence of metastability behaviour is an interesting feature of the Becker-Döring system. Another very interesting and challenging problem is to know what happens to the solution for times so large that all possible metastability regimes have elapsed. So, the problem is to understand how do the excess density $\rho - \rho_s$ spreads to larger and larger clusters once the solution "starts to move". This problem of evolution of large clusters has been studied, in the small excess density regime, by several authors [130, 166–168] striving to get rigorous proofs to the pioneering work Penrose and collaborators [181, 183] who established the relation of the large time asymptotic of small excess density solutions with solutions to the Lipschitz-Slyozov-Wagner equation of Oswald ripening [149, 214]. The detailed presentation of the existing rigorous results on this very interesting hydrodynamic limit of the Becker-Döring (as yet with no parallel in more general coagulation-fragmentation models) would take too long. In what follows we will just attempt to give an heuristic idea of the approach.

A very crude formal computation suggests a possible connection: if one considers extremely large clusters sizes j, then $j-1$ and j are extremely close to each other and, considering j as a continuous real variable, the right-hand side of the equation for the j-cluster in the Becker-Döring system is roughly equal to $-\partial_j J$, so the equation itself is $\partial_t c + \partial_j J = 0$; the assumptions about the reaction coefficients a_j and b_j determine the way the flux J depends on $c(t,j)$, and the monomer dynamics is determined by density conservation. To make more precise this idea, let us take the following situation, considered in Penrose's approach to the modelling of first-order phase transitions:

(i) $a_j = j^\alpha$, for some $\alpha \in [0,1)$.
(ii) $b_j = a_j\big(z_s + qj^{-\gamma}\big)$, for constants $z_s, q > 0$, and $\gamma \in (0,1)$.

The more precise heuristic argument is roughly the following [166, 180, 202]: since we are considering large times, consider a new time scale $\tau := \varepsilon^{1+\gamma-\alpha}t$, with a small positive parameter $\varepsilon \to 0$. Adequately choosing a separation $j_* = j_*(\varepsilon) \to +\infty$ between small and large clusters, we consider the large cluster sizes $j > j_*$ as a continuous variable $x = \varepsilon j$ and introduce the rescaled functions $\nu(\tau,x)$, $\upsilon(\tau,x)$, and $u(\tau)$, such that $c_j(t) = \varepsilon^2\nu(\tau,x)$, $J_j(c) = \varepsilon^{2+\alpha-\gamma}\upsilon(\tau,x)$, and $c_1 = z_s + \varepsilon^\gamma u(\tau)$, respectively. The equation for large cluster sizes becomes, in the limit $\varepsilon \to 0$,

$$\partial_\tau \nu + \partial_x\Big(x^\alpha\big(u - qx^{-\gamma}\big)\nu\Big) = \mathscr{o}(1). \tag{95}$$

In the other hand, density conservation and a judicious choice of j_* lead to, in the limit $\varepsilon \to 0$,

$$\int_0^\infty x\nu(\tau,x)dx = \rho - \rho_s + \mathscr{o}(1),$$

which is equivalent to

$$u(\tau) = \frac{q\displaystyle\int_0^\infty x^{\alpha-\gamma}\nu(\tau,x)dx}{\displaystyle\int_0^\infty x^\alpha \nu(\tau,x)dx},$$

at least for some relation between α and γ [166].

System (95)–(96) is the classic Lipschitz-Slyozov-Wagner model. We direct the interested reader to the references above for the precise statement and proof of this result.

3.2 *Self-Similar Behaviour of Solutions*

Contrasting to what happens with the coagulation-fragmentation system, in Smoluchowski's coagulation equations it is easy to prove that the identically zero sequence is the only non-negative equilibrium, and the proof that every solutions converges, in the weak-$*$ sense, to this equilibrium as $t \to \infty$ is also elementary (cf. below, and Sect. 3.2.1).

The zero sequence has density equal to zero and, being this density value the largest for which there is an equilibrium, it can be considered the critical density of the system and so, in a sense, all non-zero solutions are supercritical. So, the problem considered in the closing part of last subsection, namely how does the excess density is spread to larger and larger cluster sizes as $t \to \infty$, is also relevant in the Smoluchowski's equation setting, where its concretization has taken the form of the investigation of self-similar behaviour of solutions, i.e., the existence of a function (or family of functions) for which, after an appropriate rescaling of the variables, all solutions converge when $t \to \infty$. This problem, of clear scientific importance, has received a good deal of attention in the mathematical modelling community (cf., e.g. [70, 72, 96, 145–147, 150, 217] and ref. cit.) but important progresses in its rigorous analysis are much more recent. In what follows we will review some of these.

3.2.1 Similarity Behaviour in Smoluchowski's Coagulation Equations

Let us start by justifying the sentence above about the triviality of the convergence to equilibria in the Smoluchowski system:

Theorem 14 ([36]) *Let $a_{j,j} > 0$ for all j. Let c be a solution of (31) in $[0, \infty)$ with $c_0 \in X_1^+$. Then $c(t) \overset{*}{\rightharpoonup} 0$ as $t \to +\infty$.*

Sketch of proof The heuristic idea is clear enough: since the coagulation process entails the increase of the clusters' mean size (cf. Sect. 1.2) we expect the total density of clusters with sizes below any arbitrarily fixed value to decrease with time. The convergence proof is based in this monotonicity property.

Let c be a solution of (31) and, for each $n \in \mathbf{N}$, let us consider

$$p_n(t) := \sum_{j=1}^{n} jc_j(t). \tag{96}$$

This function measures the total density at time t of clusters with size not larger than n. By (96) and the definition of solution we get, for all $t, \tau \geq 0$,

$$p_n(t+\tau) - p_n(t) = -\int_t^{t+\tau} \sum_{T_{1,n}^4} ja_{j,k}c_j(s)c_k(s)ds \leq 0, \tag{97}$$

where $T^4_{1,n}$ was defined in page 119. As $c_n(t)$ and $p_n(t)$ are non-negative functions, there exists a non-decreasing positive sequence $(\bar{p}_n)$ such that $p_n(t) \to \bar{p}_n$ when $t \to +\infty$. Since $c_n(t) = \frac{p_n(t)-p_{n-1}(t)}{n}$, these functions $c_n(t)$ also converge to some constants, $\bar{c}_n := \frac{\bar{p}_n-\bar{p}_{n-1}}{n} \geq 0$, as $t \to +\infty$. By induction in the coagulation equation integrated in $[t, t+\tau]$ the conclusion that $\bar{c}_n \equiv 0$ is easily reached [36]. ■

We are now interested in knowing whether or not solutions to (31) converge to the zero solution in a self-similar way. In a slightly more precise manner: under what conditions there exists a function Φ such that, for a large class of initial conditions, the corresponding solutions to (31) satisfy

$$c_j(t) \approx \varsigma(t)^{-a}\Phi(j\varsigma(t)^{-b}), \quad \text{as } t \to +\infty \text{ and } j \to +\infty, \tag{98}$$

where $\varsigma(\cdot)$ is a positive increasing function, and a and b are positive constants?

Not much is rigorously known about the problem in this general setting. What we present next are answers for certain particular coefficients $a_{j,k}$ (constant, additive, product) for which rigorous and fairly complete answers have been obtained, and then point to very recent rigorous analysis for systems with more general classes of rate coefficients.

We start by the constant coefficient case,[6] $a_{j,k} = 2$. This case was studied by Kreer and Penrose [117] and by da Costa [47] using an idea first introduced by Lushnikov [150] which is based in exploiting the generating function

$$\varphi(z,t) := \sum_{j=1}^{\infty} c_j(t)z^j, \quad |z| \leq 1. \tag{99}$$

Observe that (99) is the discrete Laplace transform $\sum_{j=1}^{\infty} c_j(t)e^{-jw}$, ($\mathrm{Re}(w) \geq 0$), of the solution c of (31). It is easy to prove that φ is solution of the initial value problem

$$\begin{cases} \frac{d\|c\|_0}{dt} = -\|c\|_0^2 \\ \frac{\partial\varphi}{\partial t} = \varphi^2 - 2\|c\|_0\varphi \end{cases} \tag{100}$$

with $\|c(0)\|_0 = N_0 := \|c_0\|_0$ and $\varphi(z,0) = \phi(z) := \sum_{j=1}^{\infty} c_{0j}z^j$, from which we immediately conclude that

$$\varphi(z,t) = t^{-2}\frac{1}{N_0 + t^{-1}}\frac{\phi(z)}{N_0 + t^{-1} - \phi(z)}. \tag{101}$$

---

[6]The value of the constant is irrelevant for the result since it can always be transformed into another value by a time rescaling. The choice we make simplifies the computations a bit.

Since $\varphi(\cdot, t)$ is an analytic function on the unit open ball $B_1 \subset \mathbf{C}$, we use Cauchy's integral formula to write

$$t^2 c_j(t) = \frac{1}{2\pi i} \frac{1}{N_0 + t^{-1}} \oint_{\gamma_0} \frac{1}{z^{j+1}} \frac{\phi(z)}{N_0 + t^{-1} - \phi(z)} dz, \tag{102}$$

where $\gamma_0 = \{z \in \mathbf{C} : |z| = r_0 < 1\}$. In order to conclude something about the long-time behaviour of the right-hand side of (102) we need to know the behaviour of the zeros of $F(z,\tau) := N_0 + \tau - \phi(z)$ as $\tau \to 0$. With the additional hypothesis of an exponentially decaying initial condition $c_{0j} \leq A(1+\Delta)^{-j}$, for some constants $A \geq 0$ and $\Delta \in (0,1)$, it can be proved that, for all sufficiently small τ there exists q simple zeros of $F(z,\tau)$, $z_k(\tau)$, satisfying $|z_k(\tau)| > 1$ and $z_k(\tau) = \omega_q^k\big(1 + \frac{1}{\|c_0\|_1}\tau + \mathcal{O}(\tau^2)\big)$, when $\tau \to 0$, where ω_q is the qth root of unity, $e^{2\pi i/q}$; all the remaining roots of $F(z,\tau)$ are in the exterior of B_1 and keep a distance uniformly positive from B_1 when $\tau \to 0$. The positive integer constant q is given by $q = \gcd \mathcal{J}(0)$, where $\mathcal{J}(0)$ is the set of subscripts j for which $c_{0j} > 0$ (cf. statement of Theorem 1). With this result and the representation formula (101) it is not difficult to prove the following:

Theorem 15 ([47, 117]) *Let $a_{j,k} \equiv 2$ and consider a non-negative exponentially decreasing initial condition $c_0 \in X_1^+$. Let q and $\mathcal{J}(0)$ be as above. Thus, the solution c of (31) has the following self-similar behaviour:*

$$\lim_{\substack{j,t \to +\infty \\ \xi = j/t \text{ fixed} \\ j \in span_{\mathbf{N}_0} \mathcal{J}(0)}} t^2 c_j(t) = \frac{q}{\|c_0\|_1} e^{-\xi/\|c_0\|_1}. \tag{103}$$

A version of this result valid for the continuous Smoluchowski's equations, was also proved by Kreer and Penrose in [117], also for exponentially decaying initial data. The same behaviour occurs with the obvious changes of notation:

$$\lim_{\substack{j,t \to +\infty \\ \xi = j/t \text{ fixed}}} t^2 c(x,t) = \frac{1}{\|c(x,0)\|_1} e^{-\xi/\|c(x,0)\|_1}. \tag{104}$$

If the coefficients $a_{j,k}$ [or $a(x,y)$] are not constants, or if the initial condition does not decay exponentially, the above technique is not applicable. In these cases, only recently the rigorous analysis of the self-similar behaviour of solutions was achieved, in a number of notable papers by Menon and Pego [159–161] (see also [179]). These works use a more general notion of solution, in terms of measures, allowing for the simultaneous consideration of the discrete and the continuous cases of the Smoluchowski's equation. Using the modified Laplace transform $\varphi(z,t) := \int_0^\infty (1 - e^{-zx})\nu_t(dx)$, where $\nu_t(dx) = c(x,t)dx$ is the finite measure on $(0,\infty)$ that describes the cluster size distribution, Menon and Pego proved the following:

Theorem 16 ([179]) *Let $a(x,y) \equiv 2$, and $t_0 = 1$ and consider initial data satisfying $\int_0^\infty \nu_1(dx) = 1$. Let $\nu_t(dx)$, be a finite measure solution of (5)–(6), and let F_t be the probability distribution function*

$$F_t(x) := \frac{\int_0^x \nu_t(dy)}{\int_0^\infty \nu_t(dy)} = t\int_0^x c(t,y)dy. \tag{105}$$

1. *Suppose there exists a $\lambda(t) \to \infty$ and a probability distribution F_* such that $F_*(x) < 1$ for some $x > 0$, and*

$$F_t(\lambda(t)x) \to F_*(x), \quad \text{as } t \to \infty, \tag{106}$$

in its points of continuity. Then

$$\int_0^x y\nu_1(dy) \sim x^{1-\rho}L(x), \quad \text{as } x \to \infty, \tag{107}$$

for some constant $\rho \in (0,1]$ and some function L slowly varying at infinity [87, pp. 275–9].

2. *Reciprocally, suppose (107) is true. Then (106) holds, with*

$$F_*(x) := F_\rho(x) = \sum_{k=1}^{\infty} \frac{(-1)^{k+1}x^{\rho k}}{\Gamma(1+\rho k)}, \tag{108}$$

being a Mittag-Leffler distribution [87, page 453], whose Laplace transform is $\int_0^\infty e^{-zx}F_*(dx) = \frac{1}{1+q^\rho}$.

Observe that Theorem 16 provides a classification of all possible self-similar solutions of the Smoluchowski's equation with constant coefficients, which are solutions of the type

$$c(t,x) = t^{-1-1/\rho}n_\rho\big(t^{-1/\rho}x\big), \quad \rho \in (0,1], \tag{109}$$

where $n_\rho(\cdot) = F'_\rho(\cdot)$ is the Mittag-Leffler distribution density (108).

Using the Laplace transform, the self-similar *ansatz* like (98) for the transform, and separation of variables, it is easy to check that the functions (109) are in fact solutions to (5)–(6).

It is interesting to observe that if $\nu_1(dx)$ has finite density, then $\rho = 1$ and $F_1(x) = 1 - e^{-x}$, corresponding to the solution $c(t,x) = \frac{1}{t^2}e^{-x/t}$. Compare this with (104).

This result is similar to the Central Limit Theorem. The distributions F_ρ with $\rho \in (0,1)$ have infinite density and correspond to Lévy's stable distributions.

Menon and Pego also proved that the convergence to the self-similar limit is uniform in the similarity variable $\frac{x}{t}$:

Theorem 17 ([160]) *Let* $c(1,x) > 0$, *be an initial condition satisfying* $\int_0^\infty c(1,x)dx = \int_0^\infty xc(1,x)dx = 1$. *Suppose the Fourier transform of* $xc(1,x)$ *is integrable. Then*

$$\lim_{t\to+\infty} \sup_{\frac{x}{t}>0} \frac{x}{t} \left| t^2 c(t,x) - e^{-x/t} \right| = 0.$$

An improved version, establishing convergence in a stronger topology and providing an upper bound $Kt^{-\delta}$ for the rate of this convergence to zero was obtained by Cañizo et al. on [33]. The result of this theorem is also valid, *mutatis mutandis,* for the solutions of the discrete system (cf. [160, Theorem 2.2]).

Results similar to those above were also proved by Menon and Pego [159] for the two solvable types of coefficients (additive $a(x,y) = x+y$, and product $a(x,y) = xy$) for which they also obtained a complete characterization of the self-similar attractor [161]. For these kernels the rate of the convergence stated in the theorem was proved to be exponential by Srinivasan [207] also using methods based on the Laplace transform in an approach that is analogous, in spirit, to the Berry-Esséen theorem when one considers the result in theorems 16 and 17 as a Central Limit Theorem for the clusters distributions.

The first rigorous proofs of existence of self-similar solutions for non-solvable kernels are by Fournier and Laurençot [90] and Escobedo and collaborators [82], who proved the existence of (but not the convergence to) self-similar solutions to the continuous system (5)–(6). Their approach is different from those presented above, not resorting to Laplace transforms, and returning to an idea commonly used in the mathematical modelling and physics literatures [72, 145], which is to use the *ansatz* (98) directly in (5)–(6) in order to obtain an integro-differential equation for the self-similar profile $\Phi(\cdot)$, and to prove that, for certain types of coefficients, this equation has a non identically zero weak solution. This is an extremely natural approach from a mathematical perspective and was certainly attempted before; the fact that only in [82, 90] this idea could have been set on a firm basis attests to the very demanding technical difficulties its concretization involves. We will now present a very brief description of the result and approach of [90].

Write

$$c(t,x) = \varsigma(t)^{-2}\Phi(\varsigma(t)^{-1}x) \tag{110}$$

and assume the coagulation coefficients satisfy the homogeneity condition

$$a(ux,uy) = u^\lambda a(x,y), \quad \forall u,x,y \in \mathbf{R}^+, \tag{111}$$

for some real constant λ.

Substituting (110) into (5)–(6) and using (111) the following equation arises

$$\gamma \frac{d}{dx}\left(x^2\Phi(x)\right) + xQ_c(\Phi)(x) = 0 \tag{112}$$

$$\int_0^\infty x\Phi(x)dx = \rho, \tag{113}$$

for the unknown function Φ and the real positive unknowns (γ, ρ).

It is easy to conclude that, if (Φ, γ, ρ) is a solution of (112)–(113), then each of the elements of the two parameter family $\left(a\Phi(bx), a\gamma b^{-1-\gamma}, a\rho b^{-2}\right)$ is also a solution of (112)–(113). This means that, without loss of generality, we can consider $\gamma = \frac{1}{1-\lambda}, \rho = 1$.

A non-negative function $\Phi \in L^1(0,\infty, xdx)$ is a weak solution of (112) if $\Phi \in L^1(0,\infty, x^2dx)$, if $(x,y) \mapsto xya(x,y)\Phi(x)\Phi(y) \in L^1(\mathbf{R}^+ \times \mathbf{R}^+)$, and if

$$\gamma z^2\Phi(z) = \int_0^z \int_{z-x}^\infty a(x,y)x\Phi(x)\Phi(y)dydx, \tag{114}$$

for almost all $z \in \mathbf{R}^+$. Thus, if Φ is a weak solution of (112) we have, for every $\phi \in C_b^1([0,\infty))$,

$$\gamma \int_0^\infty x^2\Phi(x)\phi'(x)dx = \int_0^\infty \int_0^\infty xa(x,y)(\phi(x+y) - \phi(x))\Phi(x)\Phi(y)dydx. \tag{115}$$

From a technical viewpoint, it is more suitable to consider the unknown function $\tilde{\Phi}(x) = x\Phi(x)$ instead of $\Phi(x)$, and, instead of (115), to write the weak version of (112) as

$$\gamma \int_0^\infty x\tilde{\Phi}(x)\phi'(x)dx = \int_0^\infty \int_0^\infty \frac{a(x,y)}{y}(\phi(x+y)-\phi(x))\tilde{\Phi}(x)\tilde{\Phi}(y)dydx. \tag{116}$$

The main result of [90] is the following:

Theorem 18 ([90]) *Consider coagulation coefficients satisfying any one of the following conditions:*

(i) $a(x,y) = (x^\alpha + y^\alpha)(x^{-\beta} + y^{-\beta}),\ \alpha \in [0,1), \beta \in \mathbf{R}^+, \lambda = \alpha - \beta \in (-\infty, 1)$
(ii) $a(x,y) = (x^\alpha + y^\alpha)^\beta,\ \alpha \in [0,\infty), \beta \in \mathbf{R}^+, \lambda = \alpha\beta \in [0,1)$
(iii) $a(x,y) = x^\alpha y^\beta + x^\beta y^\alpha,\ \alpha \in (0,1), \beta \in (0,1), \lambda = \alpha + \beta \in (0,1)$

Let $\gamma = \frac{1}{1-\lambda}, \rho = 1$. *Then, there exists a positive weak solution* Φ *of (112)–(113) and the function* $c_s(x,t) := t^{-2\gamma}\Phi(xt^{-\gamma})$, *with* $x, t > 0$, *is a (self-similar) weak solution of (5)–(6) with unit density for all* $t > 0$.

The proof by Fournier and Laurençot starts by the following discretization of (116),

$$-\frac{\gamma}{n}\left(i\mathbf{1}_{1\le i\le n^2-1}f_{i+1}-(i-1)f_i\right)=\sum_{j=1}^{i-1}\frac{1}{j}a\Big(\frac{i-j}{n},\frac{j}{n}\Big)f_{i-j}f_j-\sum_{j=1}^{n^2-i}\frac{1}{j}a\Big(\frac{i}{n},\frac{j}{n}\Big)f_if_j.$$

Considering the solutions of this system of n^2 equations as the stationary solutions of an appropriate system of ordinary differential equations, one concludes the existence of a non-negative solution $f=\tilde{f}^n$ satisfying

$$\sum_{i=1}^{n^2}\tilde{f}_j^n=1,\quad \forall n.$$

Using the solutions $\tilde{f}^n$, the following sequence of probability measures indexed by n is constructed

$$\tilde{\Phi}^n(dx)=\sum_{i=1}^{n^2}\tilde{f}_i^n\delta_{i/n}(dx),\tag{117}$$

and the *a priori* estimate

$$\sup_{n\ge 1}\int_0^\infty x^\sigma\tilde{\Phi}^n(dx)<\infty,\tag{118}$$

is proved, where the domain of the parameter σ depend of the type of coefficient, (i), (ii) or (iii), considered. From (117) and (118) one deduces the tightness of the sequence $\big(\tilde{\Phi}^n(dx)\big)$ and thus the existence of a probability measure $\tilde{\Phi}(dx)$ and a subsequence $\big(\tilde{\Phi}^{n_k}(dx)\big)$ such that, for all functions $\phi\in C_b^1([0,\infty))$, it holds that

$$\lim_{k\to\infty}\int_0^\infty\phi(x)\tilde{\Phi}^{n_k}(dx)=\int_0^\infty\phi(x)\tilde{\Phi}(dx).$$

The last step in the proof is to establish that $\tilde{\Phi}$ is a weak solution of (116) and thus $\Phi(x)=\frac{\tilde{\Phi}(x)}{x}$ is a weak solution of (112)–(113).

The problem of self-similar dynamic behaviour in (5)–(6), i.e., the convergence of a "generic" solution of (5)–(6) to the self-similar ones, whose existence was proved in Theorem 18, is still largely open.

A natural approach to study this stability problem would be to consider a transformation like the following one, more general than (110),

$$c(t,x)=\varsigma(t)^{-2}\varphi(\log\varsigma(t),\varsigma(t)^{-1}x),$$

and, substituting in (5)–(6), to obtain an evolution equation for φ that would allow to prove that, for an appropriate notion of convergence, $\varphi(\log \varsigma(t), \cdot) \longrightarrow \Phi(\cdot)$, when $t \to +\infty$. This idea was successfully implemented, in the framework of weak convergence in L^1, in the case of constant coefficients $a(x, y) \equiv 1$, using Lyapunov functions whose construction were strongly dependent of the known form of the limit Φ [134], which turns the potentially promising method useless if the form of Φ is not known. The idea was also applied with success in the prove of existence and stability of self-similar solutions in the Oort-Hulst-Safronov equations with constant [127], with additive [9], and with multiplicative [128] coefficients.

A number of recent studies have greatly enhanced our understanding of the self-similar behaviour of Smoluchowski's equation with non-solvable kernels. It was established by Fournier and Laurençot in [91] that, for sum type kernels $a(x, y) = x^\lambda + y^\lambda$, with $\lambda \in (0, 1)$, the self-similar profile $\eta \mapsto \Phi(\eta)$ proved to exist in [82, 90] is continuously differentiable in $\mathbf{R}^+$ decay exponentially fast as $\eta \to \infty$ and is singular at $\eta \to 0$. Further improved results on the behaviour of the profile $\Phi(\eta)$ when $\eta \to 0$ were obtained in [32, 170] and when $\eta \to \infty$ in [173]. A very interesting, albeit formal, study of the behaviour of the density conserving self-similar profiles in the limit $\eta \to 0$ was published in [157].

The regularity of self-similar profiles in [91] was greatly improved by Cañizo and Mishler [32] who proved that, if $a(x, y) = x^\alpha y^\beta + x^\beta y^\alpha$, with $-1 < \alpha \le \beta < 1$ and $\alpha + \beta \in (-1, 1)$, then the profiles are $C^\infty(\mathbf{R}^+)$. Some results on the uniqueness of self-similar profiles have also been established [32, 172].

The existence of self-similar fat tail solutions (i.e., non-exponentially decaying profiles, as exists for solvable kernel equations) has been proved by Niethammer and Velázquez for diagonal kernels, in [169], and more recently, in [171], for homogeneous kernels satisfying $a(x, y) \le C(x^\lambda + y^\lambda)$, with $\lambda \in [0, 1)$.

Most of these papers use an *ansatz* like (110) in order to get an equation for the profile Φ that is then exploited recurring to a variety of means, comprising among other tools, rather delicate estimates and adequately carved fixed point theorems. We direct the reader to the original papers for the statement of the results and to fully appreciate the beauty and difficulty of their proofs.

With the exception of the exact case $a(x, y) = xy$, studied by Menon and Pego [159, 161] and Srinivasan [207], all the above results correspond to systems for which solutions conserve density. The general problem of conservation, or non-conservation, of density will be treated in Sect. 4. Here we just briefly refer to a recent work by Breschi and Fontelos [25] which is the first rigorous proof of existence of self-similar solutions for a non-exact kernel for which solutions do not conserve density for all times, namely $a(x, y) = (xy)^{1-\varepsilon}$, with $\varepsilon \ll 1$. With these coefficients there exists a time T_g before which all solutions conserve density, but density decrease afterwards (cf. Theorem 21 in Sect. 4). In [25] the authors study the existence of self-similar solutions for $t < T_g$; in particular they prove, among other things, that, with these coefficients, the Laplace transform of Smoluchowski equation results in a nonlocal Burgers' equation $\omega(\lambda, t) = \frac{1}{2}\partial_\lambda\big(D_\lambda^{-\varepsilon}\omega(\lambda, t)\big)^2$, where $\omega(\lambda, t) := -\int_0^\infty (e^{-\lambda x} - 1)xc(x, t)dx$, and $D_\lambda^{-\varepsilon}$ is a nonlocal operator.

If one proves that this Burgers' equation has self-similar solutions of the form $\omega(\lambda,t) = (T_g - t)^{\alpha}\psi(\xi)$, where $\xi := \lambda(T_g - t)^{-\beta}$, one can use the inverse Laplace transform to prove that the original Smoluchowski equation has a corresponding self-similar solution $\Phi(\eta) = \frac{1}{2\pi i}\frac{1}{\eta^2}\int_{-i\infty}^{i\infty} e^{\xi\eta}\psi'(\xi)d\xi$. The equation satisfied by ψ is the following ordinary differential equation

$$-((1-2\varepsilon)\beta - 1)\psi(\xi) + \beta\xi\psi'(\xi) = \frac{1}{2}\frac{d}{d\xi}\left(D_{\xi}^{-\varepsilon}\psi(\xi)\right)^2,$$

the (rather non-trivial) analysis of which, using perturbative functional analytic techniques, is one of the accomplishments of [25].

3.2.2 Similarity Behaviour in Addition Models with Input of Monomers

Another system for which the self-similar behaviour of solutions has been studied is the "addition model", referred to in page 113, with input of monomers. Its kinetics consists of the Smoluchowski's coagulation equation where the only coefficients that are eventually non-zero are those corresponding to reactions between clusters and monomers: $a_{j,k} = 0$ if $j \wedge k > 1$. This kind of models are extensively used in studies of the early stages of submonolayer epitaxial growth, where a thin layer of a material is built on the surface of a crystal by bombarding it with monomers (see, for example, [10, 16, 101]). This is an extremely important technological process and has been theoretically modelled by a variety of techniques.

Using the notation adopted for the Becker-Döring system (cf. page 94) the mean field approach to these processes by coagulation equations consists in the following addition model

$$\begin{cases} \dot{c}_1 = J_1(t) - a_1c_1^2 - c_1\sum_{j=1}^{\infty} a_jc_j \\ \dot{c}_j = a_{j-1}c_1c_{j-1} - a_jc_1c_j, \;\; j \geq 2, \end{cases} \tag{119}$$

where $J_1(t)$ is a function describing the input rate of monomers.

The first rigorous studies of the self-similar behaviour of solutions to systems like (119) are relatively recent [54, 57, 58, 62, 195] and only for the case of constant rate coefficients $a_j \equiv 1$, and for monomer input rate of polynomial-like type $J_1(t) = (1+\varepsilon(t))\alpha t^{\omega}$, where $\alpha > 0$ and ω are real constants and $\varepsilon(\cdot)$ is a continuous function converging to zero at infinity. The methods used in these works are distinct from those presented above and are based, in an essential way, in the fact that by defining the auxiliary variable

$$c_0 := \sum_{j=1}^{\infty} c_j, \tag{120}$$

the Eq. (119) can be written as

$$\begin{cases} \dot{c}_0 = (1+\varepsilon(t))\alpha t^{\omega} - c_0c_1, \\ \dot{c}_1 = (1+\varepsilon(t))\alpha t^{\omega} - c_0c_1 - c_1^2, \\ \dot{c}_j = c_1c_{j-1} - c_1c_j, \ \ j \geq 2. \end{cases} \tag{121}$$

The essential observation is that (121) can be studied by decoupling the system with the first two equations, for the variables (c_0, c_1), from the remaining infinite dimensional system for the variables $c_j(t)$ with $j \geq 2$. Furthermore, considering in this last system the change of time scale defining by $t \mapsto \varsigma(t) := \int_{t_0}^{t} c_1(s)ds$, the infinite system is transformed in the lower triangular linear system

$$\tilde{c}_j{}' = \tilde{c}_{j-1} - \tilde{c}_j, \quad j \geq 2, \tag{122}$$

where $\tilde{c}_j(\varsigma) := c_j(t(\varsigma))$. Clearly, (122) can be explicitly solved by the variation of constants formula to get

$$\tilde{c}_j(\varsigma) = e^{-\varsigma}\sum_{k=2}^{j}\frac{\varsigma^{j-k}}{(j-k)!}c_k(0) + \frac{1}{(j-2)!}\int_0^{\varsigma}\tilde{c}_1(\varsigma - s)s^{j-2}e^{-s}ds. \tag{123}$$

Hence, to study the self-similar behaviour of solutions to (119) we can exploit the representation formula (123) if the needed information about the behaviour of the component $\tilde{c}_1(\varsigma)$ of the solution is known, and this can in principle be obtained from the study of the long-time behaviour of solutions to the two-dimensional system

$$\begin{cases} \dot{c}_0 = (1+\varepsilon(t))\alpha t^{\omega} - c_0c_1 \\ \dot{c}_1 = (1+\varepsilon(t))\alpha t^{\omega} - c_0c_1 - c_1^2. \end{cases} \tag{124}$$

The result obtained in [54, 58] using this approach is the following:

Theorem 19 ([54, 58]) *Let $a_j \equiv 1$ and $J_1(t) = (1+\varepsilon(t))\alpha t^{\omega}$, with $\alpha > 0$, $\omega > -\frac{1}{2}$, and $\varepsilon(t)$ a continuous function such that $\varepsilon(t) \to 0$ when $t \to +\infty$. For $r_0 := \frac{1-\omega}{2+\omega}$, define $Q_0(\omega) := \left(\frac{3}{(1+2\omega)\alpha}\right)^{\frac{1}{2+\omega}}\left(\frac{2+\omega}{3}\right)^{r_0}$. Let (c_j) be a solution to (119) with initial condition $c_j(0) \in X_0$, and let $\varsigma(t)$ and $\tilde{c}_j(\varsigma)$ be as above. Then*

(i) $$\lim_{\substack{j,\varsigma\to+\infty\\ \eta=j/\varsigma \text{ fixed}\\ \eta\neq 1}} Q_0(\omega)\varsigma^{r_0}\,\tilde{c}_j(\varsigma) = \Phi_{1,\omega}(\eta) := \begin{cases} (1-\eta)^{-r_0}, & \text{if } 0<\eta<1 \\ 0, & \text{if } \eta>1, \end{cases}$$

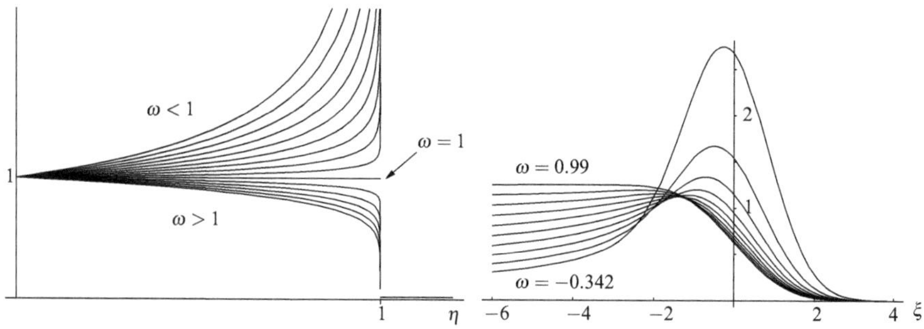

Fig. 9 Graphs of the similarity limits in Theorem 19. On the *left*: $\Phi_{1,\omega}$ for values of ω below and above 1 in steps of 0.1; On the *right*: $\Phi_{2,\omega}$ with ω from -0.342 to 0.99 in steps of 0.148

(ii) Furthermore, if $c_j(0) = 0$ when $j \geq 2$, then

$$\lim_{\substack{j,\,\varsigma \to +\infty \\ \xi = \frac{j-\varsigma}{\sqrt{\varsigma}} \text{ fixed} \\ \xi \in \mathbf{R}}} \left(\frac{\pi}{2}\right)^{\frac{1}{2}} Q_0(\omega)\, \varsigma^{\frac{1}{2}r_0}\, \tilde{c}_j(\varsigma) = \Phi_{2,\omega}(\xi)$$

$$:= e^{-\frac{1}{2}\xi^2} \int_0^{+\infty} y^{1-2r_0} e^{-\xi y^2 - \frac{1}{2}y^4}\, dy.$$

In Fig. 9 we present the graphs of some similarity profiles $\Phi_{1,\omega}$ and $\Phi_{2,\omega}$ for several values of the parameter ω. It is interesting to observe that the functions $\Phi_{2,\omega}$ can be thought of as something like an inner expansion for the jump discontinuity that exists in the functions $\Phi_{1,\omega}$ at $\eta = 1$ when $\omega \leq 1$.

Sketch of proof In order to use (123) to get the seeked for conclusions we need to know not only the limit of $\tilde{c}_1(\varsigma)$ when $\varsigma \to +\infty$, but also its rate of convergence. In order to obtain this information a detailed study of the long-time behaviour of the solutions of (124) is needed. In the autonomous case ($\omega = 0$ and $\varepsilon(t) \equiv 0$), this study can be done using invariant regions for the dynamics of (124), a change of variables suggested by Poincaré's compactification and the use of central manifolds techniques [57], or else using only elementary (but somewhat more elaborate) arguments based on invariant regions and monotonicity [52]. For the non-autonomous case all of these approaches do not seem to be applicable and the study uses an *ansatz* for a non-autonomous change of variables that is suggested by Wattis in [218]. In the new variables system (124) takes the form

$$\begin{cases} x' = (1 + \varepsilon(\tau) - xy) - A\tau^{-\frac{1}{2}}x^2 + B\tau^{-1}x \\ y' = (1 + \varepsilon(\tau) - xy)\cdot A\tau^{-\frac{1}{2}} - A^2\tau^{-1}y\,, \end{cases} \tag{125}$$

where τ is the new time, related with the variable t used in (124) by $\frac{d\tau}{dt} = \left(\frac{3\alpha^2}{1+2\omega}\right)^{1/3} t^{\frac{1+2\omega}{3}}$, the constants A and B are defined by $A := \left(\frac{1+2\omega}{4+2\omega}\right)^{\frac{1}{2}}$, and $B := \frac{1-\omega}{4+2\omega}$, respectively, and the vector (x, y) is obtained from (c_1, c_0) by the same non-autonomous change of variables. Exploiting some differential inequalities, the behaviour of the auxiliary functions $h := xy$ and $b := y - A\tau^{-\frac{1}{2}}x$ along solutions, and methods from the qualitative theory of ordinary differential equations, leads to the following [54, 58]:

$$\left(\frac{3}{\alpha(1+2\omega)}\right)^{\frac{1}{3}} t^{\frac{1-\omega}{3}} c_1(t) \overset{t\to+\infty}{\longrightarrow} 1. \tag{126}$$

Using (126) and the change of variables $t \mapsto \varsigma$ we conclude that

$$Q_0(\omega)\varsigma^{r_0}\,\tilde{c}_1(\varsigma) \overset{\varsigma\to+\infty}{\longrightarrow} 1, \tag{127}$$

and using this result and appropriate estimates for the sum and the integral of (123) we obtain the similarity limits stated in the theorem. ∎

What happens if the monomers input rate is slower than the stated in Theorem 19 is studied by Sasportes in [195], where he concludes that if $\omega = -\frac{1}{2}$ there exists a self-similar profile for the variable η, correspondent to (i) in Theorem 19, and the limit:

$$\lim_{\substack{j,\,\varsigma\to+\infty \\ \eta=j/\varsigma \text{ fixed} \\ \eta\neq 1}} (1/2)^{1/3}(3/\alpha)^{2/3}\varsigma(\log\varsigma)^{2/3}\,\tilde{c}_j(\varsigma) = \Phi_{1,-1/2}(\eta).$$

However, the limit in (ii), for that similarity variable, does not exist.

The kind of self-similar behaviour presented in Theorem 19(i) seems to be valid also when $a_j = j^p$ with $p < 1$. Although a rigorous proof is lacking, non-rigorous formal computations [51] seem to suggest that, for certain functions $Q_p(\omega)$ and $A(\omega, p)$, for $r_p = \frac{1-\omega(1-p)}{(2+\omega)(1-p)}$, and for the time scale $\tau = \left(\frac{(3-2p)A(\omega,p)}{2+\omega}\right)^{\frac{1}{1-p}} t^{\frac{2+\omega}{3-2p}}$, it could be true that

$$\lim_{\substack{j,\,\varsigma\to+\infty \\ \eta=j/\varsigma \text{ fixed} \\ \eta\neq 1}} Q_p(\omega)\varsigma^{r_p}\,\tilde{c}_j(\varsigma) = \Phi_{1,\omega,p}(\eta) := \begin{cases} \eta^{-p}\left(1-\eta^{1-p}\right)^{-r_p}, & \text{if } \eta < 1 \\ 0, & \text{if } \eta > 1. \end{cases}$$

An important feature of many physical systems that is definitely not considered in the simple model (119) is the existence of a critical cluster size below which clusters are very unstable and do not exist. The first rigorous approach to the modelling of this phenomenon in the framework we are considering was proposed recently by

Costin and co-workers in [62], where the following system, analogous to (119) but with constant input of monomers, constant reaction rates, and a critical cluster size $n > 2$, was considered:

$$\begin{cases} \dot{c}_1 = \alpha - nc_1^n - c_1 \sum_{j=n}^{\infty} c_j \\ \dot{c}_n = c_1^n - c_1 c_n \\ \dot{c}_j = c_1 c_{j-1} - c_1 c_j, \quad j > n. \end{cases} \tag{128}$$

Again, as in (119), the definition of an auxiliary variable $X(t) := \sum_{j=n}^{\infty} c_j(t)$ allows for the decoupling of (128) into a two-dimensional and an infinite dimensional that can be solved recursively. The qualitative methods used in [52, 57] for the determination of the exact long-time convergence rates of solutions of (119) do not seem to work in this case. However, a careful rigorous asymptotic analysis was possible to implement fully and the following self-similar behaviour was obtained in [62]:

$$\lim_{\substack{j,\varsigma \to +\infty \\ \eta = j/\varsigma \text{ fixed} \\ \eta \neq 1}} (n/\alpha)^{(n-1)/n} \varsigma^{(n-1)/n} \tilde{c}_j(\varsigma) = \begin{cases} (1-\eta)^{-(n-1)/n}, & \text{if } 0 < \eta < 1 \\ 0, & \text{if } \eta > 1. \end{cases}$$

Observe that if $n = 2$ the result of Theorem 19(i) is recovered.

4 Density Conservation and Gelation

More than once in this chapter we referred to problems and results related to the conservation, or non conservation, of the solution density through time evolution. The problem of characterizing the rate coefficients and the initial data for which there is conservation of density, or lack thereof, has been one of the main open problems in the mathematics of coagulation-fragmentation for many decades. Only from the late 1990s was real significant progress made, first by Jeon [110], using probabilistic methods, and afterwards by Escobedo et al. [78] and by these authors together with Laurençot [80], using purely analytic methods.

Seeing the coagulation-fragmentation equations as a model from chemical kinetics it is all too natural to expect the density of solutions to be a time invariant, due to mass conservation in each elementary reaction. In fact, proceeding in a formal way, if we substitute (4), (14) and (16) into $\sum_{j=1}^{\infty} j\dot{c}_j$ we obtain $\sum_{j=1}^{\infty} j\dot{c}_j = 0$, with an analogous formal result being valid for the continuous version of the equations.

The attempts to turn these formal computations rigorous were faced, from the beginning, with mathematical difficulties, perhaps unexpected, which resulted in that, for many years, the results available in the literature were formal studies (cf.

e.g. [77, 83, 106, 228]), rigorous studies of particular cases [26, 27], and examples of solutions that did not conserve density [144, 148]. All these studies were for Smoluchowski's equation with product type coagulation kernels growing fast with the cluster size. We will start by a brief review of the first mathematically rigorous studies.

The first results on the existence of solution to Smoluchowski's coagulation system, proved by McLeod in the beginning of the 1960s [154–156] considered coagulation coefficients $a_{j,k} = r_j r_k$ and an initial condition $c_{0j} = \delta_{j,1}$. The condition considered in those studies on the finiteness of the solutions' second moment $\|c(t)\|_2 < \infty$ implied that, when $r_j = j$, the maximal interval of existence is $[0, 1)$, when $r_j \leq j$ it contains $[0, e^{-1}]$, and when $r_j = jq_j$, with $q_j \to +\infty$, there is no solution in any non degenerate time interval. Observe that, when $a_{j,k} \leq jk$, the requirement of finite second moment easily implies density conservation of the solution [cf. (65)] and, in fact, waving this condition Leyvraz and Tschudi proved [148] that for $r_j = j$ McLeod's solution can be continued for $t > 1$, using to that end the generating function

$$G(t,z) := \sum_{j=1}^{\infty} \varphi_j(t) z^j, \quad \text{where } \varphi_j(t) := jc_j(t)e^{j\int_0^t \|c(s)\|_1 ds}.$$

This generating function is a solution of the equation

$$\frac{\partial G}{\partial t} = zG\frac{\partial G}{\partial z}, \quad z \in (0,1), \ t > 0,$$

with $G(0,z) = z$. As this problem can be integrated by the method of characteristics to obtain an explicit expression for G, from which we then obtain φ_j and hence c_j. The final result is Leyvraz-Tschudi's solution:

$$c_j(t) = \begin{cases} \dfrac{j^{j-2}}{j!} t^{j-1} e^{-jt}, & \text{if } 0 \leq t \leq 1 \\[2ex] \dfrac{j^{j-2}e^{-j}}{j!}\dfrac{1}{t}, & \text{if } t > 1, \end{cases} \tag{129}$$

for which an easy computation results in

$$\|c(t)\|_1 = \begin{cases} 1, & \text{if } 0 \leq t \leq 1 \\ t^{-1}, & \text{if } t > 1. \end{cases}$$

Thus, Leyvraz-Tschudi's solution does not conserve density for $t > 1$. This same result was later re-derived by Slemrod [200] without recourse to generating functions.

It is very interesting to observe that the nonconservation of density happens at a finite time (in Leyvraz-Tschudi's solution, at $t = 1$) and not at the long-time limit, $t \to +\infty$, as in the Becker-Döring and the coagulation-fragmentation equations with weak fragmentation. The physical interpretation is analogous: the missing density $\|c(0)\|_1 - \|c(t)\|_1$ corresponds to the runaway of part of the density to entities whose sizes are not described by subscripts $j \in \mathbf{N}$ (or $x \in \mathbf{R}^+$ in the continuous version), which means that they are, from the physical point of view, incommensurably larger than every j. This "infinite cluster" is interpreted in the physics literature as a different macroscopic phase, called a *gel,* and its occurrence is called the sol-gel transition or gelification. The smallest time $T_g \geq 0$ after which density conservation no longer holds is called the geling time.[7]

The above interpretation for the finite time break down of density conservation also suggests that such a phenomenon occurs when the coagulation coefficients grow fast with the cluster sizes. In fact, Leyvraz showed in [144] that, if $r_j = j^\alpha$, with $\alpha > \frac{1}{2}$, then there exists a solution of (31) with a specially chosen initial condition, for which $T_g = 0$. This result, as well as those above are examples of non density conserving solutions with very particular initial data; nevertheless they are historically important because for many years they were essentially the only rigorous examples known of solutions exhibiting a behaviour that was generally believed to hold for all non zero solutions to (31) when $a_{j,k} \geq (jk)^\alpha$ with $\alpha > \frac{1}{2}$.

The case $\alpha > 1$ was first studied by van Dongen [69] and rigorously proved by Carr and da Costa [35]:

Theorem 20 ([35]) *Let $C_L(j^\alpha + k^\alpha) \leq a_{j,k} \leq C_U(jk)^\beta$, with constants $C_L, C_U > 0$ and $\beta > \alpha > 1$. Let c be a solution of (31) in $[0, T)$ with $c_0 \neq 0$. Then, c does not conserve density in any time interval $[0, t_\infty)$, $\forall t_\infty \leq T$.*

Sketch of proof The basic idea of the proof is an argument by contradiction using higher moments: assuming that c is a density conserving solution in an interval $[0, t_\infty)$ and using the lower bound for the coefficients it is possible to prove that, for all $p > 1$,

$$\sum_{j=m}^{\infty} j^p c_j(t) \leq \|c_0\|_1 \sum_{j=m}^{\infty} j^{p-1} e^{-C_L\|c_0\|_1 j^{1-\alpha}(\varepsilon - t)/2}, \tag{130}$$

where $0 < t < \varepsilon < t_\infty$ and m is sufficiently large.

Obviously (130) implies that all moments $\|c(t)\|_p$ are finite in $(0, t_\infty)$ and this is the result that is at the centre of the contradiction because the lower bound on the coefficients, the hypothesis of density conservation, and Hölder's inequality imply

[7] There is an analogous behaviour of loss of mass in the pure fragmentation system, called "shattering" and interpreted as a loss of mass to infinitesimal clusters. We shall not consider this behaviour here, directing the interested reader to the literature, e.g., [7, 89, 152].

that, $\forall \delta, t, \tau \in (0, t_\infty)$ with $\delta < t \leq \tau$,

$$\|c(t)\|_p - \|c(\delta)\|_p \geq pC_L\|c_0\|^{1-\frac{\alpha-1}{p-1}} \int_\delta^t \|c(s)\|_p^{1+\frac{\alpha-1}{p-1}} ds,$$

from which we get a blowing up time for $\|c(t)\|_p$, $T^{(p)}$, with $\lim_{p\to+\infty} T^{(p)} \leq \delta$. ∎

For many decades the only rigorous results for the case $\alpha \in \left(\frac{1}{2}, 1\right]$ were the particular solutions in [144, 148], already referred to. An attempt by da Costa [49] to prove that the same behaviour would occur for all solutions, based in a dynamical systems approach, resulted in the identification of a larger family of gelling solutions but did not solve the problem, although it had some use in the numerical analysis of the gelling phenomenon [8]. For the continuous system Laurençot [125] considered $a(x,y) = r(x)r(y) + \alpha(x,y)$ with $\alpha(x,y) \leq Ar(x)r(y)$ and $r(x) \geq Rx$, and proved that all solution exhibit gelation and obtained some results about the density decay and the gelification time.

In other works fragmentation was also included. Recall that, in [46] it was proved that, with coagulation coefficients for which gelation was expected to occur, a sufficiently strong fragmentation prevents that to happen at least for solutions obtained as limits of truncated systems (cf. Theorem 8). This is also in line with the interpretation of gelation as a loss of density to an infinite size entity, because it is natural to expect that a high rate of fragmentation of big clusters inhibits the accumulation of density in larger and larger clusters, thus preventing the runaway phenomenon causing the emergence of gelation

The rigorous analytic elucidation of gelation was achieved in 2002 by Escobedo et al. in [78], and in 2003 by the same authors with Laurençot in [80]. In what follows we briefly describe those results.

Let us start by observing that the results in [78, 80] are proved for the continuous version of the coagulation-fragmentation equations but, naturally, they are also valid for the discrete case. We shall concentrate our attention in the coagulation equation:

Theorem 21 ([78]) *Let* $a(x,y) = \frac{1}{2}\left(x^\alpha y^\beta + x^\beta y^\alpha\right)$, *with* $0 \leq \alpha \leq \beta \leq 1$ *and* $\lambda := \alpha + \beta > 1$. *Let* c *be an arbitrary weak solution of (5)–(6) with a non zero initial condition*[8] $c_0 \in Y_1$. *Then, there exists a positive constant* $C_* = C_*(M_1(0), M_0(0), \lambda)$ *such that, for all* $t \geq 0$, *it holds*

$$M_1(t) \leq \frac{C_*}{(1+t)^{1/\lambda}} \tag{131}$$

and so the geling time is finite and satisfies the upper bound

$$T_g \leq T_* := \left(\frac{C_*}{M_1(0)}\right)^\lambda. \tag{132}$$

---

[8]The Banach space Y_1 was defined in page 108.

In this statement the notation $M_k(t) := \|c(t,\cdot)\|_{L^1(\mathbf{R}^+, y^k dy)}$ was used.

Sketch of proof The proof is based on some estimates for the weak solutions using carefully constructed test functions, from which it is possible to conclude that, for all $\tau \geq 0$,

$$\int_\tau^\infty M_1(t)^2 dt \leq C_\lambda M_0(0)^{\lambda-1} M_1(\tau)^{2-\lambda}, \tag{133}$$

from which we obviously see that the density cannot be constant and, with a bit more extra work, obtain (131). Let us look at this argument in more detail. The definition of weak solution of (5)–(6) is like the one in the discrete case: it is any function $c \in C([0,\infty); L^1) \cap L^\infty(0,T;Y_1)$, $\forall T > 0$, satisfying $M_1(t) \leq M_1(0)$, $\forall t \geq 0$, such that, for all $t \geq \tau \geq 0$ and $g \in L^\infty(0,\infty)$, the following holds

$$\begin{aligned} &\int_0^\infty g(x)c(t,x)dx - \int_0^\infty g(x)c(\tau,x)dx = \\ &= \frac{1}{2}\int_\tau^t \iint_{\mathbf{R}^+\times\mathbf{R}^+} (g(x+y)-g(x)-g(y))a(x,y)c(s,x)c(s,y)dxdyds. \end{aligned} \tag{134}$$

Let c be a weak solution of (5)–(6) and in (134) consider the test function $g(x) = g_A(x) := x \wedge A \in L^\infty(0,\infty)$. As $g_A(x+y) - g_A(x) - g_A(y) \leq 0$ in $\mathbf{R}^+ \times \mathbf{R}^+$ it is possible to estimate the right-hand side of (134) keeping only the contribution due to the integration on $[A,\infty)^2$, which immediately results in

$$\int_\tau^t \left(\int_A^\infty y^{\lambda/2} c(s,x)dx\right)^2 ds \leq \frac{2M_1(\tau)}{A}. \tag{135}$$

Consider now a function $\Phi : [0,\infty) \to [0,\infty)$ which is monotonic increasing, differentiable a.e., with $\Phi(0) = 0$, and $C_\Phi := \|\Phi'\|_{L^1(\mathbf{R}^+, y^{-1/2}dy)} < \infty$. Writing $\Phi(x) = \int_0^x \Phi'(A)dA$, using Fubini's theorem, Cauchy-Schwarz inequality and (135) we concluded that

$$\int_\tau^t \left(\int_0^\infty x^{\lambda/2}\Phi(x)c(s,x)dx\right)^2 ds \leq 2C_\Phi^2 M_1(\tau).$$

Taking limits as $t \to \infty$ and considering $\Phi(x) := \left(x^{1-\lambda/2} - (R/2)^{1-\lambda/2}\right)^+$, where $R > 0$ is an arbitrary constant, we conclude that

$$\int_\tau^\infty \left(\int_R^\infty xc(s,x)dx\right)^2 ds \leq CR^{1-\lambda} M_1(\tau). \tag{136}$$

Finally, (133) is obtained by using $M_1(t)^2 \le 2\big(\int_0^R xc(t,x)dx\big)^2 + 2\big(\int_R^\infty xc(t,x)dx\big)^2$, and $\big(\int_0^R xc(t,x)dx\big)^2 \le R^{2-\lambda}M_{\lambda/2}(\tau)^2$, applying (136) and (134) with the test function $g \equiv 1$, and taking $R = M_1(\tau)/M_0(\tau)$. ■

Escobedo et al. [78] also contains an extensive study of several properties of the density of weak solutions of (5)–(6), including the behaviour of the solutions at geling time T_g.

The same method was used in [78, 80] for the continuous system with fragmentation, establishing the following result:

Theorem 22 ([78, 80]) *Let* $a(x,y) = \frac{1}{2}\left(x^\alpha y^\beta + x^\beta y^\alpha\right)$, *with* $0 \le \alpha \le \beta \le 1$ *and* $\lambda := \alpha + \beta$. *Let* $b(x,y) = (1+x+y)^\gamma$, *with* $\gamma \in \mathbf{R}$. *Then, the following hold:*

(i) if $\lambda \le 1$ *or if* $\gamma > \lambda - 2$, *there exists a density conserving weak solution (17)*
(ii) if $\lambda > 1$ *and* $\gamma < \lambda - 2$, *there exists a critical density* $\rho^* > 0$ *such that, when* $c_0 \in Y_1$ *satisfies* $\|c_0\|_{L^1(\mathbf{R}^+, ydy)} > \rho^*$, *every weak solution of (17) with initial condition* c_0 *exhibits gelation.*

Note that when the condition on γ is (i) the behaviour of solutions is the same that occurred for the discrete system in the strong fragmentation case (cf. page 117), that is: even with conditions on the coagulation coefficients for which solutions to the purely coagulating dynamics have break down of density conservation, a sufficiently strong fragmentation forces density to remain constant through time evolution.

Case (ii) leaves unanswered what happens for sufficiently small densities. Formal arguments presented in [80] lead to the conjecture that, if $\gamma \in ((\lambda-3)/2, \lambda-2)$, there are weak solutions, with low density initial conditions, for which density is conserved, whereas if $\gamma < (\lambda-3)/2$, all non zero solutions have gelation.

As stated above, Theorems 21 and 22 are fundamental contributions to the problem of density conservation in coagulation-fragmentation systems, although, as stated in [78, 80], several relevant problems still wait for a proof.

Acknowledgements I would like to thank Prof. Conceição Carvalho for the invitation to give an overview talk about the mathematics of coagulation equations at the thematic session on *Non-Equilibrium Statistical Mechanics: Kinetics, Chemistry and Coagulation* she organized as part of the *International Conference on Mathematics of Energy and Climate Change,* held in Lisbon, Portugal, in March 2013 and organized by CIM-Centro Internacional de Matemática.

I am also grateful to CIM president, Prof. Alberto Pinto, who enthusiastically supported my suggestion of writing this chapter by updating and translating into English an unpublished document I wrote in Portuguese a few years ago, and was very understanding in accepting the somewhat oversized result.

This work was partially supported by FCT under Strategic Project - LA 9 - 2013–2014.

References

1. Aizenman, M., Bak, T.A.: Convergence to equilibrium in a system of reacting polymers. Commun. Math. Phys. **65**, 203–230 (1979)
2. Aldous, D.J.: Deterministic and stochastic models for coalescence (aggregation, coagulation): a review of mean-field theory for probabilists. Bernoulli **5**, 3–48 (1999)
3. Amann, H.: Coagulation-fragmentation processes. Arch. Rat. Mech. Anal. **151**, 339–366 (2000)
4. Amann, H., Walker, Ch.: Local and global strong solutions to continuous coagulation-fragmentation equations with diffusion. J. Differ. Equ. **218**, 159–186 (2005)
5. Amann, H., Weber, F.: On a quasilinear coagulation-fragmentation model with diffusion. Adv. Math. Sci. Appl. **11**, 227–263 (2001)
6. Arino, O., Rudnicki, R.: Phytoplankton dynamics. C. R. Biol. **327**, 961–969 (2004)
7. Arlotti, L., Banasiak, J.: Perturbations of Positive Semigroups with Applications. Springer Monographs in Mathematics. Springer, London (2006)
8. Babovsky, H.: On the modeling of gelation rates by finite systems. Technisch Universität Ilmenau, Institut für Mathematik, Preprint No. M11/01 (2001)
9. Bagland, V., Laurençot, Ph.: Self-similar solutions to the Oort-Hulst-Safronov coagulation equation. SIAM J. Math. Anal. **39**, 345–378 (2007)
10. Bales, G.S., Chrzan, D.C.: Dynamics of irreversible island growth during submonolayer epitaxy. Phys. Rev. B **50**, 6057–6067 (1994)
11. Ball, J.M., Carr, J.: Asymptotic behaviour of solutions to the Becker-Döring equations for arbitrary initial data. Proc. R. Soc. Edinb. **108A**, 109–116 (1988)
12. Ball, J.M., Carr, J.: The discrete coagulation-fragmentation equations: existence, uniqueness, and density conservation. J. Stat. Phys. **61**, 203–234 (1990)
13. Ball, J.M., Carr, J., Penrose, O.: The Becker-Döring cluster equations: basic properties and asymptotic behaviour of solutions. Commun. Math. Phys. **104**, 657–692 (1986)
14. Banasiak, J.: Transport processes with coagulation and strong fragmentation. Discrete Contin. Dyn. Syst. Ser. B **17**(2), 445–472 (2012)
15. Baranger, C.: Collisions, coalescences et fragmentations des gouttelettes dans un spray: écriture précise des équations relatives au modèle TAB. Prepublications du Centre de Mathématiques et de Leurs Applications, N°2001-21, Ecole Normale Supérieure de Cachan (2001)
16. Bartlet, M.C., Evans, J.W.: Exact island-size distributions in submonolayer deposition: influence of correlations between island size and separation. Phys. Rev. B **54**, R17359–R17362 (1996)
17. Becker, R., Döring, W.: Kinetische Behandlung in übersättigten Dämpfern. Ann. Phys. (Leipzig) **24**, 719–752 (1935)
18. Ben-Naim, E., Krapivsky, P.: Kinetics of aggregation-annihilation processes. Phys. Rev. E **52**, 6066–6070 (1995)
19. Bénilan, Ph., Wrzosek, D.: On an infinite system of raction-diffusion equations. Adv. Math. Sci. Appl. **7**, 349–364 (1997)
20. Berry, E.X.: A mathematical framework for cloud models. J. Atmos. Sci. **26**, 109–111 (1969)
21. Bertoin, J.: Random Fragmentation and Coagulation Processes. Cambridge Studies in Advanced Mathematics, vol. 102. Cambridge University Press, Cambridge (2006)
22. Binder, K.: Theory for the dynamics of clusters, II. Critical diffusion in binary systems and the kinetics for phase separation. Phys. Rev. B **15**, 4425–4447 (1977)
23. Blatz, P.J., Tobolsky, A.V.: Note on the kinetics of systems manifesting simultaneous polymerization-depolymerization phenomena. J. Phys. Chem. **49**, 77–80 (1945)
24. Bonilla, L., Carpio, A., Neu, J.C., Wolfer, W.G.: Kinetics of helium bubble formation in nuclear materials. Physica D **222**, 131–140 (2006)
25. Breschi, G., Fontelos, M.A.: Self-similar solutions of the second kind representing gelation in finite time for the Smoluchowski equation. Nonlinearity **27**, 1709–1745 (2014)

26. Buffet, E., Pulé, J.: Gelation: the diagonal case revisited. Nonlinearity **2**, 373–381 (1989)
27. Buffet, E., Werner, R.F.: A counter-example in coagulation theory. J. Math. Phys. **32**, 2276–2278 (1991)
28. Burton, J.J.: Nucleation theory. In: Berne, B.J. (ed.) Statistical Mechanics, Part A: Equilibrium Techniques. Modern Theoretical Chemistry, vol. 5, pp. 195–234. Plenum Press, New York (1977)
29. Cañizo, J.A.: Asymptotic behaviour of solutions to the generalized Becker-Döring equations for general initial data. Proc. R. Soc. Lond. Ser. A Math. Phys. Eng. Sci. **461**, 3731–3745 (2005)
30. Cañizo, J.A.: Convergence to equilibrium for the discrete coagulation-fragmentation equations with detailed balance. J. Stat. Phys. **129**, 1–26 (2007)
31. Cañizo, J.A., Lods, B.: Exponential convergence to equilibrium for subcritical solutions of the Becker-Döring equations. J. Differ. Equ. **255**, 905–950 (2013)
32. Cañizo, J.A., Mischler, A.: Regularity, local behaviour and partial uniqueness for self-similar profiles of Smoluchowski's coagulation equation. Rev. Mat. Iberoam. **27**(3), 803–839 (2011)
33. Cañizo, J.A., Mischler, A., Mouhot, C.: Rate of convergence to self-similarity for Smoluchowski's coagulation equation with constant coefficients. SIAM J. Math. Anal. **41**(6), 2283–2314 (2010)
34. Carr, J.: Asymptotic behaviour of solutions to the coagulation-fragmentation equations. I. The strong fragmentation case. Proc. R. Soc. Edinb. **121A**, 231–244 (1992)
35. Carr, J., da Costa, F.P.: Instantaneous gelation in coagulation dynamics. Z. Angew. Math. Phys. **43**, 974–983 (1992)
36. Carr, J., da Costa, F.P.: Asymptotic behavior of solutions to the coagulation-fragmentation equations. II. Weak fragmentation. J. Stat. Phys. **77**, 89–123 (1994)
37. Carr, J., Dunwell, R.: Kinetics of cell surface capping. Appl. Math. Lett. **12**, 45–49 (1999)
38. Carr, J., Pego, R.L.: Very slow phase separation in one dimension. In: Rascle, M., Serre, D., Slemrod, M. (eds.) PDEs and Continuum Models of Phase Transitions. Proceedings of an NSF-CNRS Joint Seminar held in Nice, France, January 18–22, 1988. Lecture Notes in Physics, vol. 344, pp. 216–226. Springer, Berlin (1989)
39. Carr, J., Pego, R.L.: Self-similarity in a coarsening model in one dimension. Proc. R. Soc. Lond. A **436**, 569–583 (1992)
40. Carr, J., Pego, R.L.: Self-similarity in a cut-and-paste model of coarsening. Proc. R. Soc. Lond. A **456**, 1281–1290 (2000)
41. Chandrasekhar, S.: Stochastic problems in physics and astronomy. Rev. Mod. Phys. **15**, 1–89 (1941)
42. Cheng, Z., Redner, S.: Scaling theory of fragmentation. Phys. Rev. Lett. **60**, 2450–2453 (1988)
43. Collet, F.: Some modelling issues in the theory of fragmentation-coagulation systems. Commun. Math. Sci. **2**(Suppl. 1), 35–54 (2004)
44. da Costa, F.P.: Studies in coagulation-fragmentation equations. Ph.D. Thesis, Heriot-Watt University, Edinburgh (1993)
45. da Costa, F.P.: Existence and uniqueness of density conserving solutions to the coagulation-fragmentation equations with strong fragmentation. J. Math. Anal. Appl. **192**, 892–914 (1995)
46. da Costa, F.P.: On the positivity of solutions to the Smoluchowski equations. Mathematika **42**, 406–412 (1995)
47. da Costa, F.P.: On the dynamic scaling behaviour of solutions to the discrete Smoluchowski equation. Proc. Edinb. Math. Soc. **39**, 547–559 (1996)
48. da Costa, F.P.: Asymptotic behaviour of low density solutions to the Generalized Becker-Döring equations. Nonlinear Differ. Equ. Appl. **5**, 23–37 (1998)
49. da Costa, F.P.: A finite-dimensional dynamical model for gelation in coagulation processes. J. Nonlinear Sci. **8**, 619–653 (1998)
50. da Costa, F.P.: Convergence to equilibria of solutions to the coagulation-fragmentation equations. In: Li, T.-t., Lin, L.-w., Rodrigues, J.F. (eds.) Nonlinear Evolution Equations and Their Applications, pp. 45–56. Luso-Chinese Symposium, Macau, 7–9 Oct 1998. World Scientific, Singapore (1999)

51. da Costa, F.P.: Convergence to self-similarity in addition models with input of monomers. Oberwolfach Rep. **4**(4), 2754–2756 (2007)
52. da Costa, F.P.: Dynamics of a differential system using invariant regions. L'Enseignement Math. **53**, 3–14 (2007)
53. da Costa, F.P., Pinto, J.T.: A nonautonomous predator-prey system arising from coagulation theory. Int. J. Biomath. Biostat. **1**(2), 129–140 (2010)
54. da Costa, F.P., Sasportes, R.: Dynamics of a nonautonomous ODE system occuring in coagulation theory. J. Dyn. Diff. Equ. **20**, 55–85 (2008)
55. da Costa, F.P., Grinfeld, M., McLeod, J.B.: Unimodality of steady size distributions of growing cell populations. J. Evol. Equ. **1**, 405–409 (2001)
56. da Costa, F.P., Grinfeld, M., Wattis, J.A.D.: A hierarchical cluster system based on Horton-Strahler rules for river networks. Stud. Appl. Math. **109**, 163–204 (2002)
57. da Costa, F.P., van Roessel, H.J., Wattis, J.A.D.: Long-time behaviour and self-similarity in a coagulation equation with input of monomers. Markov Process. Relat. Fields **12**, 367–398 (2006)
58. da Costa, F.P., Pinto, J.T., Sasportes, R.: Convergence to self-similarity in an addition model with power-like time-dependent input of monomers. In: Cutello, V., Fotia, G., Puccio, L. (eds.) Applied and Industrial Mathematics in Italy II, Selected Contributions from the 8th SIMAI Conference. Series on Advances in Mathematics for Applied Sciences, vol. 75, pp. 303–314. World Scientific, Singapore (2007)
59. da Costa, F.P., Pinto, J.T., Sasportes, R.: The Redner–Ben-Avraham–Kahng Cluster system. São Paulo J. Math. Sci. **6**(2), 171–201 (2012)
60. da Costa, F.P., Pinto, J.T., van Roessel, H.J., Sasportes, R.: Scaling behaviour in a coagulation-annihilation model and Lotka-Volterra competition systems. J. Phys. A Math. Theor. **45**, 285201 (2012)
61. da Costa, F.P., Pinto, J.T., Sasportes, R.: The Redner–Ben-Avraham–Kahng coagulation system with constant coefficients: the finite dimensional case. Z. Angew. Math. Phys. (13 August 2014, accepted for publication). arXiv:1401.3715v2
62. Costin, O., Grinfeld, M., O'Neill, K.P., Park, H.: Long-time behaviour of point islands under fixed rate deposition. Commun. Inf. Syst. **13**(2), 183–200 (2013)
63. Coutsias, E.A., Wester, M.J., Perelson, A.S.: A nucleation theory of cell surface capping, J. Stat. Phys. **87**, 1179–1203 (1997)
64. Coveney, P.V., Wattis, J.A.D.: Becker-Döring model of self-reproducing vesicles. J. Chem. Soc. Faraday Trans. **92**(2), 233–246 (1998)
65. Davidson, J.: Existence and uniqueness theorem for the Safronov-Dubovski coagulation equation, pp. 10. Z. Angew. Math. Phys. (31 August 2013, to appear). doi:10.1007/s00033-013-0360-y
66. Deaconu, M., Fournier, N., Tanré, E.: A pure jump Markov process associated with Smoluchowski's coagulation equation. Ann. Probab. **30**, 1763–1796 (2002)
67. Derrida, B., Godrèche, C., Yekutieli, I.: Scale-invariant regimes in one-dimensional models of growing and coalescing droplets. Phys. Rev. A **44**, 6241–6251 (1991)
68. Desvillettes, L., Fellner, K.: Duality and entropy methods in coagulation-fragmentation models. Riv. Mat. Univ. Parma **4**(2), 215–263 (2013)
69. van Dongen, P.G.J.: On the possible occurrence of instantaneous gelation in Smoluchowski's coagulation equation. J. Phys. A Math. Gen. **20**, 1889–1904 (1987)
70. van Dongen, P.G.J., Ernst, M.H.: Scaling solutions of Smoluchowski's coagulation equations. J. Stat. Phys. **50**, 295–329 (1988)
71. van Dongen, P.G.J.: Spatial fluctuations in reaction-limited aggregation. J. Stat. Phys. **54**, 221–271 (1989)
72. Drake, R.L.: A general mathematical survey of the coagulation equation. In: Hidy, G.M., Brock, J.R. (eds.) Topics in Current Aerosol Research (Part 2). International Reviews in Aerosol Physics and Chemistry, pp. 201–376. Pergamon Press, Oxford (1972)
73. Dreyer, W., Duderstadt, F.: On the Becker/Döring theory of nucleation of liquid droplets in solids. J. Stat. Phys. **123**, 55–87 (2006)

74. Dubovskii, P.B.: Mathematical Theory of Coagulation. Lecture Notes Series, vol. 23. Research Institute of Mathematics/Global Analysis Research Center, Seoul National University, Seoul (1994)
75. Dubovski, P.B.: A 'triangle' of interconnected coagulation models. J. Phys. A Math. Gen. **32**, 781–793 (1999)
76. Ernst, M.H., Pagonabarraga, I.: The nonlinear fragmentation equation. J. Phys. A Math. Theor. **40**, F331–F337 (2007)
77. Ernst, M.H., Ziff, R.M., Hendriks, E.M.: Coagulation processes with a phase transition. J. Colloid Interface Sci. **97**, 266–277 (1984)
78. Escobedo, M., Mischler, S., Perthame, B.: Gelation in coagulation and fragmentation models. Commun. Math. Phys. **231**, 157–188 (2002)
79. Escobedo, M., Laurençot, Ph., Mischler, S.: Fast reaction limit of the discrete diffusive coagulation-fragmentation equation. Commun. Partial Diff. Equ. **28**, 1113–1133 (2003)
80. Escobedo, M., Laurençot, Ph., Mischler, S., Perthame, B.: Gelation and mass conservation in coagulation-fragmetation models. J. Differ. Equ. **195**, 143–174 (2003)
81. Escobedo, M., Laurençot, Ph., Mischler, S.: On a kinetic equation for coalescing particles. Commun. Math. Phys. **246**, 237–267 (2004)
82. Escobedo, M., Mischler, S., Rodriguez-Ricard, M.: On self-similarity and stationary problems for fragmentation and coagulation models. Ann. Inst. H. Poincaré Anal. Non Linéaire **22**, 99–125 (2005)
83. Family, F., Landau, D.P. (eds.): Kinetics of aggregation and gelation. In: Proceedings of the International Topical Conference on Kinetics of Aggregation and Gelation, Athens, Georgia, USA, 2–4 April 1984. North-Holland, Amesterdam (1984)
84. Fasano, A., Rosso, F.: Dynamics of droplets in an agitated dispersion with multiple breakage. Part I: formulation of the model and physical consistency. Math. Meth. Appl. Sci. **28**, 631–659 (2005)
85. Fasano, A., Rosso, F.: Dynamics of droplets in an agitated dispersion with multiple breakage. Part II: uniqueness and global existence. Math. Meth. Appl. Sci. **28**, 1061–1088 (2005)
86. Fasano, A., Rosso, F., Mancini, A.: Implementation of a fragmentation-coagulation-scattering model for the dynamics of stirred liquid-liquid dispersions. Physica D **222**, 141–158 (2006)
87. Feller, W.: An Introduction to Probability Theory and Its Applications, vol. II, 2nd edn. Wiley, New York (1971)
88. Filbet, F., Laurençot, P.: Numerical simulation of the Smoluchowski coagulation equation. SIAM J. Sci. Comput. **25**, 2004–2028 (2004)
89. Filippov, A.F.: On the distribution of the sizes of particles which undergo splitting. Theory Prob. Appl. **6**, 275–294 (1961)
90. Fournier, N., Laurençot, Ph.: Existence of self-similar solutions to Smoluchowski's coagulation equation. Commun. Math. Phys. **256**, 589–609 (2005)
91. Fournier, N., Laurençot, Ph.: Local properties of self-similar solutions to Smoluchowski's coagulation equation with sum kernels. Proc. R. Soc. Edinb. Sect. A **136**, 485–508 (2006)
92. Fournier, N., Laurençot, Ph.: Markus-Lushnikov processes, Smoluchowski's and Flory's models. Stoch. Process. Appl. **119**, 167–189 (2009)
93. Fournier, N., Mischler, S.: Exponential trend to equilibrium for discrete coagulation equations with strong fragmentation and without a balance condition. Proc. R. Soc. Lond. A **460**, 2477–2486 (2004)
94. Fournier, N., Mischler, S.: On a discrete Boltzmann-Smoluchowski equations with rates bounded in the velocity variables. Commun. Math. Sci. **2**(Suppl. 1), 55–63 (2004)
95. Fournier, N., Mischler, S.: A spatially homogeneous Boltzmann equation for elastic, inelastic and coalescence collisions. J. Math. Pure Appl. **84**, 1173–1234 (2005)
96. Friedlander, S.K.: Smoke, Dust, and Haze: Fundamentals of Aerosol Dynamics. Topics in Chemical Engineering, 2nd edn. Oxford University Press, New York (2000)
97. Friedman, A., Ross, D.S.: Mathematical Models in Photographic Science. Mathematics in Industry, vol. 3. Springer, Berlin (2003)

98. Fusco, G.: A geometry approach to the dynamics of $u_t = \varepsilon^2 u_{xx} + f(x)$ for small ε. In: Kirchgässner, K. (ed.) Problems Involving Change of Type. Proceedings of a Conference Held at the University of Stuttgart, FRG, October 11–14, 1988. Lecture Notes in Physics, vol. 359, pp. 53–73. Springer, Berlin (1990)
99. Gabrielov, A., Newman, W.I., Turcotte, D.L.: Exactly soluble hierarchical clustering model: inverse cascades, self-similarity, and scaling. Phys. Rev. E **60**, 5293–5300 (1999)
100. Gallay, T., Mielke, A.: Convergence results for a coarsening model using global linearization. J. Nonlinear Sci. **13**, 311–346 (2003)
101. Gibou, F., Ratsch, C., Caflisch, R.: Capture numbers in rate equations and scaling laws for epitaxial growth. Phys. Rev. B **67**, 155403 (2003)
102. Greer, M.L., Pujo-Menjouet, L., Webb, G.F.: A mathematical analysis of the dynamics of prion proliferation. J. Theor. Biol. **242**, 598–606 (2006)
103. Grinfeld, M., Lamb, W., O'Neill, K.P., Mulheran, P.A.: Capture-zone distribution in one-dimensional sub-monolayer film growth: a fragmentation theory approach. J. Phys. A Math. Theor. **45**, 015002 (2012)
104. Großkinsky, S., Klingenberg, C., Oelschläger, K.: A rigorous derivation of Smoluchowski's equation in the moderate limit. Stoch. Anal. Appl. **22**, 113–141 (2004)
105. Guiaş, F.: Coagulation-fragmentation processes: relations between finite particle models and differential equations. PhD thesis, Ruprecht-Karls-Universität Heidelberg, SFB 359, Preprint 41/1998 (1998)
106. Hendriks, E.M., Ernst, M.H.: Critical properties for gelation: a kinetic approach. Phys. Rev. Lett. **49**(8), 593–595 (1982)
107. Herrmann, M., Naldzhieva, M., Niethammer, B.: On a thermodynamically consistent modification of the Becker-Döring equations. Physica D **222**, 116–130 (2006)
108. Ispolatov, I., Krapivsky, P.L., Redner, S.: War: the dynamics of vicious civilizations. Phys. Rev. E **54**, 1274–1289 (1996)
109. Jabin, P.-E., Niethammer, B.: On the rate of convergence to equilibrium in the Becker-Döring equations. J. Differ. Equ. **191**(2), 518–543 (2003)
110. Jeon, I.: Existence of gelling solutions for coagulation-fragmentation equations. Commun. Math. Phys. **194**, 541–567 (1998)
111. Ke, J., Wang, X., Lin, Z., Zhuang, Y.: Scaling in the aggregation process with catalysis-driven fragmentation. Physica A **338**, 356–366 (2004)
112. Kolmogorov, A.N.: Über das logarithmisch normale Verteilungsgesetz der Dimensionen der Teilchen bei Zerstückelung. Dokl. Akad. Nauk SSSR **31**, 99–101 (1941)
113. Kolmogorov, A.N., Fomin, S.V.: Introductory Real Analysis. Dover, New York (1975)
114. Kolokoltsov, V.N.: Hydrodynamic limit of coagulation-fragmentation type models of k-nary interacting particles. J. Stat. Phys. **115**, 1621–1653 (2004)
115. Krapivsky, P.: Nonuniversality and breakdown of scaling in two-species aggregation with annihilation. Physica A **198**, 135–149 (1993)
116. Kreer, M.: Cluster equations for the Glauber kinetic Ising ferromagnet: I. Existence and uniqueness. Ann. Physik **2**, 720–737 (1993)
117. Kreer, M., Penrose, O.: Proof of dynamic scaling in Smoluchowski's coagulation equations with constant kernels. J. Stat. Phys. **74**, 389–407 (1994)
118. Krivitsky, D.: Numerical solution of the Smoluchowski kinetic equation and asymptotics of the distribution function. J. Phys. A Math. Gen. **28**, 2025–2039 (1995)
119. Kumar, J., Peglow, M., Warnecke, G., Heinrich, S.: An efficient numerical technique for solving population balance equation involving aggregation, breakage, growth and nucleation. Powder Technol. **182**, 81–104 (2008)
120. Lachowicz, M., Laurençot, Ph., Wrzosek, D.: On the Oort-Hulst-Safronov coagulation equation and its relation to the Smoluchowski equation. SIAM J. Math. Anal. **34**, 1399–1421 (2003)
121. Lachowicz, M., Wrzosek, D.: A nonlocal coagulation-fragmentation model. Appl. Math. (Warsaw) **27**, 45–66 (2000)

122. Laurençot, Ph.: Uniforme integrabilite et théorème de de la Vallée Poussin, pp. 8 (unpublished note, not dated)
123. Laurençot, Ph.: Global solutions to the discrete coagulation equations. Mathematika **46**, 433–442 (1999)
124. Laurençot, Ph.: Singular behaviour of finite approximations to the addition model. Nonlinearity **12**, 229–239 (1999)
125. Laurençot, Ph.: On a class of continuous coagulation-fragmentation equations. J. Differ. Equ. **167**, 245–274 (2000)
126. Laurençot, Ph.: The discrete coagulation equations with multiple fragmentation. Proc. Edinb. Math. Soc. **45**, 67–82 (2002)
127. Laurençot, Ph.: Convergence to self-similar solutions for a coagulation equation. Z. Angew. Math. Phys. **56**, 398–411 (2005)
128. Laurençot, Ph.: Self-similar solutions to a coagulation equation with multiplicative kernel. Physica D **222**, 80–87 (2006)
129. Laurençot, Ph., Mischler, S.: The continuous coagulation-fragmentation equations with diffusion. Arch. Rational Mech. Anal. **162**, 45–99 (2002)
130. Laurençot, Ph., Mischler, S.: From the Becker-Döring to the Lifshitz-Slyozov-Wagner equations. J. Stat. Phys. **106**, 957–991 (2002)
131. Laurençot, Ph., Mischler, S.: From the discrete to the continuous coagulation-fragmentation equations. Proc. R. Soc. Edinb. **132A**, 1219–1248 (2002)
132. Laurençot, Ph., Mischler, S.: Global existence for the discrete diffusive coagulation-fragmentation equations in L^1. Rev. Mat. Iberoam. **18**, 731–745 (2002)
133. Laurençot, Ph., Mischler, S.: On coalescence equations and related models. In: Degond, P., Pareschi, L., Russo, G. (eds.) Modelling and Computational Methods for Kinetic Equations, pp. 321–356. Birkhäuser, Boston (2004)
134. Laurençot, Ph., Mischler, S.: Liapunov functional for Smoluchovski's coagulation equation and convergence to self-similarity. Monat. Math. **146**, 127–142 (2005)
135. Laurençot, P., van Roessel, H.: Nonuniversal self-similarity in a coagulation-annihilation model with constant kernels. J. Phys. A Math. Theor. **43**, 455210 (2010)
136. Laurençot, Ph., Walker, Ch.: Well-posedness for a model of prion proliferation dynamics. J. Evol. Equ. **7**, 241–264 (2006)
137. Laurençot, Ph., Wrzosek, D.: The Becker-Döring model with diffusion. I. Basic properties of solutions. Colloq. Math. **75**, 245–269 (1998)
138. Laurençot, Ph., Wrzosek, D.: The Becker-Döring model with diffusion. II. Long time behaviour. J. Differ. Equ. **148**, 268–291 (1998)
139. Laurençot, Ph., Wrzosek, D.: The discrete coagulation equation with collisional breakage. J. Stat. Phys. **104**, 193–253 (2001)
140. Lê Châu-Hoàn, Etude de la classe des opérateurs m-accrétifs de $L^1(\Omega)$ et accrétifs dans $L^\infty(\Omega)$. Thèse de troisième cycle, Université de Paris VI, Paris (1977)
141. Lécot, C., Wagner, W.: A quasi-Monte Carlo scheme for Smoluchowski's coagulation equation. Math. Comput. **73**, 1953–1966 (2004)
142. Lee, M.H.: A survey of numerical solutions to the coagulation equation. J. Phys. A Math. Gen. **34**, 10219–10241 (2001)
143. Levin, L., Sedunov, Yu.S.: A kinetic equation describing microphysical processes in clouds. Dokl. Akad. Nauk SSSR **170**, 4–7 (1966)
144. Leyvraz, F.: Existence and properties of pos-gel solutions for the kinetic equations of coagulation. J. Phys. A Math. Gen. **18**, 321–326 (1985)
145. Leyvraz, F.: Scaling theory and exactly solved models in the kinetics of irreversible aggregation. Phys. Rep. **383**, 95–212 (2003)
146. Leyvraz, F.: Rigorous results in the scaling theory of irreversible aggregation kinetics. J. Nonlinear Math. Phys. **12**(Suppl. 1), 449–465 (2005)
147. Leyvraz, F.: Scaling theory for gelling systems: work in progress. Physica D **222**, 21–28 (2006)

148. Leyvraz, F., Tschudi, H.R.: Singularities in the kinetics of coagulation processes. J. Phys. A Math. Gen. **14**, 3389–3405 (1981)
149. Lifshitz, I.M., Slyozov, V.V.: The kinetics of precipitationfrom supersaturated solid solutions. J. Phys. Chem. Solids **19**, 35–50 (1961)
150. Lushnikov, A.A.: Evolution of coagulating systems: II. Asymptotic size distributions and analytical properties of generating functions. J. Coll. Interf. Sci. **48**, 400–409 (1974)
151. Matsoukas, T., Friedlander, S.K.: Dynamics of aerosol agglomerate formation. J. Coll. Interf. Sci. **146**, 495–506 (1991)
152. McGrady, E.D., Ziff, R.M.: "Shattering" transition in fragmentation. Phys. Rev. Lett. **58**, 892–895 (1987)
153. McLaughlin, D.J., Lamb, W., McBride, A.C.: A semigroup approach to fragmentation models. SIAM J. Math. Anal. **28**, 1158–1172 (1997)
154. McLeod, J.B.: On an infinite set of non-linear differential equations. Q. J. Math. Oxford (2) **13**, 119–128 (1962)
155. McLeod, J.B.: On an infinite set of non-linear differential equations (II). Q. J. Math. Oxford (2) **13**, 193–205 (1962)
156. McLeod, J.B.: On a recurrence formula in differential equations. Q. J. Math. Oxford (2) **13**, 283–284 (1962)
157. McLeod, J.B., Niethammer, B., Velázquez, J.J.L.: Asymptotics of self-similar solutions to coagulation equations with product kernel. J. Stat. Phys. **144**, 76–100 (2011)
158. Melzak, Z.A.: A scalar transport equation. Trans. Am. Math. Soc. **85**, 547–560 (1957)
159. Menon, G., Pego, R.L.: Approach to self-similarity in Smoluchowski's coagulation equations. Commun. Pure Appl. Math. **57**, 1197–1232 (2004)
160. Menon, G., Pego, R.L.: Dynamical scaling in Smoluchowski's coagulation equations: uniform convergence. SIAM J. Math. Anal. **36**, 1629–1651 (2005)
161. Menon, G., Pego, R.L.: The scaling attractor and ultimate dynamics for Smoluchowski's coagulation equations. J. Nonlinear Sci. **18**, 143–190 (2008)
162. Menon, G., Niethammer, B., Pego, R.L.: Dynamics and self-similarity in min-driven clustering. Trans. Am. Math. Soc. **362**, 6591–6618 (2010)
163. Morgenstern, D.: Analytical studies related to the Maxwell-Boltzmann equation. J. Ration. Mech. Anal. **4**, 533–555 (1955)
164. Müller, H.: Zur allgemeinen Theorie der raschen Koagulation. Kolloidchemische Beihefte **27**, 223–250 (1928)
165. Nagai, T., Kawasaki, K.: Statistical dynamics of interactiong kinks II. Physica A **134**(3), 483–521 (1986)
166. Niethammer, B.: On the evolution of large clusters in the Becker-Döring model. J. Nonlinear Sci. **13**, 115–155 (2003)
167. Niethammer, B.: A scaling limit of the Becker-Döring equations in the regime of small excess density. J. Nonlinear Sci. **14**, 453–468 (2004)
168. Niethammer, B.: Macroscopic limits of the Becker-Döring equations. Commun. Math. Sci. **2**(Suppl. 1), 85–92 (2004)
169. Niethammer, B., Velázquez, J.J.L.: Self-similar solutions with fat tails for a coagulation equation with diagonal kernel. C.R. Acad. Sci. Paris Ser. I **349**, 559–562 (2011)
170. Niethammer, B., Velázquez, J.J.L.: Optimal bounds for self-similar solutions to coagulation equations with product kernel (11 February 2011). arXiv:1010.1857v2
171. Niethammer, B., Velázquez, J.J.L.: Self-similar solutions with fat tails for Smoluchowski's coagulation equations with locally bounded kernels. Commun. Math. Phys. **318**(2), 505–532 (2013) (erratum: same volume 533–534)
172. Niethammer, B., Velázquez, J.J.L.: Uniqueness of self-similar solutions to Smoluchowski's coagulation equations for kernels that are close to constant (18 September 2013). arXiv:1309.4621v1
173. Niethammer, B., Velázquez, J.J.L.: Exponential tail behaviour of self-similar solutions to Smoluchowski's coagulation equation (17 October 2013). arXiv:1310.4732v1
174. Niwa, H.-S.: School size statistics of fish. J. Theor. Biol. **195**, 351–361 (1998)

175. Norris, J.R.: Smoluchowski's coagulation equation: uniqueness, nonuniqueness and a hydrodynamic limit for the stochastic coalescent. Ann. Appl. Prob. **9**, 78–109 (1999)
176. Norris, J.R.: Notes on Brownian coagulation. Markov Process. Relat. Fields **12**, 407–412 (2006)
177. Oort, J.H., van de Hulst, H.C.: Gas and smoke in interstellar space. Bull. Astron. Inst. Neth. **10**, 187–204 (1946)
178. Oshanin, G.S., Burlatsky, S.F.: Fluctuation-induced kinetics of reversible coagulation. J. Phys. A Math. Gen. **22**, L973–L976 (1989)
179. Pego, R.L.: Lectures on dynamics in models of coarsening and coagulation. In: Bao, W., Liu, J.-G. (eds.) Dynamics in Models of Coarsening, Coagulation, Condensation and Quantization. Lecture Notes Series, Institute for Mathematical Sciences, National University of Singapore, vol. 9, pp. 1–61. World Scientific, Singapore (2007)
180. Penrose, O.: Metastable states for the Becker-Döring cluster equations. Commun. Math. Phys. **124**, 515–541 (1989)
181. Penrose, O.: The Becker-Döring equations at large times and their connection with the LSW theory of coarsening. J. Stat. Phys. **89**, 305–320 (1997)
182. Penrose, O., Lebowitz, J.L.: Towards a rigorous molecular theory of metastability. In: Montroll, E.W., Lebowitz, J.L. (eds.) Studies in Statistical Mechanics VII: Fluctuation Phenomena, pp. 321–375. North-Holland, Amesterdam (1987)
183. Penrose, O., Lebowitz, J.L., Marro, J., Kalos, M.H., Sur, A.: Growth of clusters in a first-order phase transition. J. Stat. Phys. **19**(3), 243–267 (1978)
184. Perthame, B.: Transport Equations in Biology. Frontiers in Mathematics. Birkhäuser Verlag, Basel (2007)
185. Pesz, K., Rodgers, G.J.: Kinetics of growing and coalescing droplets. J. Phys. A Math. Gen. **25**, 705–713 (1992)
186. Piskunov, V.N., Petrov, A.M.: Condensation/coagulation kinetics for mixture of liquid and solid particles: analytical solutions. Aerosol Sci. **33**, 647–657 (2002)
187. Pöschel, T., Brilliantov, N.V., Frömmel, C.: Kinetics of prion growth. Biophys. J. **85**, 3460–3474 (2003)
188. Pruppacher, H.R., Klett, J.D.: Microphysics of Clouds and Precipitation. Atmospheric and Oceanographic Sciences Library, vol. 18, 2nd edn. Kluwer, Dordrecht (1997)
189. Ranjbar, M., Adibi, H., Lakestani, M.: Numerical solution of homogeneous Smoluchowski's coagulation equation. Int. J. Comput. Math. **87**(9), 2113–2122 (2010)
190. Rao, M.M., Ren, Z.D.. Theory of Orlicz Spaces. Pure and Applied Mathematics, vol. 146. Marcel Dekker, New York (1991)
191. Redner, S., Ben-Avraham, D., Kahng, B.: Kinetics of 'cluster eating'. J. Phys. A Math. Gen. **20**, 1231–1238 (1987)
192. Rezakhanlou, F.: The coagulating brownian particles and Smoluchowski's equation. Markov Process. Relat. Fields **12**, 425–445 (2006)
193. Roquejoffre, J.-M., Villedieu, Ph.: A kinetic model for droplet coalescence in dense sprays. Math. Meth. Models Appl. Sci. **11**, 867–882 (2001)
194. Safronov, V.: Evolution of the Protoplanetary Cloud and Formation of the Earth and the Planets. Israel Program for Scientific Translations, Jerusalem (1972)
195. Sasportes, R.: Long time behaviour and self similarity in an addition model with slow input of monomers. In: Bourguignon, J.P. (eds.) Mathematics of Energy and Climate Change. International Conference and Advanced School Planet Earth, Portugal, March 21-28, 2013. Springer, Heidelberg (2015)
196. Scott, W.T.: Analytic studies of cloud droplet coalescence I. J. Atmos. Sci. **25**, 54–65 (1968)
197. Simha, R.: Kinetics of degradation and size distribution of long chain polymers. J. Appl. Phys. **12**, 569–578 (1941)
198. Simonett, G., Walker, Ch.: On the solvability of a mathematical model for prion proliferation. J. Math. Anal. Appl. **234**, 580–603 (2006)

199. Slemrod, M.: Coagulation-diffusion systems: derivation and existence of solutions for the diffuse interface structure equations. Physica D **46**, 351–366 (1990)
200. Slemrod, M.: A note on the kinetic equations of coagulation. J. Integr. Equ. Appl. **3**, 167–173 (1991)
201. Slemrod, M.: Metastable fluid flow described via a discrete-velocity coagulation-fragmentation model. J. Stat. Phys. **83**, 1067–1108 (1996)
202. Slemrod, M.: The Becker-Döring equation. In: Bellomo, N., Pulvirenti, M. (eds.) Modelling in Applied Sciences, A Kinetic Theory Approach; Modelling and Simulation in Science, Engineering and Technology, pp. 149–171. Birkhäuser, Boston (2000)
203. Slemrod, M., Qi, A., Grinfeld, M., Stewart, I.: A discrete velocity coagulation-fragmentation model. Math. Meth. Appl. Sci. **18**, 959–993 (1995)
204. von Smoluchowski, M.: Drei Vorträge über Diffusion, Brownsche Molekularbewegung und Koagulation von Kolloidteilchen. Physik. Zeitschr. **17**,557–571, 587–599 (1916)
205. von Smoluchowski, M.: Versuch einer mathematischen Theorie der Koagulationskinetik kolloider Lösungen. Zeitschrift f. physik. Chemie **92**, 129–168 (1917)
206. Spouge, J.: An existence theorem for the discrete coagulation-fragmentation equations. Math. Proc. Camb. Phil. Soc. **96**, 351–357 (1984)
207. Srinivasan, R.: Rates of convergence for Smoluchowski's coagulation equation. SIAM J. Math. Anal. **43**(4), 1835–1854 (2011)
208. Srivastava, R.C.: Parametrization of raindrop size distributions. J. Atmos. Sci. **35**, 108–117 (1978)
209. Srivastava, R.C., A simple model of particle coalescence and breakup. J. Atmos. Sci. **39**, 1317–1322 (1982)
210. Stewart, I.W.: A global existence theorem for the general coagulation-fragmentation equation with unbounded kernels. Math. Meth. Appl. Sci. **11**, 627–648 (1989)
211. Stewart, I.W.: A uniqueness theorem for the coagulation-fragmentation equation. Math. Proc. Camb. Phil. Soc. **107**, 573–578 (1990)
212. Straube, R., Falcke, M.: Reversible clustering under the influence of a periodically modulated biding rate. Phys. Rev. E **76**, 010402(R) (2007)
213. Vigil, R.D., Vermeersch, I., Fox, R.O.: Destructive aggregation: aggregation with collision-induced breakage. J. Colloid Interf. Sci. **302**, 149–158 (2006)
214. Wagner, C.: Theorie der alterung von niederschlägen durch umlösen. Z. Electrochemie **65**, 581–594 (1961)
215. Walker, Ch.: Coalescence and breakage processes. Math. Meth. Appl. Sci. **25**, 729–748 (2002)
216. Walker, Ch.: Prion proliferation with unbounded polymerizaton rates. Electr. J. Differ. Equ. **15**, 387–397 (2007)
217. Wang, C., Friedlander, S.K.: The self-preserving particle size distribution for coagulation by Brownian motion. J. Coll. Interf. Sci. **22**, 126–132 (1966)
218. Wattis, J.A.D.: Similarity solutions of a Becker-Döring system with time-dependent monomer input. J. Phys. A Math. Gen. **37**, 7823–7841 (2004)
219. Wattis, J.A.D.: An introduction to mathematical models of coagulation-fragmentation processes: a discrete deterministic mean-field approach. Physica D **222**, 1–20 (2006)
220. Wattis, J.A.D.: Exact solutions for cluster-growth kinetics with evolving size and shape profiles. J. Phys. A Math. Gen. **39**, 7283–7298 (2006)
221. Webb, G.F.: Theory of Nonlinear Age-Dependent Population Dynamics. Marcel Dekker, New York (1985)
222. Wilkins, D.: A geometrical interpretation of the coagulation equation. J. Phys. A Math. Gen. **15**, 1175–1178 (1982)
223. White, W.H.: A global existence theorem for Smoluchowski's coagulation equations. Proc. Am. Math. Soc. **80**, 273–276 (1980)
224. Wrzosek, D.: Existence and uniqueness for the discrete coagulation-fragmentation model with diffusion. Topol. Meth. Nonlin. Anal. **9**, 279–296 (1997)

225. Wrzosek, D.: Mass-conserving solutions to the discrete coagulation-fragmentation with diffusion. Nonlinear Anal. **49**, 297–314 (2002)
226. Yıldırım, A., Koçak, H.: Series solution of the Smoluchowski's coagulation equation. J. King Saud Univ. - Science **23**(2), 183–189 (2011)
227. Ziff, R.M., McGrady, E.D.: Kinetics of polymer degradation. Macromolecules **19**, 2513–2519 (1986)
228. Ziff, R.M., Stell, G.: Kinetics of polymer gelation. J. Chem. Phys. **73**(7), 3492–3499 (1980)

Resampling-Based Methodologies in Statistics of Extremes: Environmental and Financial Applications

M. Ivette Gomes, Lígia Henriques-Rodrigues, and Fernanda Figueiredo

Abstract Resampling computer intensive methodologies, like the *jackknife* and the *bootstrap* are important tools for a reliable semi-parametric estimation of parameters of extreme or even rare events. Among these parameters we mention the *extreme value index*, ξ, the primary parameter in *statistics of extremes*. Most of the semi-parametric estimators of this parameter show the same type of behaviour: nice asymptotic properties, but a high variance for small k, the number of upper order statistics used in the estimation, a high bias for large k, and the need for an adequate choice of k. After a brief reference to some estimators of the aforementioned parameter and their asymptotic properties we present an algorithm that deals with an adaptive reliable estimation of ξ. Applications of these methodologies to the analysis of environmental and financial data sets are undertaken.

1 A Brief Introduction

Let us assume that we have access to a sample $(X_1, \ldots, X_n)$ of independent, identically distributed (i.i.d.), or even stationary and weakly dependent, random variables (r.v.'s) from an underlying model F, and let us denote by $(X_{1:n} \leq \cdots \leq X_{n:n})$ the sample of associated ascending order statistics (o.s.'s). Let us further assume that it is possible to normalize the sequence of maximum values, $\{X_{n:n}\}_{n\geq 1}$ so that we get a non-degenerate limit. Then (Gnedenko [16]), that limiting r.v. has a distribution function (d.f.) of the type of the *general extreme value* (GEV) d.f.,

M. Ivette Gomes (✉)
Universidade de Lisboa, FCUL, DEIO, and CEAUL, Lisboa, Portugal
e-mail: ivette.gomes@fc.ul.pt

L. Henriques-Rodrigues
Universidade de São Paulo, IME, and CEAUL, São Paulo, Brasil
e-mail: ligiahr@ime.usp.br

F. Figueiredo
Universidade do Porto, FEP, and CEAUL, Porto, Portugal
e-mail: otilia@fep.up.pt

J.-P. Bourguignon et al. (eds.), *Mathematics of Energy and Climate Change*,
CIM Series in Mathematical Sciences 2, DOI 10.1007/978-3-319-16121-1_6

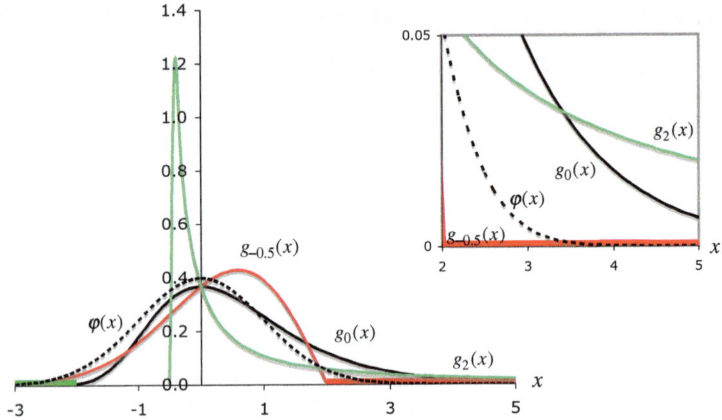

Fig. 1 Probability density function (p.d.f.) $g_\xi(x) = dG_\xi(x)/dx$, for $\xi = -0.5$, $\xi = 0$ and $\xi = 2$, together with the normal p.d.f., $\varphi(x) = \exp(-x^2/2)/\sqrt{2\pi}$, $x \in \mathbb{R}$

given by

$$G_\xi(x) = \begin{cases} \exp\left(-(1+\xi x)^{-1/\xi}\right), \ 1+\xi x > 0, & \text{if } \xi \neq 0, \\ \exp(-\exp(-x)), \ x \in \mathbb{R}, & \text{if } \xi = 0, \end{cases} \tag{1}$$

and ξ is the so-called *extreme value index* (EVI), the primary parameter in *statistics of univariate extremes* (SUE). We then say that F is in the max-domain of attraction of G_ξ, in (1), and use the notation $F \in \mathscr{D}_{\mathscr{M}}\left(G_\xi\right)$.

The *extreme value index* ξ measures essentially the weight of the right tail-function $\overline{F} := 1 - F$, as illustrated in Fig. 1.

- If $\xi < 0$, the right tail is light, and F has a finite *right endpoint*, i.e. $x^F := \sup\{x : F(x) < 1\} < +\infty$;
- If $\xi > 0$, the right tail is heavy, of a negative polynomial type, and F has an infinite *right endpoint*;
- If $\xi = 0$, the right tail is of an exponential type. The *right endpoint* can then be either finite or infinite.

Slightly more restrictively than the full max-domain of attraction of the GEV d.f., we now consider a positive EVI, i.e. we work with heavy-tailed models F in $\mathscr{D}_{\mathscr{M}}\left(G_\xi\right)_{\xi>0} =: \mathscr{D}^+_{\mathscr{M}}$. As usual, we shall further use the notations, $F^{\leftarrow}$ for the *generalized inverse* function of F, i.e. $F^{\leftarrow}(y) := \inf\{x : F(x) \geq y\}$, and $\mathscr{R}_{\mathrm{a}}$ for the class of *regularly varying* functions at infinity with an index of regular variation a, i.e. positive Borel measurable functions $g(\cdot)$ such that $g(tx)/g(t) \to x^{\mathrm{a}}$, as $t \to \infty$, for all $x > 0$. Let us further use the notation

$$U(t) := \left(\frac{1}{1-F}\right)^{\leftarrow}(t) = F^{\leftarrow}(1 - 1/t),$$

for the tail quantile function, defined for $t > 1$.

Equivalently to say that $F \in \mathscr{D}^+_{\mathscr{M}}$, we can say (Gnedenko [16]) that the tail function

$$\overline{F} := 1 - F$$

belongs to $\mathscr{R}_{-1/\xi}$ or that $U \in \mathscr{R}_\xi$ (de Haan [6]), i.e. for heavy-tailed models we have the validity of the so-called *first-order conditions*,

$$F \in \mathscr{D}^+_{\mathscr{M}} \quad \Longleftrightarrow \quad \overline{F} \in \mathscr{R}_{-1/\xi} \quad \Longleftrightarrow \quad U \in \mathscr{R}_\xi. \tag{2}$$

For these heavy-tailed models, and given a sample $\underline{\mathbf{X}}_n = (X_1, \ldots, X_n)$, the classical EVI-estimators are Hill estimators (Hill [34]), with the functional expression

$$H_{k,n} \equiv H(k; \underline{\mathbf{X}}_n) := \frac{1}{k} \sum_{i=1}^{k} V_{ik},$$
$$V_{ik} := \ln X_{n-i+1:n} - \ln X_{n-k:n}, \quad 1 \le i \le k < n. \tag{3}$$

The Hill EVI-estimators are thus the average of the k log-excesses above a random level $X_{n-k:n}$, that compulsory needs to be an *intermediate* o.s., i.e.

$$k = k_n \to \infty \quad \text{and} \quad k/n \to 0, \text{ as } n \to \infty, \tag{4}$$

so that we have consistent EVI-estimation in the whole $\mathscr{D}^+_{\mathscr{M}}$.

Under adequate second-order conditions that rule the rate of convergence in any of the first-order conditions in (2), Hill estimators, $H_{k,n}$, in (3), have usually a high asymptotic bias, i.e., $\sqrt{k}\,(H_{k,n} - \xi)$ is asymptotically normal with variance ξ^2 and a non-null mean value for the moderate k-values that lead to minimal mean square error (MSE), as sketched in Sect. 2.2. This non-null asymptotic bias and a rate of convergence of the order of $1/\sqrt{k}$ lead to sample paths with a high variance for small k, a high bias for large k, and a very peaked MSE pattern. Recently, several authors have considered different ways of reducing bias in the area of SUE (see the overviews in Gomes et al. [24], Chap. 6 of Reiss and Thomas [36]; Gomes et al. [25]; Beirlant et al. [3]). A simple class of *minimum-variance reduced-bias* (MVRB) EVI-estimators is the class studied in Caeiro et al. [4], to be introduced in Sect. 2.1. These MVRB EVI-estimators depend on the estimation of second-order parameters, and their asymptotic behaviour is presented in Sect. 2.2. Both the Hill and the MVRB EVI-estimators are invariant to changes in scale, but they are not invariant to changes in location. And particularly the Hill EVI-estimators can suffer drastic changes when we induce an arbitrary shift in the data. This was one of the reasons that led Araújo Santos et al. [1] to introduce the so-called *peaks over random threshold* (PORT) methodology, to be sketched in Sect. 2.3.

Resampling methodologies, introduced in Sect. 3, have recently revealed to be quite fruitful in the field of SUE. We mention the importance of the *generalized jackknife* (GJ), detailed in Gray and Schucany [33], in the reduction of bias, revisited recently in the field of extremes by Gomes et al. [32]. We further refer the relevance of the *bootstrap* (Efron [11]) in the estimation of a crucial tuning parameter in the area, the number k of top order statistics involved in the estimation of the tails. Together, these two resampling procedures enable the obtention of reliable semi-parametric estimates of any parameter of extreme or even rare events, like a *high quantile*, the *expected shortfall*, the *return period* of a high level or the two primary parameters of extreme events, the *extreme value index* (EVI) and the *extremal index*, related to the degree of local dependence in the extremes of a stationary sequence. In order to illustrate such topics, we essentially consider the GJ EVI-estimators in Gomes et al. [32], associated with the simplest class of MVRB estimators of a positive EVI introduced and studied in Caeiro et al. [4].

In Sect. 4, an application of these methodologies to the analysis of an environmental data set, related to the number of hectares, exceeding 100 ha, burnt during wildfires recorded in Portugal during 14 years (1990–2003), is undertaken. To enhance the relevance of the PORT methodology, we further consider an application to financial data.

2 Second-Order Reduced-Bias (SORB), MVRB and PORT EVI-Estimators

As mentioned above, for consistent semi-parametric EVI-estimation, in the whole $\mathscr{D}^+_{\mathscr{M}}$, we merely need to work with adequate functionals, dependent on an *intermediate tuning* parameter k, the number of top o.s.'s involved in the estimation, i.e. (4) should hold. To obtain full information on the non-degenerate asymptotic behaviour of semi-parametric EVI-estimators, we need further assuming a *second-order condition*, ruling the rate of convergence in the *first-order condition*, or even a *third* or *fourth-order condition*. Whenever dealing with reduced-bias estimators of parameters of extreme events, like the EVI, and essentially due to technical reasons, we slightly restrict the domain of attraction, $\mathscr{D}^+_{\mathscr{M}}$, and consider a Pareto-type class of models, assuming that, with $C > 0, \xi > 0, \rho < 0$, and $\beta \neq 0$,

$$U(t) = Ct^{\xi}\Big(1 + A(t)/\rho + o(t^{\rho})\Big), \quad A(t) := \xi\beta t^{\rho}, \tag{5}$$

as $t \to \infty$, i.e. we assume that the slowly varying function $L_U(t) = t^{-\xi}U(t)$ tends to a finite non-null constant. To obtain information on the bias of MVRB EVI-estimators it is even common to slightly restrict our class of models, further assuming the following third-order condition,

$$U(t) = Ct^{\xi}\Big(1 + A(t)/\rho + \beta' t^{2\rho} + o(t^{2\rho})\Big), \tag{6}$$

as $t \to \infty$, with $\beta' \neq 0$. And if we deal with GJ-MVRB EVI-estimators, to be detailed in Sect. 3.2, and also want to obtain full information on their asymptotic bias, we can further assume, in the lines of Taylor series, that

$$U(t) = Ct^{\xi}\Big(1 + A(t)/\rho + \beta' t^{2\rho} + \beta'' t^{3\rho} + o(t^{3\rho})\Big), \tag{7}$$

as $t \to \infty$, with $\beta'' \neq 0$.

More generally than (5), it is often assumed that there exists a function $A(\cdot)$, such that

$$\lim_{t\to\infty} \frac{\ln U(tx) - \ln U(t) - \xi \ln x}{A(t)} = \psi_\rho(x) := \begin{cases} (x^\rho - 1)/\rho, & \text{if } \rho \neq 0, \\ \ln x, & \text{if } \rho = 0. \end{cases} \tag{8}$$

Then, we compulsory have $|A| \in \mathscr{R}_\rho$. Moreover, if the limit in the left hand-side of (8) exists, it is compulsory equal to the above defined $\psi_\rho(\cdot)$ function (Geluk and de Haan [15]). Further note that the validity of (8) with $\rho < 0$ is equivalent to (5). Additional details on second and higher-order conditions can be found in de Haan and Ferreira [7].

As mentioned above, and provided that (4) and (8) hold, Hill EVI-estimators, $H_{k,n}$, have usually a high asymptotic bias. The adequate accommodation of this bias has recently been extensively addressed. We mention the pioneering papers by Peng [35], Beirlant et al. [2], Feuerverger and Hall [12], and Gomes et al. [20], among others. In these papers, authors are led to SORB EVI-estimators, with asymptotic variances larger than or equal to $(\xi\ (1-\rho)/\rho)^2$, where $\rho(<0)$ is the aforementioned 'shape' second-order parameter, ruling the rate of convergence of the distribution of the normalized sequence of maximum values towards the limiting law G_ξ, in (1).

2.1 MVRB EVI-Estimation

Recently, Caeiro et al. [4] and Gomes et al. [23, 27] have been able to *reduce the bias without increasing the asymptotic variance*, kept at ξ^2, just as happens with the Hill EVI-estimators. Those estimators, called MVRB EVI-estimators, are all based on an adequate 'external' and a bit more than consistent estimation of the pair of second-order parameters, $(\beta, \rho) \in (\mathbb{R}, \mathbb{R}^-)$, in (5), done through adequate estimators denoted by $(\hat\beta, \hat\rho)$, and outperform the classical estimators for all k. Different algorithms for the estimation of (β, ρ) can be found in Gomes and Pestana [19], among others.

Among the most common MVRB EVI-estimators, we now consider the class in Caeiro et al. [4], used for Value-at-Risk (VaR) estimation in the aforementioned seminal paper by Gomes and Pestana [19]. Such a class, denoted by $\overline{H} \equiv \overline{H}_{k,n}$, has the functional form

$$\overline{H}_{k,n} \equiv \overline{H}_{\hat{\beta},\hat{\rho}}(k;\underline{\mathbf{X}}_n) := H_{k,n}\Big(1 - \hat{\beta}(n/k)^{\hat{\rho}}/(1-\hat{\rho})\Big), \tag{9}$$

where $(\hat{\beta},\hat{\rho})$ is an adequate consistent estimator of (β,ρ), with $\hat{\beta}$ and $\hat{\rho}$ based on a number of top o.s.'s k_1 usually of a higher order than the number of top o.s.'s k used in the EVI-estimation. Further details on such estimation are given in Sect. 3.3.

2.2 A Brief Asymptotic Comparison of Classical and MVRB EVI-Estimators

The Hill estimators reveal usually a high asymptotic bias. Indeed, from the results of de Haan and Peng [8], and with $\mathcal{N}_{\mu,\sigma^2}$ denoting a normal r.v. with mean value μ and variance σ^2,

$$\sqrt{k}\,(H_{k,n} - \xi) \stackrel{d}{=} \mathcal{N}_{0,\xi^2} + b_H\sqrt{k}A(n/k) + o_p\big(\sqrt{k}A(n/k)\big), \tag{10}$$

where the bias $b_H\sqrt{k}A(n/k)$ can be very large, moderate or small (i.e. go to ∞, constant or 0) as $n \to \infty$. Under the same conditions as before, $\sqrt{k}\,\big(\overline{H}_{k,n} - \xi\big)$ is asymptotically normal with variance also equal to ξ^2 but with a null mean value. Indeed, under the validity of the aforementioned third-order condition in (6), related to Pareto-type class of models, we can then adequately estimate the vector of second-order parameters, (β,ρ) so that $\overline{H}_{k,n}$ outperforms $H_{k,n}$ for all k. Indeed, we can write (Caeiro et al. [5])

$$\sqrt{k}\,\big(\overline{H}_{k,n} - \xi\big) \stackrel{d}{=} \mathcal{N}_{0,\xi^2} + b_{\overline{H}}\sqrt{k}A^2(n/k) + o_p\big(\sqrt{k}A^2(n/k)\big). \tag{11}$$

And when we try answering the question whether it is still possible to improve the performance of these MVRB EVI-estimators through the use of resampling methods, we are led to a positive answer, as provided in Sect. 3.2.

2.3 PORT EVI-Estimation

The estimators in (3) and (9) are scale invariant but not location invariant. In order to achieve location invariance, Araújo Santos et al. [1] introduced the so-called PORT EVI-estimators, functionals of a sample of excesses over a random level $X_{n_q:n}$, $n_q :=$

$\lfloor nq \rfloor + 1$, with $\lfloor x \rfloor$ denoting the integer part of x, i.e. functionals of the sample,

$$\underline{\mathbf{X}}_n^{(q)} := \left(X_{n:n} - X_{n_q:n}, \ldots, X_{n_q+1:n} - X_{n_q:n}\right). \tag{12}$$

Generally, we can have $0 < q < 1$, for any $F \in \mathscr{D}_{\mathscr{M}}^{+}$ (*the random level is an empirical quantile*). If the underlying model F has a finite *left endpoint*, $x_F := \inf\{x : F(x) \geq 0\}$, we can also use $q = 0$ (*the random level can then be the minimum*).

If we think, for instance, on Hill EVI-estimators, in (3), the new classes of PORT-Hill EVI-estimators, theoretically studied in Araújo Santos et al. [1], and for finite samples in Gomes et al. [26], are given by

$$\begin{aligned} H_{k,n}^{(q)} &:= H(k; \underline{\mathbf{X}}_n^{(q)}) \\ &= \frac{1}{k} \sum_{i=1}^{k} \left\{ \ln \frac{X_{n-i+1:n} - X_{n_q:n}}{X_{n-k:n} - X_{n_q:n}} \right\}, \quad 0 \leq q < 1. \end{aligned} \tag{13}$$

Similarly, if we think on the MVRB EVI-estimators, in (9), the new classes of PORT-MVRB EVI-estimators, studied for finite samples in Gomes et al. [28, 31], are given by

$$\begin{aligned} \overline{H}_{k,n}^{(q)} &:= \overline{H}_{\hat{\beta}_q, \hat{\rho}_q}(k; \underline{\mathbf{X}}_n^{(q)}) \\ &= H_{k,n}^{(q)} \left(1 - \hat{\beta}(n_q/k)^{\hat{\rho}}/(1-\hat{\rho})\right), \quad 0 \leq q < 1, \end{aligned} \tag{14}$$

with $H_{k,n}^{(q)}$ in (13), and $\hat{\beta} \equiv \hat{\beta}_q := \hat{\beta}(\underline{\mathbf{X}}_n^{(q)})$, $\hat{\rho} \equiv \hat{\rho}_q := \hat{\rho}(\underline{\mathbf{X}}_n^{(q)})$ any adequate estimator of (β, ρ) based on the sample $\underline{\mathbf{X}}_n^{(q)}$, in (12).

These PORT EVI-estimators are thus dependent on a *tuning parameter* q, $0 \leq q < 1$, that makes them highly flexible. Moreover, they are invariant to changes in both location and scale. We shall further use the notation $X_{n+1:n} \equiv 0$, and work with $0 \leq q \leq 1$, so that with $\overline{H}$ and $\overline{H}^{(q)}$, given in (9) and (14), respectively, we can consider that $\overline{H} = \overline{H}^{(q)}$ for $q = 1$ ($n_1 = n + 1, \hat{\beta}_1 = \hat{\beta},\ \hat{\rho}_1 = \hat{\rho}$).

We get to know that the second-order MVRB EVI-estimators in (9) are not location invariant, but they are approximately location invariant. Almost equivalent to the PORT-MVRB EVI-estimators in (14), we can consider, in the lines of Figueiredo et al. [13], quasi-PORT-MVRB EVI-estimators, with a functional expression similar to the one in (14) but where for all $0 \leq q < 1$, $(\hat{\beta}, \hat{\rho}) = (\hat{\beta}_1, \hat{\rho}_1)$ are the (β, ρ)-estimators based on the original sample, i.e.

$$\overline{\overline{H}}_{k,n}^{(q)} := H_{k,n}^{(q)} \overline{H}_{k,n} / H_{k,n} = H_{k,n}^{(q)} \left(1 - \hat{\beta}(n/k)^{\hat{\rho}}/(1-\hat{\rho})\right), \tag{15}$$

with $H_{k,n}$, $\overline{H}_{k,n}$ and $H_{k,n}^{(q)}$ given in (3), (9) and (13), respectively.

3 Resampling Methodologies in SUE

The use of resampling methodologies (Efron [11]) has revealed to be promising in the estimation of the nuisance parameter k, or equivalently, in the estimation of the optimal sample fraction (OSF), k/n, as well as in the reduction of bias of any estimator of a parameter of extreme events. If we ask how to choose the tuning parameter k in the EVI-estimation, either through $H_{k,n}$ or $\overline{H}^{(q)}_{k,n}$ or $\overline{\overline{H}}^{(q)}_{k,n}$, $0 \leq q \leq 1$, generally denoted $E_{k,n}$, we usually consider the estimation of

$$k_{0|E}(n) := \arg\min_k \text{MSE}(E_{k,n}). \tag{16}$$

3.1 OSF-Estimation and the Bootstrap Methodology

To obtain estimates of $k_{0|E}(n)$, in (16), one can use a *double-bootstrap* method applied to an adequate *auxiliary statistic* like

$$T_{k,n} \equiv T_{k,n|E} := E_{\lfloor k/2 \rfloor,n} - E_{k,n}, \quad k = 2, \ldots, n-1, \tag{17}$$

which tends to the well-known value zero and has an asymptotic behaviour similar to the one of $E_{k,n}$ (see Gomes and Oliveira [18], among others, for the estimation through $H_{k,n}$ and Gomes et al. [30], for the estimation through MVRB EVI-estimators). See also Sect. 3.3 of this article. At such optimal levels, we have a non-null asymptotic bias, and if we still want to remove such a bias, we can then make use of the GJ methodology.

3.2 The GJ Methodology and Bias Reduction

The main objectives of the *jackknife methodology* are:

1. Bias and variance estimation of a certain statistic, only through manipulation of observed data $\underline{x}$.
2. The building of estimators with bias and MSE smaller than those of an initial set of estimators.

The jackknife or the GJ are resampling methodologies, which usually give a positive answer to the question: '*May the combination of information improve the quality of estimators of a certain parameter or functional?*' The pioneering SORB EVI-estimators are, in a certain sense, *generalized jackknife* estimators, i.e., affine combinations of well-known estimators of ξ.

The *generalized jackknife* statistic was introduced by Gray and Shucany [33], and the main objective of the method is related to bias reduction. Let $E_n^{(1)}$ and $E_n^{(2)}$ be two biased estimators of ξ, with similar bias properties, i.e.,

$$\mathrm{Bias}(E_n^{(i)}) = \phi(\xi)d_i(n), \quad i = 1, 2.$$

Then, and trivially, if

$$p = p_n = d_1(n)/d_2(n) \neq 1,$$

the affine combination

$$E_n^{GJ} := \left(E_n^{(1)} - pE_n^{(2)}\right)/(1-p)$$

is an unbiased estimator of ξ.

3.2.1 GJ-MVRB EVI-Estimation

Given $\overline{H}$, in (9), the most natural *GJ* r.v. is the one associated with the random pair $\left(\overline{H}_{k,n}, \overline{H}_{\lfloor\theta k\rfloor,n}\right)$, $0 < \theta < 1$, i.e.

$$\overline{H}_{k,n}^{GJ(p,\theta)} := \frac{\overline{H}_{k,n} - p\,\overline{H}_{\lfloor\theta k\rfloor,n}}{1-p}, \quad 0 < \theta < 1,$$

with

$$p = p_n = \frac{Bias_\infty[\overline{H}_{k,n}]}{Bias_\infty[\overline{H}_{\lfloor\theta k\rfloor,n})]} = \frac{A^2(n/k)}{A^2(n/\lfloor\theta k\rfloor)} \underset{n/k\to\infty}{\longrightarrow} \theta^{2\rho}.$$

It is thus sensible to consider $p = \theta^{2\rho}$, $\theta = 1/2$ (see Gomes et al. [21], for further details on the choice of θ), and, with $\hat{\rho}$ a consistent estimator of ρ, the GJ-MVRB EVI-estimators,

$$\overline{\overline{H}}_{k,n} \equiv \overline{H}_{k,n}^{GJ} := \frac{2^{2\hat{\rho}}\,\overline{H}_{k,n} - \overline{H}_{\lfloor k/2\rfloor,n}}{2^{2\hat{\rho}} - 1}. \tag{18}$$

Then, and provided that $\hat{\rho} - \rho = o_p(1)$,

$$\sqrt{k}\left(\overline{\overline{H}}_{k,n} - \xi\right) \overset{d}{=} \mathcal{N}_{0,\sigma_{GJ}^2} + o_p\big(\sqrt{k}A^2(n/k)\big),$$

with

$$\sigma_{GJ}^2 = \xi^2\big(1 + 1/(2^{-2\rho} - 1)^2\big),$$

just as proved in Gomes et al. [32]. More precisely, and under the fourth-order framework in (7), we can write

$$\sqrt{k}\left(\overline{\overline{H}}_{k,n}-\xi\right) \stackrel{d}{=} \mathcal{N}_{0,\sigma^2_{GJ}} + b_{GJ}\sqrt{k}A^3(n/k) + o_p\big(\sqrt{k}A^3(n/k)\big). \tag{19}$$

We have thus again a trade-off between variance and bias. The bias decreases, but the variance increases. Anyway, we are able to reach a better performance at optimal levels, as desired.

Consequently, even if

$$\sqrt{k}\,A(n/k) \to \infty, \quad \text{with} \quad \sqrt{k}\,A^2(n/k) \to \lambda_A, \text{ finite},$$

the type of levels k where the MSE of $\overline{H}_{k,n}$ is minimized,

$$\sqrt{k}\,\left(\overline{H}_{k,n}-\xi\right) \underset{n\to\infty}{\overset{d}{\longrightarrow}} \mathcal{N}_{\lambda_A b_{\overline{H}},\sigma^2_{\overline{H}}} \quad \text{and} \quad \sqrt{k}\,\left(\overline{\overline{H}}_{k,n}-\xi\right) \underset{n\to\infty}{\overset{d}{\longrightarrow}} \mathcal{N}_{0,\sigma^2_{GJ}}.$$

$$\sqrt{k}\,\left(\overline{\overline{H}}_{k,n}-\xi\right) \underset{n\to\infty}{\overset{d}{\longrightarrow}} \mathcal{N}_{\lambda_A b_{GJ},\sigma^2_{GJ}}.$$

If $\sqrt{k}\,A^3(n/k) \to \lambda_A$, finite, the type of levels where the MSE of $\overline{\overline{H}}_{k,n} \equiv \overline{H}^{GJ}_{k,n}$ is minimized.

Let E denote either H or $\overline{H}$ or $\overline{\overline{H}} \equiv \overline{H}^{GJ}$. We then get, on the basis of (10), (11) and (19),

$$\begin{aligned} k_{A|E}(n) &:= \arg\min_k \text{AMSE}\big(E_{k,n}\big) \\ &= \arg\min_k \begin{cases} \sigma_E^2/k + b_E^2\,A^2(n/k), \text{ if } E = H, \\ \\ \sigma_E^2/k + b_E^2\,A^4(n/k), \text{ if } E = \overline{H}, \\ \\ \sigma_E^2/k + b_E^2\,A^6(n/k), \text{ if } E = \overline{\overline{H}} \end{cases} \\ &= k_{0|E}(n)(1+o(1)), \end{aligned}$$

with $k_{0|E}(n)$ defined in (16). See Theorem 1 of Draisma et al. [9], for a proof of this result, in the case of H. The proof is similar for the cases of $\overline{H}$ and $\overline{\overline{H}}$. Things work more intricately for the PORT-MVRB and quasi-PORT-MVRB EVI-estimators, and we shall consider an algorithm similar to the one devised for the Hill EVI-estimators in case we are working with either $\overline{H}^{(q)}$ or $\overline{\overline{H}}^{(q)}$, $0 \le q < 1$, since only for specific values of q will these estimators be second-order reduced-bias. The bootstrap methodology enables us to estimate the OSF, $k_{0|E}(n)/n$, on the basis of a consistent estimator of $k_{0|E}(n)$, in (16), in a way similar to the one used for the classical EVI-estimators, now through the use of an auxiliary statistic like the one in (17), a method detailed in Gomes et al. [29, 30] for the MVRB EVI-estimation. Indeed, under the above-mentioned fourth-order framework in (7), we get

$$T_{k,n}^E \stackrel{d}{=} \frac{\xi\, P_k^E}{\sqrt{k}} + \begin{cases} b_E(2^{\rho}-1)\, A(n/k)(1+o_p(1)), & \text{if } E = H, \\ b_E(2^{2\rho}-1)\, A^2(n/k)(1+o_p(1)), & \text{if } E = \overline{H}, \\ b_E(2^{3\rho}-1)\, A^3(n/k)(1+o_p(1)), & \text{if } E = \overline{\overline{H}}, \end{cases}$$

with P_k^E asymptotically standard normal.

Consequently, denoting $k_{0|T}(n) := \arg\min_k \mathrm{MSE}(T_{k,n})$, we have

$$k_{0|E}(n) = k_{0|T}(n) \times \begin{cases} (1-2^{\rho})^{\frac{2}{1-2\rho}}\,(1+o(1)), & \text{if } E = H, \\ \left(1-2^{2\rho}\right)^{\frac{2}{1-4\rho}}\,(1+o(1)), & \text{if } E = \overline{H}, \\ \left(1-2^{3\rho}\right)^{\frac{2}{1-6\rho}}\,(1+o(1)), & \text{if } E = \overline{\overline{H}}. \end{cases}$$

3.3 *Adaptive EVI-Estimation*

In the following Algorithm, and with the notation $X_{n+1:n} = 0$, we consider that $\overline{H} \equiv \overline{H}_{k,n}^{(q)} \equiv \overline{\overline{H}}_{k,n}^{(q)}$, for $q = 1$, i.e. we include the MVRB EVI-estimators in the overall selection. Moreover, whenever dealing with $0 \le q < 1$ replace n by $n - n_q$, $n_q = \lfloor nq \rfloor + 1$.

3.3.1 Algorithm: Adaptive Bootstrap Estimation of ξ

1. Given the sample $(x_1, \ldots, x_n)$, compute for the tuning parameters $\tau = 0$ and $\tau = 1$, the observed values of $\hat\rho_\tau(k)$, the most simple class of estimators in Fraga Alves et al. [14]. Such estimators have the functional form

$$\hat\rho_\tau(k) := -\left|3(W_{k,n}^{(\tau)} - 1)/(W_{k,n}^{(\tau)} - 3)\right|,$$

dependent on the statistics

$$W_{k,n}^{(\tau)} := \begin{cases} \dfrac{\left(M_{k,n}^{(1)}\right)^{\tau} - \left(M_{k,n}^{(2)}/2\right)^{\tau/2}}{\left(M_{k,n}^{(2)}/2\right)^{\tau/2} - \left(M_{k,n}^{(3)}/6\right)^{\tau/3}}, & \text{if } \tau \neq 0, \\[2ex] \dfrac{\ln M_{k,n}^{(1)} - \ln\left(M_{k,n}^{(2)}/2\right)/2}{\ln\left(M_{k,n}^{(2)}/2\right)/2 - \ln\left(M_{k,n}^{(3)}/6\right)/3}, & \text{if } \tau = 0, \end{cases}$$

where

$$M_{k,n}^{(j)} := \frac{1}{k}\sum_{i=1}^{k}\Big(\ln X_{n-i+1:n} - \ln X_{n-k:n}\Big)^{j},\ j = 1,2,3.$$

2. Consider $\mathscr{K} = \big(\lfloor n^{0.995}\rfloor,\ \lfloor n^{0.999}\rfloor\big)$. Compute the median of $\{\hat\rho_\tau(k)\}_{k\in\mathscr{K}}$, denoted χ_τ, and compute $I_\tau := \sum_{k\in\mathscr{K}}\left(\hat\rho_\tau(k)-\chi_\tau\right)^2$, $\tau = 0,1$. Next choose the *tuning parameter* $\tau^* = 0$ if $I_0 \le I_1$; otherwise, choose $\tau^* = 1$.

3. Work with $\hat\rho \equiv \hat\rho_{\tau^*} = \hat\rho_{\tau^*}(k_1)$ and $\hat\beta \equiv \hat\beta_{\tau^*} := \hat\beta_{\hat\rho_{\tau^*}}(k_1)$, with $k_1 = \lfloor n^{0.999}\rfloor$, being $\hat\beta_{\hat\rho}(k)$ the estimator in Gomes and Martins [17], given by

$$\hat\beta_{\hat\rho}(k) := \left(\frac{k}{n}\right)^{\hat\rho}\frac{d_k(\hat\rho)\,D_k(0) - D_k(\hat\rho)}{d_k(\hat\rho)\,D_k(\hat\rho) - D_k(2\hat\rho)},$$

dependent on the estimator $\hat\rho = \hat\rho_{\tau^*}(k_1)$, and where, for any $\alpha \le 0$,

$$d_k(\alpha) := \frac{1}{k}\sum_{i=1}^{k}(i/k)^{-\alpha} \quad \text{and} \quad D_k(\alpha) := \frac{1}{k}\sum_{i=1}^{k}(i/k)^{-\alpha}\,U_i,$$

with $U_i = i\left(\ln X_{n-i+1:n} - \ln X_{n-i:n}\right)$, $1 \le i \le k < n$, the *scaled log-spacings*.

4. For $k = 1,2,\ldots$, compute the observed values of $H_{k,n}$, $\overline{H}_{k,n}$ and $\overline{\overline{H}}_{k,n} \equiv \overline{H}^{GJ}_{k,n}$, in (3), (9) and (18), respectively.

5. For $q = 0(0.1)0.9$, execute steps **1.**, **2.** and **3.** for the observed value of the sample of excesses in (12), and compute the observed values of $\overline{H}^{(q)}_{k,n}$, in (14), $\big($or alternatively $\overline{\overline{H}}^{(q)}_{k,n}$, in (15)$\big)$, for all admissible k.

6. Consider sub-sample sizes $m_1 = o(n)$ and $m_2 = \lfloor m_1^2/n\rfloor + 1$, having n the same meaning as $n - \lfloor nq\rfloor - 1$ if $0 \le q < 1$.

7. For l from 1 until $B = 250$, independently generate from the empirical d.f. associated with the underlying sample $(x_1, x_2, \ldots, x_n)$, B bootstrap samples

$$(x_1^*, \ldots, x_{m_2}^*) \quad \text{and} \quad (x_1^*, \ldots, x_{m_2}^*, x_{m_2+1}^*, \ldots, x_{m_1}^*),$$

with sizes m_2 and m_1, respectively.

8. Again generally denoting $E_{k,n}$ any of the aforementioned EVI-estimators, let us denote by $T^*_{k,n|E}$ the bootstrap counterpart of the auxiliary statistic in (17), and obtain $t^*_{k,m_1,l|E}$, $1 < k < m_1$, $t^*_{k,m_2,l|E}$, $1 < k < m_2$, $1 \le l \le B$, the observed values of the statistics $T^*_{k,m_i|E}$, $i = 1,2$, and compute

$$\mathrm{MSE}^*_E(m_i,k) = \frac{1}{B}\sum_{l=1}^{B}\left(t^*_{k,m_i,l|E}\right)^2,\ \ k = 1,2,\ldots,m_i-1,\ \ i = 1,2.$$

9. Obtain $\hat{k}^*_{0|E}(m_i) := \arg\min_{1\le k\le m_i-1} \mathrm{MSE}^*_E(m_i,k)$, $i=1,2$.
10. Compute

$$\hat{k}_{0|E} := \min\left(n-1, \left\lfloor \frac{c_{\hat{\rho}}\left(\hat{k}^*_{0|E}(m_1)\right)^2}{\hat{k}^*_{0|E}(m_2)} \right\rfloor + 1\right),$$

with

$$c_\rho = \begin{cases} (1-2^\rho)^{\frac{2}{1-2\rho}}, & \text{if } E = H \text{ or } \overline{H}^{(q)},\ 0 \le q < 1, \\ \left(1-2^{2\rho}\right)^{\frac{2}{1-4\rho}}, & \text{if } E = \overline{H}, \\ \left(1-2^{3\rho}\right)^{\frac{2}{1-6\rho}}, & \text{if } E = \overline{\overline{H}}, \end{cases}$$

and the OSF's estimates, $\hat{k}_{0|E}/n$.
11. Obtain $H^* = H_{\hat{k}_{0|H},n}$, $\overline{H}^* = \overline{H}_{\hat{k}_{0|\overline{H}},n}$, $\overline{\overline{H}}^* = \overline{\overline{H}}_{\hat{k}_{0|\overline{\overline{H}}},n}$ and $\overline{H}^{*(q)}_{n,m_1} := \overline{H}^{(q)}_{\hat{k}^{(q)}_0,n}$, with $\hat{k}^{(q)}_0 := \hat{k}_{0|\overline{H}^{(q)}}$.
12. With $B^*_q(m_i,k) = \frac{1}{B}\sum_{l=1}^{B} t^*_{k,m_i,l|\overline{H}^{(q)}}$, $k=1,2,\ldots,m_i-1$, $i=1,2$, consider

$$\widehat{\mathrm{AMSE}}(k;q) := \frac{\left(\overline{H}^{*(q)}_{n,m_1}\right)^2}{k} + \left(\frac{\left(B^*_q(m_1,k)\right)^2}{(2^\rho-1)B^*_q(m_2,k)}\right)^2, \quad q \ne 1,$$

with the previously obtained values $\hat{\rho} = \hat{\rho}_q$, and $\overline{H}^{*(q)}_{n,m_1}$.
13. Compute $\hat{q} := \arg\min_q \widehat{\mathrm{AMSE}}(\hat{k}^{(q)}_0;q)$.
14. Obtain the final adaptive EVI-estimate,

$$\overline{H}^{**} \equiv \overline{H}^{**}|\hat{q} \equiv \overline{H}^{*(\hat{q})}_{n,m_1} := \overline{H}^{(\hat{q})}_{\hat{k}^{(\hat{q})}_0,n}.$$

Remark 1 An analogue procedure can be used for any other parameter of extreme events.

Remark 2 A few practical questions may be again raised under the set-up developed: How does the asymptotic method work for moderate sample sizes? What is the type of the sample path of the new estimator for different values of m_1? What is the dependence of the method on the choice of m_1? What is the sensitivity of the method with respect to the choice of ρ-estimators? Although aware of the need of $m_1 = o(n)$, what happens if we choose $m_1 = n$? Answers to these questions were given in Gomes and Oliveira [18] for the estimation of ξ through the Hill EVI-estimators, can be addressed here, but are beyond the scope of this article.

Remark 3 Note that bootstrap confidence intervals associated with the adaptive EVI-estimates are easily computed on the basis of the replication of the *Algorithm* R times, for an adequate R.

4 Applications to Real Data

4.1 An Environmental Application

The first set of data, already considered in Gomes et al. [30], is related to the number of hectares, exceeding 100 ha, burnt during wildfires recorded in Portugal during 14 years (1990–2003). Most of the wildfires are extinguished within a short period of time, with almost negligible effects. However, some wildfires go out of control, burning hectares of land and causing significant and negative environmental and economical impacts. The data (a sample of size $n = 2627$) do not seem to have a significant temporal structure, and we have used it as a whole. A box-and-whiskers plot of the data provides evidence on the heaviness of the right tail, as can be seen in Fig. 2.

Let us have a look at the behaviour of the adaptive EVI-estimators under consideration for this data set. We have been led to the ρ-estimate, $\hat{\rho} \equiv \hat{\rho}_0 = -0.388$, obtained at the level $k_1 = \lfloor n_0^{0.999} \rfloor = 2606$. The associated β-estimate is $\hat{\beta} \equiv \hat{\beta}_0 = 0.470$. Note that the sample paths of the ρ-estimates associated with $\tau = 0$ and $\tau = 1$ lead us indeed to choose, on the basis of any stability criterion for

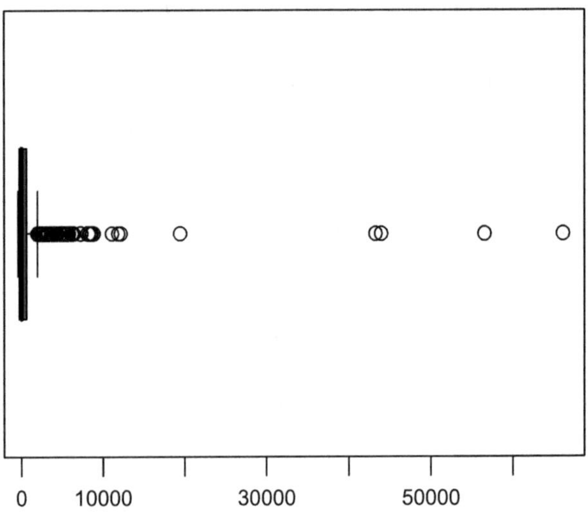

Fig. 2 Box-and-whiskers plot associated with the burnt areas in Portugal, above 100 ha, in the period 1990–2003

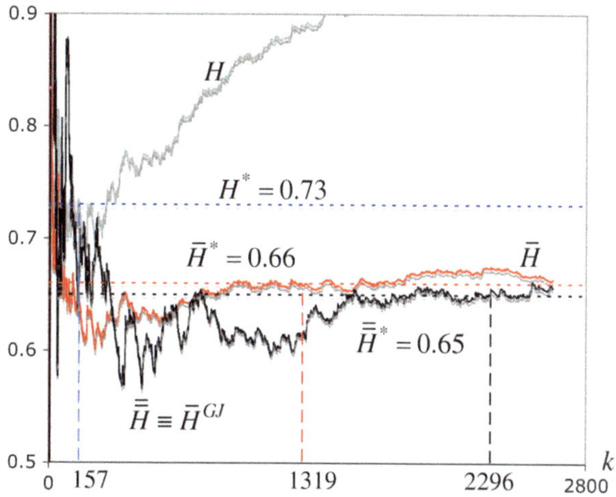

Fig. 3 EVI-estimates for the burned areas

large k, the estimate associated with $\tau = 0$. The aforementioned double-bootstrap algorithm (until Step 11., and with $q = 1$, so that we are working with $\overline{H}$ only, among the $\overline{H}^{(q)}$ EVI-estimators) depends very weakly on the choice of a subsample size $m_1 = o(n)$ (see Gomes et al. [30]). For $m_1 = \lfloor n^{0.955} \rfloor = 1843$, and B=250 bootstrap replications, we have got

- $\hat{k}_0^H = 157$ ($\hat{k}_0^H/n = 0.060$) and the Hill EVI-estimate, $H^* = 0.73$,
- $\hat{k}_0^{\overline{H}} = 1319$ ($\hat{k}_0^{\overline{H}}/n = 0.502$) and the MVRB EVI-estimate, $\overline{H}^* = 0.66$,
- $\hat{k}_0^{\overline{H}^{GJ}} = 2296$ ($\hat{k}_0^{\overline{H}^{GJ}}/n = 0.874$) and the GJ-MVRB EVI-estimate, $\overline{\overline{H}}^* = 0.65$,

the values presented in Fig. 3, together with sample paths of the EVI-estimates under consideration.

For the PORT-MVRB EVI-estimation, illustrated in Fig. 4, and with the exclusion of the value $q = 1$, we have been led to the choice $q = 0$, $\hat{k}_0^{(0)} = 242$ ($\hat{k}_0^{(0)}/n_0 = 0.092$, $n_0 = n - 1$) and $\overline{H}^{**}|0 = 0.670$. With the inclusion of $q = 1$ in the algorithm, we have been led to $\overline{H}$, as expected, due to the data characteristics (positive values only). For this type of data we have thus no particular gain in terms of efficiency when we use the PORT methodology.

4.2 An Application in the Area of Finances

To enhance the importance of the PORT-MVRB EVI-estimation, we shall further consider an application to the analysis of the log-returns associated with one of the four sets of finance data considered in Gomes and Pestana [19]. Such data, collected over the period from January 4, 1999, until November 17, 2005, and with a size $n = 1762$, were the daily closing values of the Microsoft Corp. (MSFT). Note that

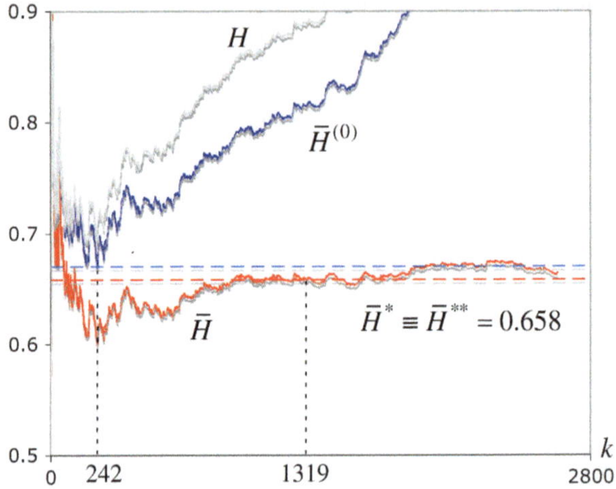

Fig. 4 PORT-MVRB EVI-estimation for burned areas

these MSFT data have also been analysed in Gomes et al. [29, 31], through the use of different algorithms. Although there is some increasing trend in the volatility of all these log-returns, stationarity and weak dependence is often assumed, under the same considerations as in Drees [10].

The underlying model has heavy left and right tails. We have thus eliminated the estimators associated with $q = 0$, due to their inconsistency (see Gomes et al. [26], for details). The number of positive elements in the available sample of MSFT log-returns is $n_0 = 882$. We have been led to the ρ-estimate $\hat{\rho} \equiv \hat{\rho}_0 = -0.72$, obtained at the level $k_1 = \lfloor n_0^{0.999} \rfloor = 876$. The associated β-estimate is $\hat{\beta} \equiv \hat{\beta}_0 = 1.02$. Just as above, the sample paths of the ρ-estimates associated with $\tau = 0$ and $\tau = 1$ lead us indeed to choose, on the basis of any stability criterion for large k, the estimate associated with $\tau = 0$.

In Fig. 5, we present the adaptive and non-adaptive estimates of ξ, provided by H, $\overline{H}^{(q)}$, $q = 0.1$, 0.3 and 1 ($\overline{H}^{(1)} = \overline{H}$), with H and $\overline{H}^{(q)}$ given in (3) and (14), respectively. Note that the Hill estimators $H_{k,n}$, in (3), are unbiased for the EVI estimation only when the underlying model is a strict Pareto model. Otherwise, i.e. when we have only Pareto-like tails, as surely happens here and can be seen from Fig. 5 (as well as from Figs. 3 and 4), it exhibits a quite relevant bias. The PORT-MVRB estimators, $\overline{H}^{(q)}$, in (14), which are expected to be 'asymptotically unbiased' for adequate values of q, have a smaller bias, exhibit more stable sample paths as functions of k, and enable us to take a decision upon the estimate of ξ and other parameters of extreme events to be used, even with the help of any heuristic stability criterion, like the '*largest run*' method suggested in Gomes et al. [22], and the ones provided in Gomes et al. [31], among others.

The *Algorithm* in Sect. 3.3, for $0 < q \leq 1$, led us to the choice $\hat{q} = 0.1$, $\hat{k}_0^{(0.1)} = 449$ ($\hat{k}_0^{(0.1)}/1585 = 0.283$, $n_{(0.1)} = 1585$), and $\overline{H}^{**} \equiv \overline{H}^{**}|0.1 = 0.241$,

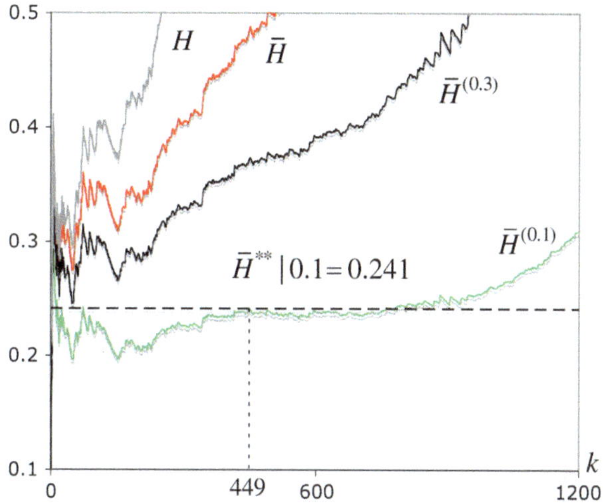

Fig. 5 Adaptive and non-adaptive EVI-estimates for the MSFT data set

as shown in Fig. 5. Indeed, the MVRB EVI-estimators, despite of 'asymptotically unbiased' reveal a relevant bias for models like the Student-t, one of the most common candidates in a parametric estimation of log-returns. The PORT-MVRB EVI-estimates are then serious candidates to a reliable EVI-estimation.

5 Some Overall Conclusions

- The double-bootstrap algorithm, despite of computationally intensive, is quite reliable for the estimation of OSFs.
- The most attractive features of the GJ EVI-estimators are their stable sample paths (for a wide region of k or k/n values).
- The GJ-MVRB EVI-estimate is quite close to the MVRB EVI-estimate, but with a higher OSF-estimate.
- Due to stability reasons we advise for positive data sets the use of the GJ-MVRB or the MVRB EVI-estimators rather than the PORT-MVRB EVI-estimators.
- For the MSFT data set, or for any data set with negative values, we advise the use of the PORT-MVRB EVI-estimators, due to their stable sample paths as functions of k or k/n for an adequate q. Moreover, note that these estimates are more reliable since they involve a larger number of top o.s.'s due to the increase of the positive elements in the sample from n_0 to $n - \lfloor nq \rfloor - 1$.

Acknowledgements Research partially supported by National Funds through FCT—Fundação para a Ciência e a Tecnologia, projects PEst-OE/MAT/UI0006/2011, 2014 (CEAUL), EXTREMA, PTDC/MAT/101736/2008, and grant SFRH/BPD/77319/2011.

References

1. Araújo Santos, P., Fraga Alves, M.I., Gomes, M.I.: Peaks over random threshold methodology for tail index and quantile estimation. Revstat **4**(3), 227–247 (2006)
2. Beirlant, J., Dierckx, G., Goegebeur, Y., Matthys, G.: Tail index estimation and an exponential regression model. Extremes **2**, 177–200 (1999)
3. Beirlant, J., Caeiro, F., Gomes, M.I.: An overview and open research topics in the field of statistics of univariate extremes. Revstat **10**(1), 1–31 (2012)
4. Caeiro, F., Gomes, M.I., Pestana, D.: Direct reduction of bias of the classical Hill estimator. Revstat **3**(2), 111–136 (2005)
5. Caeiro, F., Gomes, M.I., Henriques-Rodrigues, L.: Reduced-bias tail index estimators under a third order framework. Commun. Stat. Theory Methods **38**(7), 1019–1040 (2009)
6. de Haan, L.: Slow variation and characterization of domains of attraction. In: de Oliveira, T. (ed.) Statistical Extremes and Applications, pp. 31–48. D. Reidel, Dordrecht (1984)
7. de Haan, L., Ferreira, A.: Extreme Value Theory: An Introduction. Springer Science+Business Media, LLC, New York (2006)
8. de Haan, L., Peng, L.: Comparison of tail index estimators. Stat. Neerl. **52**, 60–70 (1998)
9. Draisma, G., de Haan, L., Peng, L., Themido Pereira, T.: A bootstrap-based method to achieve optimality in estimating the extreme value index. Extremes **2**(4), 367–404 (1999)
10. Drees, H.: Extreme quantile estimation for dependent data, with applications to finance. Bernoulli **9**(4), 617–657 (2003)
11. Efron, B.: Bootstrap methods: another look at the jackknife Ann. Stat. **7**(1), 1–26 (1979)
12. Feuerverger, A., Hall, P.: Estimating a tail exponent by modelling departure from a Pareto distribution. Ann. Stat. **27**, 760–781 (1999)
13. Figueiredo, F., Gomes, M.I., Henriques-Rodrigues, L., Miranda, C.: A computational study of a quasi-PORT methodology for VaR based on second-order reduced-bias estimation. J. Stat. Comput. Simul. **82**(4), 587–602 (2012)
14. Fraga Alves, M.I., Gomes M.I., de Haan, L.: A new class of semi-parametric estimators of the second order parameter. Port. Math. **60**(2), 194–213 (2003)
15. Geluk J., de Haan L.: Regular variation, extensions and tauberian theorems. Technical Report CWI Tract 40. Centre for Mathematics and Computer Science, Amsterdam (1987)
16. Gnedenko, B.V.: Sur la distribution limite du terme maximum d'une série aléatoire. Ann. Math.**44**, 423–453 (1943)
17. Gomes, M.I., Martins M.J.: Asymptotically unbiased estimators of the tail index based on external estimation of the second order parameter. Extremes **5**(1), 5–31 (2002)
18. Gomes, M.I., Oliveira, O.: The bootstrap methodology in statistics of extremes: choice of the optimal sample fraction. Extremes **4**(4), 331–358 (2001)
19. Gomes, M.I., Pestana, D.: A sturdy reduced-bias extreme quantile (VaR) estimator. J. Am. Stat. Assoc. **102**(477), 280–292 (2007)
20. Gomes, M.I., Martins, M.J., Neves, M.: Alternatives to a semi-parametric estimator of parameters of rare events: the Jackknife methodology. Extremes **3**(3), 207–229 (2000)
21. Gomes, M.I., Martins, M.J., Neves, M.: Generalized Jackknife semi-parametric estimators of the tail index. Port. Math. **59**(4), 393–408 (2002)
22. Gomes, M.I., Figueiredo, F., Mendonça, S.: Asymptotically best linear unbiased tail estimators under a second order regular variation condition. J. Stat. Plann. Infer. **134**(2), 409–433 (2005)
23. Gomes, M.I., Martins, M.J., Neves, M.: Improving second order reduced bias extreme value index estimation. Revstat **5**(2), 177–207 (2007)
24. Gomes, M.I., Reiss, R.-D., Thomas, M.: Reduced-bias estimation. In: Reiss, R.-D., Thomas, M. (eds.) Statistical Analysis of Extreme Values with Applications to Insurance, Finance, Hydrology and Other Fields, 3rd edn, pp. 189–204. Birkhäuser, Basel (2007)
25. Gomes, M.I., Canto e Castro, L., Fraga Alves, M.I., Pestana, D.: Statistics of extremes for iid data and breakthroughs in the estimation of the extreme value index: Laurens de Haan leading contributions. Extremes **11**(1), 3–34 (2008)

26. Gomes, M.I., Fraga Alves, M.I., Araújo Santos, P.: PORT Hill and moment estimators for heavy-tailed models. Commun. Stat. Simul. Comput. **37**, 1281–1306 (2008)
27. Gomes, M.I., de Haan, L., Henriques Rodrigues, L.: Tail index estimation through accommodation of bias in the weighted log-excesses. J. R. Stat. Soc. B **70**(1), 31–52 (2008)
28. Gomes, M.I., Henriques-Rodrigues, L., Miranda, C.: Reduced-bias location-invariant extreme value index estimation: a simulation study. Commun. Stat. Simul. Comput. **40**(3), 424–447 (2011)
29. Gomes, M.I., Mendonça, S., Pestana, D.: Adaptive reduced-bias tail index and VaR estimation via the bootstrap methodology. Commun. Stat. Theory Methods **40**(16), 2946–2968 (2011)
30. Gomes, M.I., Figueiredo, F., Neves, M.M.: Adaptive estimation of heavy right tails: resampling-based methods in action. Extremes **15**, 463–489 (2012)
31. Gomes, M.I., Henriques-Rodrigues, L., Fraga Alves, M.I., Manjunath, B.G.: Adaptive PORT-MVRB estimation: an empirical comparison of two heuristic algorithms. J. Stat. Comput. Simul. **83**(6), 1129–1144 (2013)
32. Gomes, M.I., Martins, M.J., Neves, M.M.: Generalised jackknife-based estimators for univariate extreme-value modelling. Commun. Stat. Theory Methods **42**(7), 1227–1245 (2013)
33. Gray, H.L., Schucany, W.R.: The Generalized Jackknife Statistic. Marcel Dekker, New York (1972)
34. Hill, B.M.: A simple general approach to inference about the tail of a distribution. Ann. Stat. **3**, 1163–1174 (1975)
35. Peng, L.: Asymptotically unbiased estimator for the extreme value index. Stat. Probab. Lett. **38**(2), 107–115 (1998)
36. Reiss, R.-D., Thomas, M.: Statistical Analysis of Extreme Values with Applications to Insurance, Finance, Hydrology and Other Fields, 3rd edn. Birkhäuser, Basel (2007)

On the Optimal Control of Flow Driven Dynamic Systems

Teresa Grilo, Sílvio M.A. Gama, and Fernando Lobo Pereira

Abstract The objective of this work is to develop a mathematical framework for the modeling, control and optimization of dynamic control systems whose state variable is driven by interacting ODE's (ordinary differential equations) and solutions of PDE's (partial differential equations). The ultimate goal is to provide a sound basis for the design and control of new advanced engineering systems arising in many important classes of applications, some of which may encompass, for example, underwater gliders and mechanical fishes. For now, the research effort has been focused in gaining insight by applying necessary conditions of optimality for shear flow driven dynamic control systems which can be easily reduced to problems with ODE dynamics. In this article we present and discuss the problem of minimum time control of a particle advected in a Couette and Poiseuille flows, and solve it by using the maximum principle.

1 Introduction

The development a mathematical framework for the modeling, control and optimization of dynamic control systems whose state variable is driven by interacting ODE's and PDE's is still a significant challenge. In [5], it is presented some earlier work aiming at the development of a theory of optimal control of dynamic systems, [4, 6], whose state evolves due to the interaction of ordinary differential equations with partial differential equations in which the later part is replaced by some known particular solution. Underwater gliders and robotic fishes, Fig. 1, are two examples of the class of applications whose currently available models we intend to improve.

An underwater glider is a winged autonomous underwater vehicle (AUV) that moves by modulating its buoyancy and attitude in the velocity vector fields of its environment. This vehicles are used for long-term, large-scale oceanographic

T. Grilo (✉) • S.M.A. Gama
FCUP, Rua Campo Alegre 687, 4169-007 Porto, Portugal
e-mail: tgrilo@fc.up.pt; smgama@fc.up.pt

F.L. Pereira
FEUP, Rua Dr. Roberto Frias s/n, 4200-465 Porto, Portugal
e-mail: flp@fe.up.pt

J.-P. Bourguignon et al. (eds.), *Mathematics of Energy and Climate Change*,
CIM Series in Mathematical Sciences 2, DOI 10.1007/978-3-319-16121-1_7

Fig. 1 Underwater glider (*left*), robotic fish (*right*)

monitoring, undersea surveillance and other applications. The kinetic and dynamic equations that described the vehicle motion can be found in [8, 9]. In [7], the motion of the robotic fish is approximated by a model featuring several components. The key advantage of this model is the fact that, instead of being considered a rigid body, the structure of the fish is composed of three parts: head, body and tail.

While the optimal control of systems with dynamics given by ordinary differential equations only has been making great strides in the twentieth and twenty-first centuries (see, among others, [1, 3, 10]), such a theory for hybrid—in the sense that the controlled dynamics involve ordinary and partial differential equations—systems is still at its infancy.

Here, we formulate and solve two optimal control problems, where linear and parabolic velocity profiles are considered. Each one of these velocity profiles corresponds to a particular solution of the incompressible two-dimensional Navier-Stokes equation, respectively, the Couette and Poiseuille steady flows.

The Couette flow is the steady laminar unidirectional and two dimensional flow due to the relative motion of two infinite horizontal and parallel rigid plates [2]. The liquid between these two plates is driven by the viscous drag force originated by the uniform motion of the upper plate which moves in the x-direction with velocity v_0 (the lower plate is at rest). In this case, the velocity of such a flow has a linear profile and is given by

$$v(x, y) = (my, 0)\,, \quad x \in \mathbb{R}, \quad y \in [0, L]$$

with $m = v_0/L$, the plates being L distance units apart (Fig. 2, left).

The Poiseuille flow is the steady flow due to the presence of a pressure gradient between two fixed (i.e., with zero relative velocity) rigid plates, [2]. In this case, a parabolic velocity profile is of the form

$$v(x, y) = \left(a - \frac{a}{L^2} y^2, 0\right), \quad x \in \mathbb{R}, \quad y \in [-L, L]\,,$$

where, now, L is half the distance between the upper and lower plates (Fig. 2, right).

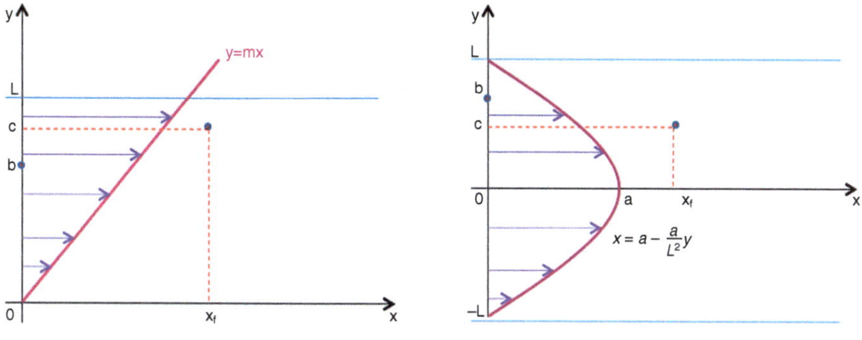

Fig. 2 Linear (*left*) and quadratic velocity field (*right*)

2 Minimum Time Control Problem

Let us consider a particle placed in a flow, contained in a channel of width L, with a given velocity field. $(0, b)$ is the initial position of the particle, with $0 \leq b \leq L$.

Our problem consists in moving the particle in minimum time along a path in the channel connecting a given initial position to a given end point (x_f, c), with $0 \leq c \leq L$. Since the particle is subject to the flow field, we must determine the value of the control function $u(\cdot) = (u_x(\cdot), u_y(\cdot))$ to be applied so that the conditions of the proposed control problem are satisfied. Let $X(t) = (x(t), y(t))$ be the position of the particle at time t, the control problem can be formulated as follows:

$$\begin{cases} \text{Minimize } T \\ \text{subject to } \dot{X}(t) = F(X(t), u(t)) \\ \qquad X(0) = (0, b) \\ \qquad X(T) = (x_f, c) \\ \qquad y(t) \in [0, L] \\ \qquad \|u(t)\|_\infty \leq 1 \end{cases} , \forall t \in [0, T]. \tag{1}$$

Remark From now on, for simplicity of notation, we will not indicate de time t as an independent variable of the others variables, although this is the case we are considering.

The maximum principle, [10], allows us to determine the optimal control $u^* = (u_x^*, u_y^*)$ by using the maximization of the Pontryagin's function $H(X, P, u)$, where $P = (p_x, p_y)$ is the adjoint variable satisfying $-\dot{P} = \nabla_X H(X, P, u)$, being ∇_X the gradient of H with respect to X, almost everywhere with respect to the Lebesgue measure (from here onwards, functions are specified in this sense), together with the satisfaction of the appropriate boundary conditions.

2.1 Couette Flow

Consider the case of linear flow, with slope $m = v_0/L > 0$, whose velocity field is given by $v(x, y) = (\frac{y}{m}, 0)$. So, the dynamics of this control system is

$$F(X, u) = (\frac{y}{m} + u_x, u_y)$$

and the position of the particle at time t is given by

$$\begin{cases} x(t) = \frac{b}{m}t + \int_0^t u_x(\tau)\, d\tau + \frac{1}{m}\int_0^t (t-\tau)u_y(\tau)\, d\tau \\ y(t) = b + \int_0^t u_y(\tau)\, d\tau\ . \end{cases}$$

So, the Pontryagin's function is given by

$$H(X, P, u) = p_x(\frac{y}{m} + u_x) + (p_y + \gamma)u_y\ ,$$

where γ is a certain function which reflects the activity of the state constraints of the variable y, it follows from the maximum principle that

$$\begin{cases} -\dot{p}_x = 0 \\ -\dot{p}_y = \frac{p_x}{m} \end{cases} \Leftrightarrow \begin{cases} p_x = K_x \\ p_y = K_y - \frac{K_x}{m}t \end{cases}$$

for some constants $K_x, K_y > 0$.

By taking into account the position of the particle at each instant t, we conclude by the maximization of the Pontryagin's function that $u_x^*(t) = 1$ (Fig. 2) and u_y^* depends the final position of the particle, $X(T)$.

If $x_f \leq 2(L - b)$ the state constraint of the variable y remains inactive and

$$u_y^*(t) = \begin{cases} 1, & t \in [0, \frac{c-b+t^*}{2}[\\ -1, & t \in]\frac{c-b+t^*}{2}, t^*] \end{cases}.$$

By substituting in the equations of the particle's position, we conclude that the optimum time for (1) is given by

$$t^* = \sqrt{(c + b + 2m)^2 + (c - b)^2 + 4mx_f} - (c + b + 2m)\,.$$

For the case of $x_f > 2(L - b)$ the state constraint of variable y is active and

$$u_y^*(t) = \begin{cases} 1, & t \in [0, t_1[\\ 0, & t \in]t_1, t^* - t_2[\\ -1, & t \in]t^* - t_2, t^*] \end{cases},$$

where $t_1 = \sqrt{(b+m)^2 + 2m(L-b)} - (b+m)$ is the time when the particle is on the boundary of the channel, and $t_2 = \sqrt{(c+m)^2 + 2m(L-c)} - (c+m)$ is the time when the particle leaves the boundary. Now the minimum time is

$$t^* = \sqrt{(b+m)^2 + 2m(L-b)} - (b+m) + \frac{(c+m)^2 + m(b-c) + mx_f}{\sqrt{(b+m)^2 + 2m(L-b)}} - (c+m)A\,,$$

where $A = \sqrt{\frac{(c+m)^2+2m(L-c)}{(b+m)^2+2m(L-b)}}$.

2.2 *Poiseuille Flow*

Let us consider a flow with a parabolic velocity vector field, with vertex at $(a, 0)$. In this case the velocity field is given by $v(x, y) = (a - \frac{a}{L^2}y^2, 0)$. Then, the dynamics of the control system is

$$F(X, u) = (a - \frac{a}{L^2}y^2 + u_x, u_y)$$

and the Pontryagin's function is given by

$$H(X, P, u) = p_x(a - \frac{a}{L^2}y^2 + u_x) + (p_y + \gamma)u_y\,.$$

It follows from the maximum principle that

$$\begin{cases} -\dot{p}_x = 0 \\ -\dot{p}_y = -\frac{2a}{L^2}p_x y \end{cases} \Leftrightarrow \begin{cases} p_x = K_x \\ p_y = K_y + \frac{2aK_x}{L^2}\int_0^t y(\tau)\,d\tau \end{cases}$$

We remark that there is symmetry with respect to the axis $y = 0$ and, that the state constraint will be inactive along the optimal trajectory. By using these observations in the application of the maximum principle, as well as the fact that the position of the particle is given by

$$\begin{cases} x(t) = at - \frac{a}{L^2}\int_0^t y^2(\tau)\,d\tau + \int_0^t u_x(\tau)\,d\tau \\ y(t) = b + \int_0^t u_y(\tau)\,d\tau, \end{cases}$$

we conclude that $u_x^*(t) = 1$ (Fig. 2), and that u_y^* is defined by

$$u_y^*(t) = \begin{cases} -1, & t \in [0, \frac{b-c+t^*}{2}[\\ 1, & t \in]\frac{b-c+t^*}{2}, t^*] \end{cases},$$

if $x_f \leq 2b$, being, in this case, the minimum time t^* a root of the polynomial

$$t^{*3} - 3(b+c)t^{*2} + 3\left((c+b)^2 - 4L^2(1+\frac{1}{a})\right)t^* + 3A = 0, \qquad (2)$$

with $A = b^3 - b^2c - bc^2 + c^3 + \dfrac{4L^2}{a}x_f$.

In the case of $x_f > 2b$ we have

$$u_y^*(t) = \begin{cases} -1, & t \in [0, t_1[\\ 0, & t \in]t_1, t^* - t_2[\\ 1, & t \in]t^* - t_2, t^*] \end{cases},$$

being the optimal time given by

$$t^* = \frac{a(2t_1^3 - 3b(t_1^2 + t_2^2) + 3t_1t_2^2 - t_2^3) - 3L^2x_f}{3a(b-t_1)^2 - 3aL^2 - 3L^2},$$

where t_1 and t_2 is a half of the value of the t^* obtained in (2) with $x_f = 2b$ and $x_f = 2c$, respectively.

3 Conclusions and Future Work

The cases we have presented and discussed here were introduced for the first time in [5]. These two examples are very simple and differ only in the profile of the fluid velocity field. Not only the dynamics of the control system are defined by a set of ODE's, but also the conditions resulting from the application of the maximum principle can be easily solved in an explicit way. The next step consists in deriving optimality conditions in the form of a maximum principle leading to the computation of the solution to optimal control problems for which the above simplifications can not be exploited. This study suggests that the optimality conditions to be developed will require an adjoint variable satisfying a mixed system with ODE's and PDE's, so that the optimal control can be obtained by maximizing an appropriated Pontryagin's function, coupled with appropriate boundary conditions.

Acknowledgements The first and the third authors gratefully acknowledge the financial support of the FCT funded the doctoral grant POPH/FSE/SFRH/BD/94131/2013, and the R&D projects PEST-OE/EEI/UI0147/2014 and SYSTEC R&D Unit ref. UID/EEA/00147/2013, respectively. The second author was partially supported by CMUP (UID/MAT/00144/2013), which is funded by FCT (Portugal) with national (MEC) and European structural funds through the programs FEDER, under the partnership agreement PT2020.

References

1. Arutyunov, A., Karamzin, D., Lobo Pereira, F.: The maximum principle for optimal control problems with state constraints by R.V. Gamkrelidze: revisited. J. Optim. Theory Appl. **149**, 474–493 (2011)
2. Batchelor, G.: An Introduction to Fluid Dynamics. Cambridge University Press, Cambridge (1992)
3. Clarke, F.: The maximum principle in optimal control, then and now. J. Control Cybern. **34**, 709–722 (2005)
4. Clarke, F., Ledyaev, Y., Stern, R., Wolenski, P.: Nonsmooth Analysis and Control Theory. Springer, New York (1998)
5. Grilo, T., Lobo Pereira, F., Gama, S.: Optimal control of particle advection in Couette and Poiseuille flows. J. Conf. Papers Math. **2013**, Art. ID 783510 (2013). http://dx.doi.org/10.1155/2013/783510
6. Lions, J.: Optimal Control of Systems Governed by Partial Differential Equations. Springer, New York (1971)
7. Liu, J., Hu, H.: Biological inspiration: from carangiform fish to multi-joint robotic fish. J. Bionic Eng. **7**, 35–48 (2010)
8. Mahmoudian, N., Woolsey, C.: Dynamics and control of underwater gliders II: motion planning and control. Virginia Center for Autonomous Systems, Technical Report No. VaCAS-2010-02 (2010)
9. Mahmoudian, N., Geisbert, J., Woolsey, C.: Dynamics and control of underwater gliders I: steady motions. Virginia Center for Autonomous Systems, Technical Report No. VaCAS-2007-01 (2009)
10. Pontryagin, L., Boltyanskiy, V., Gamkrelidze, R., Mishchenko, E.: Mathematical Theory of Optimal Processes. Interscience Publishers, New York (1962)

An Overview of Network Bifurcations in the Functionalized Cahn-Hilliard Free Energy

Noa Kraitzman and Keith Promislow

Abstract The functionalized Cahn-Hilliard (FCH) free energy models interfacial energy in amphiphilic phase-separated mixtures. Its minimizers and quasi-minimizers encompass rich classes of network morphologies with detailed inner layers incorporating bilayers, pore, pearled pore, and micelle type structures. We present an overview of the stability of the network morphologies as well as the competitive evolution of bilayer and pore morphologies under a gradient flow in three space-dimensions.

1 Amphiphilic Materials

Traditionally, an amphiphilic molecule is one which finds its energetically favorable interaction at the interface of two disparate fluids, such as soap in oily water. Indeed, early studies of amphiphilic materials concerned emulsions formed from two immiscible fluids combined with an amphiphilic surfactant. Lipids, formed of a hydrophilic head group and a hydrophobic tail also belong to the class of amphiphilic molecules. More recently, developments in synthetic chemistry, such as atom transfer radical polymerization, have simplified the process of attaching charge groups to polymers, greatly expanding the possible classes of amphiphilic polymers that can be readily synthesized, see [4, 18]. Amphiphilic blends typically phase separate, however the propensity of the amphiphilic molecules to form monolayers leads to an energetic preference for thin interfaces. As a result the interfaces are often co-dimension one bilayers, or co-dimension two pore structures—morphologies that are often referred to collectively as networks. And these networks have significant value: they self-assemble at the nano-scale, yielding huge densities of solvent-accessible surface area and are often charge-lined, which renders them effective as selective ionic conductors. Due to these traits, amphiphilic materials have found use in many types of energy conversion devices, forming the ionomer membranes in fuel cells, the photo-active collecting matrix in bulk-heterojunction solar cells, and the separator membrane in Lithium ion batteries.

N. Kraitzman (✉) • K. Promislow
Department of Mathematics, Michigan State University, East Lansing, MI 48824, USA
e-mail: kraitzm1@msu.edu; kpromisl@math.msu.edu

J.-P. Bourguignon et al. (eds.), *Mathematics of Energy and Climate Change*,
CIM Series in Mathematical Sciences 2, DOI 10.1007/978-3-319-16121-1_8

The casting of blends of amphiphilic mixtures and di-block polymers presents a rich array of distinct morphologies, however control of the end-state morphology is experimentally challenging due to the delicate roles played by solvent type, salt concentration and counter-ion type, di-block composition and polydispersity, temperature, and pH. It has been shown [8], that changing the concentration of water in a water-dioxane solvent blend induces bifurcations in amphiphilic di-blocks yielding micelle, micelle-pore, pore, pore-vesicle, and vesicle network morphologies. Similar bifurcation were obtained in PEO-PB amphiphilic di-blocks by changing the density of charge groups in the hydrophilic portion [15], or by varying the length of the hydrophobic portion of the di-block [21]. Morphological reconfigurations can also be achieved through varying temperature [12, 25], and concentrations of counter-ions [29].

We pay particular attention to the experimental investigation in [2, 28] of the division of primitive lipid membranes, for which a particularly simple method was devised to induce the bilayer to micelle bifurcations discussed above. Szostak's group formed a suspension of spherical vesicles of 10 % phospholipid and found that increasing the concentration of free oleo-lipids dispersed in the bulk solvent induced a fingering instability in spherical phospholipid vesicles, depicted in the three horizontally arranged panels on the left side of Fig. 1. The end state consisted of long, co-dimension two pore morphologies. In a subsequent experiment, the charge density of cylindrical pores was suddenly increased through a photo-oxidation process; the jump in charge density induced a pearling bifurcation causing the

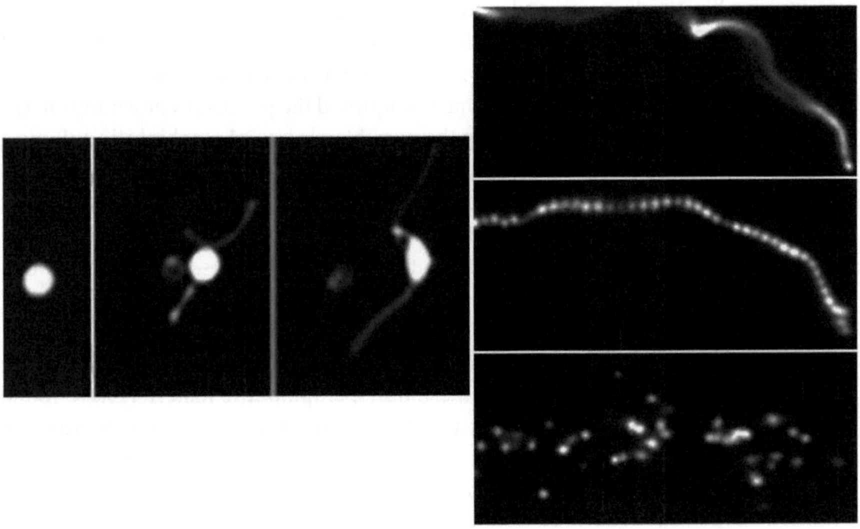

Fig. 1 Szostak's mechanism for division of primitive cell membrane: (*left*) raising the background concentration of lipids induces the vesicle to grow worm-like (co-dimension two) protrusions over a 74 ns time period [2], (*right*) changing the density of charged groups on the surface via a photochemically induced redox reaction incites the pore to pearl and break into micelles [28]. Reprint permission grunted by Proceedings of National Academy of Science

pore structures to break into individual micelles, as depicted in the three vertically arranged panels on the right side of Fig. 1. The goal of this overview is to present an analysis of related bifurcations within the context of the Functionalized Cahn-Hilliard free energy.

2 The Functionalized Cahn-Hilliard Free Energy

The Cahn-Hilliard free energy [3], describes the spinodal decomposition of a binary mixture. For a fixed domain, $\Omega \subset \mathbf{R}^3$, a phase function $u \in H^1(\Omega)$ describes the volume fraction of one component of the binary mixture, and the free energy is modeled by a function of the density u weakly perturbed by the spatially isotropic gradients

$$\mathscr{E}(u) = \int_\Omega f(u, \varepsilon^2|\nabla u|^2, \varepsilon^2 \Delta u)\, dx. \tag{1}$$

Expanding the free energy in orders of ε and truncating at $O(\varepsilon^4)$, yields an expression of the form

$$\mathscr{E}(u) = \int_\Omega f(u, 0, 0) + \varepsilon^2 A(u)|\nabla u|^2 + \varepsilon^2 B(u)\Delta u\, dx. \tag{2}$$

To obtain a generic normal form for the free energy, Cahn and Hilliard integrated by parts on the $B(u)$ term, set the resulting coefficient of $|\nabla u|^2$ to $\frac{1}{2}$, and renamed the potential $f(u, 0, 0)$ to $W(u)$. The result is the Cahn-Hilliard free energy

$$\mathscr{E}(u) = \int_\Omega \frac{\varepsilon^2}{2}|\nabla u|^2 + W(u)\, dx. \tag{3}$$

The corresponding H^{-1} gradient flow, the Cahn-Hilliard equation, takes the form

$$u_t = \Delta \frac{\delta \mathscr{E}}{\delta u} = \Delta(-\varepsilon^2 \Delta u + W'(u)). \tag{4}$$

Subject to traceless boundary conditions the Cahn-Hilliard equation preserves the total mass

$$\frac{d}{dt}\int_\Omega u(x,t)\, dx = 0, \tag{5}$$

and dissipates the Cahn-Hilliard free energy

$$\frac{d}{dt}\mathscr{E}(u) = \left\langle u_t, \frac{\delta \mathscr{E}}{\delta u} \right\rangle_{L^2} = -\left\| \nabla \frac{\delta \mathscr{E}}{\delta u} \right\|_{L^2}^2 \le 0. \tag{6}$$

To model amphiphilic mixtures, such as emulsions formed by adding a minority fraction of an oil and soap mixture to water, Teubner and Strey [24] and Gompper and Schick [13] were motivated by small-angle X-ray scattering (SAXS) data to include a higher-order term in the usual Cahn-Hilliard expansion,

$$\mathscr{F}(u) := \int_{\Omega} f(u,0,0) + \varepsilon^2 A(u)|\nabla u|^2 + \varepsilon^2 B(u)\Delta u + \overbrace{C(u)}^{\geq 0} (\varepsilon^2 \Delta u)^2 \, dx. \tag{7}$$

The full form of this system supports too many possible regimes to permit a systematic study. It is important to find the simplest mathematical framework that supports the network morphologies typical of amphiphilic mixtures; we need new normal form. With this goal we first shift all the differential terms to powers of Laplacians; specifically, letting $\overline{A}$ denote the primitive of A, we replace $A(u)\nabla u$ with $\nabla \overline{A}(u)$ and integrate by parts on the term $\nabla \overline{A} \cdot \nabla u$ to obtain

$$\mathscr{F}(u) = \int_{\Omega} f(u,0,0) + (B(u) - \overline{A}(u))\varepsilon^2 \Delta u + C(u)(\varepsilon^2 \Delta u)^2 \, dx. \tag{8}$$

The energy density is a quadratic polynomial in $\varepsilon^2 \Delta u$, which suggests that we complete the square

$$\mathscr{F}(u) = \int_{\Omega} C(u) \left(\varepsilon^2 \Delta u - \frac{\overline{A} - B}{2C} \right)^2 + f(u,0,0) - \frac{(\overline{A} - B)^2}{4C(u)} \, dx. \tag{9}$$

For simplicity we replace $C(u)$ with $\frac{1}{2}$, and relabel the potential within and outside the squared term by $W'(u)$ and $P(u)$, respectively. The key point is that the first term is the square of the variational derivative of a Cahn-Hilliard type free energy, consequently the case $P \equiv 0$, when the energy is a perfect square, has the special property that its global minimizers are precisely the *critical points* of the corresponding Cahn-Hilliard energy. Indeed, a variant of this case was proposed as a target for Γ-convergence analysis by De Giorgi, see [22]. The normal form of the network is obtained by unfolding the perfect square via a scaled perturbation

$$\mathscr{F}(u) = \int_{\Omega} \frac{1}{2} \left(\varepsilon^2 \Delta u - W'(u) \right)^2 + \delta P(u) \, dx, \tag{10}$$

where $\delta \ll 1$. The function $W(u)$ is assumed to be a double-well potential with two minima at $u = b_{\pm}$ whose unequal depths are normalized so that $W(b_-) = 0 > W(b_+)$. Typically $b_- = 0$, however it is helpful to give this value a specific name. Thus $u = b_-$ is associated to a bulk solvent phase, while the size of $u - b_- > 0$ is proportional to the density of the amphiphilic phase. The parameter $\varepsilon \ll 1$ determines the interfacial width and corresponds to the ratio of the typical length of an amphiphilic molecule to the domain size.

The Functionalized Cahn-Hilliard free energy is a class of two distinguished limits and a particular choice for p,

$$\mathscr{F}_{\rm CH}(u) := \int_{\Omega} \frac{1}{2}\left(\varepsilon^2 \Delta u - W'(u;\tau)\right)^2 - \varepsilon^p \left(\frac{\varepsilon^2 \eta_1}{2}|\nabla u|^2 + \eta_2 W(u)\right) dx. \tag{11}$$

The functionalization terms, parameterized by $\eta_1 > 0$ and $\eta_2 \in \mathbb{R}$, are analogous to the surface and volume energies typical of models of charged solutes in confined domains, see [23] and particularly equation (67) of [1]. The minus sign in front of η_1 is of considerable significance—it incorporates the propensity of the amphiphilic surfactant phase to drive the creation of interface. Indeed, experimental tuning of solvent quality shows that morphological instability in amphiphilic mixtures is associated to (small) negative values of surface tension [26, 27]. In the FCH energy the gradient term, $-\eta_1|\nabla u|^2 < 0$, is localized on interfaces, associated to single-layers of surfactant molecules, whose growth lowers overall system energy—however the effect is *perturbative* and unrestricted growth is arrested by the penalty nature of the square term which keeps u close to the critical points of $\mathscr{E}_{\rm CH}$. The two distinguished limits correspond to difference choices for the exponent p in the functionalization terms. In the Strong Functionalization, $p = 1$, the functional terms dominate the Willmore corrections from the squared variational term. The Weak Functionalization, corresponding to $p = 2$, is the natural scaling for the Γ-limit as the curvature-type Willmore terms appear at the same asymptotic order as the functional terms.

The well-posedness of the minimization problem for the FCH, including the existence of global minimizers for fixed values of $\varepsilon > 0$ was established in [20] for a more general functional form over various natural function spaces. Depending upon the application, the volume-type η_2 functionalization perturbation incorporates the impact of counter-ion entropy (PEM fuel cells), capillary pressure, or entropic effects from constraint of tail groups (lipid bilayers) [11]. The form $\eta_2 W(u)$ is chosen primarily for convenience, as integrals of $W(u)$ evaluated at critical points of $\mathscr{E}_{\rm CH}$ grow increasingly negative with increasing interfacial co-dimension. We remark that the surface term $\eta_1|\nabla u|^2$ is equivalent to an $\eta_1 u W'(u)$ functional-form since an integration by parts on $-\eta_1|\nabla u|^2$ yields $\eta_1 u \Delta u$ which can be absorbed into the squared variation with a perturbed form of W.

The goal of this survey is to present an overview of the stability and dynamics of classes of quasi-minimizer network morphologies $\mathscr{N}$ of $\mathscr{F}_{CH}$, which we define to be functions $u \in H^2(\Omega)$ which have an asymptotically small minority of amphiphilic phase, satisfy assigned boundary conditions, and render the free energy sufficiently small. Specifically for each fixed $C > 0$ we define the set of quasi-minimizing network morphologies

$$\mathscr{N}_C := \left\{ u \in H^2(\Omega) \Big| \int_{\Omega} |u - b_-|\, dx \le C\varepsilon \text{ and } \mathscr{F}_{CH}(u) \le C\varepsilon^{p+1} \right\}, \tag{12}$$

where the value of $\mathscr{F}_{CH}(u)$ can be negative.

3 Critical Points of the Functionalized Cahn-Hilliard Free Energy

For simplicity, we focus on the strong FCH, whose critical points, subject to a total mass constraint, are the solutions of the associated Euler-Lagrange equation

$$\frac{\delta \mathscr{F}}{\delta u} := \left(\varepsilon^2 \Delta - W''(u)\right)\left(\varepsilon^2 \Delta u - W'(u)\right) - \varepsilon\left(-\varepsilon^2 \eta_1 \Delta u + \eta_2 W'(u)\right) = \lambda, \tag{13}$$

where $\lambda \in \mathbb{R}$ is the Lagrange multiplier. Intuitively, solutions of the critical point equation which are close to global minima of the FCH should also be close to critical points of the Cahn-Hilliard free energy, solving

$$\frac{\delta \mathscr{E}}{\delta u} := -\varepsilon^2 \Delta u + W'(u) = O(\varepsilon). \tag{14}$$

This observation further suggests that the Lagrange multiplier should scale with ε, that is $\lambda = \varepsilon \hat{\lambda}$, and we may rewrite the critical point equation as two, coupled second order systems

$$\begin{aligned} \varepsilon^2 \Delta u - W'(u) &= \varepsilon v, \\ \left(\varepsilon^2 \Delta - W''(u)\right) v &= \left(-\varepsilon^2 \eta_1 \Delta u + \eta_2 W'(u)\right) + \hat{\lambda}. \end{aligned} \tag{15}$$

The singularly perturbed nature of the Euler-Lagrange system makes it amenable to dimensional reduction, yielding localized solutions build upon immersions in $\mathbb{R}^3$ of different co-dimensions, see Fig. 2. We first consider co-dimension one immersions, which we dress into bilayer morphologies which are quasi-minimizer of $\mathscr{F}_{CH}$. The first step in the dressing process is to produce a local coordinate system. More specifically, given a smooth, closed two-dimensional manifold Γ_b embedded

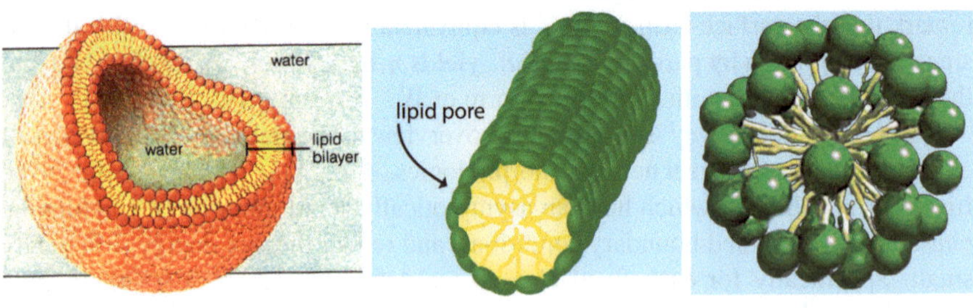

Fig. 2 Depiction of bilayer (*left*, *source*: academic.brooklyn.cuny.edu), pore (*center*), and micelle (*right*) morphologies of lipids. The co-dimension associated to the morphology is the difference between the space dimension and the number of tangent directions of the minimal manifold whose normal bundle locally foliates the morphology. In $\mathbb{R}^3$ bilayers are co-dimension one, pores are co-dimension two, and micelles are co-dimensional three

in $\Omega \subset \mathbb{R}^3$, the "whiskered" coordinate system is defined in a tubular neighborhood of Γ_b via the mapping

$$x = \rho(s, z) := \zeta_b(s) + \varepsilon\nu(s)z, \tag{16}$$

where $\zeta_b : S \mapsto \mathbb{R}^3$ is a local parameterization of Γ_b and $\nu(s)$ is the outward unit normal to Γ. The variable z is often called the ε-scaled, signed distance to Γ, while the variables $s = (s_1, s_2)$ parameterize the tangential directions of Γ. In general it is the number of normal directions, the co-dimension, of the manifold which determines the stability properties of the associated dressing. We define an admissible class of co-dimension one interfaces, whose dressings will be quasi-minimizer of $\mathscr{F}_{CH}$.

Definition 1 ([14]) For fixed $K, \ell > 0$ the family, $\mathscr{G}_{K,\ell}$, of "admissible interfaces" is comprised of closed (compact and without boundary), oriented two dimensional manifolds Γ embedded in $\mathbb{R}^3$, which are far from self-intersection and with a smooth second fundamental form. More precisely,

(i) The $W^{4,\infty}(S)$ norm of the second fundamental form of Γ and its principal curvatures are bounded by K.
(ii) The whiskers of length $3\ell < 1/K$, in the unscaled distance, defined for each $s_0 \in S$ by, $w_{s_0} := \{x : s(x) = s_0, |z(x)| < 3\ell/\varepsilon\}$, neither intersect each-other nor $\partial\Omega$ (except when considering periodic boundary conditions).
(iii) The surface area, $|\Gamma|$, of Γ is bounded by K.

We will denote an admissible co-dimension one manifold by Γ_b, the 'b' is for bilayer. The associated change of variables $x \to \rho(s, z)$ is a C^4 diffeomorphism on the "reach",

$$\Gamma_b^\ell := \left\{\rho(s, z) \in \mathbb{R}^d \middle| s \in S, -\ell/\varepsilon \le z \le \ell/\varepsilon\right\} \subset \Omega, \tag{17}$$

of Γ_b. The white region in Fig. 3 (right) depicts the reach of the associated immersion Γ_b.

In the whiskered coordinates the Cartesian Laplacian takes the form

$$\varepsilon^2 \Delta_x = \partial_z^2 + \varepsilon\partial_z J/(\varepsilon J)\partial_z + \varepsilon^2 J^{-1} \sum_{i,j=1}^{2} \frac{\partial}{\partial s_i} G^{ij} J \frac{\partial}{\partial s_j} = \partial_z^2 + \varepsilon H(s, z)\partial_z + \varepsilon^2 \Delta_G, \tag{18}$$

where J is the Jacobian of the change of variables, $H = \partial_z J/(\varepsilon J)$ is the extended curvature, and $\mathbf{G} = G_{ij}$ is the metric tensor, see Sect. 6 of [14] for details tailored to the context of the FCH free energy. In particular, at leader order $H(s, z) = H_0(s) + O(\varepsilon z)$ where H_0 is the mean curvature of Γ_b at $\rho(s, 0)$ and $\Delta_G = \Delta_s + O(\varepsilon z)$ where Δ_s is the usual Laplace-Beltrami operator on Γ_b.

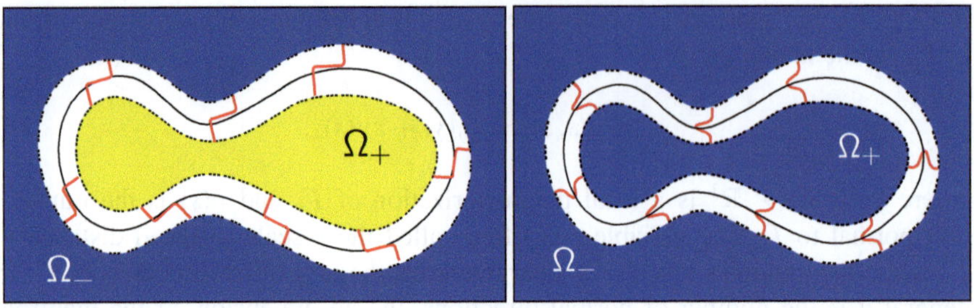

Fig. 3 Single layer and Bilayer solutions of the Euler-Lagrange equation associated to the interface Γ. For the single layer solution Γ separates regions $u = b_-$ from $u = b_+$, while the bilayer solution corresponds to $u = b_-$ on either side of the bilayer, with a brief excursion $u > b_-$ near Γ

In the whiskered coordinates the first equation of (15) reduces, at leading order to a second-order ODE in z, for the one-dimension profile $\phi(z)$,

$$\partial_z^2 \phi(z) = W'(\phi), \tag{19}$$

defined for $|z| \leq \ell/\varepsilon$. Since the double-well W is assumed to have unequal depth wells $0 = W(b_-) > W(b_+)$, a simple phase-plane analysis shows that this equation supports a unique solution ϕ_b which is *homoclinic* to b_-, that is $\phi_b(z) \to b_-$ as $z \to \pm\infty$, see [16] for a detailed analysis of the existence and linear analysis stability for homoclinic waves. We define the leading-order structure of the bilayer critical point, $u_b = u_b(x; \Gamma_b)$ as the bilayer "dressing" of Γ_b with ϕ_b,

$$u_b(x) := \phi_b(z(x)) + O(\varepsilon), \tag{20}$$

for $x \in \Gamma_b^\ell$ and smoothly extend u_b to equal b_- off of Γ_b^ℓ, see Fig. 3. We remark that if $g = g(z)$ decays exponentially to zero in z then $g \in L^2(\mathbb{R})$ and g has an extension $\tilde{g} \in L^2(\Omega)$ defined by $\tilde{g}(x) = g(z(x))$ on the reach of Γ_b and smoothly extended to zero off the reach. By abuse of notation, we use g to denote both the original function and its extension, in particular using both $\|g\|_{L^2(\mathbb{R})}$ and $\|g\|_{L^2(\Omega)}$ where the meaning is made clear by choice of inner product.

The $O(\varepsilon)$ correction, $u_{b,1}$ to u_b also plays a fundamental role, it is straight forward to see that it should solve

$$L_0 u_{b,1} = L_0^{-1}\left(-\eta_1 \phi_b'' + \eta_2 W'(\phi_b) + \hat{\lambda}\right) + H_0(s)\phi_b', \tag{21}$$

where we introduce the Sturm-Liouville operator

$$L_0 := \partial_z^2 - W''(\phi_b), \tag{22}$$

which is the linearization of (19) about ϕ_b. However there is a complication, whose resolution requires an understanding of the spectral properties of L_0 acting on an "infinite" whisker, that is on $L^2(\mathbb{R})$. The translational invariance of the critical point equation (19) forces $L_0\phi_b' = 0$, and since ϕ_b is homoclinic its derivative has a zero at $z = 0$. By the Sturm-Liouville theory, $\psi_1 := \phi_b'$ is the first excited state (eigenmode) of L_0 acting on $L^2(\mathbb{R})$ with eigenvalue $\lambda_1 = 0$, and there exists a ground state eigenmode ψ_0 with no zeros, and eigenvalue $\lambda_0 > 0$. The remainder of the spectrum of L_0 is strictly negative. It is easy to see that the terms acted upon by L_0^{-1} in (21) are $L^2(\mathbb{R})$ orthogonal to ψ_1, and hence lie in the range of L_0. However for each fixed value of s, the term $H_0(s)\phi_b'$ is not orthogonal to ψ_1. Nevertheless, in [9] it is shown that exact critical points can be constructed for flat interfaces, where $H_0 \equiv 0$ and for constant curvature interfaces if λ is appropriately tuned. However we may construct quasi-minimizers of $\mathscr{F}_{CH}$ by dropping the curvature term in the construction of u_b. It will be accounted for later as a driving force for the geometric evolution of the underlying co-dimension one immersion Γ_b. Proceeding, we drop the curvature term, invert L_0 and decompose $u_{b,1}$ into a local term $\phi_{b,1}$ which decays exponentially to zero in z, and is smoothly extended to be zero off of Γ_b^ℓ, and a constant term

$$\gamma_1 = \frac{\hat{\lambda}}{\alpha_-^2}, \tag{23}$$

where we introduce the far-field well coercivity $\alpha_- := W''(b_-) > 0$. Consequently u_b admits the quasi-steady expansion

$$u_b(x) = \phi_b(z) + \varepsilon\left(\gamma_1 + \phi_{b,1}(z)\right) + O(\varepsilon^2). \tag{24}$$

The local term $\phi_{b,1}$ corrects the structure of ϕ_b within the reach, while the spatial constant γ_1 adjusts the far-field behavior of u_b, which is now $b := b_- + \varepsilon\gamma_1 + O(\varepsilon^2)$. It is γ_1 that plays a key role in the evolution and bifurcation of the quasi-steady interfaces. Indeed this is the parameter Szostak tweaked when adding oleo-lipids to the bulk solvent phase.

Momentarily setting aside the mass constraint, there are two classes of free parameters in our construction of u_b, the spatially constant background correction, γ_1, and the interface shape Γ_b. It is instructive to examine the value of the FCH energy over the associated families of bilayer dressings, u_b, in particular the relation between the interface size and the total mass of available lipid. We first evaluate the free energy, which takes the form

$$\mathscr{F}(u_b) = \int_\Omega \frac{1}{2}\left(\varepsilon^2\Delta u_b - W'(u_b)\right)^2 - \varepsilon\left(\frac{\varepsilon^2\eta_1}{2}|\nabla u_b|^2 + \eta_2 W(u_b)\right) dx, \tag{25}$$

and break the integral over the near-field Γ_b^ℓ and far-field $\tilde{\Gamma}_\ell := \Omega \backslash \Gamma_b^\ell$. Denoting the near-field integral by $\mathscr{F}_\ell(u_b)$ we change to local coordinates

$$\begin{aligned}\mathscr{F}_\ell(u_b) &= \int_{\Gamma_b^\ell} \frac{1}{2}\left(\varepsilon^2 \Delta u_b - W'(u_b)\right)^2 - \varepsilon\left(\frac{\varepsilon^2\eta_1}{2}|\nabla u_b|^2 + \eta_2 W(u_b)\right) dx,\\ &= \int_{\Gamma_b}\int_{-\ell/\varepsilon}^{\ell/\varepsilon} \frac{1}{2}\left(\partial_z^2\phi_b - W'(\phi_b) + \varepsilon H_0(s)\partial_z\phi_b\right)^2\\ &\quad -\varepsilon\left(\frac{\eta_1}{2}|\partial_z\phi_b|^2 + \eta_2 W(\phi_b)\right) J(s,z)\, dz\, ds,\end{aligned} \tag{26}$$

where the Jacobian takes the form $J = \varepsilon + \varepsilon^2 z H_0(s) + O(\varepsilon^3 z^2)$. Expanding the Jacobian and keeping only leading order terms we find

$$\begin{aligned}\mathscr{F}_\ell(u_b) = \varepsilon \int_{\Gamma_b}\int_{-\ell/\varepsilon}^{\ell/\varepsilon} &\frac{\varepsilon^2}{2}\left(L_0(\gamma_1 + \phi_{b,1}) + H_0(s)\phi_b'\right)^2\\ &-\varepsilon\left(\frac{\eta_1}{2}|\phi_b'|^2 + \eta_2 W(\phi_b(z))\right) dz\, ds.\end{aligned} \tag{27}$$

The localized functions in the squared term will yield $O(\varepsilon^3)$ integrals which are negligible. Moreover integrating (19) we see that $(\phi_b')^2 = 2W(\phi_b)$. Together these two observations allow us to rewrite the localized component of the free energy as,

$$\mathscr{F}_\ell(u_b) = \varepsilon^2|\Gamma_b|\left(\ell\gamma_1^2\alpha_-^2 - \frac{\eta_1+\eta_2}{2}\sigma_b\right), \tag{28}$$

where we introduced the bilayer 'surface tension' $\sigma_b := \|\phi_b'\|^2_{L^2(\mathbb{R})}$. In the far-field region u_b takes the spatially constant value

$$u_b(x) = b := b_- + \varepsilon\gamma_1 + O(\varepsilon^2), \qquad x \in \tilde{\Gamma}_\ell, \tag{29}$$

for which value $W'(b) = \varepsilon\alpha_-\gamma_1 + O(\varepsilon^2)$, and $W(b) = O(\varepsilon^2)$. The far-field contribution to the energy thus reduces to the leading order expression,

$$\tilde{\mathscr{F}}_\ell(u_b) = \varepsilon^2(|\Omega| - 2\ell|\Gamma|)\frac{1}{2}\gamma_1^2\alpha_-^2 + O(\varepsilon^3). \tag{30}$$

Combining the near- and far-field expressions, the total energy takes the form

$$\mathscr{F}(u_b) = \varepsilon^2\left(\frac{\alpha_-^2|\Omega|}{2}\gamma_1^2 - |\Gamma_b|\frac{\eta_1+\eta_2}{2}\sigma_b\right). \tag{31}$$

A similar decomposition of the integrals shows that the total mass of amphiphilic material is

$$M := \int_{\Omega} u_b(x) - b_- \, dx = \int_{\tilde{\Gamma}_{b,\ell}} \varepsilon \frac{\mu_1}{\alpha_-^2} \, dx + \int_{\Gamma_b} \int_{-\ell/\varepsilon}^{\ell/\varepsilon} (U_b + \varepsilon \frac{\mu_1}{\alpha_-^2}) J_b \, dz \, ds$$
$$= \varepsilon |\Omega| \frac{\mu_1}{\alpha_-^2} + \varepsilon |\Gamma_b| m_b, \tag{32}$$

where $m_b := \int_{\mathbb{R}} \phi_b(z) - b_- \, dz > 0$, is the mass of amphiphilic material per unit length of bilayer. Typically the amphiphilic component is scarce within the bulk, so that $M = \varepsilon \hat{M}$ (don't put too much soap in the washing machine!), and since Γ_b is admissible its interfacial area $|\Gamma_b|$ is $O(1)$. These assumptions render u_b a quasi-minimizer of $\mathscr{F}_{CH}$, moreover a prescribed value of $\hat{M}$ and γ_1 determines the area $|\Gamma_b|$ of the bilayer interface. Consequently, the minimization of $\mathscr{F}(u_b)$ over Γ_b and γ_1, subject to the mass constraint reduces to the optimization of a quadratic polynomial in γ_1 which yields the optimal value

$$\gamma_b^* = -\frac{\eta_1 + \eta_2}{2} \frac{\sigma_b}{m_b \alpha_-^2}, \tag{33}$$

of amphiphilic material in the bulk region. For the strong functionalization only the area of an admissible co-dimension one interface, and not its curvature, enter into the leading-order determination of the free energy of its bilayer dressing. Moreover bilayers prefer an optimal far-field value of lipid, γ_b^* which is independent of the scaled mass constraint $\hat{M}$ and hence the area of the bilayer—it is a universal property of the system as determined by the shape of the well W throughout m_b, σ_b, and α_- and through the functionalization parameters η_1 and η_2. For the weak functionalization the Willmore term, the integral of the square of the mean curvature over Γ_b, enters into the free energy at leading order, and the optimization is more subtle.

There are critical points of $\mathscr{F}$ for which λ is $O(1)$, in particular the *single-layer* solutions, which correspond to heteroclinic orbits of (19) that connect two equilibrium values, see Fig. 3 (left). For the Cahn-Hilliard free energy single-layers form the dominant global minimizers, however they are generically saddle points of the FCH, and are susceptible to meander instabilities in the gradient flow, as discussed below. It is important to emphasize that single-layers and bilayers are distinct morphologies—single-layers separate phase A from phase B while bilayers separate phase A into two regions by a thin layer of phase B, see Fig. 3. In particular bilayers can rupture, re-uniting the two regions of phase A, as when a lipid bilayer opens a pore, or tears. In addition, the interfacial component is a conserved quantity for bilayers, and when the bilayer is stretched the interface must thin, which naturally increases its free energy as it deforms from its equilibrium profile ϕ_b—bilayers can support non-zero tangential stresses.

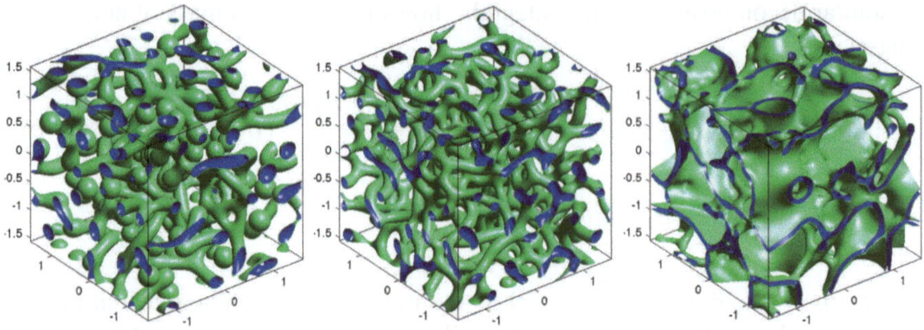

Fig. 4 Level sets $u = 0.4$ (*green*) and $u = 0.45$ (*blue*) of quasi-minimizers of the Functionalized Cahn-Hilliard free energy obtained from the mass preserving gradient flow (63) from identical initial data. The parameter values are $\varepsilon = 0.03$, $b_\pm = \pm 1$, the well satisfies $W(-1) = 0 > W(1) = -0.3$, $\eta_1 = 5$ and $\eta_2 = -2, 1$, and 5 in the panels from *left to right*, yielding micelles—pore, pore, and bilayer dominate networks respectively. Images courtesy of Andrew Christlieb and Jaylan Jones

The FCH critical point equation also possesses co-dimension two solutions in $\mathbb{R}^3$, see Fig. 4. These are based upon a foliation of a neighborhood of a smooth, closed, non-self intersecting one dimensional manifold Γ_p immersed in Ω. The local coordinate system takes the form

$$x = \rho_p(s, z_1, z_2) = \zeta_p(s) + \varepsilon \left(z_1 N_1(s) + z_2 N_2(s)\right), \tag{34}$$

where N_1 and N_2 are orthogonal unit vectors which are also orthogonal to the tangent vector $\zeta_p'(s)$. Within the reach Γ_p^ℓ of Γ_p the Laplacian admits the local form

$$\Delta_s = \Delta_R + \varepsilon \boldsymbol{\kappa}(s, z) \cdot \nabla_z + \varepsilon^2 D_s^2, \tag{35}$$

where Δ_R is the usual cylindrical Laplacian in (R, Θ) which correspond to the scaled normal distances $\mathbf{z} = (z_1, z_2)$, $\boldsymbol{\kappa} = (\kappa_1, \kappa_2)^T$ are the two curvatures of Γ_p at $\zeta_p(s)$, and D_s^2 reduces to the line diffusion operator on Γ_p when $\mathbf{z} = 0$, see [7] for details. Assuming axial symmetry, the leading order pore profile associated to the critical point equation (13) satisfies co-dimension two critical point equation

$$\partial_R^2 \phi_p + \frac{1}{R} \partial_R \phi_p = W'(\phi_p), \tag{36}$$

subject to $\partial_R \phi_p(0) = 0$ and $\phi_p \to b = b_- + \varepsilon \gamma_1 + O(\varepsilon^2)$ as $R \to \infty$. The leading order form for the pore quasi-minimizer network arises from the pore profile dressing of a co-dimension two interface Γ_p,

$$u_p(x) := \phi_p(R(x)) + \varepsilon \left(\gamma_1 + \phi_{p,1}(R)\right) + O(\varepsilon^2), \tag{37}$$

It is also possible to combine bilayer and pore quasi-minimizer, so long as the associated manifolds have non-intersecting reaches, *and* the far-field constant γ_1

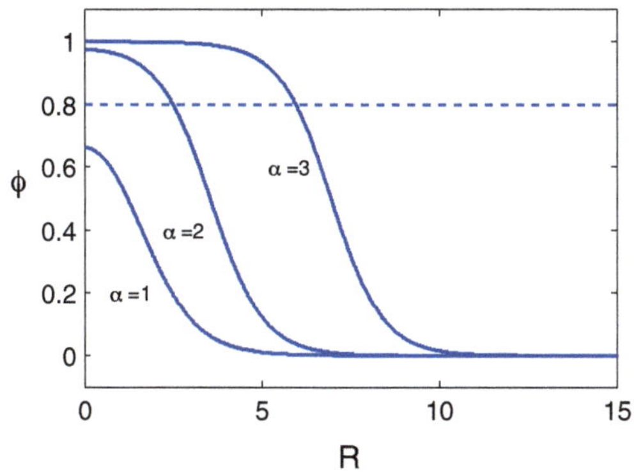

M_n^{core} (g/mol)	2500 ± 40	5850 ± 204
bilayer	8.7 ± 1.2	15.8 ± 2.8
pore	14.3 ± 1.6	25.4 ± 3.3
micelle (nm)	18.4 ± 2.6	38.8 ± 10.2

Fig. 5 (*Left*) a comparison of co-dimension $\alpha = 1, 2$, and 3 profiles computed from (19), (36), and (38) respectively. The profile is most sensitive to the difference in depths of the two wells: $W(b_-) - W(b_+) > 0$. (*Right*) a table of experimental data, from [15], indicating radii of bilayer, pore, and micelle morphologies obtained by varying the hydrophilic length of polymer in PEO-PB amphiphilic di-blocks with fixed hydrophobic (core) molecule weight, M_n^{core}, as indicated

takes a common value. Indeed, the quasi-steady evolution between co-existing co-dimension one and co-dimension two interfaces is driven by the competition between this common far-field value b. If the optimal far-field values associated to distinct co-dimensional morphologies differ, then the morphologies will not coexist over long time periods; one will grow on a slow time scale at the expense of the other, as described in Sect. 5.

Micelle, or co-dimension three solutions of the critical point equation reduce, in $\mathbb{R}^3$, to solutions of the usual spherical Laplacian. Assuming rotational symmetry, the leading order micelle profile is the unique solution of

$$\partial_R^2 \phi_m + \frac{2}{R}\partial_R \phi_m = W'(\phi_m), \tag{38}$$

subject to $\partial_R \phi_m(0) = 0$ and $\phi_m \to b$ as $R \to \infty$. A key prediction of the FCH free energy is that bilayers must be thinner than pores, which in turn are thinner than micelles. This observation is born out by experimental data, Fig. 5 (right).

4 Network Bifurcation in the FCH

The quasi-minimizer network morphologies developed in Sect. 3 are, at leading order, critical points of the Cahn-Hilliard, however these structures are not close to local minima but are rather quasi-saddle points of the CH free energy. An essential feature of the functional form of the FCH is its facility to build local minima out of the saddle points of the simpler CH free energy. This process is best understood by examining the second variational derivative of the FCH free energy at

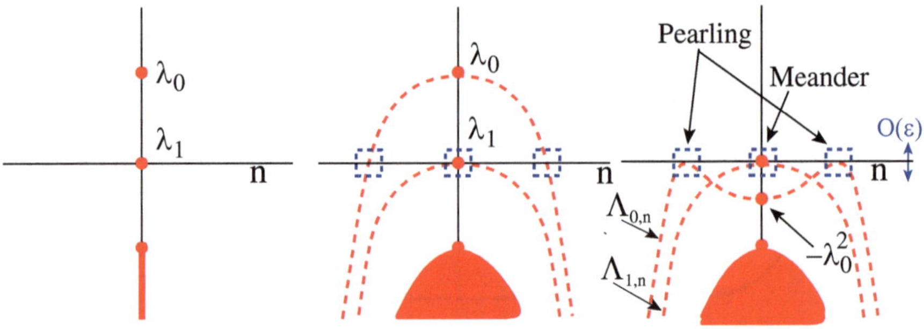

Fig. 6 The structure of the real spectrum of $-\mathbb{L}_b$ plotted verses Laplace-Beltrami wavenumber n. (*Left*) the Sturm-Liouville operator L_0, defined in (22), has one positive ground state eigenvalue, $\lambda_0 > 0$ and a one dimensional kernel, denoted λ_1. (*Center*) the extension of L_0 to $\mathscr{L} = L_0 + \varepsilon^2 \Delta_s$ adds side-bands in n, the Laplace-Beltrami index which bend back negatively at the rate $-(\lambda_0 - \varepsilon^2 \beta_k)^2$. (*Right*) the spectrum of the operator $-\mathbb{L} = -\mathscr{L}^2 + O(\varepsilon)$, (*minus sign* chosen to preserve orientation of images) is, to $O(\varepsilon)$, the negative square of the spectrum of $\mathscr{L}$. The side-band associated to λ_0 has a quadratic tangency at leading order, which may be raised or lowered by the functional terms, η_1 and η_2, the crossing of this spectrum through zero is the mechanism of the pearling instability

a smooth critical point, u_c, of the Cahn-Hilliard free energy. For traceless boundary conditions, such as periodic boundary conditions, see [20] for a detailed discussion of appropriate boundary conditions, the second variation takes the form

$$\mathbb{L}_{u_c} := \frac{\delta^2 \mathscr{F}}{\delta u^2}(u_c) = \left(\varepsilon^2 \Delta - W''(u_c)\right)^2 - \varepsilon^p \left(\eta_1 \varepsilon^2 \Delta + \eta_2 W''(u_b)\right). \tag{39}$$

For the bilayer quasi-minimizer, u_b, associated to an admissible, co-dimension one interface Γ_b, the second variational derivative $\mathbb{L}_b := \mathbb{L}_{u_b}$, takes a simplified form when acting on functions $u \in H^4(\Omega)$ whose support lies within the reach, Γ_b^ℓ, of Γ_b. On this subspace the operator admits the asymptotic expansion

$$\mathbb{L}_b = \left(L_0 + \varepsilon H \partial_z + \varepsilon^2 \Delta_s\right)^2 - \varepsilon^p \left(\eta_1 \partial_z^2 + \eta_2 W''(\phi_b)\right) + O(\varepsilon^{p+1}), \tag{40}$$

and the leading order structure of $\mathbb{L}_b$ is controlled by $\mathscr{L}^2$ where $\mathscr{L} := L_0 + \varepsilon^2 \Delta_s$ is the dominant part of the second variation of $\mathscr{E}$ at u_b. The remaining parts of $\mathbb{L}_b$ are relatively bounded and asymptotically small in comparison to $\mathscr{L}^2$.

The spectrum of the operator $\mathbb{L}_b$ can be built from the spectrum of its constituents L_0 and Δ_s. The spectral properties of L_0 are described in Sect. 3 and depicted in Fig. 6 (left). The Laplace-Beltrami operator Δ_s is self-adjoint over $L^2(\Gamma_b)$ where, for each admissible interface Γ_b, the inner product is defined by

$$\langle f, g \rangle_{L^2(\Gamma)} := \int_\Gamma f(s) g(s) J_0(s)\, ds, \tag{41}$$

where $J_0 = \sqrt{\mathbf{g}}$ is the square root of the determinant of the first fundamental form of Γ_b. The eigenvalues $\{\beta_n\}_{n=0}^\infty$ of $-\Delta_s$ may be enumerated in increasing size with

$\beta_0 = 0$ and $\beta_1 > 0$. The associated Laplace-Beltrami eigenmodes $\{\Theta_n\}_{n=0}^{\infty}$ are orthonormal in the $L^2(\Gamma)$ norm.

Theorem 1 (Weyl Asymptotics [5]) *Let Γ_b be an admissible co-dimension one interface immersed in $\mathbb{R}^d$, and let $N(x)$ denote the number of eigenvalues of $-\Delta_s$, counted according to multiplicity, that are smaller than x, then*

$$N(x) \sim Cx^{\frac{d-1}{2}}. \tag{42}$$

In particular, $\beta_n \sim \tilde{C} n^{\frac{2}{d-1}}$.

Indeed, in [14] it was shown that for each admissible class, $\mathcal{G}_{k,\ell}$, of co-dimension one interfaces there exists $U > 0$, which may be chosen independent of $\varepsilon > 0$ such that the eigenfunctions associated to $\mathbb{L}_b$ with corresponding eigenvalues $\lambda < U$ comprise two sets, the *pearling eigenmodes* $\{\Psi_{0,n}\}_{n=N_1}^{N_2}$ and the *meander eigenmodes* $\{\Psi_{1,n}\}_{n=0}^{N_3}$ and moreover these eigenmodes admit the asymptotic form

$$\Psi_{j,n} = \psi_j(z)\Theta_n(s) + O(\varepsilon), \tag{43}$$

for $j = 0, 1$ and n running over the corresponding indices. For $j = 0, 1$ we introduce Σ_j, the set of indices n for which $\mathbb{L}_b$ acting on $\psi_j\Theta_n$ is small, i.e.,

$$\Sigma_j := \{n \mid (\lambda_j - \varepsilon^2\beta_n) \sim O(\sqrt{\varepsilon})\}. \tag{44}$$

From Weyl's asymptotic formula we deduce that $|\Sigma_0| \sim O(\varepsilon^{3/2-d}) \gg 1$. For the strong functionalization, $p = 1$, we look for an expression of the pearling eigenvalues in order to determine a condition for pearling stability. For Γ_b an admissible co-dimension one interface we consider the eigenvalue problem

$$\mathbb{L}_b\Psi_{0,n} = \Lambda_{0,n}\Psi_{0,n}, \tag{45}$$

associated to the second variation of $\mathcal{F}_{CH}$ about the bilayer dressing u_b. The spectrum of $\mathbb{L}_b$ cannot be localized by a regular perturbation expansion since the eigenvalues are asymptotically close together. We need bounds on the spectrum that are uniform in $\varepsilon \ll 1$. To this end we introduce the $L^2(\Omega)$ orthogonal projection Π onto the space

$$X_b := \text{span}\{\psi_j(z)\Theta_n(s) \big| j = 0, 1, \text{ and } n \in \Sigma_j \text{ respectively}\}, \tag{46}$$

which approximates the eigenspaces of $\mathbb{L}_b$ corresponding to pearling ($j = 0$) and meander ($j = 1$) eigenmodes. Functionally, Π, acts on $f \in L^2(\Omega)$ by

$$\Pi f := \sum_{k\in\Sigma_0} \langle f, \psi_0\Theta_k\rangle_{L^2(\Omega)}\psi_0\Theta_k + \sum_{k\in\Sigma_1} \langle f, \psi_1\Theta_k\rangle_{L^2(\Omega)}\psi_1\Theta_k, \tag{47}$$

with its complementary projection denoted $\tilde{\Pi} := I - \Pi$. We decompose the operator $\mathbb{L}_b$ into a 2×2 block form using the projections

$$\tilde{\mathbb{L}}_b := \begin{bmatrix} \Pi\mathbb{L}_b\Pi & \Pi\mathbb{L}_b\tilde{\Pi} \\ \tilde{\Pi}\mathbb{L}_b\Pi & \tilde{\Pi}\mathbb{L}_b\tilde{\Pi} \end{bmatrix}. \tag{48}$$

The upper-left element $\Pi\mathbb{L}_b\Pi$ can be written as a matrix $M \in \mathbb{R}^{N\times N}$ where $N \approx \varepsilon^{3/2-d}$ has entries

$$M_{j,k} := \left\langle \mathbb{L}_b\psi_0\Theta_j, \psi_0\Theta_k \right\rangle_{L^2(\Omega)}. \tag{49}$$

The off-diagonal terms are small, in norm, and the spectrum of the fully infinite dimensional piece, $\tilde{\Pi}\mathbb{L}_b\tilde{\Pi}$, is bounded from below by the aforementioned $U > 0$. We will show that the spectrum of $\mathbb{L}_b$ sufficiently below U is controlled by the spectrum of the matrix M which we characterize.

The matrix M has large dimension and hence care must be taken to distinguish between the size of the entries of M and the size of M as an operator. Indeed, the norm of a matrix as an operator from $l^2(\mathbb{R}^N)$ to $l^2(\mathbb{R}^N)$ generically scales like $\sqrt{N}$ times the l^∞ norm of its entries. However uniform norm bounds are possible, via a convolution style argument, if the off-diagonal elements decay sufficiently quickly.

Lemma 1 ([17]) *Fix $c > 0$ and assume that the entries of $A \in \mathbb{R}^{N\times N}$ satisfy the bound*

$$|A_{j,k}| \leq \frac{c}{1 + (k-j)^2}. \tag{50}$$

Then the matrix A is uniformly bounded from $l^2(\mathbb{R}^N)$ to $l^2(\mathbb{R}^N)$ independent of N.

Our goal is to show that the matrix M admits an asymptotic decomposition

$$M = M^0 + \varepsilon^q\tilde{M}. \tag{51}$$

For values of $q > \frac{1}{4} + \frac{d}{2}$, if the entries of $\tilde{M}$ are uniformly bounded then there exists a constant $C > 0$, independent of ε such that $\varepsilon^q\|\tilde{M}\|_{L^2} \ll \varepsilon$. That is, it is sufficient to uniformly bound the entries of $\tilde{M}$ to see that $\varepsilon^q\tilde{M}$ acts as a lower-order perturbation on the eigenvalues of M. We will handle the matrix M^0 using Lemma 1, so long as the interface Γ_b is sufficiently smooth, as is guaranteed by its admissibility. For simplicity we focus only on the pearling modes $j = 0$, neglecting the meander terms associated to $j = 1$. Using the expansion (40) of $\mathbb{L}_b$ and recasting the inner product in (49) in terms of the whiskered coordinates, the diagonal terms of the matrix M^0, at leading order, take the form

$$M^0_{k,k} = (\lambda_0 - \varepsilon^2\beta_k)^2 - \varepsilon(\gamma_1\alpha_-^2 S + \lambda_0(\eta_1 - \eta_2)\|\psi_0\|_2^2), \tag{52}$$

where the sign of the "shape factor"

$$S := \int_{\mathbb{R}} W'''(\phi_b)\psi_0^2(z)L_0^{-1}1\,dz, \tag{53}$$

determines if the pearling bifurcation absorbs amphiphilic material from the bulk or releases it. For $k \in \Sigma_0$, the quadratic term is bounded by $O(\varepsilon)$ but becomes dominant as k approaches the boundary of the set Σ_0 of pearling indices. The off-diagonal terms of M^0 are formally lower order, admitting the expansion

$$M^0_{j,k} = -\varepsilon^2 \int_\Gamma (\sigma_b H_0^2 + S_1 H_1)\Theta_k\Theta_j J_0(s)\,ds, \tag{54}$$

where the coefficient $S_1 := \int_{\mathbb{R}} W'''(\phi_b)\phi_b'\psi_0^2 z\,dz$ and $H_1 = k_1^2 + k_2^2$ is the sum of the squares of the curvatures of Γ_b. However they form a lower order operator on $l^2(\mathbb{R}^N)$ only if we can apply Lemma 1. To characterize the diagonal terms of M^0 we apply the following theorem

Theorem 2 ([17]) *Let $\Gamma \subset R^d$ be an admissible interface, with curvatures $\mathbf{k} = (k_1, \cdots, k_{d-1})$ in $W^{4,\infty}(\Gamma_b)$ and $f : R^{d-1} \to R$ a smooth function. Then there exist constants $c_1, c_2, c > 0$ such that for every $k, j \in N$, $k \neq j$,*

$$\begin{aligned}\left|\int_\Gamma f(\mathbf{k})\Theta_k\Theta_j J_0 ds\right| &\le \frac{1}{\beta_k+\beta_j}\left(c_1 + c_2\int_\Gamma \left|\Delta_s^{-1}(\nabla_s\Theta_k\nabla_s\Theta_j)\right| J_0 ds\right)\\ &\le \frac{c}{1+|k-j|^2}.\end{aligned} \tag{55}$$

As a consequence, the matrix M^0 can be written as

$$M^0 = D + \varepsilon^2 A, \tag{56}$$

where D is a diagonal matrix with entries

$$D_{k,k} = (\lambda_0 - \varepsilon^2\beta_k)^2 - \varepsilon(\gamma_1\alpha_-^2 S + \lambda_0(\eta_1-\eta_2)||\psi_0||_2^2), \tag{57}$$

and A is uniformly bounded as an operator on $l^2(\mathbb{R}^N)$. Since $\varepsilon^q\tilde{M}$ is also lower order as an operator, the eigenvalues of M take the form

$$\Lambda_{0,n} = (\lambda_0 - \varepsilon^2\beta_n)^2 - \varepsilon(\gamma_1\alpha_-^2 S + \lambda_0(\eta_1-\eta_2)\|\psi_0\|_2^2) + o(\varepsilon^2). \tag{58}$$

Weyl asymptotics imply that the separation between Laplace-Beltrami eigenvalues scales like $O(\sqrt{\varepsilon})$ for $\beta_n \approx \varepsilon^{-\frac{1}{2}}$, hence $\lambda_0 - \varepsilon^2\beta_k$ can be made as small as $O(\varepsilon)$. This shows that the squared term is lower order near the turning point of the pearling spectrum, see Fig. 6 (right). Assuming that the shape factor $S < 0$, which is true for a genetic class of double-wells, W, see Sect. 5 of [9], the spectrum of M will contain negative eigenvalues if and only if γ_1 satisfies the *pearling condition*

$$P_* := -\frac{\lambda_0(\eta_1 - \eta_2)\|\psi_0\|_{L^2}^2}{\alpha_-^2 S} > \gamma_1. \tag{59}$$

To connect the spectrum of M to that of $\mathbb{L}_b$, we must bound the interaction between the projection Π and the operator $\mathbb{L}_b$. If Π where a spectral projection associated to $\mathbb{L}_b$ then the two operators would commute, and since $\Pi\tilde{\Pi} = 0$, the off-diagonal terms would be zero. However X_b only approximates a spectral subset of $\mathbb{L}_b$, and the estimates $\|\Pi\mathbb{L}_b\tilde{\Pi}\|_{L^2(\Omega)} = \|\tilde{\Pi}\mathbb{L}_b\Pi\|_{L^2(\Omega)} \leq C\varepsilon$, are sharp. However the restricted operator $\tilde{\Pi}\mathbb{L}_b\tilde{\Pi}$ is uniformly coercive on L^2 and its spectrum is bounded from below by $U > 0$ which may be chosen independent of sufficiently small $\varepsilon > 0$. In this case, for any $\lambda < U$ we introduce $B := \Pi\mathbb{L}_b\tilde{\Pi}$ and $C := \tilde{\Pi}\mathbb{L}_b\tilde{\Pi}$ and reduce the 2×2 representation of the eigenvalue problem

$$\begin{bmatrix} M & B \\ B^T & C \end{bmatrix} \begin{bmatrix} v_1 \\ v_2 \end{bmatrix} = \lambda \begin{bmatrix} v_1 \\ v_2 \end{bmatrix} \tag{60}$$

to a finite dimensional system for the component v_1, which solves

$$(M - \lambda)v_1 = -B(C - \lambda)^{-1}B^T v_1. \tag{61}$$

In particular, taking the l^2 norm of both sides and using the estimates on B, B^T, and the distance of λ to $\sigma(C)$ to estimate the norm of the resolvent, we have

$$\|(M - \lambda)v_1\|_{l^2} \leq \frac{c\varepsilon^2}{|U - \lambda|}\|v_1\|_{l^2}. \tag{62}$$

For λ an order of one distance from U this estimate implies that $\mathrm{dist}(\lambda, \sigma(M)) = O(\varepsilon^2)$, so that the spectrum of $\mathbb{L}_b$ below U lies within $O(\varepsilon^2)$ of the spectrum of M. In particular, if the spectrum of M is bounded from below by a positive $O(\varepsilon)$ quantity, then so is the spectrum of $\mathbb{L}_b$. Conversely, since $M_{k,k} = \langle \mathbb{L}_b\psi_0\Theta_k, \psi_0\Theta_k\rangle_{L^2(\Omega)}$ it follows from the variational characterization of the spectrum of $\mathbb{L}_b$ that the smallest eigenvalue of $\sigma(\mathbb{L}_b)$ is smaller (more negative) than the smallest diagonal element of M. We deduce from these calculations that the pearling condition (59) applies to $\mathbb{L}_b$. Moreover, this is in agreement with Szostak's experiment, the photo-induced increase in charge on the lipid heads induced a pearling bifurcation and drove pores to micelles. The increase in charge corresponds, within the FCH, to an instantaneous increase in η_1; a sufficiently large increase, for a fix value of γ_1, will trigger the pearling condition (59). Figure 7 depicts the pearling of a co-dimension one sphere.

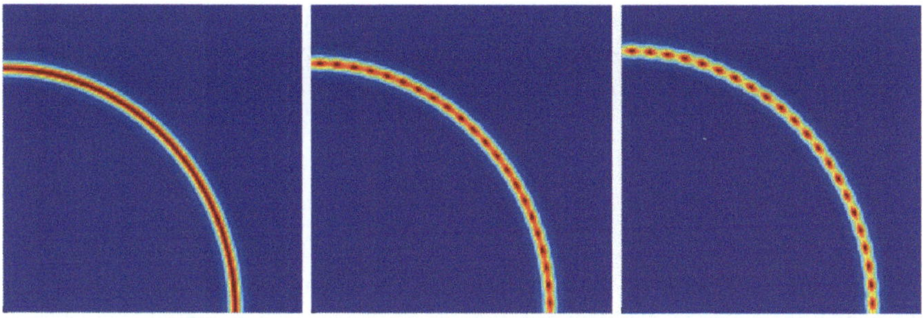

Fig. 7 Time evolution of a circular, co-dimension one bilayer under the FCH gradient flow (63) for values $\varepsilon = 0.1$ and $\eta_1 = \eta_2 = 2$. The times depicted correspond to $t = 0$, $t = 114$, and $t = 804$ and show the onset of the pearling bifurcation

5 Competitive Geometric Evolution of Bilayers and Pores

The over damped dynamics of amphiphilic polymer suspensions can be received from the Functionalized Cahn-Hilliard free energy via its gradient flows whose evolution preserves the volume fraction of the constituent species and lowers the free energy. Similar to the Cahn-Hilliard gradient flow given in (4), the simplest mass preserving gradient flow of the FCH is generated by the H^{-1} gradient,

$$u_t = \Delta \frac{\delta \mathscr{F}}{\delta u} = \Delta\left[\left(-\varepsilon^2\Delta + W''(u) - \varepsilon\eta_1\right)\left(-\varepsilon^2\Delta u + W'(u)\right) + \varepsilon(\eta_1 - \eta_2)W'(u)\right]. \tag{63}$$

The quasi-minimizer network morphologies constructed in Sect. 3 are not stationary solutions of the FCH gradient flow, but generate slow dynamics which may be locally parameterized by the interfacial sub-manifolds of bilayers and pores, respectively Γ_b and Γ_p. Indeed, when the pearling condition does not hold, then meander eigenvalues associated to the bilayer morphologies span the tangent plane to the manifold of bilayer configurations, parameterized by the admissible interfaces. The flow of the underlying interfacial structure can be obtained by projecting the residual $\frac{\delta \mathscr{F}}{\delta u}(u_b)$ of the critical point equation (13) onto this tangent plane. The method of matched asymptotic expansion provides a more accessible, but formal method to derive the interfacial motion. For a bilayer morphology, the ansatz (24) for u_b is augmented by taking the signed distance z to the interface Γ_b and the background state γ_1 to be functions of the slow scaled time $t_1 = t/\varepsilon$, and the gradient flow is solved by matching fluxes, particularly across the interfacial layers. For single layer morphologies, under the Cahn-Hilliard gradient flow this results in a Mullins-Sekerka flow for the interface, see [19]. For the FCH gradient flow (63) reduces, at leading order, to

$$\varepsilon\phi_b'(z)\frac{\partial z}{\partial t_1} + \varepsilon\frac{d\gamma_1}{dt_1} = \Delta\frac{\delta \mathscr{F}}{\delta u}(u_b) = \varepsilon\Delta H_0(s)\phi_b'(z) + O(\varepsilon^2). \tag{64}$$

The leading order residual arises from the mean-curvature term which was neglected in the construction of the bilayer, u_b. This term now becomes a driving force for the evolution of the interface Γ_b through the change in the signed distance function. Indeed, the quantity

$$V_b(s) := -\frac{\partial z}{\partial t_1}, \tag{65}$$

is the normal velocity of the interface Γ_b. The asymptotic reduction does lead to a Mullins-Sekerka problem for the far-field chemical potential, however its driving force is given by the interfacial mean curvature times the derivative of the bilayer profile at the interface, $H_0(s)\phi_b'(0)$. Since the bilayer is symmetric across the interface its derivative is zero, $\phi_b'(0) = 0$, and the Mullins-Sekerka problem is trivial. The outer chemical potential reduces to a spatial constant, and the far-field is characterized by amphiphilic density, $\gamma_1(t_1)$, whose value in determined by conservation of total mass, see [6] for details for bilayers under the weak functionalization. For the strong functionalization the resulting system takes the form

$$\begin{aligned} V_b &= \nu_b(\gamma_1 - \gamma_b^*)H_0, \\ \frac{d\gamma_1}{dt_1} &= -\nu_b m_b(\gamma_1 - \gamma_b^*)\int_{\Gamma_b} H_0^2\, dS, \end{aligned} \tag{66}$$

where $\nu_b := \frac{m_b}{\int_{\mathbb{R}}(\phi_b - b_-)^2\, dz} > 0$ and γ_b^* is the optimal far-field amphiphilic density derived by the optimization process in (33). The H^{-1} gradient flow drives pure bilayer interfaces by a quenched mean-curvature flow. While the flow drives γ_1 to its optimal value γ_b^*, the sign of the difference $\gamma_1 - \gamma_b^*$ is consequential. Indeed, in two space dimension, modulo reparamerization of the evolving interface, the curvature driven flow can be recast as an evolution equation of the single curvature H_0,

$$\frac{\partial H_0}{\partial t_1} = -(\partial_s^2 + H_0^2)V_b = -\nu_b(\gamma_1 - \gamma_b^*)(\partial_s^2 + H_0^2)H_0, \tag{67}$$

see Sect. 3.3 of [10] for details. If $\gamma_1 > \gamma_b*$, that is if the bulk value of amphiphilic material is in excess then the curvature driven flow is a backwards-heat equation in the curvatures. This is the nature of the fingering instability induced in [2] when oleo-lipids were added to the bulk of the spherical bilayer suspension. The fingering instability corresponds to a backward heat flow in the curvature. The resulting singularity is associated to the development of the pore type growth from the bilayer surface. Moreover, in [9] the condition $\gamma_1 > \gamma_b^*$ was identified as the point of bifurcation to linear instability of the meander eigenvalues associated to spherical bilayers. For $\gamma_1 < \gamma_b^*$ the curvature driven flow is locally well-posed but is subject to finite-time blow-up due to the cubic driving force, H_0^3. This is the familiar finite-type extinction of droplets under curvature driven flow. However, for the quenched flow (66), the relaxation of γ_1 to its equilibrium value precludes the blow-up if the initial curvatures are not too large.

A similar reduction can be performed for co-dimension two pore structures, parametrized by the one-dimensional immersion Γ_p. The result is a similar quenched curvature driven flow for the vector valued normal velocity $\mathbf{V}_p = -(\frac{\partial z_1}{\partial_{t_1}}, \frac{\partial z_2}{\partial_{t_1}})^T$,

$$
\begin{aligned}
\mathbf{V}_p &= \nu_p(\gamma_1 - \gamma_p^*)\boldsymbol{\kappa}(s), \\
\frac{d\gamma_1}{dt_1} &= -\varepsilon m_p(\gamma_1 - \gamma_p^*) \int_{\Gamma_p} |\boldsymbol{\kappa}|^2 \, ds,
\end{aligned}
\tag{68}
$$

where $\nu_p := \frac{m_p}{\pi \int_0^\infty (\phi_p')^2 R\, dR} > 0$, $\boldsymbol{\kappa}$ is the vector curvature of Γ_p, $m_p := 2\pi \int_0^\infty (\phi_p - b_-) R\, dR$ is the mass of amphiphilic material per unit length of pore structure and the equilibrium value

$$
\gamma_p^* := -\frac{\eta_1}{\alpha_-^2} \frac{\int_0^\infty (\phi_p')^2 R\, dR}{\int_0^\infty (\phi_p - b_-)^2 R\, dR},
\tag{69}
$$

is again independent of Γ_p. Most intriguingly, initial data corresponding to spatially separated pores and bilayers yields a competitive evolution that can be understood as a fight for surfactant, mediated through the common value of the bulk amphiphilic density γ_1, whose evolution is determined to impose the conservation of total mass,

$$
\begin{aligned}
V_n &= \nu_b(\gamma_1 - \gamma_b^*)H, \\
\mathbf{V}_p &= \nu_p(\gamma_1 - \gamma_p^*)\boldsymbol{\kappa}, \\
\frac{d\gamma_1}{dt_1} &= -\nu_b m_b(\gamma_1 - \gamma_b) \int_{\Gamma_b} H_0^2 dS - \varepsilon \nu_p m_p(\gamma_1 - \gamma_p^*) \int_{\Gamma_p} |\boldsymbol{\kappa}|^2 \, ds.
\end{aligned}
\tag{70}
$$

The competitive evolution of the bilayers and pores couples through curvature-weighted surface area. However, generically, the two morphologies seek differing equilibria values, which typically satisfy $\gamma_b^* > \gamma_p^*$, making coexistence of bilayers and pores impossible under the strong functionalization, unless one of the structures is flat, since zero curvature interfaces are at equilibrium independent of bulk value of amphiphile. For curved interfaces, the range $\gamma_1 \in [\gamma_p^*, \gamma_b^*]$ is invariant under the flow, and once γ_1 enters this range the bilayers will shrink, while the pore morphologies will grow. Moreover, if the pearling threshold P_* lies within the invariant range $[\gamma_p^*, \gamma_b^*]$ then the value of γ_1 may transiently decrease through the pearling threshold for bilayers (59), causing the bilayers to pearl as they shrink. Various realizations of this transient interaction are depicted in Fig. 8, for double wells W with increasing well tilt.

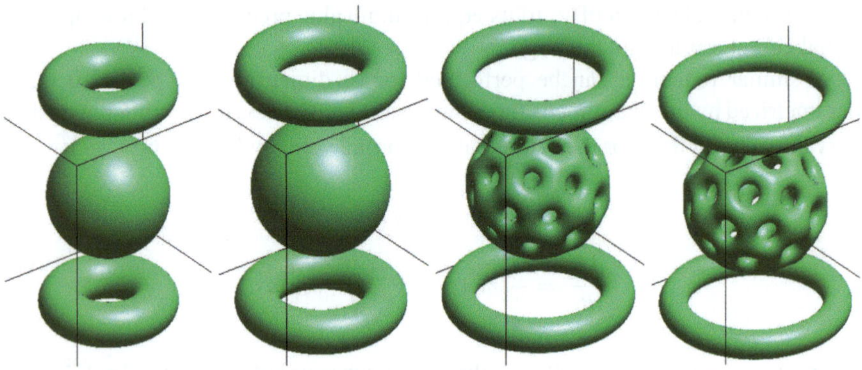

Fig. 8 Competition for the amphiphilic phase between a spherical bilayer (beach ball) and circular solid pore (hula hoop) as a function of the well tilt $W(b_-) - W(b_+)$. The image shows $t = 100$ end states of the FCH gradient flow (63) from identical initial data but with increasing values of the well tilt. Small tilt prefers bilayers, larger tilt prefers pores by increasing γ_b^* and the pearling threshold, P_*, which drives bilayers to pearl. Images courtesy of Andrew Christlieb and Jaylan Jones

6 Conclusion

The Functionalized Cahn-Hilliard free energy provides a compact description of the energy landscape driving morphological selection in amphiphilic mixtures, such as lipid bilayers. We have shown that the strength of the interactions of the hydrophilic units with the solvent phase, parameterized by $\eta_1 > 0$, the packing entropy of the hydrophobic tails, parameterized by η_2, and the pressure jump between amphiphilic and hydrophobic phases, characterized by the difference in self energies, $W(b_\pm)$ of the amphiphilic and bulk phases, can trigger a range of bifurcations. Specifically the fingering and pearling instabilities observed experimentally in [2, 28] by adjusting the bulk values of lipids and the charge density of the lipids, respectively, can be induced in the FCH framework by varying the corresponding control parameters. There are however many avenues to explore, for example the pearling bifurcation induces a periodic dimpling of a bilayer surface which can lead to perforation. Within the biological context of cell membranes, it is of particular interest to understand the energy required to open a single hole. Can a local adjustment of parameter values, such as a spatial variation in η_1, induce the opening of isolated holes in the membrane?

Acknowledgements The second author acknowledges the support of the DMS of the National Science Foundation through grant 1109127. Both authors thank Andrew Christlieb and Jaylan Jones for the computations which resulted in the images presented in Figs. 4 and 8.

References

1. Andreussi, O., Dabo, I., Marzari, N.: Revised self-consistent continuum solvation in electronic-structure calculations. J. Chem. Phys. **136**, 064102 (2012)
2. Budin, I., Szostak, J.: Physical effects underlying the transition from primitive to modern cell membranes. Proc. Natl. Acad. Sci. **108**, 5249–5254 (2011)
3. Cahn, J.W., Hilliard, J.E.: Free energy of a nonuniform system, I. Interfacial free energy. J. Chem. Phys. **28**, 258–267 (1958)
4. Charleux, B., Delaittre, G., Rieger, J., D'Agosto, F.: Polymerization-induced self-assembly: from soluble macromolecules to block copolymer nano-objects in one step. Macromolecules **45**, 6753–6765 (2012)
5. Chavel, I.: Eigenvalues in Riemannian Geometry. Academic, London (1984)
6. Dai, S., Promislow, K.: Geometric evolution of bilayers under the functionalized Cahn-Hilliard equation. Proc. R. Soc. Lond. Ser. A **469**, (2013)
7. Dai, S., Promislow, K.: Competitive geometric evolution of amphiphilic interfaces. SIAM Journal on Mathematical Analysis, **47**(1), 347–380 (2015)
8. Discher D., Eisenberg, A.: Polymer vesicles. Science **297**, 967–973 (2002)
9. Doelman, A., Hayrapetyan, G., Promislow, K., Wetton, B.: Meander and Pearling of Single-Curvature Bilayer Interfaces in the Functionalized Cahn–Hilliard Equation. SIAM Journal on Mathematical Analysis, **46**(6), 3640–3677 (2014)
10. Gavish, N., Hayrapetyan, G., Promislow, K., Yang, L.: Curvature driven flow of bi-layer interfaces. Phys. D **240**, 675–693 (2011)
11. Gavish, N., Jones, J., Xu, Z., Christlieb, A., Promislow, K.: Variational models of network formation and ion transport: applications to perfluorosulfonate ionomer membranes. Polymers **4**, 630–655 (2012)
12. Gomez, E., Rappl, T., Agarwal, V., Bose, A., Schmutz, M., Marques, C., Balsara, N.: Platelet self-assembly of an amphiphilic A-B-C-A tetrablock copolymer in pure water. Macromolecules **38**, 3567–3570 (2005)
13. Gompper, G., Schick, M.: Correlation between structural and interfacial properties of amphiphilic systems. Phys. Rev. Lett. **65**, 1116–1119 (1990)
14. Hayrapetyan, G., Promislow, K.: Spectra of functionalized operators arising from hypersurfaces. Zeitschrift für angewandte Mathematik und Physik, 1–32 (2014)
15. Jain, S., Bates, F.: Consequences of nonergodicity in aqueuos binary PEO-PB micellar dispersions. Macromolecules **37**, 1511–1523 (2004)
16. Kapitula, T., Promislow, K.: Spectral and Dynamical Stability of Nonlinear Waves. Springer, New York (2013)
17. Kraitzman, N.: Bifurcation and competitive evolution of network morphologies in the strong functionalized Cahn-Hilliard equation. Ph.D. thesis (2015)
18. Matyjaszewski, K.: Atom transfer radical polymerization (ATRP): current status and future perspectives. Macromolecules **45**, 4015–4039 (2012)
19. Pego, R.L.: Front migration in the nonlinear Cahn-Hilliard equation. Proc. R. Soc. Lond. Ser. A **422**, 261–278 (1989)
20. Promislow, K., Zhang, H.: Critical points of functionalized Lagranagians. Discrete Cont. Dyn. Syst. A **33**, 1–16 (2013)
21. Ratcliffe, L., Ryan, A., Armes, S.: From a water-immiscible monomer to block copolymer nano-objects via a one-pot RAFT aqueous dispersion polymerization formulation. Macromolecules **46**, 769–777 (2013)
22. Röger, M., Schätzle, R.: On a modified conjecture of De Giorgi. Math. Z. **254**, 675–714 (2006)
23. Scherlis, D.A., Fattebert, J.L., Gygi, F., Cococcioni, M., Marzari, N.: A unified electrostatic and cavitation model for first-principles molecular dynamics in solution. J. Chem. Phys. **124**, 074103 (2006)
24. Teubner, M., Strey, R.: Origin of scattering peaks in microemulsions. J. Chem. Phys. **87**, 3195–3200 (1987)

25. Zare, P., Stojanovic, A., Herbst, F., Akbarzadeh, J., Peterlik, H., Binder, W.H.: Hierarchically nanostructured polyisobutylene-based ionic liquids. Macromolecules **45**, 2074–2084 (2012)
26. Zhu, J., Hayward, R.C.: Interfacial tension of evaporating emulsion droplets containing amphiphilic block copolymers: effects of solvent and polymer composition. J. Colloid Interf. Sci. **365**, 275–279 (2012)
27. Zhu, J., Ferrer, N., Hayward, R.C.: Tuning the assembly of amphiphilic block copolymers through instabilities of solvent/water interfaces in the presence of aqueous surfactants. Soft Matter **5**, 2471–2478 (2009)
28. Zhu, T.F., Adamala, K., Zhang, N., Szostak, J.W.: Photochemically driven redox chemistry induces protocell membrane pearling and division. Proc. Natl. Acad. Sci. **109**, 9828–9832 (2012)
29. Zhulina, E.B., Borisov, O.V.: Theory of block polymer micelles: recent advances and current challenges. Macromolecules **45**, 4429–4440 (2012)

The Economics of Ethanol: Use of Indirect Policy Instruments

Charles B. Moss, Andrew Schmitz, and Troy G. Schmitz

Abstract General equilibrium models typically ignore environmental goods because it is assumed that they have zero price. In the United States the Renewable Fuel Standard was introduced to offset the carbon emissions from by burning ethanol. The model in this study extends the standard general equilibrium approach to consider both positive and negative externalities. The negative externality is due to gasoline consumption while the positive externality is from substitution of ethanol for gasoline.

1 Introduction

Policy in the twenty-first century is complicated by a variety of factors including the complexity of transactions in the modern economy and the collaboration across sometimes diverse support groups to generate political support. In the United States there is continued support for the Renewable Fuel Standards (RFS) as part of the Energy Independence and Security Act (EISA) of 2007. Further, the RFS is a policy that was created to meet an array of policy goals including fuel security and carbon recycling. These policy goals are met indirectly—the policy does not increase the price of either fuel security or carbon recycling. Instead it operates through secondary markets—the market for gasoline and ethanol. This paper examines the effect of indirect policy instruments on the level of non-priced environmental goods. It develops a general equilibrium model to analyze the use of indirect policies to affect changes in the levels of environmental goods. The general equilibrium model

C.B. Moss (✉)
Food and Resource Economics Department, University of Florida, 1175 McCarty Hall, Gainesville, FL 32611-0240, USA
e-mail: cbmoss@ufl.edu

A. Schmitz
University of Florida, Gainesville, FL, USA
e-mail: aschmitz@ufl.edu

T.G. Schmitz
Arizona State University, Tempe, AZ, USA
e-mail: troy.schmitz@asu.edu

J.-P. Bourguignon et al. (eds.), *Mathematics of Energy and Climate Change*, CIM Series in Mathematical Sciences 2, DOI 10.1007/978-3-319-16121-1_9

demonstrates how policy instruments in the fuel or ethanol market changes the level of environmental goods (i.e., affects the either the positive or negative externality).

The diversity of policy goals can be seen in the five major goals of the National Institute of Food and Agriculture (NIFA) (the United States Department of Agriculture's competitive funding program):

- Food Security and Hunger—NIFA supports science to boost domestic agricultural production, improve capacity to meet the growing food demand, and foster innovation in fighting hunger and food insecurity in vulnerable populations.
- Climate Change—NIFA funded projects help producers adapt to changing weather patterns and sustain economic vitality while also reducing greenhouse gas emissions and increasing carbon sequestration in agricultural and forest production.
- Sustainable Energy—NIFA contributes to the President's goal of energy independence with a portfolio of grant programs to develop optimum biomass, forests, and crops for bioenergy production; and produce value-added, bio-based industrial products.
- Childhood Obesity—NIFA supported programs ensure that nutritious foods are affordable and available and that individuals and families are able to make informed, science-based decisions about their health and well-being.
- Food Safety—NIFA food safety programs work to provide a safer food supply and reduce the incidence of food-borne illness by addressing the causes of microbial contamination and anti-microbial resistance, educating consumer and food safety professionals, and developing enhanced food processing technologies [8].

Many of these goals are conflicting. For example, the sustainable energy goal implies using some of the U.S.'s agricultural potential to increase biofuels. This implicitly reduces the amount of farmland used to produce food crops. Hence, sustainable energy conflicts with food security. Other policy goals such as sustainable energy and climate change may be weakly complementary.

The second section of this paper develops an unconstrained general equilibrium model including both priced and unpriced (i.e., environmental goods). The following section changes the formulation slightly by introducing taxes on gasoline and subsidies on ethanol production. These variables (i.e., the tax on gasoline and subsidy on ethanal) are the policy variables used to change the level of the environmental good. The fourth section of the paper then considers a more complex set of environmental objectives. Finally, we offer conclusions about the implications of the general equilibrium framework.

2 Unconstrained and Policy Constrained General Equilibrium

As a starting point, we consider the general equilibrium solution for prices and quantities in an economy based on four traded goods. The equilibrium prices and quantities are determined by that set of prices where all the excess demands ($\xi_l(.)$) are less than or equal to zero

$$\begin{aligned} \xi_i(p,w) &= y_i^D(p,w,k) - y_i^S(p,w,k) \leq 0 \text{ for } i = 1,2,3,4 \\ \xi_j(p,w) &= k_j^D(p,w,k) - k_j \leq 0 \; \forall j \end{aligned} \quad (1)$$

where p is the vector of output (consumption) prices, w is a vector of input (factor) prices, $y_i^D(.)$ is the amount of output i demanded (as a function of input prices and income as determined by each household's factor endowments $[k_m]$ and the equilibrium prices for those factors $[w_m]$), $y_j^S(.)$ is the supply of output i offered to the market as a function of input and output prices, $k_j^D(.)$ is the derived demand for factors of production and k_j is the factor endowment for each factor of production ($k_j = \sum_m k_{jm}$ where m households are initially endowed with the factors of production). In this equilibrium the amount of output consumed by each household is determined by the household's utility maximization

$$\left.\begin{array}{c} \max\limits_{y} U_m(y_1,y_2,y_3,y_4) \\ y_1p_1 + y_2p_2 + y_3p_3 + y_4p_4 \leq \sum_j k_{jm}w_j \end{array}\right\} \Rightarrow \left\{\begin{array}{c} y_{im}^D(p,w,k_m),\; i = 1,2,3,4 \\ y_i^D(p,w,k) = \sum_m y_{im}^D(p,w,k_m),\; i = 1,2,3,4 \end{array}\right. . \quad (2)$$

Similarly, the output supply and input demands are determined by the optimizing behavior of the firms

$$\left.\begin{array}{c} \max\limits_{x,y} y_1p_1 + y_2p_2 + y_3p_3 + y_4p_4 - w'k \\ \{x,y,k\} \in T \end{array}\right\} \Rightarrow \left\{\begin{array}{c} y_i^S(p,w,k),\; i = 1,2,3,4 \\ k_j^D(p,w,k)\; \forall j \end{array}\right. \quad (3)$$

where $\{x,y,k\} \in T$ denotes a convex technology set.

Under the standard Arrow-Debreu formulation, a vector of prices p exists so that the conditions in Eq. (1) are satisfied given a normalization condition. In this analysis, we use the normalization condition $p_1 = 1$ so that p_2, p_3, p_4, and w_j (all non-negative) are prices relative to the first consumption good [1, 7]. For subsequent discussions, we note that these prices follow the complementary slackness conditions

$$\begin{aligned} \xi_i(p,w)\,p_i &= \left[y_i^D(p,w,k) - y_i^S(p,w,k)\right]p_i = 0,\ i = 1,2,3,4 \\ \xi_j(p,w)\,w_j &= \left[k_j^D(p,w,k) - k_j\right]w_j = 0,\ \forall j \end{aligned} \quad . \tag{4}$$

Hence, by choice of the numeraire good, we assume that $y_1^D(p,w,k) - y_1^S(p,w,k) = 0$ (so that $p_1^* = 1 \gg 0$). We denote the vector of input prices that solves Eq. (4) as $p^* = \left(1, p_2^*, p_3^*, p_4^*\right)$ and the vector of input prices that solves Eq. (4) as w^*.

We assume that good 1 is a general consumption good, good 2 is gasoline, good 3 is ethanol, and good 4 is an environmental good affected by the combination of gasoline and ethanol produced and consumed. At this stage, we allow for a fairly flexible definition of this environmental good.[1] The benefits to bioenergy have been rather loosely defined to include fuel security and carbon reduction.[2] For our purpose, we simply define good 4 as environmental goods.

In the general equilibrium formulation, given a functioning market for each good, the price vector generates a Pareto best solution. Specifically, no change in resource use or consumption allocation can be made to make one individual better off without making another worse off. However, certain difficulties associated with environmental goods distort this solution. For example, non-exclusivity of consumption causes difficulties in the specification of demand for many environmental goods. Given the failure of price signals in the market for good 4 the consumer's decision in Eq. (2) is rewritten as

$$\left.\begin{array}{c} \max\limits_{y} U_m(y_1, y_2, y_3, y_4) \\ y_1\tilde{p}_1 + y_2\tilde{p}_2 + y_3\tilde{p}_3 \le \sum_j k_{jm}\tilde{w}_j \end{array}\right\} \Rightarrow \\ \left\{\begin{array}{c} y_{im}^D(\tilde{p}, \tilde{w}, k_m),\ i = 1,2,3 \\ y_i^D(\tilde{p}, \tilde{w}, k) = \sum_m y_{im}^D(\tilde{p}, \tilde{w}, k_m),\ i = 1,2,3 \end{array}\right. \tag{5}$$

where $\tilde{p} = (1, \tilde{p}_2, \tilde{p}_3)$ and the vector of new input prices becomes $\tilde{w}$ (again normalizing on the vector of prices on the first consumption good). Similarly, the producer's choice in Eq. (3) is rewritten as

$$\left.\begin{array}{c} \max\limits_{x,y} y_1\tilde{p}_1 + y_2\tilde{p}_2 + y_3\tilde{p}_3 - \tilde{w}'k \\ \{x, y, k\} \in T \end{array}\right\} \Rightarrow \left\{\begin{array}{c} y_i^S(\tilde{p}, \tilde{w}, k),\ i = 1,2,3 \\ k_j^D(\tilde{p}, \tilde{w}, k)\ \forall j \end{array}\right. . \tag{6}$$

[1]The definition of environmental goods tends to be rather all encompassing. For example, Wikipedia states that *Environmental goods* are a sub-category of public goods which includes— clean air, clean water, landscape, scenic towns, green transportation infrastructure (footpaths, cycleways, greenways, etc.), a diverse flora, a diverse fauna, public parks, town squares, urban parks, rivers, mountains, forests, and beaches.

[2]Early studies such as Hill [2] suggested that ethanol could have significant benefits, later studies focus primarily on the potential for cellulosic ethanol.

Finally, equilibrium prices are determined by the first three ($i = 1, 2, 3$) excess demand curves in Eq. (4).

Does the introduction of environmental goods associated with negative or positive externalities distort the market? To examine this question, we focus on the fourth excess demand equation for consumption. If the excess demand at the market clearing price (p^*, w^*) is less than zero

$$\xi_4 \left(p^*, w^*\right) = y_4^D \left(p^*, w^*, k\right) - y_i^S \left(p^*, w^*, k\right) \leq 0 \tag{7}$$

the general equilibrium conditions for the environmental goods are met—the environmental goods are being optimally produced. However, under standard assumptions the original market clearing price yields a positive excess demand for environmental goods

$$\xi_4 \left(p^*, w^*\right) = y_4^D \left(p^*, w^*, k\right) - y_i^S \left(p^*, w^*, k\right) \gg 0 \tag{8}$$

so that the economy is producing a level of environmental quality less than the optimal. In the most general terms

$$\begin{aligned} &U_m \left(y_{1m}^D \left(\tilde{p}, \tilde{w}, k_m\right), y_{2m}^D \left(\tilde{p}, \tilde{w}, k_m\right), y_{3m}^D \left(\tilde{p}, \tilde{w}, k_m\right), y_{4m}^D \left(\tilde{p}, \tilde{w}, k_m\right)\right) \\ &\ll U_m \left(y_{1m}^D \left(p^*, w^*, k_m\right), y_{2m}^D \left(p^*, w^*, k_m\right), y_{3m}^D \left(p^*, w^*, k_m\right), y_{4m}^D \left(p^*, w^*, k_m\right)\right) \end{aligned} \tag{9}$$

for some collection of consumers ($m \in M$)—at least some consumers are made worse off by the failure to price environmental outputs.

3 Policy Response

Given the failure of the market to price environmental quality, we consider three different policy scenarios: pricing the environmental quality through a tax, placing a tax on other goods that affect environmental quality (i.e., the Volumetric Ethanol Excise Tax Credit [VEETC]), and imposing regulatory mandates on the production of other goods (i.e., the Renewable Fuels Standard [RFS]). In order to simplify the mathematics, we express the general equilibrium formulation in the preceding section using a volume index

$$\bar{y}\left(p, w\right) = y_1 + y_2 p_2 + y_3 p_3 \tag{10}$$

where we normalized the output prices on the price of general consumption goods in the general equilibrium formulation (i.e., we divide the prices by p_1—we make the first consumption good the numeraire good).

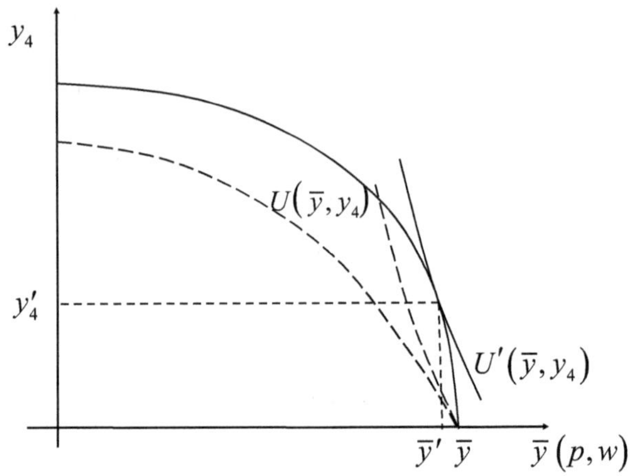

Fig. 1 Effect of non-priced environmental good

As depicted in Fig. 1, the tradeoff between consumption goods and environmental quality produces a concave production surface $(y_4 \bar{y})$ similar to the unrestricted production possibility frontier in the two output case. The difference is that each point on the surface may imply different relative prices among the consumption goods in Eq. (10) (as relative prices change in the restricted general equilibrium model). The aggregate output of the priced goods $(\bar{y})$ assumes that the appropriate (general equilibrium) combinations of y_1, y_2, and y_3 are selected by the economy for a given level of y_4 (assumed in Fig. 1 to be initially zero). Next, we restrict the new level of y_4 to be higher than the original value $y_4' \gg y_4 = 0$. Given this restriction, we can solve for a new set of equilibrium price $p' = 1, p_2', p_3'$ and quantities $y' = y_1', y_2', y_3'$. If the production of y_4 is binding (i.e., y_4 is scarce) the new level of the aggregated good from Eq. (10) will be less than the original level $\bar{y}'(p', w') \ll \bar{y}(p, w)$. Thus, the tradeoff between the aggregate good and the environmental good is downward sloping. In addition, given that the solution is an interior point (i.e., not on either boundary—there exists a point such that both $y_4 > 0$ and $\bar{y} > 0$), the surface is concave. Put slightly differently, the combination of optimal volume indices and environmental quantities resembles the traditional production possibilities frontier.

The dashed line curves in Fig. 1 represent the scenario where the original excess demand for environmental quality is less than zero. Under this scenario, the original equilibrium is Pareto optimal—environmental quality has a zero price in the general equilibrium solution. The solid frontier represents the scenario where the original excess demand curve for the environmental good is greater than zero. Under this scenario the original equilibrium is not Pareto optimal. Reallocating resources to increase environmental quality would shift the utility outward from $U(\bar{y}, y_4)$ to $U'(\bar{y}, y_4)$—improving the utility of at least one consumer. The possibility of shifting to a Pareto improving position raises questions concerning the magnitude of the

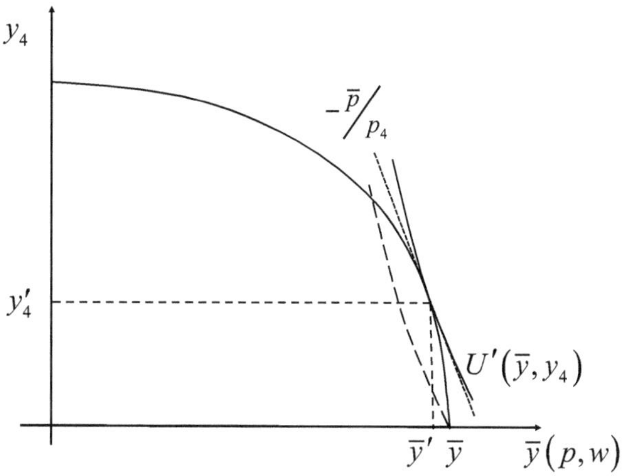

Fig. 2 Market price for non-priced good

production shift and the allocation of the changes across producers and consumers. Figure 2 depicts the scenario where a market price for environmental quality could be established. Under this scenario, production occurs where the production surface is tangent to the negative of the price ratio ($-\bar{p}/p_4$; the dotted line in Fig. 2). This price ratio yields an optimal price ratio for the other consumption goods in Eq. (10).

We now consider two difficulties with the market equilibrium conditions. First, as discussed above, there is a market failure in the demand for environmental quality. Second, environmental quality is hypothesized to be a byproduct of other production activities

$$F : \{y_2, y_3\} \rightarrow y_4 \tag{11}$$

so that the choice of both gasoline and ethanol production and consumption determines the level of environmental quality observed. Figure 3 presents a graphical depiction of the physical relationship in Eq. (11). Environmental quality is a decreasing function of the quantity of gasoline (y_2) produced and consumed and an increasing function of the quantity of ethanol (y_3) produced and consumed. One intuitive justification for this relationship is to consider y_4 as atmospheric carbon. As the quantity of gasoline consumed increases, the level of atmospheric carbon increases. While it may be somewhat controversial, we assume that generating fuel using ethanol reduces the level of carbon through the production of corn. Thus, as the required blend of ethanol to gasoline increases, the overall level of environmental quality increases.

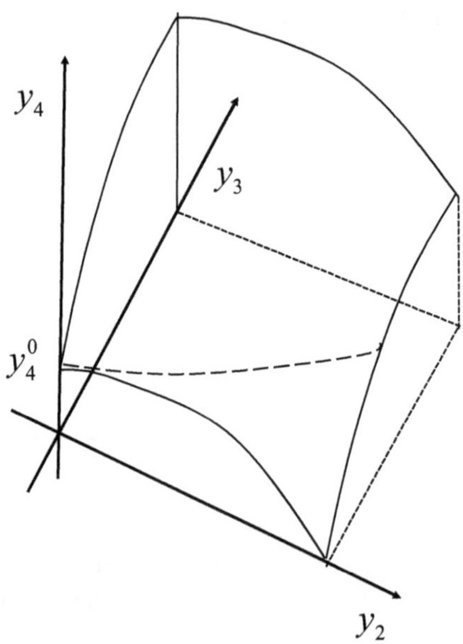

Fig. 3 Technological tradeoff between gasoline, environmental quality, and ethanol

Integrating the production relationship for environmental quality into the production problem yields

$$\left.\begin{array}{c}\max\limits_{x,y} y_1\tilde{p}_1 + y_2(\tilde{p}_2+\tau_2) + y_3(\tilde{p}_3+\tau_3) - \tilde{w}'k \\ \{x,y,k\} \in T \\ F:\{y_2,y_3\} \to y_4 \end{array}\right\} \Rightarrow \left\{\begin{array}{c} y_i^S(\tilde{p},\tau,\tilde{w},k),\ i=1,2,3 \\ y_4\left(y_2^S(\tilde{p},\tau,\tilde{w},k), y_2^S(\tilde{p},\tau,\tilde{w},k)\right) \\ k_j^D(\tilde{p},\tilde{w},k)\ \forall j \end{array}\right. \tag{12}$$

where τ_2 is a subsidy (tax if $\tau_2 < 0$) on gasoline production and τ_3 is a subsidy on ethanol production. Hence, the VEETC was designed to increase the quantity of environmental good produced (i.e, enhance carbon recycling), but does not directly introduce a price for environmental quality. In Fig. 4, the VEETC as a subsidy for the production of ethanol that increases the environmental quality. But the tradeoff between environmental quality and other goods is determined by Eq. (11). Mathematically, consider the case where $\tau_2 = 0$ and policy makers attempt to increase the quality of the environmental good by increasing the subsidy on ethanol. The tradeoff between the environmental and consumption good becomes

$$\frac{dy_4/d\bar{y}}{d\tau_3} \equiv \frac{\frac{\partial F}{\partial y_2}\frac{dy_2}{d\tau_3} + \frac{\partial F}{\partial y_3}\frac{dy_3}{d\tau_3}}{\frac{dy_1}{d\tau_3} + \frac{\partial y_2}{\partial p_2}\frac{dp_2}{d\tau_3} + \frac{dy_2}{d\tau_3} + \frac{\partial y_3}{\partial p_3}\frac{dp_3}{d\tau_3} + \frac{dy_3}{d\tau_3}} \tag{13}$$

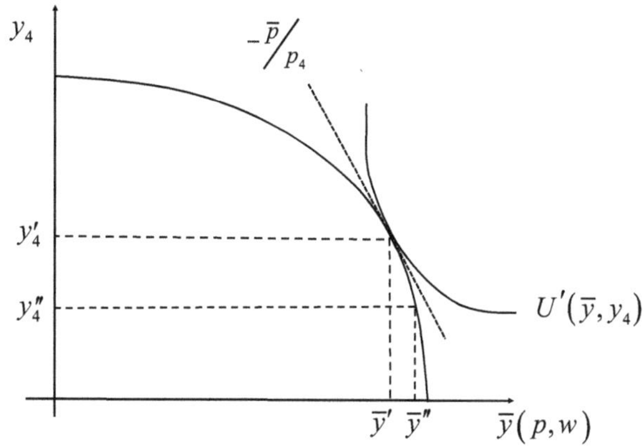

Fig. 4 Using ethanol tax credit to increase environmental quality

where some $d\tau_3^*$ exists such that

$$\left.\frac{dy_4}{d\bar{y}}\right|_{d\tau_3^*} = \left.\frac{\frac{\partial F}{\partial y_2}\frac{dy_2}{d\tau_3} + \frac{\partial F}{\partial y_3}\frac{dy_3}{d\tau_3}}{\frac{dy_1}{d\tau_3} + \frac{\partial y_2}{\partial p_2}\frac{dp_2}{d\tau_3} + \frac{dy_2}{d\tau_3} + \frac{\partial y_3}{\partial p_3}\frac{dp_3}{d\tau_3} + \frac{dy_3}{d\tau_3}}\right|_{d\tau_3 = d\tau_3^*} = -\frac{\bar{p}}{p_4} = \frac{\partial U/\partial \bar{y}}{\partial U/\partial y_4}. \tag{14}$$

If society's marginal tradeoff between consumption goods and environmental benefits is known along with the tradeoff between environmental quality and fuels (gasoline and ethanol), we can derive an optimal subsidy on ethanol that maximizes social welfare (i.e., the point ($\bar{y}'$, y_4') in Fig. 4).

Figure 4 presents the scenario where the original equilibrium provides some environmental (i.e., the solution $(\bar{y}'', y_4'')$ depicts the scenario where some level of environmental quality is being produced in the original equilibrium). For example, we assume that $y_3 > 0$ in the original equilibrium so that some level of ethanol is being produced and used. However, this original equilibrium does not allow for the effect of ethanol production on the unpriced environmental quality variable. The point $(\bar{y}', y_4')$ denotes the optimal point—the point where the marginal utility of the environmental good is equal to the marginal disutility of lost consumption goods. This equilibrium solution is generated by a specific ethanol subsidy determined in Eq. (14).

The partial equilibrium equivalent of Eqs. (13) and (14) are presented in Fig. 5. Panel (a) of Fig. 5 depicts the market for gasoline while Panel (b) of Fig. 5 presents the market for ethanol. We start by assuming that the ethanol market is subsidized by $\tau_4 = p_E' - p_E''$. The result is a classical deadweight loss of *fgh*. The additional quantity of ethanol consumed shifts the demand curve for gasoline back from D_G to D_G'. The price of gasoline declines from p_G^0 to p_G' while the quantity of gasoline consumed falls from q_G^0 to q_G'. Consumers in the gasoline market lose *abcd*. Producers lose a surplus of $p_G^0 bep_G'$.

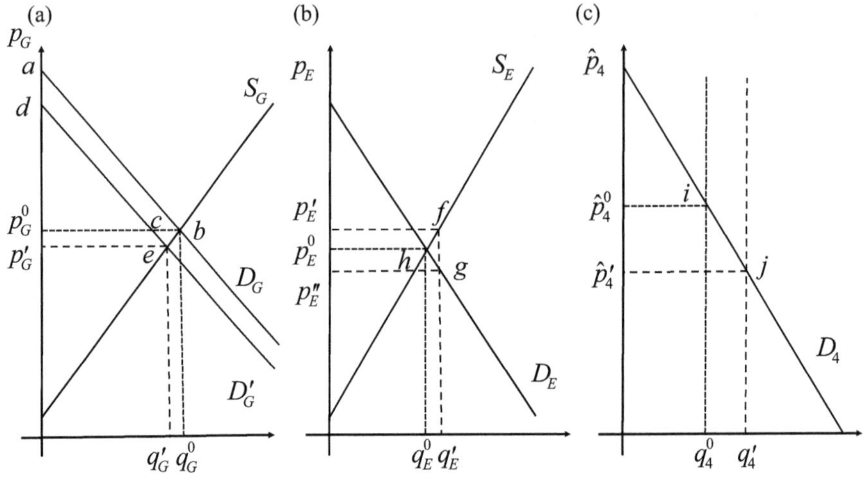

Fig. 5 Partial equilibrium model of ethanol subsidy. (**a**) Gasoline; (**b**) ethanol; (**c**) environment

Panel c of Fig. 5 presents the changes in benefits from the environmental goods. It is not a market in a classical sense because it does not represent an exchange of goods for money. While we have denoted the vertical axes as a price, it is probably better conceptualized as a willingness to pay for environmental services. Thus, in the original equilibrium q_4^0 units of an environmental index are being produced and consumed. Based on this production, consumers would be willing to pay a price of $\hat{p}_4^0$ for these services. If the quantity of ethanol used in the economy is expanded from q_E^0 to q_E', the quantity of environmental services produced increases from q_4^0 to q_4'. Associated with this increase, there is a reduction in the consumer's willingness to pay for these environmental services from $\hat{p}_4^0$ to p_4'. The shift from gasoline to ethanol produces an increase in the consumer surplus from environmental quality (as depicted in Eq. (11)) measured as $ijq_4'q_4^0$. The net gain in welfare is then

$$ijq_4'q_4^0 - fgh - \left(p_G^0 bep_G' + abcd\right). \tag{15}$$

The graphical depiction allows us to determine that the overall improvement in environmental quality is downward sloping. Specifically, note that as the amount of the subsidy increases from τ_4 to τ_4' the area of the triangular deadweight loss (fgh) increases at an increasing rate while the producer loss in the gasoline market ($ijq_4'q_4^0$) increases at a decreasing rate. The real cause of the relative rate of change is that the gain in the environmental market increases at a decreasing rate, eventually becoming zero at some point (i.e., where the demand curve reaches zero—the additional unit of environmental good has zero value). Intuitively, there is a point where the increase in ethanol subsidy yields no net social benefit (i.e., the result of Eq. (15) is zero).

Previous studies have estimated the economic costs of the ethanol tax credit in the fuel and food market. Schmitz et al. [5] a net gain from ethanol of 1,281 million dollars, but a large portion of this gain was due to the reduction in government payments to agriculture of 4,084 million dollars. They did not attempt to value the environmental benefits of substituting ethanol for gasoline. Partially as a result of the growth in ethanol demand, corn prices from 2008 through 2010 largely eliminated price supports for corn. Moss and Schmitz [3] evaluated the economic cost of ethanol subsidies using a computable general equilibrium model focusing on the effects of the VEETC on the food and fuel markets. In general, they found a small but negative cost to ethanol subsidies.

While Schmitz et al. [5] found a net benefit to the ethanol subsidy, the economic gains resulted from the reduction in the distortion from agricultural policies. This study takes a slightly different approach. Instead of the benefits resulting from the reduction in the distortion of agricultural policies, we consider economic benefits resulting from increased environmental quality. However, it is important to note that estimates of the value of improved environmental quality are subject to considerable difficulties. For example, Schmitz et al. [6] evaluate the economic benefits and cost to removing sugar land from production on the environmental quality of Florida's everglades. In general, they show that "…the benefit-cost ratios form the proposed U.S. Sugar land buyout are very low by any standards regardless of the supply price elasticities used…" (p. 81). Hence, as an alternative to traditional cost-benefit analysis Schmitz et al. [6] propose an *Environmental Equivalent* as that amount or value that "…accrue to society that would bring the net benefit of this buyout to at least a breakeven point…" (p. 82). This environmental equivalent is related to the value of the non-priced environmental good presented in Fig. 4 ($p_4\left(y'_4 - y''_4\right)$). Moss and Schmitz [4] apply the environmental equivalent approach to demonstrate how biofuel mandates can be used to generate values of environmental goods within a general equilibrium model.

4 Complex Environmental Benefits

The preceding discussion demonstrates how a policy instrument in a secondary market (i.e., a subsidy on ethanol production) can generate a welfare improvement given a single unpriced good in the economy. The ability to determine an optimal level of a single policy instrument can be linked to the unique policy mapping in Eq. (11) (i.e., by choosing one instrument we can control the level of the single output). The problem becomes more complex if two unpriced goods exist. This scenario is particularly important in the historical support for the renewable fuels standard in the United States. Specifically, the support from a variety of special interest groups is necessary for the renewable fuel standards including: (1) farm groups whose support of the renewable fuels standard can be traced the use of corn to manufacture ethanol, (2) industrial and consumer groups who support renewable fuels as a means to reduce the country's dependence on imported oil (fuel security or sufficiency), and (3) environmental groups who support the renewable fuels standards as a means to increase carbon recycling (reduction in greenhouse gases). The diversity of goals is embedded in the act. For example, while environmental groups support growth of renewable fuels to reduce carbon emissions, they oppose the environmental consequences of large scale commercial agriculture that come from expanded ethanol production. As a result, the original renewable fuels standards passed in the Energy Independence and Security Act (EISA) of 2007 included two sets of mandates—one for ethanol and another for cellulosic biofuels. The goal was to develop technology to transform grasses (such as switch grass) and other cellulosic crops into biofuels. The concept was that such a technology would generate environmental benefits. However, the technology for large scale cellulosic biofuels has been slow to develop. Hence, the overall environmental consequences of the renewable fuels standards are mixed as high corn prices have resulted in potentially fragile land drawn into production to meet the increased demand for corn.

To develop the complexity in policy design with multiple unpriced goods, we modify Eq. (11)

$$\tilde{F} : \{y_2, y_3\} \rightarrow \{y_{4A}, y_{4B}\} \tag{16}$$

where y_{4A} is the value of fuel security and y_{4B} is the environmental benefit of reduced atmospheric carbon. The contribution of this paper is the conjecture that the combination of unpriced goods responds differently to a combination of tax/subsidy incentives. For example, taxing gasoline reduces the overall emission of carbon while subsidizing ethanol affects both carbon and fuel security.

5 Conclusions

The policy process is typically a complicated system of compromise between groups with different goals. Further, policy goals are typically accomplished indirectly through secondary markets. This paper demonstrates how gasoline taxes and ethanol subsidies can be used to meet possibly multiple policy goals. The Energy Independence and Security Act of 2007 imposes certain blend targets for biofuels under the Renewable Fuel Standards. The standards foresaw two sources of biofuel supplements for gasoline—ethanol and cellulosic. Implicitly the standards were imposed to meet two general policy goals: increase the availability of fuel and improve the environmental quality. The role of biofuels in the first objective is clear, but whether biofuels actually improved the environmental quality is an open debate. Some of this debate is tied up in technical discussions of carbon fate models. However, regardless of whether biofuels reduce carbon emissions this mechanism of improving the environmental quality through a secondary market introduces a variety of mathematical assumptions. The fact that these goals can be achieved in theory should not be confused with the efficiency of such policy in practice. It is not clear that the policy has had its intended effect. One of the basic problems is measuring carbon emissions. The level of carbon emitted is typically measured by the inputs into the process and not the outputs—the current level of atmospheric carbon. In addition, the increased level of U.S. biofuels may be swamped by the overall global increase in the use of carbon through energy consumption. In addition to these practical difficulties, we assumed several functional relationships such as the relationship between gasoline and ethanol consumption and environmental quality is known and well behaved. This assumption is probably suspect.

References

1. Debreu, G.: Theory of Value: An Axiomatic Analysis of Economic Equilibrium. Yale University Press, London (1959)
2. Hill, J., Nelson, E., Tilman, D., Polasky, S., Tiffany, D.: Environmental, economic, and energetic costs and benefits of biodiesel and ethanol biofuels. Proc. Natl. Acad. Sci. USA **103**(30), 11206–11210 (2006)
3. Moss, C.B., Schmitz, T.G.: Potential effect of ethanol on the U.S. Economy: a general equilibrium approach, Chap. 4. In: The Economics of Alternative Energy Sources and Globalization, pp. 35–48. Bentham Science. (2011). http://ebooks.benthamscience.com/book/9781608052332/
4. Moss, C.B., Schmitz, A.: Valuing carbon recycling through ethanol: zero prices for environmental goods. Theor. Econ. Lett. **4**(3), 235–240 (2014)
5. Schmitz, A., Moss, C.B., Schmitz, T.G.: Ethanol: no free lunch. J. Agric. Food Ind. Organ. **5**(2), 34 (2007)
6. Schmitz, A., Kennedy, P.L., Hill-Gabriel J.: Restoring the Florida Everglades through a sugar land bayout: benefits, costs, and legal challenges. Environ. Econ. **3**(1), 74–85 (2012)
7. Shoven, J.B., Whalley, J.: Applying General Equilibrium. Cambridge University Press, Cambridge (1992)
8. USDA/NIFA: NIFA Fact sheet. Webpage: http://www.nifa.usda.gov/newsroom/factsheet.pdf (2002)

Geostatistical Analysis in Extremes: An Overview

M. Manuela Neves

Abstract Classical statistics of extremes is very well developed in the univariate context for modeling and estimating parameters of rare events. Whenever rain, snow, storms, hurricanes, earthquakes, and so on, happen the analysis of extremes is of primordial importance. However such rare events often present a temporal aspect, a spatial aspect or both. Classical geostatistics, widely used for spatial data, is mostly based on multivariate normal distribution, inappropriate for modeling tail behavior. The analysis of spatial extreme data, an active research area, lies at the intersection of two statistical domains: extreme value theory and geostatistics. Some statistical tools are already available for the spatial modeling of extremes, including Bayesian hierarchical models, copulas and max-stable random fields. The purpose of this chapter is to present an overview of basic spatial analysis of extremes, in particular reviewing max-stable processes. A real case study of annual maxima of daily rainfall measurements in the North of Portugal is slightly discussed as well the main functions in R environment for doing such analysis.

1 Motivation and Introduction

This chapter is related to a talk presented in a session of the "International Conference Planet Earth. Mathematics of Energy and Climate Change" entitled *The role of Statistics of Extremes in Society*. Through the ages natural disasters and catastrophes have happened, causing deaths and destruction. Remember, for example, the 1755 Lisbon earthquake and tsunami and the 2005 hurricane Katrina, New Orleans, see Fig. 1.

The scientific community has been worried about whether anything could be done for society to be at least better prepared for those occurrences.

On 18.01.12, Nicolas Guerin [22], from École Polytechnique Fédérale de Lausanne, wrote in *News Mediacom:*

…*"The problem of extremes is that there are so few events, by definition… explains EPFL mathematician Anthony Davison. It's thus necessary to*

M.M. Neves (✉)
CEAUL, and ISA, University of Lisbon, Lisbon, Portugal
e-mail: manela@isa.ulisboa.pt

J.-P. Bourguignon et al. (eds.), *Mathematics of Energy and Climate Change*, CIM Series in Mathematical Sciences 2, DOI 10.1007/978-3-319-16121-1_10

Fig. 1 The 1755 Lisbon earthquake and tsunami (*left*) and the 2005 hurricane Katrina, New Orleans (*right*)

create specific models that are different from those that use innumerable mean values...

For several years now, the scientists have noted that the increase in extreme events associated with climate change appears to be having much more of an impact on society than the increase in mean temperatures. Natural disasters are accompanied by a significant human and economic cost... "

Extreme value theory (EVT) is the branch of probability and statistics dedicated to characterizing the behavior of the extreme observations. An extreme observation is a datum that has low probability of occurrence, but which can be very large (or small).

EVT has its beginnings in the early to middle part of the last century. It formally began with the paper by Dodd [14], followed by papers of Fréchet [18], Fisher and Tippett [17] Gumbel [23] and von Mises [42], to cite the pioneering and most relevant works.

Rare events such as the risk of flooding, potential crop damage from drought, health effects of extreme air pollution, storms, and so on, may cause severe impacts on human life as well as on ecosystems. In weather and climate studies as well as in other fields, often what one wants to characterize is not the usual behavior, but the extreme events.

Assessing the behavior of rare events such as wind speed, precipitation and temperature presents unique statistical challenges, and requires one to characterize the tail of the distribution of the quantity of interest.

Emil Gumbel (1891–1966) was the pioneer in the application of statistics of extremes. He wrote " *...The aim of a statistical theory of extreme values is to analyze observed extremes and to forecast further extremes*". Gumbel [24] presents several applications of EVT on real world problems in engineering and in meteorological phenomena. There appear first applications in hydrology. "*...It seems that the rivers know the theory...*" is a remarkable expression from Gumbel.

Recently, a special issue of the journal *Extremes* (2010) **13**:2 on Statistics of Extremes in Weather and Climate, shows the relevance of EVT in the area.

Nowadays, extreme value analysis appears in quite diversified areas revealing the importance of statistics of extremes in applications. Many excellent books, presenting both the methodological basis and a great emphasis in the applications must be referred to. Besides Gumbel [24], we can mention Tiago de Oliveira (ed.) [12] that is still a reference today with a wide range of contributions and applications of statistics of extremes. It was the result of a remarkable meeting that took place thirty years ago in Vimeiro and fortunately was remembered in 2013, celebrating that conference and also dedicated to Ivette Gomes, a highly recognized international researcher in Statistics of Extremes. More recently, other books emphasizing the applications appeared and deserve to be mentioned: Coles [4], Finkenstadt and Rootzén (eds.) [16], Castillo et al. [3], Beirlant et al. [1], Reiss and Thomas, [33] and Gomes et al. [21].

Statistical modeling of extremes was based initially on limiting families of distributions for maxima of a sequence, $X_1, \ldots, X_n$, of independent and identically distributed (i.i.d.) random variables from an unknown distribution function (d.f.), F. Given that the distribution of the maxima is highly dependent of the unknown form of F, similar to the central limit theory, researchers tried to obtain sequences $\{a_n > 0\}$ and $\{b_n\} \in \mathbf{R}$ such that $M_n := \max\{X_1, \ldots, X_n\}$, linearly normalized by those constants, had a non-degenerate limiting distribution.

Univariate EVT is well developed, but is it well recognized that many extreme events, particularly in the environment, environmental health, climate, hydrology or meteorology occur in a place and/or in a time. In those areas of application we are faced with the task of analyzing data that are geographically referenced and show a correlated structure that needs to be adequately modelled .

Spatial data are measurements or observations taken at specific locations or within specific regions. The dependence structure of those data needs to be adequately captured. Geostatistics is an active area of research with important applications in the environment, agriculture and public health. Classical methods are well known and explored under the Gaussian model.

Regarding extreme values there was thus a need to develop methods for analyzing and characterizing spatial extreme data. Spatial extreme theory is an area that lies at the intersection of EVT and geostatistics. While classical geostatistics is usually applied to situations where only one realization of the process is taken, the spatial extreme approach needs multiple realizations underlying the subset of extreme data analyzed.

After this general motivation and introduction, Sect. 2 introduces the basic notions in EVT and in Standard Geostatistics. In Sect. 3 a background in spatial extremes is given as well as some most common statistical models in the max-stable process approach for spatial extremes. Section 4 is devoted to slightly discuss an application to annual maxima of daily rainfall data, with particular emphasis to the use of R software. Applications to rainfall data have been done recently by Smith and Stephenson [39], Padoan et al. [31] and Davison et al. [8].

2 Fundamental Notions and Basic Results

Classical EVT has been one of the most fast developing areas in the last decades. The underlying mathematical basis is well established, see e.g. Leadbetter et al. [26], Embrechts et al. [15] and de Haan and Ferreira [11], to cite only a few books. To model the tail of a distribution, where extreme events often occur, has been a challenge for researchers. There is now a well established set of methods to model the tail of a distribution.

2.1 Main Limiting Results in EVT

For the univariate EVT the main and well known result is: Let $X_1, \ldots, X_n$ be independent replications from an unknown d.f., F, and define $M_n := \max\{X_1, \ldots, X_n\}$. Fréchet [18], Fisher and Tippet [17], Gumbel [23], and von Mises [42], in Fig. 2, obtained the first results concerning the existence of a non-degenerate law of the maximum of that series, suitably normalized.

Gnedenko [20] and later on de Haan [9] gave necessary and sufficient conditions for the existence of sequences $\{a_n > 0\}$ and $\{b_n\} \in \mathbf{R}$ such that,

$$\lim_{n\to\infty} P\left(\frac{M_n - b_n}{a_n} \leq x\right) = \lim_{n\to\infty} F^n(a_n x + b_n) = EV_\xi(x),$$

Fig. 2 Fréchet (1878–1973), Gumbel (1891–1966), von Mises (1883–1953) and Weibull (1887–1979) (from *left to right*)

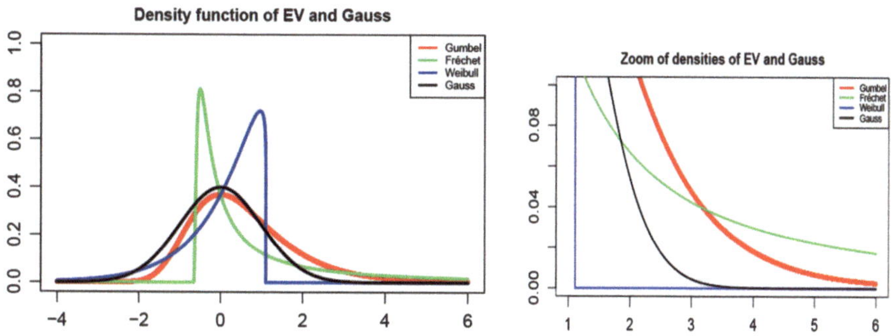

Fig. 3 Gumbel, Fréchet and Weibull p.d.f. (*left*) and zoom of Gumbel, Fréchet, Weibull and Gauss p.d.f. (*right*)

$\forall x \in \mathbf{R}$, where EV_ξ is a nondegenerate distribution function. This function, called Extreme Value d.f., is given by

$$EV_\xi(x) = \begin{cases} \exp[-(1+\xi x)^{-1/\xi}], & 1+\xi x > 0 \quad \text{if } \xi \neq 0 \\ \exp[-\exp(-x)], & x \in \mathbf{R} \quad \text{if } \xi = 0. \end{cases} \tag{1}$$

In applications, the d.f. in (1) can also present a more general form with a location parameter $\lambda \in \mathbf{R}$ and a scale parameter $\delta \in \mathbf{R}^+$.

The $\mathbf{EV}_\xi$ incorporates the three (Fisher-Tippett) types: Gumbel type: $\Lambda(x) = \exp(-\exp(-x)) \equiv EV_0(x)$, $x \in \mathbf{R}$, $(\xi = 0)$, the limit for exponential tailed distributions; Fréchet type: $\Phi_\xi(x) = \exp(-(x)^{-1/\xi}) \equiv EV_\xi(\frac{x-1}{\xi})$, $x > 0$, $\xi > 0$, the limit for heavy tailed distribution and Weibull type: $\Psi_\xi(x) = \exp(-(-x)^{1/\xi}) \equiv EV_\xi(\frac{-x-1}{\xi})$, $x < 0$, $\xi < 0$, the limit for short tailed distributions. These three families of models were combined as the d.f. EV_ξ in (1) by von Mises [42].

If $\xi = 0$, the *right endpoint*, $x^* := \sup\{x : F(x) < 1\}$, can then be either finite or infinite. If $\xi > 0$, F has an infinite *right endpoint*. If $\xi < 0$, F has a finite *right endpoint*, x^*. The shape parameter, ξ, is then directly related to the weight of the right tail, $\overline{F} := 1 - F$, of the underlying model F. As ξ increases the right tail becomes heavier. Figure 3 shows the behavior of the right-tails for the three different types of EV models, together with the Gauss model for comparison.

Jointly with the knowledge of methodological aspects of extreme values theory, the interest for having free, accurate and simple software has increased tremendously, motivated by the wide range of areas of application of EVT.

R is an environment and a programming language for statistical computing and graphics. It is a free and open source project. Several packages for extreme value analysis are already available, with a large set of functions, among which we mention: `evd`, Stephenson [40], with functions for statistical analysis of extremes, including multivariate extremes and Bayesian methods; `ismev`, Stephenson [41], with functions for classical extreme value analysis, fitting extreme value distribution to "block maxima", as well as generalized Pareto distribution to excesses over a high threshold and `extRemes`, with a graphical user interface based on `ismev`, are perhaps the most well known. Other packages are nevertheless available, see Gilleland, et al. [19]. They published a very nice software review, comparing the available statistical software and presenting the main characteristics of the main packages for extreme value analysis.

2.2 Standard Geostatistics

If the quantity of interest, e.g. the rainfall level, is observed at different locations spread over a region, it is necessary to know how to take into account the spatial pairwise dependence among sites, for an adequate analysis of data. There are three types of spatial data:

- Geostatistical data or point referenced data are measurements taken at fixed locations $\mathbf{s} \in \mathbf{K} \subset \mathbf{R}^d$. These locations are generally spatial continuous. Data may be modeled as values from a spatial process.
- Lattice data or areal referenced data where $\mathbf{K}$ is again a fixed subset but observations are associated with spatial regions and with well defined boundaries.
- Spatial points patterns where locations are now the variable of interest, so $\mathbf{K}$ is itself random and observation sites may be treated as random.

We shall consider geostatistical data modeled by a stochastic process $\{\mathbf{Y(s)}\}$ where $\mathbf{s} \in \mathbf{K}$ and $\mathbf{K}$ is a compact subset in $\mathbf{R}^d$. Data are observed at $D = \{\mathbf{s}_1, \ldots, \mathbf{s}_D\} \subset \mathbf{K} \subset \mathbf{R}^d$. Usually $d = 2$ and we shall assume it throughout. Essential elements for exploring and modeling spatial data are: stationarity, isotropy and the variogram, key elements of the "Matheron school", see Cressie [7].

Let us consider a spatial process with mean, $\mu(\mathbf{s}) = E[\mathbf{Y(s)}]$ and variance $Var[\mathbf{Y(s)}]$ finite for all $\mathbf{s} \in \mathbf{K} \subset \mathbf{R}^2$.

Usually it is assumed second-order stationarity what implies that covariance relationship between values of the process at any two locations can be summarized by the covariance function $C(\mathbf{h}) = Cov(Y(\mathbf{s}), Y(\mathbf{s}+\mathbf{h}))$, for all $\mathbf{h} \in \mathbf{R}^2$, such that s and $\mathbf{s}+\mathbf{h}$ both lie within $\mathbf{K}$. That function depends only on the separation vector $\mathbf{h}$.

Assuming also that $E\left[\mathbf{Y}(\mathbf{s}+\mathbf{h}) - \mathbf{Y}(\mathbf{s})\right] = 0$, with $\mathbf{s}$ and $\mathbf{s}+\mathbf{h} \in \mathbf{K}$, the variogram is defined as:

$$E\left[\mathbf{Y}(\mathbf{s}+\mathbf{h}) - \mathbf{Y}(\mathbf{s})\right]^2 = Var\left[\mathbf{Y}(\mathbf{s}+\mathbf{h}) - \mathbf{Y}(\mathbf{s})\right] = 2\gamma(\mathbf{h}),$$

where $\gamma(\mathbf{h})$ is designated as the semivariogram and is a crucial measure for quantifying spatial dependence in the data. If the semivariogram depends only on the length of $\mathbf{h}$, and not on the orientation, $\gamma(\mathbf{h}) = \gamma(\|\mathbf{h}\|)$, the process is said to be isotropic, i.e., roughly speaking "as it looked the same in all directions". Table 1 summarizes the main models for isotropic semivariograms.

Table 1 Theoretical semivariograms

Model	Semivariogram ($\gamma(t)$)
Linear	$\gamma(t) = \begin{cases} \tau^2 + \sigma^2 t, & t > 0; \\ 0, & \text{otherwise.} \end{cases}$
Spherical	$\gamma(t) = \begin{cases} \tau^2 + \sigma^2, & t \geq 1/\phi; \\ \tau^2 + \sigma^2\left(\frac{3}{2}\phi t - \frac{1}{2}(\phi t)^3\right), & 0 < t \leq 1/\phi; \\ 0, & \text{otherwise.} \end{cases}$
Exponential	$\gamma(t) = \begin{cases} \tau^2 + \sigma^2\left(1 - \exp(-\phi t)\right), & t > 0; \\ 0, & \text{otherwise.} \end{cases}$
Powered exponential	$\gamma(t) = \begin{cases} \tau^2 + \sigma^2\left(1 - \exp(-\lvert\phi t\rvert^p)\right), & t > 0; \\ 0, & \text{otherwise.} \end{cases}$
Matérn at $\nu = 3/2$	$\gamma(t) = \begin{cases} \tau^2 + \sigma^2\left(1 - (1 + \phi t)\exp(-\phi t)\right), & t > 0; \\ 0, & \text{otherwise.} \end{cases}$

In an exploratory phase of a geostatistical analysis, the dependence is investigated via an empirical covariogram or empirical semivariogram.

Assuming stationarity and isotropy the simplest semivariogram estimator is the *moments estimator*, Matheron [27],

$$\hat{\gamma}(h) = \frac{1}{2|N(h)|} \sum_{(\mathbf{s}_i,\mathbf{s}_j)\in N(h)} \{Y(\mathbf{s}_i) - Y(\mathbf{s}_j)\}^2,$$

where $N(h) = \{(\mathbf{s}_i, \mathbf{s}_j) : ||\mathbf{s}_i - \mathbf{s}_j|| = h\}$ and $|N(h)|$ denotes the cardinality of $N(h)$.

A few steps can be summarized for basic geostatistical analysis: remove trends in mean and (perhaps) in variance; transform residuals to standard normal margins; use graphical techniques to assess likely form for the semivariogram function; fit a suitable semivariogram model; make inferences using weighted least squares (kriging), likelihood or Bayes procedures; make predictions using the fitted correlation to obtain a map of predictions, based on a fitted normal model.

In R environment several packages are available for geostatistical analysis. Exploratory analysis, modeling and extrapolating are possible through several functions in `spatial`, `gstat`, `sp`, `MASS` and `geoR`, for example.

- `gstat` includes functions as: `variogram`—calculates sample (experimental) variograms; `plot.variogram`—plots an experimental variogram with automatic detection of lag spacing and maximum distance; `fit.variogram`—iteratively fits an experimental variogram; `krige`—a generic function to make predictions by inverse distance interpolation, ordinary kriging, OLS regression, regression-kriging and co-kriging; `krige.cv`—runs krige with cross-validation, see Pebesma [32] and Bivand et al. [2] for a complete overview of `gstat` functions and examples.

- `geoR`—extensively described by Diggle and Ribeiro Jr. [13] and Ribeiro Jr. et al. [35], with a series of tutorials.

Usually it is supposed that $\{\mathbf{Y}(\mathbf{s})\}$ follows a Gaussian process, so likelihood inference is realized under this assumption.

3 Geostatistical Analysis in Extremes

As mentioned in the previous section, Gaussian processes play a central role in modeling spatial processes. In this approach relevance is given to studying the central tendencies of the distribution rather than the distribution tails. However in many events already pointed out behind, the extremes are of main interest. The generalization of classical multivariate extreme value distributions to the spatial case is done through max-stable processes. Other statistical approaches have been developed for the spatial modeling of extremes, such as Bayesian hierarchical models and copulas, topics that will not be considered in this brief overview. A recent and very good survey on spatial extremes is Cooley et al. [6].

The purpose of this section is to review some max-stable processes, and mainly to show packages and functions already available in the R environment for the analysis of spatial extremes. First results date back to de Haan [10] and were developed by several authors, such as Smith [38], Schlather [36] and Kabluchko et al. [25], to mention only a few.

Max-stable processes follow a similar asymptotic motivation to the univariate EV distribution, providing a general approach to modeling process extremes incorporating temporal or spatial dependence. For the region $\mathbf{K}$ under study and the location $\mathbf{s}$, we will assume $Y_1(\mathbf{s}), Y_2(\mathbf{s}), \ldots$ as independent replicas of a stochastic process.

Consider that we have daily rainfall data collected in some stations and we are interested in considering the maximum of that quantity over a period of time (e.g. a year) in order to:

- Assess the dependence of the extreme precipitation levels between stations.
- Predict values at unobserved locations.
- Elaborate a map of the distribution of the maximum precipitation levels.

Interest is now to model extremes of a process $\{\mathbf{Y}(\mathbf{s})\}$ over spatial domain $\mathbf{K}$, where data are observed at sites $\mathbf{s}_d \in \{\mathbf{s}_1, \ldots, \mathbf{s}_D\}$ within $\mathbf{K}$ and at times $\mathbf{T} = \{t_1, \ldots, t_n\}$. A difference between geostatistics and spatial extremes is that much of geostatistics is applied to situations where one has only one realization of the process $\{\mathbf{Y}(\mathbf{s})\}$. However to perform an extreme value analysis it is necessary that multiple realizations $\{Y_i(\mathbf{s})\}$ underlie the subset of extreme data which are analyzed.

An overview on some main max-stable models is here considered and briefly applied to a case study.

3.1 Some Models for Max-Stable Processes

Let us recall the definition of max-stable process by de Haan [10]. Let $\{Y_i(\mathbf{s})\}$, $\mathbf{s} \in \mathbf{D}, i = 1, \ldots, n$, be independent replicas of a stochastic process $\{\mathbf{Y}(\mathbf{s})\}$ defined in $\mathbf{D} \subset \mathbf{R}^2$.

Definition 1 The stochastic process $\{\mathbf{Y}(\mathbf{s})\}$ is called max-stable if, for all $\mathbf{s} \in \mathbf{D}$ there exist normalizing sequences $\{a_n(\mathbf{s}) > 0\}$ and $\{b_n(\mathbf{s})\}$ such that, as $n \to \infty$,

$$\left\{ \frac{\max_{i=1,\ldots,n} Y_i(\mathbf{s}) - b_n(\mathbf{s})}{a_n(\mathbf{s})} \right\} \to_d \{\mathbf{Y}^*(\mathbf{s})\}_{\mathbf{s} \in \mathbf{R}^2}$$

where $\{\mathbf{Y}^*(\mathbf{s})\}$ is identical in law to $\{\mathbf{Y}(\mathbf{s})\}$.

To characterize max-stable processes is a very difficult task. For max-stable processes with Fréchet unit margins de Haan [10] introduced a very useful representation that allowed the construction of parametric models for spatial extremes. Such a general representation is as follows: let $\{Y_i(\mathbf{s})\}_{i \in \mathbf{N}}$, be independent realizations of a stochastic process $\{\mathbf{Y}(\mathbf{s})\}$ with $E[\mathbf{Y}(\mathbf{s})] = 1$ and let ζ_i be points of a Poisson process $\prod$ with intensity $d\zeta/\zeta^2$ on $(0, \infty)$. Then

$$Z(\mathbf{s}) = \max_{i \geq 1} \zeta_i Y_i(\mathbf{s}), \quad \mathbf{s} \in \mathbf{D}, \tag{2}$$

is a max-stable process with unit-Fréchet margins. and the distribution function is determined by

$$P\big(Z(\mathbf{s}) \leq z(\mathbf{s}), \mathbf{s} \in \mathbf{D}\big) = \exp\left(- E\left[\sup_{\mathbf{s} \in \mathbf{D}} \left\{ \frac{Y(\mathbf{s})}{z(\mathbf{s})} \right\} \right] \right).$$

Different choices of the process $Y_i(\mathbf{s})$ lead to different models of max-stable processes. As examples let us mention the following models.

- Smith [38] proposed to take $Y_i(\mathbf{s}) = \varphi(\mathbf{s} - \mathbf{s}_i)$, where φ is a zero mean multivariate normal density with covariance matriz Σ, where the joint distribution at two sites is given by

$$P\left[Z(\mathbf{s}_1) \leq z_1, Z(\mathbf{s}_2) \leq z_2\right] = \exp\left[-\frac{1}{z_1} \Phi\left(\frac{a}{2} + \frac{1}{a} \log \frac{z_2}{z_1} \right) - \frac{1}{z_2} \Phi\left(\frac{a}{2} + \frac{1}{a} \log \frac{z_1}{z_2} \right) \right], \tag{3}$$

  where $a = \sqrt{(\mathbf{s}_1 - \mathbf{s}_2)^T \Sigma^{-1} (\mathbf{s}_1 - \mathbf{s}_2)}$, is a dependence parameter, and Φ is the standard normal d.f.

- Schlather [36] proposed a more flexible class of max-stable processes by taking $Y_i(\mathbf{s})$ to be any stationary Gaussian random field with finite expectation. He considered

$$Z(\mathbf{s}) = \max_{i \geq 1} \zeta_i \max\{0, Y_i(\mathbf{s})\}$$

where $\mu = E\big[\max\{0, Y_i(\mathbf{s})\}\big] < \infty$, being the joint distribution at two sites given by

$$P\left[Z(\mathbf{s}_1) \leq z_1, Z(\mathbf{s}_2) \leq z_2\right] = \exp\left[-\frac{1}{2}\left(\frac{1}{z_1} + \frac{1}{z_2}\right)\left(1 + \sqrt{1 - \frac{2(\rho(h)+1)z_1 z_2}{(z_1+z_2)^2}}\right)\right], \tag{4}$$

where $h = ||\mathbf{s}_1 - \mathbf{s}_2||$ and $\rho(h)$ is chosen from Whittle-Matérn, Cauchy and Powered Exponential, see Schlather [36].

- As an example of another model let us consider a more recent proposal, the so-called Brown-Resnick process, studied in Kabluchko et. al. [25], who proposed an alternative specification for the $Y_i(\cdot)$ process, $Y(\mathbf{s}) = \exp\left(\epsilon_i(\mathbf{s}) - \sigma^2(\mathbf{s})/2\right)$, where $\epsilon(\mathbf{s})$ is a Gaussian process with stationary increments, being $\sigma^2(\mathbf{s})$ the variance of $\epsilon(\mathbf{s})$.

3.2 *Spatial Dependence of Extremes*

To use max-stable models we need to have information on how the dependence between two locations decreases, when the distance increases. It would be nice to have a *kind of variogram* for extremes of a stochastic process. However if we assume that Z is a unit Fréchet max-stable process, the variance (and even the mean) might be infinite.

A new function is now needed to reflect how evolves the spatial dependence of extremes. It provides sufficient information about extremal dependence and it is called *the extremal coefficient function*, Schlather and Tawn [37].

Definition 2 If $Z(\cdot)$ is a max-stable process with unit Fréchet margins the extremal coefficient function $\theta()$ is defined by

$$P\left[Z(\mathbf{s}_1) \leq z, Z(\mathbf{s}_2) \leq z\right] = \exp\left(-\frac{\theta(||\mathbf{s}_1 - \mathbf{s}_2||)}{z}\right),$$

where $1 \leq \theta(||\mathbf{s}_1 - \mathbf{s}_2||) \leq 2$.

The extremal coefficient function has the following meaning: If $\theta(||\mathbf{s}_1 - \mathbf{s}_2||) = 1$, we have perfect dependence; if $\theta(||\mathbf{s}_1 - \mathbf{s}_2||) = 2$, we have independence.

Extremal coefficient functions for the above models are:

- $\theta(||\mathbf{s}_1 - \mathbf{s}_2||) = 2\Phi\left(\frac{\sqrt{(\mathbf{s}_1-\mathbf{s}_2)^T\Sigma^{-1}(\mathbf{s}_1-\mathbf{s}_2)}}{2}\right)$, for the Smith model covering the whole range of dependence;
- $\theta(||\mathbf{s}_1 - \mathbf{s}_2||) = 1 + \sqrt{\frac{1-\rho(||\mathbf{s}_1-\mathbf{s}_2||)}{2}}$, for Schlather model that has upper bound of $1 + \sqrt{1/2}$. A drawback of this model is that independence of extremes can not be attained because $\theta \in [1; 1.8333]$ and $\theta = 2$ is never attained;
- $\theta(||\mathbf{s}_1 - \mathbf{s}_2||) = 2\Phi\left(\sqrt{\gamma(||\mathbf{s}_1 - \mathbf{s}_2||)/2}\right)$, for the Brown-Resnick process. As $\gamma(||\mathbf{s}_1-\mathbf{s}_2||) \to 0$, we have $\theta(||\mathbf{s}_1-\mathbf{s}_2||) \to 1$, while if $\gamma(||\mathbf{s}_1-\mathbf{s}_2||)$ is unbounded, then $\theta(||\mathbf{s}_1 - \mathbf{s}_2||) \to 2$ as $||\mathbf{s}_1 - \mathbf{s}_2|| \to \infty$.

Another measure of dependence between two locations is given by a "kind" of variogram. The first idea was to consider the madogram, a tool in classical geostatistics, Matheron [28], defined as:

$$\upsilon(||\mathbf{s}_1 - \mathbf{s}_2||) = E\left[|Z(\mathbf{s}_1) - Z(\mathbf{s}_2)|\right].$$

The madogram requires the finiteness of the first-moment and for stationary max-stable processes with unit Fréchet margins, mean and variance may be not finite and that mean value does not exist theoretically. Consequently, variogram-based approaches, specially designed for extremes, have been proposed:

- the F-Madogram, Cooley et al. [5],

$$\upsilon_F(||\mathbf{s}_1 - \mathbf{s}_2||) = \frac{1}{2}E\left[|F\{Z(\mathbf{s}_1)\} - F\{Z(\mathbf{s}_2)\}|\right].$$

  The F-madogram is related to the extremal coefficient function as:

$$\upsilon_F(||\mathbf{s}_1 - \mathbf{s}_2||) = \frac{\theta(||\mathbf{s}_1 - \mathbf{s}_2||) - 1}{\theta(||\mathbf{s}_1 - \mathbf{s}_2||) + 1}.$$

- λ-Madogram, Naveau et al. [29].

$$\upsilon_\lambda(||\mathbf{s}_1 - \mathbf{s}_2||) = \frac{1}{2}E\left[|F^{\lambda}\{Z(\mathbf{s}_1)\} - F^{1-\lambda}\{Z(\mathbf{s}_2)\}|\right], \quad 0 \le \lambda \le 1,$$

  where $F(z) = \exp(-1/z)$ is the unit Fréchet d.f.

The F-Madogram is similar to the λ-Madogram when $\lambda = 0.5$. The F-Madogram has the advantage of suggesting an estimator directly from its definition, see Cooley et al. [5].

Naveau et al. [29] discussed the estimation of the madogram. For the F-Madogram it was considered the plug in, $\hat{F}$, an estimate of the d.f., at the specified location and the binned estimate of the F-madogram. For the λ−Madogram, the binned λ−Madogram estimator and the adjusted estimator have been proposed.

4 A Case Study

For 21 different stations, in the North of Portugal, daily precipitation has been recorded. Our data refers to the maximum annual values for 20 years (1977–1996), in each station. A preliminary analysis of these data was done in Neves and Prata Gomes [30]. It would be nice to have more years of observations but surprisingly in recent years some missing values were found.

Figure 4 shows the region of Portugal where data have been collected on the left, and locations of the stations pointed out on the right. In each site and location, the maxima annual values of daily precipitation were considered in our study.

We began our analysis by performing marginal analysis and transformation. Daily values of precipitation at a given station x are dependent, however, the maxima values at each hydrological year can be considered almost independent.

As an illustration of the graphical diagnostic GEV fitting, graphics in two locations were chosen and displayed in Figs. 5 and 6.

The transformation of the data at each station to the unit Fréchet distribution was then performed, through the function `gev2frech()` of the package `SpatialExtremes`, Ribatet [34].

To estimate the spatial dependence structure, estimates of values of the extremal coefficient at two stations $\mathbf{s}_1$ and $\mathbf{s}_2$ are evaluated. For this, the `fitextcoeff()` was used considering both Smith, [38], and Schlather-Tawn, [37], estimators. As far as we know only these estimators are available in R package. Work is

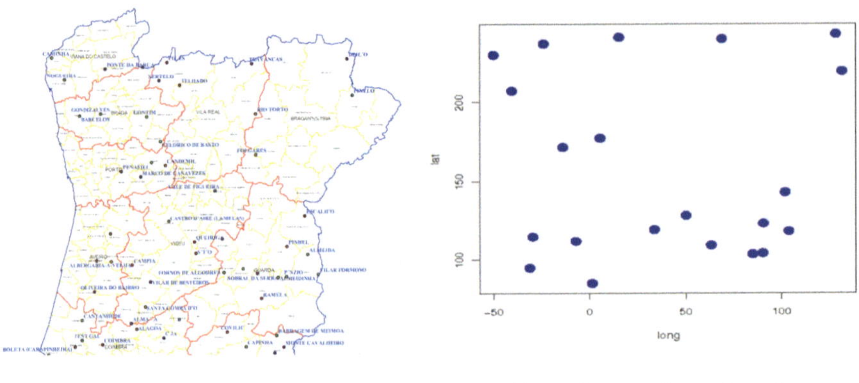

Fig. 4 Map of the North of Portugal (*left*) and the positions of the stations where data were recorded (*right*)

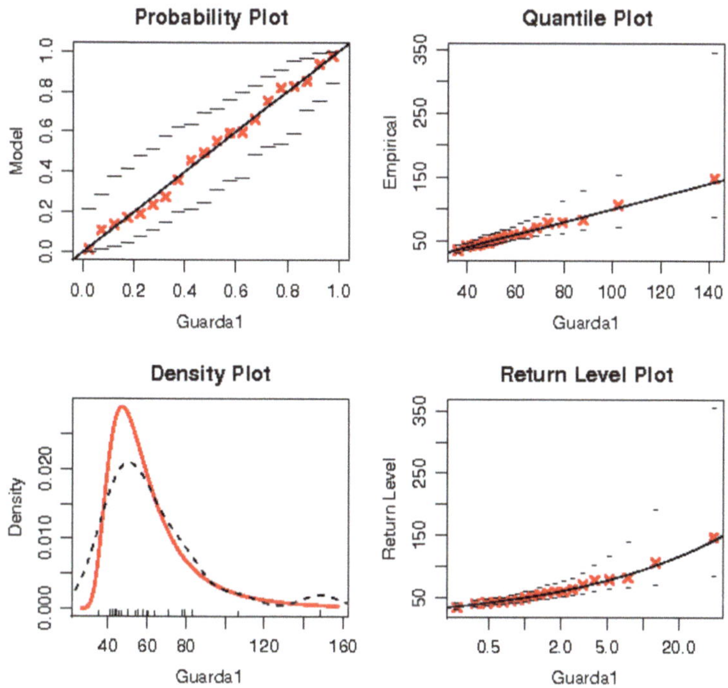

Fig. 5 GEV model diagnostic for data from Guarda

now in progress for including other functions in R environment. Figure 7 shows pairwise extremal coefficient estimates and lowess curves for Smith and Schlather-Tawn estimators and also the F-madogram, here obtained through `fmadogram()` function.

Trend surfaces were tried to be estimated in order to capture the spatial dependence structure by describing the marginal parameters as:

$$\lambda(x) = \beta_{o,\lambda} + \beta_{1,\lambda} lon(x) + \beta_{2,\lambda} lat(x)$$
$$\delta(x) = \beta_{0,\delta} + \beta_{1,\delta} lon(x) + \beta_{2,\delta} lat(x)$$
$$\xi(x) = \beta_{o,\xi}$$

where $lon(x)$ and $lat(x)$ denote the longitude and the latitude of the stations.

However, for our data, values of $\lambda(x)$, $\delta(x)$ and $\xi(x)$ showed a weak relationship with $lon(x)$ and $lat(x)$. Other trend surfaces need to be considered, but work is now in progress. Even though, considering the estimated matrix for Smith model as well as estimated sill and range for Schlather model, with the powered exponential correlation function, several simulations were performed on a 21×21 grid. Figure 8 displays one of those simulations for the Smith model and for the Schlather model.

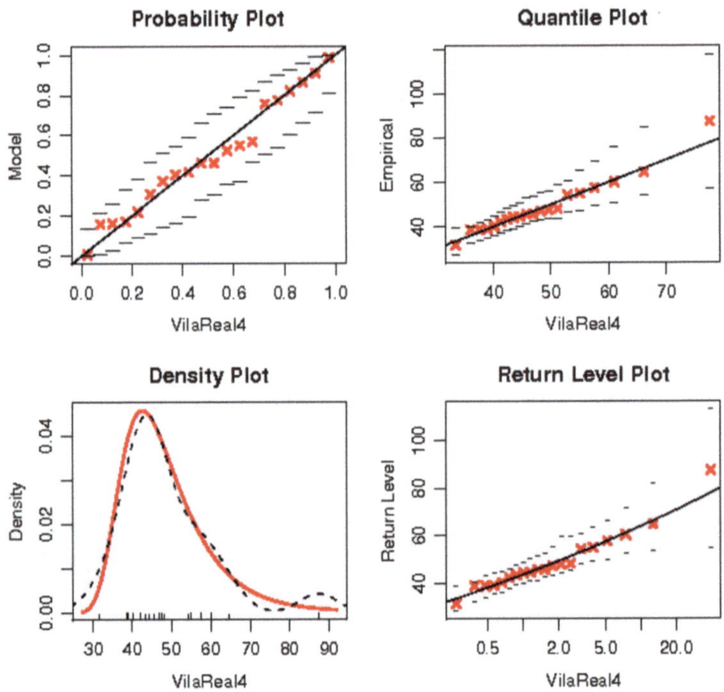

Fig. 6 GEV model diagnostic for data from Vila Real

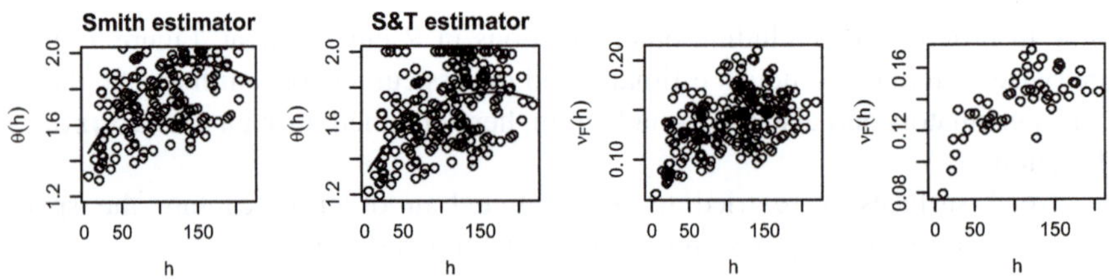

Fig. 7 Pairwise extremal coefficient estimates and lowess curves: Smith and Schlather-Tawn (*left*); F-madogram and binned madogram (*right*)

5 Concluding Remarks

This chapter was intended to introduce spatial models for extreme value analysis as well as to present an overview of functions available in the R software for doing that analysis. Max-stable processes are a natural generalization of multivariate extreme models and the most common way to deal with extreme value data in spatial statistics. Some procedures for estimating the spatial dependence available in the R

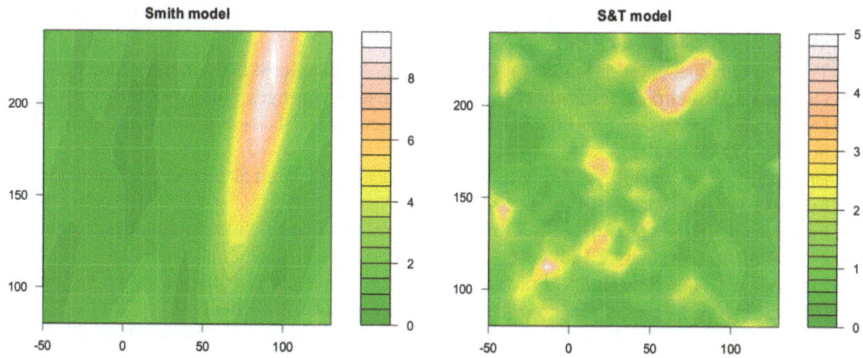

Fig. 8 One realization of the Smith (*left*) and Schlather and Tawn (*right*) models

environment were shown and some simulations using Smith and Schlather models were also performed in the R package `SpatialExtremes`.

Work is already in progress for including more max-stable models in R, as well as other approaches for modeling spatial dependence.

Some difficulties related to the amount of data available still remain. How to deal with missing values, in a given real situation, is another challenging point.

Acknowledgements Research partially supported by National Funds through **FCT**—Fundação para a Ciência e a Tecnologia, projects PEst-OE/MAT/UI0006/2014 (CEAUL)

References

1. Beirlant, J., Goegebeur, Y., Teugels, J., Segers, J.: Statistics of Extremes: Theory and Applications. Wiley, England (2005)
2. Bivand, R., Pebesma, E., Gomez-Rubio, V.: Applied Spatial Data Analysis with R, 2nd edn. Springer, New York (2013)
3. Castillo, E., Hadi, A.S., Balakrishnan, N., Sarabia, J.M.: Extreme Value and Related Models in Engineering and Science Applications. Wiley, New York (2005)
4. Coles, S.: An Introduction to Statistical Modeling of Extreme Values. Springer, London (2001)
5. Cooley D., Naveau P., Poncet P.: Variograms for spatial max-stable random fields. In: Springer (ed.) Dependence in Probability and Statistics. Lecture Notes in Statistics Edition, vol. 187, pp. 373–390. Springer, New York (2006)
6. Cooley, D., Cisewski, J., Erhardt, R.J., Jeon, S., Mannshardt, E., Omolo, B.O., Sun, Y.: A survey of spatial extremes: measuring spatial dependence and modeling spatial effects. Revstat Stat. J. **10**, 135–165 (2012)
7. Cressie, N.A.C.: Statistics for Spatial Data. Wiley, New York (1993)
8. Davison, A.C., Padoan, S.A., Ribatet, M.: Statistical modelling of spatial extremes (with discussion). Stat. Sci. **27**, 161–186 (2012)
9. de Haan, L.: On regular variation and its application to the weak convergence of sample extremes. Thesis, University of Amsterdam/Mathematical Centre Tract 32 (1970)
10. de Haan, L.: A spectral representation for max-stable processes. Ann. Probab. **12**(4), 1194–1204 (1984)

11. de Haan, L., Ferreira, A.: Extreme Value Theory: An Introduction Springer Science+Business Media, LLC, New York (2006)
12. Tiago de Oliveira, J. (ed.): Statistical Extremes and Applications. D. Reidel, Dordrecht (1983)
13. Diggle, P.J., Ribeiro, Jr. P.J.: Model-Based Geostatistics. Springer, New-York (2007)
14. Dodd, E.L.: The greatest and the least variate under general laws of error. Trans. Am. Math. Soc. **25**, 525–539 (1923)
15. Embrechts, P., Kluppelberg, C., Mikosch, T.: Modelling Extremal Events for Insurance and Finance, 3rd edn. Springer, Berlin (2001)
16. Finkenstadt, B., Rootzen, H. (eds.): Extreme Values in Finance, Telecommunications and the Environment. Chapman & Hall/CRC, London (2004)
17. Fisher R.A., Tippett, L.H.C.: Limiting forms of the frequency distribution of the largest and smallest member of a sample. Proc. Camb. Philos. Soc. **24**, 180–190 (1928)
18. Fréchet, M.: Sur la loi de probabilité de l'écart maximum. Ann. Soc. Polon. Math. **6**, 93–116 (1927)
19. Gilleland, E., Ribatet, M., Stephenson, A.G.: A software review for extreme value analysis. Extremes **16**(1), 103–119 (2013)
20. Gnedenko, B.V.: Sur la distribution limite dune série aléatoire. Ann. Math. **44**, 423–453 (1943)
21. Gomes, M.I., Fraga Alves, M.I., Neves, C.: Análise de Valores Extremos: uma Introdução. Edições SPE, Lisboa (2013)
22. Guerin, N.: Climate and the statistics of extremes. Available in http://actu.epfl.ch/news/climate-and-the-statistics-of-extremes/ (2012). Cited 25 June 2014
23. Gumbel, E.J.: Les valeurs extrêmes des distributions statistiques. Ann. Inst. Henri Poincaré **5**(2), 115–158 (1935)
24. Gumbel, E.J.: Statistics of Extremes. Columbia University Press/Dover Publications, New York (1958)
25. Kabluchko, Z., Schlather, M., de Haan, L.: Stationary max-stable fields associated to negative definite functions. Ann. Prob. **37**, 2042–2065 (2009)
26. Leadbetter, M., Lindgren, G., Rootzén, H.: Extremes and related properties of random sequences and series. Springer, New York (1983)
27. Matheron, G.: Traité de géostatistique appliquée. Tome I: Mémoires du Bureau de Recherches Géologiques et Minières, no. 14. Editions Technip, Paris (1962)
28. Matheron, G.: Suffit-il, pour une covariance, d' être de type positive. Sciences de la Terre, série informatique géologique **26**, 51–66 (1987)
29. Naveau, P., Guillou, A., Cooley, D., Diebolt, J.: Modeling pairwise dependence of maxima in space. Biometrika **96**(1) 1–17 (2009)
30. Neves, M. and Prata Gomes, D.: Geostatistics for spatial extremes. A case study of maximum annual rainfall in Portugal. Procedia Environ. Sci. **7**, 246–251 (2011)
31. Padoan, S.A., Ribatet, M., Sisson, S.A.: Likelihood-based inference for max-stable processes. J. Am. Stat. Assoc. **105**, 263–277 (2010)
32. Pebesma, E.J.: Multivariable geostatistics in S: the `gstat` package. Comput. Geol. **30**, 683–691 (2004)
33. Reiss, R.-D., Thomas, M.: Statistical Analysis of Extreme Values: From Insurance, Finance, Hydrology and other Fields. Birkhauser, Basel (2007)
34. Ribatet, M.: A User's Guide to the SpatialExtremes Package. École Polytechnique Fédérale de Lausanne, Switzerland (2011)
35. Ribeiro, Jr., P.J., Christensen, O.F., Diggle, P.J.: geoR and geoRglm: software for model-based geostatistics. In: Proceedings of DSC, vol. 2 (2003)
36. Schlather, M.: Models for stationary max-stable random fields. Extremes **5**(1) 33–44 (2002)
37. Schlather, M., Tawn, J.: A dependence measure for multivariate and spatial extremes: properties and inference. Biometrika **90**(1) 139–156 (2003)
38. Smith, R.: Max-stable processes and spatial extremes. Unpublished manuscript (1990)
39. Smith, E.L., Stephenson, A.G.: An extended Gaussian max-stable process model for spatial extremes. J. Stat. Plann. Inference **139**, 1266–1275 (2009)

40. Stephenson, A.G.: `evd`: Extreme value distributions. R News **2**(2), 31–32 (2002). http://CRAN.R-project.org/doc/Rnews/
41. Stephenson, A.G.: ismev: an introduction to statistical modeling of extreme values. R package version 1.38 (2012). http://CRAN.R-project.org/package=ismev
42. von Mises, R.: La distribution de la plus grande de n valeurs. Revue Math. Union Interbalcanique **1**, 141–160 (1936) [Reprinted in Selected Papers Volumen II, pp. 271–294. American Mathematical Society, Providence (1954)]

Reducing the Minmax Regret Robust Shortest Path Problem with Finite Multi-scenarios

Marta M.B. Pascoal and Marisa Resende

Abstract The minmax regret robust shortest path problem is a combinatorial optimization problem that can be defined over networks where costs are assigned to arcs under a given scenario. This model can be continuous or discrete, depending on whether costs vary within intervals or within discrete sets of values. The problem consists in finding a path that minimizes the maximum deviation from the shortest paths over all scenarios. This work focuses on designing tools to reduce the network, in order to make easier the search for an optimum solution. With this purpose, methods to identify useless nodes to be removed and to detect arcs that surely belong to the optimum solution are developed. Two known algorithms for the robust shortest path problem are tested on random networks with and without these preprocessing rules.

1 Introduction

The robust shortest path problem is a network optimization problem that consists in finding a path between two given nodes that minimizes the worst case for the scenarios considered for the arc costs. Two types of objective functions are commonly adopted. One is an absolute version of the problem that finds the path with the minimum maximum cost over all scenarios. This is called the minmax shortest path problem or the absolute robust shortest path problem [7, 9]. The other is concerned with the relative version of the problem, which considers the regret costs of each path towards the shortest paths among all scenarios and that minimizes the maximum of these deviation costs. This is the so called minmax regret robust shortest path problem or robust deviation shortest path problem [8, 9]. This paper addresses this latter version of the problem, considering that each arc cost is associated with a finite number of scenarios.

M.M.B. Pascoal (✉) • M. Resende
Department of Mathematics, University of Coimbra, Apartado 3008, EC Santa Cruz, 3001-501 Coimbra, Portugal

Institute for Systems Engineering and Computers – Coimbra (INESCC), Rua Antero de Quental, 199, 3000-033 Coimbra, Portugal
e-mail: marta@mat.uc.pt; mares@mat.uc.pt

J.-P. Bourguignon et al. (eds.), *Mathematics of Energy and Climate Change*,
CIM Series in Mathematical Sciences 2, DOI 10.1007/978-3-319-16121-1_11

Murthy and Her [7] were the first to approach the absolute version of the problem and introduced a labeling algorithm to solve it. Later, Yu and Yang [9] extended the study to the relative version of the problem, proposing a dynamic programming strategy for solving it and devising a particular method for layered networks. Besides, they also showed that the problem is strongly NP-hard if the number of scenarios is unlimited. More recently, Pascoal and Resende [8] developed three algorithms for the relative robust shortest path problem; the first is a labeling method where labels are pruned according to cost lower and upper bounds; the second is a ranking based method and the third is a hybrid version of the two previous methods.

An alternative to deal with costs uncertainty consists of assuming that each cost ranges within known intervals. Karasan et al. [3] were the first to address the robust shortest path problem with interval data, focusing on the case of acyclic networks. These results were extended and new methods were proposed for general networks in the works by Montemanni et al. [4–6]. Another contribution of [3] was to introduce rules to identify a priori, and later to delete, arcs that do not belong to the optimum solution. This allows to reduce the network before a robust shortest path algorithm is applied. In a recent work Catanzaro et al. [2] further developed these techniques and established new sufficient conditions to identify both nodes and arcs that cannot be part of the solution but also nodes and arcs that are certainly part of the solution.

To our knowledge no preprocessing techniques have been studied or applied to models with discrete data. Therefore, the main purpose of the current work is to exploit new preprocessing strategies for the finite multi-scenario model, inspired by the results presented in [2]. These rules were implemented together with the labeling and the hybrid approaches in [8]. The impact of the developed rules is evaluated by means of empirical tests in random instances.

The remainder of this work is organized into five other sections. The next one is concerned with the definition of the minmax regret robust shortest path problem and with introducing notation. Section 3 is dedicated to establishing theoretical results which allow to remove unnecessary nodes and identify arcs that obligatory belong to the robust shortest path. The correspondent algorithms are outlined and their time computational complexity orders are presented. Section 4 shows an example of the application of such preprocessing rules. In Sect. 5 computational experiments over randomly generated networks are presented and their results are analyzed. Finally, conclusions are discussed and future topics of investigation are suggested.

2 Problem Definition and Notation

Hereinafter a finite multi-scenario model is represented as $G(V, A, T_k)$, where G is a directed graph with a set of nodes $V = \{1, \ldots, n\}$, a set of m arcs $A \subseteq \{(i,j) : i,j \in V \text{ and } i \neq j\}$ and a finite set of acceptable parameters $T_k = \{t_l : l \in I_k\}$, with $I_k := \{1, \ldots, k\}$, $k > 1$. It is assumed that each node $j \in V$ can be reached from node 1 and that G contains no parallel arcs. Given the set T_k, a scenario $l \in I_k$ is determined according to the costs assigned under t_l. For each arc $(i,j) \in A$, i and j are named the tail and the head nodes, respectively. The associate cost function is defined by $c_{ij}^k : T_k \longrightarrow \mathbb{R}_0^+$, where $c_{ij}^{l,k} := c_{ij}^k(t_l)$ represents the cost of arc (i,j) in scenario l, or under parameter t_l.

Let A' be a nonempty subset of A. Then, $G - A'$ denotes the subgraph of G with set of arcs $A \backslash A'$. In particular, $G - \{(i,j)\}$ is represented by G_{ij}^*.

A path from i to j, $i,j \in V$, in graph G, also called an (i,j)-path, is an alternating sequence of nodes and arcs of the form

$$p = \langle v_1, (v_1, v_2), v_2, \ldots, (v_{r-1}, v_r), v_r \rangle,$$

with $v_1 = i$, $v_r = j$ and where $v_s \in V$, for $s = 2, \ldots, r-1$, and $(v_s, v_{s+1}) \in A$, for $s = 1, \ldots, r-1$. The sets of arcs and of nodes in a path p are denoted by $A(p)$ and $V(p)$, respectively. Given two paths p, q, such that the destination node of p is also the initial node of q, the concatenation of p and q is the path formed by p followed by q, and is denoted by $p \diamond q$. In the following, paths will be represented simply by their sequence of nodes.

The cost of a path p in scenario l, or under t_l, $l \in I_k$, is defined by

$$v(p, t_l) = \sum_{(i,j) \in A(p)} c_{ij}^{l,k}. \tag{1}$$

With no loss of generality, 1 and n denote the origin and the destination nodes of the graph G, respectively. For simplicity of presentation, it will also be assumed that there are no arcs arriving at 1 and no arcs starting at n in G. The set of all $(1,n)$-paths in G is denoted by $P(G)$.

Let p_{ij}^l represent the shortest (i,j)-path in G, $i,j \in V$, for a given scenario $l \in I_k$. In order to simplify notation, p_1^l is used to denote p_{1n}^l and LB_i^l is used to denote $v(p_{in}^l, t_l)$, $i \in V$.

The minmax regret robust shortest path problem corresponds to determining a path in $P(G)$ with a least maximum robust deviation, i.e. satisfying

$$\arg \min_{p \in P(G)} RC(p), \tag{2}$$

where $RC(p)$ is the robustness cost of p in G, defined by

$$RC(p) := \max_{l \in I_k} RD(p, t_l), \tag{3}$$

and $RD(p, t_l)$ represents the robust deviation of a path p under parameter t_l, $l \in I_k$, in G, given by

$$RD(p, t_l) := v(p, t_l) - LB_1^l. \tag{4}$$

Any optimum solution of (2) is called a robust shortest path of G.

Given a path $p \in P(G)$, the set of scenarios in which $RC(p)$ occurs corresponds to the set of indices of the parameters under which the robust deviation of p is maximized and will be denoted by $I(p) := \{\arg\max_{l \in I_k} RD(p, t_l)\}$.

3 Preprocessing Techniques

As mentioned in the introduction, Karasan et al. [3] addressed the robust shortest path problem with interval data and introduced preprocessing techniques to reduce the size of a problem before it is solved. Namely, rules were defined with the goal of identifying in advance arcs which do not belong to any robust shortest path. Later on, Catanzaro et al. [2] developed a similar idea, in order to identify nodes that can be known in advance not to belong to any optimum solution. The first of these results is based on the shortest paths for at least one realization of the arc costs for scenarios that result from the lower and the upper limits of the cost intervals. The second is based on the shortest path under the scenario associated with the upper bounds on the cost intervals. The cost of this path is compared with the cost of the shortest path that contains node i, for any $i \in V$, under the scenarios associated with the lower limits of the cost intervals. In this section similar ideas are explored when considering a discrete set of possible scenarios. Conditions for reducing the network while not discarding the optimum solution are introduced and algorithms for implementing such conditions are outlined. Finally, the time complexity order of these methods is analyzed.

An arc/node of a robust shortest path is called robust 1-persistent, otherwise it is denominated robust 0-persistent.

The following result presents a sufficient condition for detecting robust 0-persistent nodes, valid for any scenario.

Proposition 1 *Consider a path $q \in P(G)$ and a node $r \notin V(q)$. If $v(p^{\hat{l}}_{1r} \diamond p^{\hat{l}}_{rn}, t_{\hat{l}}) > RC(q) + LB^{\hat{l}}_1$ for some $\hat{l} \in I_k$, then node r is robust 0-persistent.*

Proof Let $r \in V \backslash V(q)$ and q' be any path in $P(G) \backslash \{q\}$ such that $r \in V(q')$. Let q'_{1r} and q'_{rn} represent the $(1, r)$-subpath and the (r, n)-subpath of q', respectively. Then, by definition of robustness cost of a given path, one has

$$RC(q') = \max_{l \in I_k} RD(q', t_l) = \max_{l \in I_k} \left\{ v(q'_{1r}, t_l) + v(q'_{rn}, t_l) - LB^l_1 \right\}.$$

Given that p^l_{1r} and p^l_{rn} are the shortest $(1, r)$-path and the shortest (r, n)-path under t_l, $l \in I_k$, in G, respectively, then

$$RC(q') \geq \max_{l \in I_k} \left\{ v(p^l_{1r}, t_l) + LB^l_r - LB^l_1 \right\} = \max_{l \in I_k} RD(p^l_{1r} \diamond p^l_{rn}, t_l). \tag{5}$$

Consequently, if $v(p_{1r}^{\hat{l}} \diamond p_{rn}^{\hat{l}}, t_{\hat{l}}) > RC(q) + LB_1^{\hat{l}}$ for some $\hat{l} \in I_k$, then $RD(p_{1r}^{\hat{l}} \diamond p_{rn}^{\hat{l}}, t_{\hat{l}}) > RC(q)$. Therefore,

$$\max_{l \in I_k} RD(p_{1r}^{l} \diamond p_{rn}^{l}, t_l) > RC(q).$$

From (5) it follows that $RC(q') > RC(q)$, which means that any path in G that contains node r cannot be a robust shortest path. Therefore, node r is robust 0-persistent. □

The detection of a robust 0-persistent node is more effective than the identification of a single robust 0-persistent arc, given that the removal of a node from a network implies the elimination of all its incoming and outgoing arcs. Thus, the latter case will not be considered.

In [2] the identification of robust 1-persistent arcs in an interval data model was also presented. Computational tests showed that using this result, together with the robust 0-persistency, led to an actual reduction of the network and, besides that, it allowed to find an optimum solution more efficiently. Specifically, by considering the shortest path under the scenario associate to the upper bounds of the interval data, the new result was based on the choice of the arcs of such path to be evaluated. For the scenario that attributed the upper limits of the interval costs to the correspondent arcs of that path and the lower limits of the interval costs to the correspondent remaining arcs of the network, the rule could be derived by determining the shortest path on the subnetwork resultant from removing the arc under analysis at the original network, in case node n was still reachable from node 1.

The following result has the same motivation and introduces a broader rule for detecting robust 1-persistent arcs, which are restricted to the shortest $(1, n)$-paths for the scenarios of the adopted finite multi-scenario model. Provided that a path and its robustness cost are known, it deals with the scenarios for which the associate shortest $(1, n)$-paths contain the arc under evaluation:

Proposition 2 *Let $q \in P(G)$ and $(i,j) \in A(q) \cap \{A(p_1^l) : l \in I_k\}$ be an arc such that node n is reachable from node 1 in G_{ij}^*. Let $S(i,j) = \{l \in I_k : (i,j) \in p_1^l\}$ be the set of scenarios for which the associate shortest $(1, n)$-paths contain arc (i,j). Let p_1^{*l} denote the shortest $(1, n)$-path in $P(G_{ij}^*)$ under t_l, $l \in I_k$. If $v(p_1^{*\hat{l}}, t_{\hat{l}}) > RC(q) + LB_1^{\hat{l}}$ for some $\hat{l} \in S(i,j)$, then arc (i,j) is robust 1-persistent.*

Proof Let $q \in P(G)$ contain some arc in the set $\{A(p_1^l) : l \in I_k\}$, $(i,j) \in A(q)$ be an arc in that set and $p \in P(G_{ij}^*)$. Then, by definition of robustness cost of a path and because p_1^{*l} is the shortest path under scenario l that does not contain arc (i,j), $l \in I_k$, one gets

$$RC(p) = \max_{l \in I_k} RD(p, t_l) \geq \max_{l \in I_k} RD(p_1^{*l}, t_l), \tag{6}$$

with

$$\max_{l \in I_k} RD(p_1^{*l}, t_l) = \max\left\{\max_{l \in I_k \setminus S(i,j)} RD(p_1^{*l}, t_l), \max_{l \in S(i,j)} RD(p_1^{*l}, t_l)\right\}. \tag{7}$$

For every $l \in I_k \setminus S(i,j)$, one has $p_1^{*l} = p_1^l$ and therefore

$$RD(p_1^{*l}, t_l) = RD(p_1^l, t_l) = 0.$$

This means that $\max_{l \in I_k \setminus S(i,j)} RD(p_1^{*l}, t_l) = 0$. Since any robust deviation of a path is non-negative, then (6) and (7) imply $RC(p) \geq \max_{l \in S(i,j)} RD(p_1^{*l}, t_l)$. Then, if

$$v(p_1^{*\hat{l}}, t_{\hat{l}}) > RC(q) + LB_1^{\hat{l}}$$

for some $\hat{l} \in S(i,j)$, i.e., $RD(p_1^{*\hat{l}}, t_{\hat{l}}) > RC(q)$, one gets

$$RC(p) > RC(q).$$

This means that any path in $P(G)$ which does not contain arc (i,j) cannot be a robust shortest path. Therefore, arc (i,j) is robust 1-persistent. □

It should be noticed that when looking for a robust shortest path, the identification of a robust 1-persistent arc (i,j) allows to delete from the network all the other arcs with tail node i or head node j. This is more effective than detecting a robust 1-persistent node, given that this information does not exclude the arcs connected to it. Hence, the first technique is adopted rather than the second.

Since the calculation of the shortest paths for all scenarios is needed to calculate the robustness cost of a given path, one can take a shortest path with the minimum robustness cost as a candidate for the optimum solution and apply Propositions 1 and 2. Let Q denote the set of shortest $(1,n)$-paths for the scenarios of the model with the least robustness cost, i.e.

$$Q = \arg\min\{RC(p_1^l) : l \in I_k\}. \tag{8}$$

Given any path $q \in Q$, the sets of possible robust 0-persistent nodes and robust 1-persistent arcs will be denoted by *Snod* and *Sarc*, respectively, and are given by

$$Snod = V - \left\{r \in V(q) : q \in Q\right\} \tag{9}$$

and

$$Sarc = \left\{(i,j) \in A(q) : q \in Q \text{ and node } n \text{ is reachable from node 1 in } G_{ij}^*\right\}. \tag{10}$$

Moreover, the inequalities stated in Propositions 1 and 2 can be particularized. Namely, for any node $r \in Snod$, since

$$v(p^l_{1r} \diamond p^l_{rn}, t_l) = v(p^l_{1r}, t_l) + v(p^l_{rn}, t_l) = v(p^l_{1r}, t_l) + LB^l_r,$$

one concludes that if

$$\exists \hat{l} \in I_k \; : \; v(p^{\hat{l}}_{1r}, t_l) > \min\{RC(p^l_1) : l \in I_k\} + LB^{\hat{l}}_1 - LB^{\hat{l}}_r \tag{11}$$

holds, then node r is robust 0-persistent. Similarly, for any arc $(i,j) \in Sarc$, if condition

$$\exists \hat{l} \in S(i,j) \; : \; v(p^{*\hat{l}}_1, t_l) > \min\{RC(p^l_1) : l \in I_k\} + LB^{\hat{l}}_1 \tag{12}$$

is satisfied, then arc (i,j) is robust 1-persistent.

3.1 Algorithms

The algorithms to identify robust 0-persistent nodes and robust 1-persistent arcs first need to determine the tree of the shortest (j,n)-paths under t_l, denoted by $\mathcal{T}^l$, $j \in V$, $l \in I_k$, and to calculate their costs LB^l_j. Any shortest path tree algorithm can be used with such purpose, for instance with a labeling method to find the shortest path, like Bellman-Ford's or Dijkstra's algorithms (see [1]). The variable $RCmin$ represents the least robustness cost of the paths in set Q.

The following pseudo-code summarizes the procedure for determining robust 0-persistent nodes which are stored in the list $P0nod$. Testing (11) for a given scenario l, implies computing the tree of the shortest $(1,j)$-paths under t_l, denoted by $\tilde{\mathcal{T}}^l$, $l \in I_k, j \in V$. Computational effort can be avoided if the smallest $l \in I_k$ for which condition (11) is fulfilled is known, for each node r selected in $Snod$. In fact, if l_r denotes that scenario, then node r is a robust 0-persistent node and its analysis can halt. The tests for scenarios $l_r + 1, \ldots, k$ if $l_r \neq k$ can thus be skipped. Moreover, when $\max\{l_r : r \in Snod\} \neq k$, the computation of the trees $\tilde{\mathcal{T}}^l$ can be skipped for $l = \max\{l_r : r \in Snod\}, \ldots, k$, which can be useful when k is large.

In terms of the worst case computational time complexity of Algorithm 1, two phases should be considered. The first corresponds to determining the costs LB^l_j, $j \in V$, $l \in I_k$ and the robustness cost, $RCmin$, which is of $\mathcal{O}^a_1 = \mathcal{O}(km + k^2 n)$ for acyclic networks and of $\mathcal{O}^c_1 = \mathcal{O}(k(m+n\log n)+k^2 n)$ for general networks [8]. The second concerns the search for robust 0-persistent nodes. The computation of the tree $\tilde{\mathcal{T}}^l$ is similar to the computation of $\mathcal{T}^l$. However, for the former only the costs for the scenario where the paths are the shortest are needed. Thus, such procedure has time of $\mathcal{O}(m + n)$ for acyclic networks and $\mathcal{O}(m + n\log n)$ in the general case for each scenario [1]. Then, for each node selected in $Snod$, (11) is checked in $\mathcal{O}(1)$

Algorithm 1: Finding robust 0-persistent nodes

for $l = 1, \ldots, k$ **do**
  Compute the tree $\mathcal{T}^l$;
  for $j = 1, \ldots, n$ **do** $LB_j^l \leftarrow$ cost of the shortest (j, n)-path under t_l;
$RCmin \leftarrow \min\{RC(p_1^l) : l \in I_k\}$;
$Q \leftarrow \{p_1^l : l \in I_k$ and $RC(p_1^l) = RCmin\}$;
$Snod \leftarrow V - \{r \in V(q) : q \in Q\}$;
$P0nod \leftarrow \emptyset$;
Compute the tree $\tilde{\mathcal{T}}^1$;
for $r \in Snod$ **do**
  for $l = 1, \ldots, k$ **do**
    if $l \neq 1$ *and tree* $\tilde{\mathcal{T}}^l$ *was not yet determined* **then** Compute the tree $\tilde{\mathcal{T}}^l$;
    $p_{1r}^l \leftarrow$ shortest $(1, r)$-path under t_l;
    if $v(p_{1r}^l, t_l) > RCmin + LB_1^l - LB_r^l$ **then**
      $P0nod \leftarrow P0nod \cup \{r\}$;
      break;
return $P0nod$

time. The analysis of the nodes in $Snod \subseteq V \backslash \{1, n\}$ considers at most k scenarios and at most $n - 2$ nodes, therefore, it can be done in $\mathcal{O}_2^a = \mathcal{O}(k(m + n))$ time for acyclic networks and in $\mathcal{O}_2^c = \mathcal{O}(k(m + n \log n))$ time for general networks. Consequently, Algorithm 1 has a time complexity of $\mathcal{O}_1^a + \mathcal{O}_2^a = \mathcal{O}(km + k^2 n)$ for acyclic networks and of $\mathcal{O}_1^c + \mathcal{O}_2^c = \mathcal{O}(km + kn \log n + k^2 n)$ for general networks.

The following pseudo-code summarizes the procedure for searching for robust 1-persistent arcs. The list $P1arc$ stores such arcs. When an arc (i, j) is selected in $Sarc$ to be scanned only the shortest $(1, n)$-path in the reduced network G_{ij}^* under the scenarios $l \in S(i, j)$ for which $(i, j) \in p_1^l$ has to be determined, since for the remaining scenarios $p_1^{*l} = p_1^l$. Thus, the inequality (12) only has to be tested for the scenarios $S(i, j)$ in I_k, and, moreover, the algorithm can halt the search when a scenario satisfies that inequality.

In a worst case, Algorithm 2 has the same time complexity order as Algorithm 1 to determine $RCmin$. Then, the second phase is concerned with the analysis of the arcs in $Sarc$, which implies the shortest paths computation, and since at most $k(n-1)$ arcs are used, one obtains $\mathcal{O}_2^a = \mathcal{O}(k^2 n(n + m)) = \mathcal{O}(k^2 mn)$ time for acyclic networks and $\mathcal{O}_2^c = \mathcal{O}(k^2 n(m + n \log n))$ time for general networks. Consequently, Algorithm 2 has a polynomial time complexity of $\mathcal{O}_1^a + \mathcal{O}_2^a = \mathcal{O}(k^2 mn)$ for acyclic networks, and $\mathcal{O}_1^c + \mathcal{O}_2^c = \mathcal{O}(k^2 mn + k^2 n^2 \log n)$ for general networks.

Algorithm 2: Finding robust 1-persistent arcs

for $l = 1, \dots, k$ **do**
  $p_1^l \leftarrow$ shortest path under t_l;
  $LB_1^l \leftarrow$ cost of p_1^l under t_l;
$RCmin \leftarrow \min\{RC(p_1^l) : l \in I_k\}$;
$Q \leftarrow \{p_1^l : l \in I_k$ and $RC(p_1^l) = RCmin\}$;
$Sarc \leftarrow \{(i,j) \in A(q) : q \in Q$ and node n is reachable from node 1 in $G_{ij}^*\}$;
$P1arc \leftarrow \emptyset$;
for $(i,j) \in Sarc$ **do**
  $S(i,j) \leftarrow \{l \in I_k : (i,j) \in p_1^l\}$;
  for $l \in S(i,j)$ **do**
    $p_1^{*l} \leftarrow$ shortest path under t_l in G_{ij}^*;
    if $v(p_1^{*l}, t_l) > RCmin + LB_1^l$ **then**
      $P1arc \leftarrow P1arc \cup \{(i,j)\}$;
      break;
return $P1arc$

4 Example

In this section the preprocessing techniques for finding robust 0-persistent nodes and robust 1-persistent arcs introduced previously are exemplified.

Let $G(V, A, T_2)$ be the network represented in Fig. 1. Figure 2 shows the trees of the shortest paths in this network from every node to node 7 under scenario 1—Fig. 2a—and under scenario 2—Fig. 2b.

After computing these trees, Q is set to $\{p_1^1, p_1^2\}$, with $p_1^1 = \langle 1, 2, 7\rangle$, $LB_1^1 = 2$, and $p_1^2 = \langle 1, 4, 6, 7\rangle$, $LB_1^2 = 7$. Since $v(p_1^1, t_2) = 10$ and $v(p_1^2, t_1) = 8$, one has $RC(p_1^1) = 3$ and $RC(p_1^2) = 6$. Hence, $p_1^1 = \langle 1, 2, 7\rangle$ is the path with the minimum robustness cost in Q, and $RCmin = 3$.

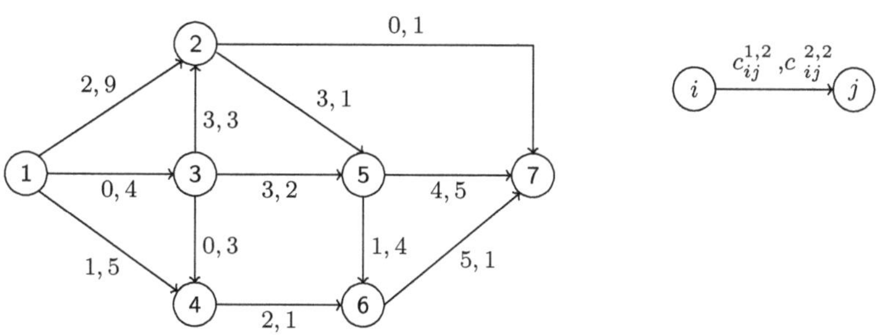

Fig. 1 Network $G(V, A, T_2)$

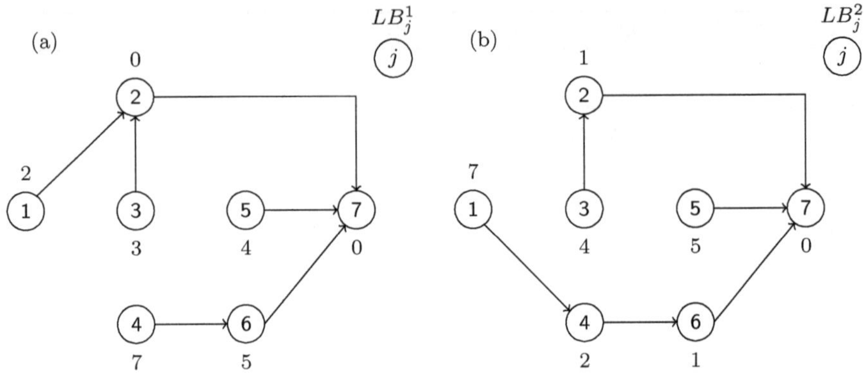

Fig. 2 Shortest path trees rooted at node $n = 7$ in $G(V, A, T_2)$. (**a**) Under scenario 1; (**b**) under scenario 2

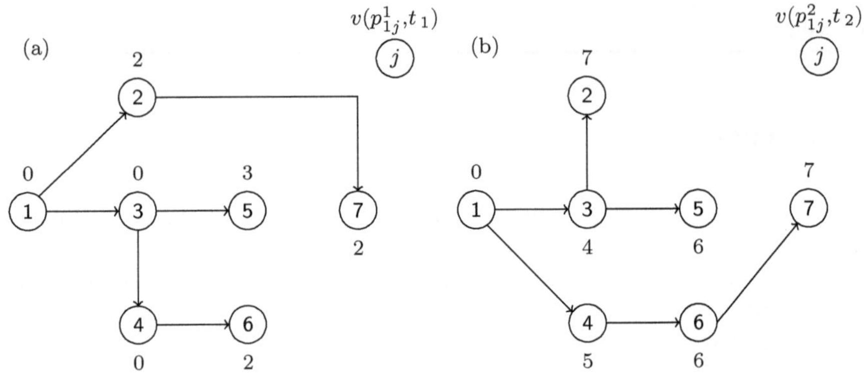

Fig. 3 Shortest path trees rooted at node 1 in $G(V, A, T_2)$. (**a**) Under scenario 1; (**b**) under scenario 2

4.1 Identifying Robust 0-Persistent Nodes

Because $q = p_1^1 = \langle 1, 2, 7 \rangle$, the set of nodes to be scanned is $Snod = \{3, 4, 5, 6\}$. Figure 3 shows the trees of the shortest paths from node 1 to any node under scenario 1—Fig. 3a—and under scenario 2—Fig. 3b.

According to Algorithm 1, one starts by checking inequality (11) in scenario 1. For nodes 4, 5 and 6 that condition is satisfied. In fact, for node 4,

$$v(p_{14}^1, t_1) = 0 > RCmin + LB_1^1 - LB_4^1 = -2;$$

for node 5,

$$v(p_{15}^1, t_1) = 3 > RCmin + LB_1^1 - LB_5^1 = 1$$

and for node 6,

$$v(p_{16}^1, t_1) = 2 > RCmin + LB_1^1 - LB_6^1 = 0.$$

Consequently, nodes 4, 5 and 6 are robust 0-persistent. For node 3 and the same scenario, this condition is not satisfied, because $v(p_{13}^1, t_1) = 0$ and $RCmin + LB_1^1 - LB_3^1 = 2$. The same happens for scenario 2, because $v(p_{13}^2, t_2) = 4$ and $RCmin + LB_1^2 - LB_3^2 = 6$. Consequently, $P0nod = \{4, 5, 6\}$ and node 3 cannot be deleted from the network when searching for an optimum solution.

4.2 Identifying Robust 1-Persistent Arcs

Because $q = p_1^1 = \langle 1, 2, 7\rangle$, the set of arcs to be scanned is $Sarc = \{(1, 2), (2, 7)\}$. According to Algorithm 2, one first considers arc $(1, 2)$ for which $S(1, 2) = \{1\}$. Node 7 is reachable from node 1 in G_{12}^* and $p_1^{*1} = \langle 1, 3, 2, 7\rangle$, with $v(p_1^{*1}, t_1) = 3$. Nevertheless, $RCmin + LB_1^1 = 5$, therefore condition (12) is not satisfied and one can not conclude anything about arc $(1, 2)$. Afterward, arc $(2, 7)$ is selected and $S(2, 7) = \{1\}$. Now, node 7 is reachable from node 1 in G_{27}^* and $p_1^{*1} = \langle 1, 3, 5, 7\rangle$, with $v(p_1^{*1}, t_1) = 7$ under scenario 1. Since $7 > RCmin + LB_1^1 = 5$, condition (12) is satisfied for this case. Consequently, arc $(2, 7)$ is robust 1-persistent and belongs to an optimum solution, $P1arc = \{(2, 7)\}$.

4.3 Computing a Robust Shortest Path After Preprocessing

After the preprocessing a robust shortest path can be computed in a reduced network, represented in Fig. 4. The robust 0-persistent nodes, 4, 5 and 6, are also removed from G, as well as all the arcs that start or end in these nodes. Arc $(2, 7)$ must be contained in the optimum solution since it is robust 1-persistent. Thus, the

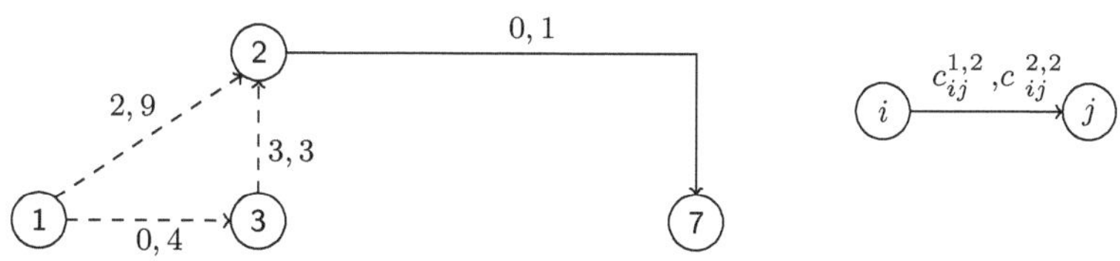

Fig. 4 Reduced network after preprocessing

reduced network results from removing from G all the remaining arcs that start in node 2, $(2, 5)$, and all the remaining arcs that end in node 7, $(5, 7)$ and $(6, 7)$. Nothing can be said at this moment about the other arcs in G, represented with a dashed line in Fig. 4.

There are only two $(1, 7)$-paths containing $(2, 7)$ in the reduced network, $q = \langle 1, 2, 7 \rangle$, with $RC(q) = 3$, and $q' = \langle 1, 3, 2, 7 \rangle$, with $v(q', t_1) = 3$ and $v(q', t_2) = 8$. Then, $RC(q') = 1 < RC(q)$, and therefore, q' is the $(1, 7)$-path with the minimum robustness cost in the reduced network, i.e. q' is the robust shortest path in G.

5 Computational Experiments

In this section a computational study of the performance of the preprocessing techniques introduced earlier and of their impact on the resolution of the robust shortest path problem when combined with the labeling and the hybrid algorithms introduced in [8] are presented.

In order to apply the preprocessing techniques described in Sect. 3, Algorithms 1 and 2 were implemented in Matlab 7.12 and ran on a computer equipped with an Intel Pentium Dual CPU T2310 1.46 GHz processor and 2GB of RAM. The codes use Dijkstra's algorithm [1] to solve the single destination shortest path problem for a given scenario. As mentioned above, the preprocessing techniques were combined with the labeling algorithm (LA) and the hybrid algorithm (HA). The robust shortest path problem was solved with and without preprocessing.

The benchmarks used in the experiments correspond to randomly generated directed graphs with n nodes, m arcs and k scenarios. For each scenario, each arc cost is assigned with a random integer number in $U(0, 100)$. The computational tests were performed for $k \in \{2, 10\}$, $n \in \{250, 500, 750, 1000\}$ and $d \in \{5, 10, 20\}$, where $d = m/n$ represents the network density.

For each network dimension, 10 instances were generated. For each instance, Algorithms 1 and 2 were applied and the associate robust shortest path problems were solved by LA and HA after preprocessing. Alternatively, LA and HA were applied to solve the same instances without preprocessing.

5.1 *Results*

In order to analyze the performance of the algorithms for each dimension, the average total running times (in seconds) are calculated. Let P be the CPU time to preprocess the network, NP be the CPU time for solving the robust shortest path without any preprocessing and AP be the CPU time for solving the same problem after preprocessing. Let $\mu(P)$, $\mu(NP)$ and $\mu(AP)$ denote the corresponding averages. The indices 0 and 1 are used to distinguish the preprocessing of robust 0-persistent nodes and the preprocessing of robust 1-persistent arcs, respectively. In

Table 1 Average results for preprocessing robust 0-persistent nodes

					HA			LA		
n	d	k	P_0	Nod_0	NP	AP_0	$P_0 + AP_0$	NP	AP_0	$P_0 + AP_0$
250	5	2	0.210	134	0.324	0.304	0.514	0.473	0.353	0.563
		10	1.033	111	1.193	1.164	2.197	1.358	1.209	2.242
	10	2	0.227	20	0.318	0.331	0.558	0.830	0.499	0.726
		10	1.218	201	1.531	1.471	2.689	1.534	1.473	2.691
	20	2	0.220	4	0.342	0.342	0.562	1.173	0.521	0.741
		10	1.185	8	1.562	1.594	2.779	2.588	1.648	2.833
500	5	2	0.479	380	0.673	0.648	1.127	0.926	0.688	1.167
		10	2.404	309	2.995	3.041	5.445	3.153	3.037	5.441
	10	2	0.537	145	0.732	0.725	1.262	1.438	1.075	1.612
		10	2.641	101	3.593	3.646	6.287	4.109	3.601	6.242
	20	2	0.553	147	0.744	0.699	1.252	2.082	1.186	1.739
		10	2.705	3	3.400	3.439	6.144	6.037	4.013	6.718
750	5	2	0.741	319	1.127	1.281	2.022	2.284	1.638	2.379
		10	5.899	370	6.797	7.263	13.162	8.221	6.860	12.759
	10	2	0.859	281	1.218	1.329	2.188	2.491	1.934	2.793
		10	5.816	90	6.920	7.433	13.249	10.610	8.264	14.080
	20	2	0.865	27	1.294	1.390	2.255	4.233	2.720	3.585
		10	6.470	0	8.775	9.189	16.659	22.426	10.581	18.051
1000	5	2	1.088	714	1.721	2.003	3.091	2.557	1.703	2.791
		10	7.881	631	8.805	10.079	17.960	10.424	9.089	16.970
	10	2	1.192	403	1.817	2.896	4.088	3.645	2.831	4.023
		10	8.313	457	9.449	11.661	19.974	11.546	10.775	19.088
	20	2	1.203	185	1.855	1.853	3.056	7.094	3.926	5.129
		10	7.568	1	9.825	9.556	17.124	29.270	11.975	19.543

addition, let Nod_0 and Arc_1 be the number of robust 0-persistent nodes and robust 1-persistent arcs, respectively, and $\mu(Nod_0)$ and $\mu(Arc_1)$ be their averages.

The averages for the preprocessing of robust 0-persistent nodes are reported in Table 1, where μ was omitted to simplify notation. The plots in Fig. 5 show the average CPU times considering the density of the network and the number of scenarios.

Since robust 1-persistent arcs were rarely detected with the performed tests, the values of $\mu(Arc_1)$ were always 0 and therefore $\mu(NP)$ and $\mu(AP_1)$ were similar. Thus, preprocessing robust 1-persistent arcs did not bring any advantage and in Table 2 only $\mu(P_1)$ is shown. These results were close to the correspondent for $\mu(P_0)$ in Table 1 when the number of scenarios was small ($k = 2$). Nevertheless, when $k = 10$, preprocessing robust 0-persistent nodes became more demanding than preprocessing robust 1-persistent arcs. This can be explained by the fact that the set of possible scenarios for checking condition (11) is generally larger than the number of scenarios involved to check (12), which implies an increased effort in

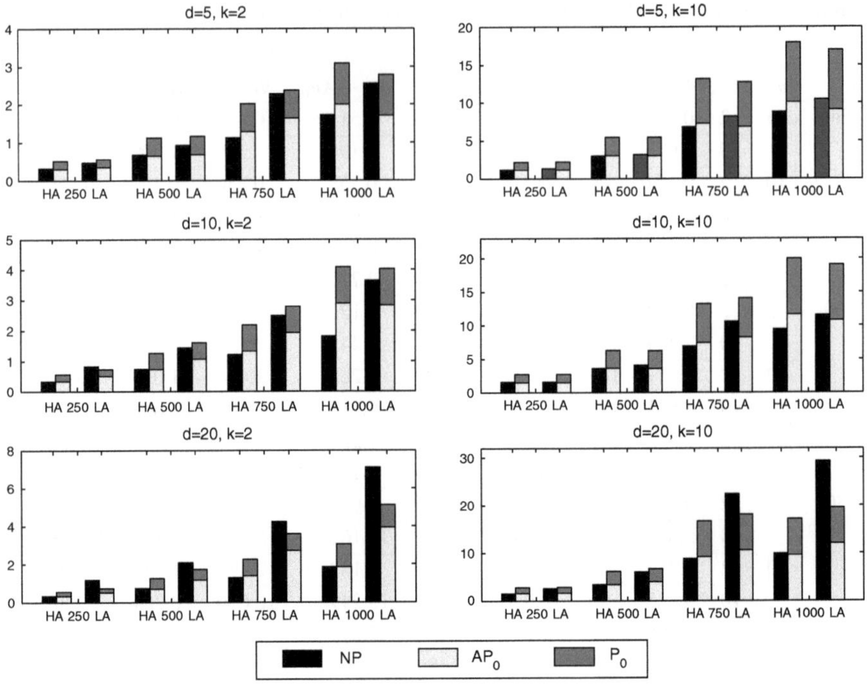

Fig. 5 Average CPU times for preprocessing robust 0-persistent nodes and for algorithms HA and LA with and without preprocessing

Table 2 Average CPU times for preprocessing robust 1-persistent arcs

d	k	$n = 250$	$n = 500$	$n = 750$	$n = 1000$
5	2	0.239	0.477	0.970	1.159
	10	0.228	0.567	1.196	1.641
10	2	0.337	0.517	0.763	1.232
	10	0.256	0.701	0.959	1.152
20	2	0.286	0.635	0.897	1.143
	10	0.231	0.473	1.257	1.368

solving shortest path problems. The results show that such difference is more clear when the networks are bigger ($n \in \{750, 1000\}$).

Table 1 shows that finding robust 0-persistent nodes was always faster than finding the robust shortest path without preprocessing ($\mu(P_0) < \mu(NP)$), both when HA and LA are considered. Moreover, the smaller the density, the higher the number of robust 0-persistent nodes. In fact, when $d = 5$, more than half of the nodes in the network were identified as being robust 0-persistent.

The preprocessing of robust 0-persistent nodes was quite effective when solving the robust shortest path problem with LA for the networks with the highest density. The reason is that the denser the network, the bigger the average number of arcs that emerge from each node, and, consequently, the bigger the number of labels that can be discarded when detecting robust 0-persistent nodes. For $d = 20$, Table 1 and the plots in Fig. 5 confirm such results ($\mu(NP) > \mu(P_0) + \mu(AP_0)$), specially for the two largest networks ($n \in \{750, 1000\}$). Exceptions were observed when $d = 20$ and $k = 10$ for the two smallest networks ($n \in \{250, 500\}$), possibly due to a major effort in calculating the robustness costs after preprocessing given that more scenarios are involved. Globally, the CPU time used by LA to solve the problem after preprocessing was always inferior to the correspondent time without preprocessing ($\mu(AP_0) < \mu(NP)$).

For HA, $\mu(P_0)$ is close to $\mu(NP)$ and then the difference between both can be easily compensated by the time necessary to solve the problem after preprocessing ($\mu(AP_0)$). This value exceeded $\mu(NP)$ for HA in many of the large networks ($n \in \{750, 1000\}$), generally when only a small number of robust 0-persistent nodes could be identified. In both cases the number of paths ranked in HA [8] did not decrease enough to spare enough computational effort for the overall performed tests. Hence, even though for some cases finding the robust shortest path with HA after preprocessing was faster than without preprocessing, in general this approach did not seem to react well to preprocessing.

Figure 5 shows that the performances of HA and LA are similar for the networks with the lowest densities ($d \in \{5, 10\}$). Globally, for the same density and the same number of nodes, $|\mu(NP) - (\mu(P_0) + \mu(AP_0))|$ grows with the number of scenarios.

6 Conclusions

This work approached preprocessing techniques for the minmax regret robust shortest path problem with a finite number of scenarios. The research concerned the identification of nodes/arcs that participate or do not participate in an optimum solution, before this is actually determined. Based on [2], sufficient conditions were derived to search for robust 0-persistent nodes and for robust 1-persistent arcs. The developed rules can be implemented in polynomial time and depend on the shortest paths for the scenarios with the least robustness cost.

Computational experiments were performed over randomly generated networks in order to study the impact of preprocessing on finding a robust shortest path. The labeling and hybrid approaches in [8] were used to compute the robust shortest path with and without preprocessing. The strategy for identifying robust 0-persistent nodes was the most useful, specially in sparse networks, for which the problem size could be significantly reduced at most of the cases. Moreover, for the networks with the highest density, the labeling method applied after detection of robust 0-persistent nodes outperformed its application without preprocessing. The results were different when using the same preprocessing followed by the hybrid method. In this case the preprocessing made some of the original instances easier to solve. Nevertheless the CPU time demanded for both preprocessing and computing the robust shortest path exceeded the CPU times of the hybrid method when applied without preprocessing. For the considered instances preprocessing robust 1-persistent arcs was not advantageous, given that very few of those arcs could have been identified.

Future research on this subject can be guided to develop new techniques that allow more nodes to be discarded or that can identify more arcs belonging to an optimum solution. One of the possible techniques to investigate is a dynamic approach for identifying such nodes or arcs based on the robust shortest path candidates determined along the algorithm.

Acknowledgements This work has been partially supported by the Portuguese Foundation for Science and Technology under project grants PEst-OE/ EEI/UI308/2014 and SFRH/BD/51169/2010.

References

1. Ahuja, R.K., Magnanti, T.L., Orlin, J.B.: Network Flows: Theory, Algorithms and Applications. Prentice Hall, Englewood Cliffs (1993)
2. Catanzaro, D., Labbé, M., Salazar-Neumann, M.: Reduction approaches for robust shortest path problems. Comput. Oper. Res. **38**, 1610–1619 (2011)
3. Karasan, O.E., Pinar, M.C., Yaman, H.: The robust shortest path problem with interval data. Technical Report, Bilkent University, Ankara (2001)
4. Montemanni, R., Gambardella, L.: An exact algorithm for the robust shortest path problem with interval data. Comput. Oper. Res. **31**, 1667–1680 (2004)
5. Montemanni, R., Gambardella, L.: The robust shortest path problem with interval data via Benders decomposition. 4OR: Q. J. Belg. Fr. Ital. Oper. Res. Soc. **3**, 315–328 (2005)
6. Montemanni, R., Gambardella, L., Donati, V.: A branch and bound algorithm for the robust shortest path problem with interval data. Oper. Res. Lett. **32**, 225–232 (2004)
7. Murthy, I., Her, S.-S.: Solving min-max shortest path problems on a network. Nav. Res. Logist. **39**, 669–683 (1992)
8. Pascoal, M., Resende, M.: Minmax regret robust shortest path problem in a finite multi-scenario model. Appl. Math. Comput. **241**, 88–111 (2014)
9. Yu, G., Yang, J.: On the robust shortest path problem. Comput. Oper. Res. **25**, 457–468 (1998)

Mathematics of Energy and Climate Change: From the Solar Radiation to the Impacts of Regional Projections

Mário Gonzalez Pereira

Abstract This chapter focuses on the natural and anthropogenic drivers of climate change and on the assessment of potential impacts of regional projections for different scenarios of future climate. Internal and external forcing factors of climate change are associated to changes in the most important processes of energy transfer with influence on the energy balance of the climate system. The role of the solar activity, regular variations in the orbital parameters of the Earth and the radiative forcing which comprises the changes in the chemical composition of the atmosphere and the characteristics of the radiative processes that occur in the atmosphere and on the surface of the Earth will be discussed. Recent evidences of climate change and the general characteristics of the climate models used in climate projection will be presented. The chapter ends with results of some case studies of potential impacts of regional climate change projections in Portugal, namely in forest fire regime, extreme precipitation intensity and in the design of storm water drainage infrastructures.

Abbreviations

20C3M	Twentieth century model simulations
AOGCMs	Atmosphere–Ocean General Circulation Models
BA	Burnt area
BAM	Burned Area Model
BUI	Buildup Index
CFFBPS	Canadian Forest Fire Behaviour Prediction System
CFFDRS	Canadian Forest Fire Danger Rating System
CFFWIS	Canadian Forest Fire Weather Index System
COSMO-CLM	COnsortium for Small-scale MOdelling and Climate Limited-area Modelling Community

M.G. Pereira (✉)
CITAB, Universidade de Trás-os-Montes e Alto Douro, Quinta de Prados, 5000-801 Vila Real, Portugal
e-mail: gpereira@utad.pt

J.-P. Bourguignon et al. (eds.), *Mathematics of Energy and Climate Change*,
CIM Series in Mathematical Sciences 2, DOI 10.1007/978-3-319-16121-1_12

DMC	Duff Moisture Code
DC	Drought Code
DSR	Daily Severity Rating
ECMWF	European Centre for Medium-Range Weather Forecasts
FFMC	Fine Fuel Moisture Code
FWI	Fire Weather Index
GCM	General Circulation Model
IDF	Intensity Duration Frequency Curve
IPCC	Intergovernmental Panel on Climate Change
IPMA	Instituto Português do Mar e da Atmosfera
ISI	Initial Spread Index
LWR	Long wave radiation
MIROC	Model for Interdisciplinary Research on Climate
OLR	Outgoing Long wave radiation
PRFD	Portuguese Rural Fire Database
RCM	Regional Climate Model
SWR	Shortwave radiation
TOA	Top of the Atmosphere
UNFCCC	United Nations Framework Convention on Climate Change
WMO	World Meteorological Organization

1 Introduction

The participation in the thematic session "Energy Transfer and Management" of the "International Conference and Advanced, School Planet Earth, Mathematics of Energy and Climate Change" was considered as an opportunity to promote the proximity and interaction between mathematicians, climatologists and engineers working in the energy transfer and management scientific areas. The participants were challenged to present their work from the standpoint of mathematics, attracting and motivating the colleagues from this area of science to improve the methodologies used in climate research and/or to propose different approaches usually used in their areas of study. This contribution is intended to respond to this request by focusing on the concepts and processes in the physics of climate as well as the assessment of climate change and its impacts. It aims to contribute to clarify some key definitions and phenomena, focuses mainly on the laws and principles governing these concepts and processes as well as on the adopted methodology and the presentation of the main results obtained in specific case studies.

2 Natural and Anthropogenic Drivers of Climate Change

This section is devoted to present the concepts and processes of the physics of climate that are fundamental to understand and assess future climate changes. Various definitions of weather and climate can be found in the literature. Weather is a detailed description of the state of the Atmosphere and daily evolution for a short period of time (of just a few days). The state of the Atmosphere is defined by the values of a complete set of thermo-hydro-dynamic properties, usually named climatic elements. These properties may be extensive (e.g. volume, mass, energy, entropy) whose values are proportional to the size of the system or intensive (e.g. temperature, pressure, wind) with local character and value defined in each point and instant [60]. Climate comes from Ancient Greek word *klima*, which means inclination in a clear mention to the different amount of energy per unit area that reaches the Earth's surface depending to the direction of the incident solar rays. Climate is commonly defined as the "expected weather" [9], the "average weather" or the "the state, including a statistical description, of the climate system" [40]. More rigorously, climate may be defined as a set of statistics (e.g., averaged quantities, higher moment statistics and information on the occurrence of extreme events) of the climate elements that characterizes the structure and behavior of the Atmosphere, Hydrosphere and Cryosphere [60], over a period of time that may range from a few months to thousands or millions of years [40], but it is common to adopt the classical period of 30 years recommended by the World Meteorological Organization [85]. A discussion of the statistical descriptors of climate may be found in [26] and variations in these statistics, on all spatial and temporal scales may be defined as the climate variability. This variability may be decomposed into the internal variability, caused by natural internal processes within the climate system (such as internal instabilities and feedbacks, leading to nonlinear interactions among various components of the climate system), or external variability, caused by natural or anthropogenic variations in external forcing namely astronomical (e.g., changes in orbital parameters, in the intensity solar radiance and rate of rotation of the Earth) and terrestrial (e.g. changes in the composition of the Atmosphere due to human activity or volcanic activity, in the land use/land cover and in long-term tectonic factors) forcings [40, 60]. In summary, the climate variability result from the complex interactions between forced (external) and free (internal) variations within the dissipative and highly nonlinear Climate System with many sources of instabilities [60]. In a recent past, the United Nations Framework Convention on Climate Change (UNFCCC) makes a distinction between climate change attributable to human activities altering the atmospheric composition, and climate variability attributable to natural causes. However, the most recent IPCC definition of climate change [40] is adopted here as the change in the state of the climate that can be identified (e.g., by using statistical tests) by changes in the mean and/or the variability of its properties, and that persists for an extended period, typically decades or longer, independently of the cause of the change.

Climate change is usually presented in direct or indirect association with the increase of global temperature. In fact, the most recent Fifth Assessment Report of the Intergovernmental Panel on Climate Change [41], provides a list of undeniable observable changes in climate since 1950 including: the warming of the Atmosphere and the Oceans; reduction of the amount of snow and ice; rising of sea level and concentrations of greenhouse gases. Most of these recent observable evidences of climate change are associated to changes (increases) of air temperature. From the thermodynamics point of view, the temperature of a system is a measure of its energy content. The object of this study is the Climate System (Fig. 1) which may be defined as an open (due to the sporadic exchanges of usually small amounts of mass with the outside) and non-isolated system (for the energy exchange with the Sun and the outer space). The Climate System is composed by the following subsystems: Atmosphere (the gas layer that surrounds the planet), Hydrosphere (the water in liquid phase deposited on the surface forming the oceans, rivers and lakes), Cryosphere (water in the solid phase—snow and ice), Lithosphere (the solid, inorganic, mineral, rocky crust covering entire planet) and the Biosphere (all living organisms from one-celled organisms to plants and animals). These subsystems interact with each other through many different processes (e.g. hydrological and carbon cycles) exchanging mass, energy and momentum.

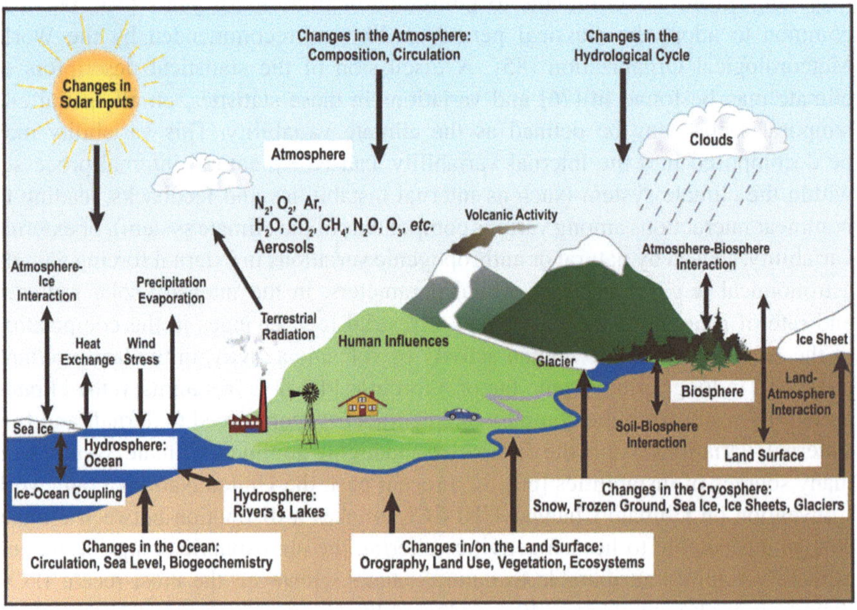

Fig. 1 The Climate System, their components, processes and interactions. Image from the [38]

Thus, to understand the changes of climate it is necessary to study all the processes of energy transfer within and between each subsystem but essentially between the Climate System and its neighborhood. The Climate System receives energy from interior of the Earth and other stars but it is mainly powered by sun energy. This energy reaches the top of the Atmosphere in the form of radiation in a process that can be described by the laws of radiation.

In 1791, Pierre Prévost, a Genevan philosopher and physicist, showed that all bodies radiate heat independent of its temperature [65]. However, it is the Planck's law that describes the distribution of the energy radiated by a black body in thermal equilibrium. This special body is an ideal model, which consist of a perfect absorber that emits the maximum possible amount of energy at a given temperature in all directions (i.e., isotropically). The Planck's law states that the amount and quality of energy emitted by blackbody is determined solely by its absolute temperature i.e. the emitted radiation has a spectrum that is only determined by the temperature alone and not by the shape or composition of the body, which can be written in terms of the wavelength or frequency (spectral radiance), using $c = \lambda\nu$, by

$$E(\lambda, T)d\lambda = \frac{2hc^2}{\lambda^5(e^{ch/k\lambda T} - 1)}d\lambda \quad \text{or} \quad E(\nu, T)d\nu = \frac{2h\nu^3}{\lambda^5(e^{ch/k\lambda T} - 1}d\nu$$

where $h = 6.63 \times 10^{-34}$ J is the Planck constant, c is the speed of the light in the propagation medium, and $k = 1.38 \times 10^{-23}\,\mathrm{J\,K^{-1}}$ is the Boltzman constant. The integration of the spectral radiance (Planck's law) for all possible wavelengths and angles of a hemisphere covering a horizontal surface leads to the total radiance,

$$\pi E(T) = \sigma T^4$$

where $\sigma = 5.670 \times 10^{-8}\,\mathrm{W\,m^{-2}\,K^{-4}}$ is the Stefan-Boltzman constant. The above equation is the Stefan-Boltzman law that states that the flux density emitted by a blackbody is proportional to the fourth power of the temperature (T^4). The wavelength of maximum emission is the solution of a simple extreme problem solved by differentiating $E(\lambda, T)$ with respect to λ and setting the derivative equal to zero,

$$T\lambda_{\max} = 2898\,\mu\mathrm{m\,K}$$

which constitutes the Wien law. These radiation laws are well illustrated in Fig. 2 where the intensity of radiation is plotted for different values of blackbody's temperature. The shape of the thick black lines are defined by Plank's law, the area below those lines are a measure of the total energy emitted by a blackbody (Stefan-Boltzman's law) and the wavelength of maximum emission decrease with the increase of temperature which helps to understand why Wien law is also known as Wien displacement law.

Two other radiation laws should be mentioned: the Kirchhoff's law and the Beer-Bouger-Lambert law. The first takes into account that, in general, a medium do not

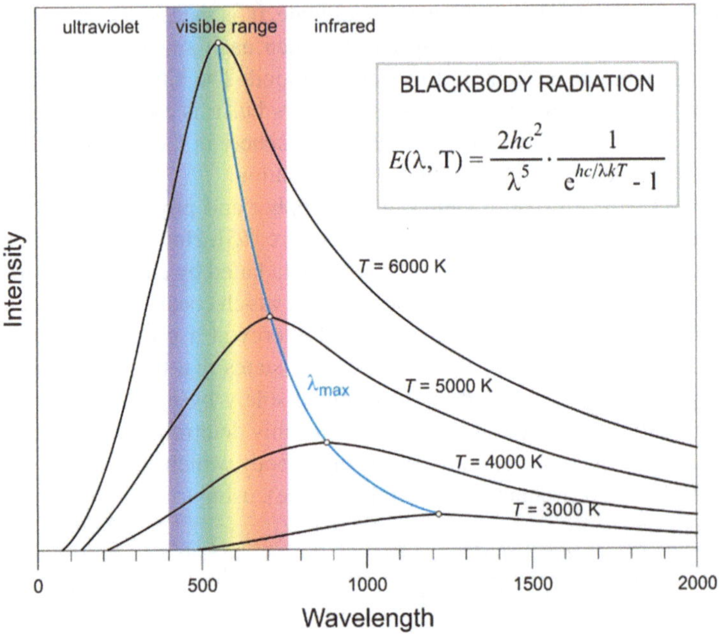

Fig. 2 Blackbody radiation (Planck Radiation Law) which relates the intensity of radiation emitted by unit surface area into a fixed direction (*solid angle*) from the blackbody as a function of wavelength for a specific temperature. Wien displacement Law ($T.\lambda_{\max} = const.$) is also illustrated. Image from the Croatian-English Chemistry Dictionary & Glossary. Credits to E. Generalic (http://glossary.periodni.com/glossary.php?en=blackbody+radiation)

absorb all but also reflect and transmit part of the incident energy so that, for each wavelength,

$$\alpha_\lambda + r_\lambda + t_\lambda = 1$$

where α_λ is the absorptivity defined as the quotient between the absorbed and total monochromatic intensity, $\alpha_\lambda = I_\alpha(\lambda)/I(\lambda)$, $r_\lambda = I_r(\lambda)/I(\lambda)$ is the reflectivity and $t_\lambda = I_t(\lambda)/I(\lambda)$ is the transmissivity of the layer [60].

The amount of energy that reaches the Climate System (top of the Atmosphere) depends on two factors: the amount of energy emitted by the Sun and the orbital radius of the Earth (distance Earth-Sun). In fact, the Sun do not emit energy at the same rate (Fig. 3) and the changes in solar activity and its relationship with climate is a contemporary topic of research (cf. [6, 11, 24, 27, 31, 46, 47, 50, 83]).

On the other hand, the distance to the sun is primarily determined by the orbital parameters of the Earth, in particular the eccentricity of the orbit, the obliquity of the Earth's axis, and the precession of the Earth. These three factors combined determine the flux of incoming solar radiation to the Climate System and the temporal and spatial distribution of that energy over the Earth Surface.

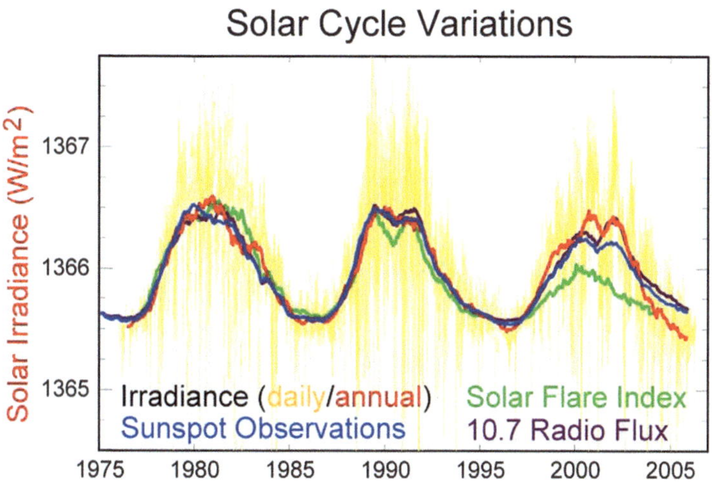

Fig. 3 Several indicators of solar variability recorded during the last three solar cycles, namely solar irradiance (i.e. direct solar power at the top of the Earth's atmosphere), sunspot numbers, solar flare activity, and 10.7 cm radio flux. All data are depicted as the annual average value except the solar irradiance which is also depicted as both a daily measurement and a moving annual average. Image created by Robert A. Rohde/Global Warming Art (http://www.globalwarmingart.com/wiki/File:Solar_Cycle_Variations_png)

The relative importance of these orbital parameters is not always evident. For example, presently, the summer season on the northern Hemisphere occurs when Earth is on the aphelion (the point of the orbit where the Earth is farthest from the Sun) as a consequence of small eccentricity of orbit and the obliquity of the Earth's axis. However, these orbital parameters do not remain constant over time but vary periodically (Fig. 4). The Earth's eccentricity determines the shape of the Earth's orbit around the Sun and is constantly fluctuating between circular and more elliptical (0–5 % ellipticity) with two main periodicities ~100,000 and ~413,000 years [12]. These oscillations, from more elliptic to less elliptic, are of prime importance to glaciation in that it alters the distance from the Earth to the Sun, thus changing the distance the Sun's short wave radiation must travel to reach the Earth, subsequently reducing or increasing the amount of radiation received at the Earth's surface in different seasons. These oscillations are termed the Milankovitch cycles due to the work of Milutin Milankovitch, the Serbian astronomer, geophysicist and mathematician who contributes to the explanation of Earth's long-term climate changes (e.g. ice ages) caused by the changes in the position of the Earth in comparison to the Sun [34] but this theory was largely advanced by Hays et al. [29]. Berger [7] provides a review on the Milankovitch theory and climate.

For relative long periods of time, the Earth's surface temperature remains constant which implies that the incoming solar radiation must be balanced by the outgoing terrestrial radiation [16]. Due to the temperature of their surfaces, the

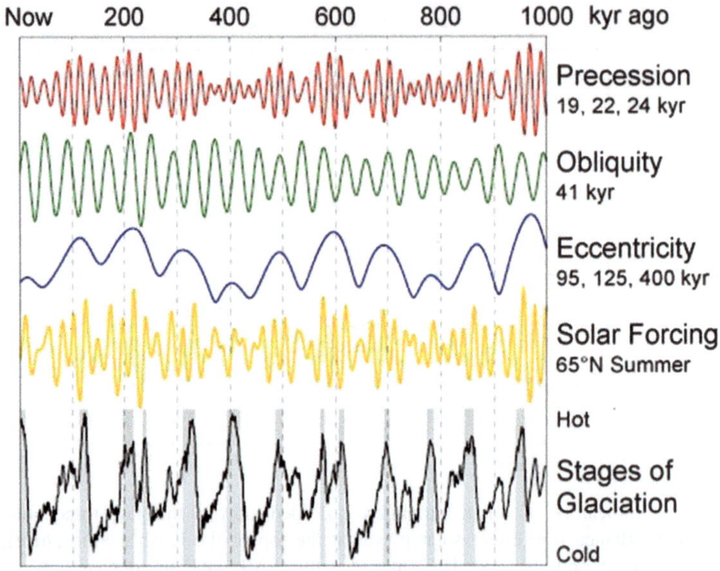

Fig. 4 Variations in Earth's orbit, the resulting changes in solar energy flux at high latitude, and the observed glacial cycles. Principal frequencies for each of the three kinds of variations are also labeled. Orbital data from [66], glacial data is from [51], solar forcing curve data (insolation) is derived from July 1st sunlight at 65° N latitude according to Jonathan Levine's insolation calculator. The *gray bars* indicate interglacial periods, defined here as deviations in the 5,000 year average of at least 0.8 standard deviations above the mean. Image created by Robert A. Rohde/Global Warming Art (http://www.globalwarmingart.com/wiki/File:Milankovitch_Variations_png)

maximum spectral intensity of the Sun and Earth radiation is located in the visible and infrared parts of the spectrum, respectively (Fig. 5).

After reaching the top of the Atmosphere (TOA) the solar radiation suffers a set of *accidents* before reaching the Earth's surface (Fig. 6). From the total incoming solar shortwave radiation (SWR) that reaches the TOA only about 50 % is absorbed by Earth's surface because: (a) about 6 % is backscattered by the atmospheric gases and 24 % reflected back to outer space by the aerosols and clouds (20 %) as well as by the Earth's surface (4 %); and (b) about 20 % is absorbed by clouds (3 %) and air gases (16 %) mainly water vapor, carbon dioxide and dust (Fig. 5). The energy that reaches the Earth's surface is then used to heat the lower levels of the Atmosphere (7 %), to feed the hydrological cycle (23 %) and emitted (20 %) as long wave radiation (LWR). The LWR emitted from the Earth's surface is strongly absorbed by the clouds and specific atmospheric constituents—e.g. water vapor, carbon dioxide, methane, nitrous oxide—usually named as greenhouse gases for reemitting LWR into all directions and in particular to the Earth's surface, reheating the lower layers of the atmosphere (greenhouse effect).

The Climate System loses energy (infrared radiation) from the middle layers of the troposphere while the Sun provides an excess of energy primarily in the

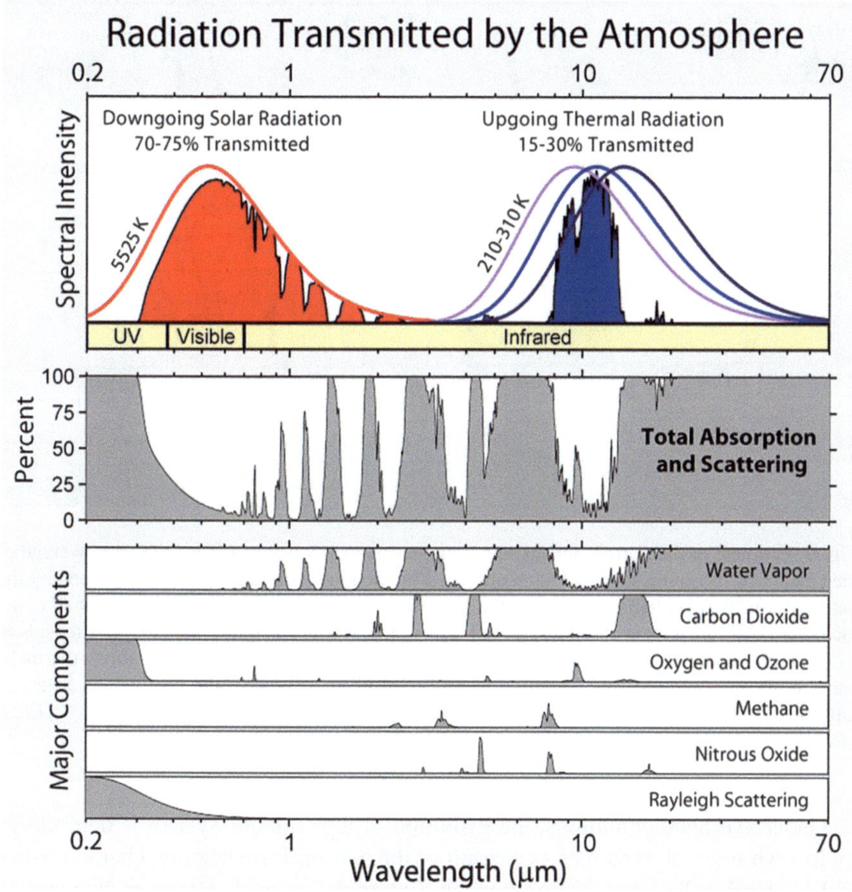

Fig. 5 Individual absorption spectrum for major greenhouse gases and total absorption plus Rayleigh scattering bands effects on both downgoing solar and upgoing terrestrial radiation. Image created by Robert A. Rohde/Global Warming Art (http://www.globalwarmingart.com/wiki/File:Atmospheric_Transmission_png)

tropics and the subtropics. The energy is then partially redistributed to middle and high latitudes by atmospheric winds and oceanic currents in complex energy transport processes. From the thermodynamics perspective, the Climate System may be viewed as a giant heat engine, where the high-temperature reservoir is the subtropical region, the low-temperature reservoir are the middle layers of the Atmosphere, and the energy (heat) of the sun is converted into the mechanical energy of the ocean and the atmosphere (here viewed as the working fluids of the heat engine) to transport an almost unimaginably large amount of heat from the tropics to the poles, largely carried out by the mid-latitude storms, and the work performed used to maintain the kinetic energy of the circulations against the continuous drain by friction [5, 60].

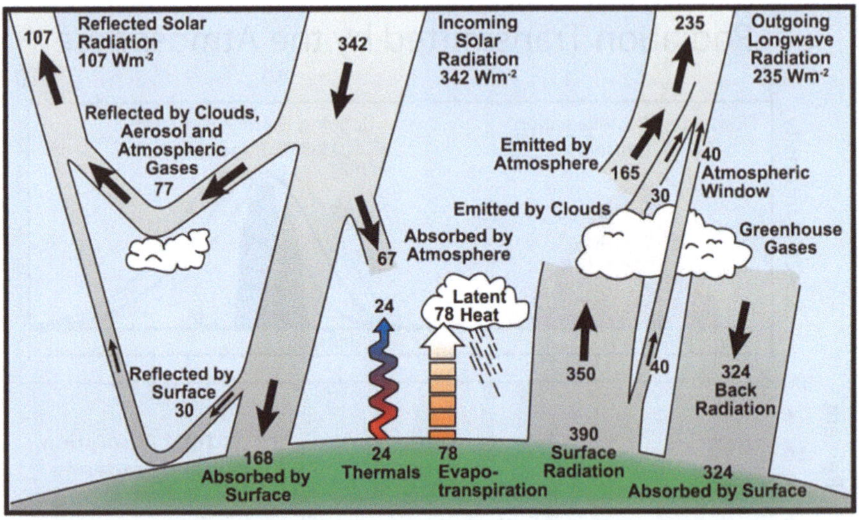

Fig. 6 Estimate of the Earth's annual and global mean energy balance. The long term average of the amount of incoming solar radiation absorbed by the Earth and atmosphere is balanced by the outgoing long wave radiation emitted by the Earth's surface and Atmosphere. Only about half of the incoming solar radiation is absorbed by the Earth's surface which is transported to the Atmosphere by warming the air in contact with the surface (sensible heat), by evapotranspiration (thermals, latent heat) and by long wave radiation partially absorbed by clouds and greenhouse gases. In turn, the Atmosphere radiates long wave energy out to space but also downward back to Earth (greenhouse effect). Image from [38]

Under averaged conditions, the evolution of the Climate System is determined by its own internal dynamics as a result of the non-uniform heating, characteristics and interaction between its components (internal forcings). However, changes in the natural (e.g. volcanic eruptions and solar radiation) and human induced (e.g. atmospheric composition, vegetation types) external forcings (Fig. 7) have the ability to affect the climate, disrupting the radiation balance of the Earth (Fig. 6): (1) by changing the incoming solar radiation (e.g., through changes in Earth's or Sun's orbit, solar activity/cycles); (2) by changing the albedo, i.e. the fraction of solar radiation that is reflected (e.g., though changes in cloud cover, atmospheric particles, land use and vegetation cover); and (3) by changing the concentration of gases and aerosols involved in atmospheric chemical reactions, able to act as cloud condensation nuclei (modifying the properties of cloud droplets and potentially affecting precipitation regime), to absorb, scatter and reflect SWR and LWR and changing the longwave radiation from Earth back towards space [16, 38]. The way the Climate System responds to these changes is not linear due to the large number of feedback mechanisms.

The drivers of climate change are all the natural or anthropogenic factors that can disrupt the climate system and cause a statistically significant change in, at least, one feature of the statistical distribution of any climatic element, beyond the limits set

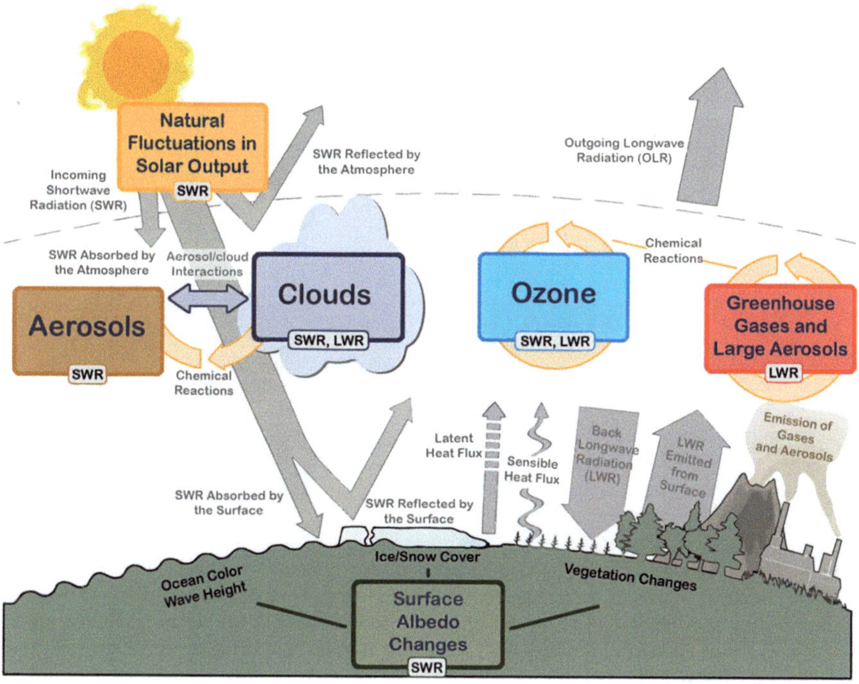

Fig. 7 Main drivers of climate change affecting the radiative balance between incoming solar shortwave radiation (SWR) and outgoing longwave radiation (OLR). Image from [16]

by the climate variability. To assess future climate change, it is necessary to have projections of the climatic elements which is achieved with climate models. These primary tools for climate research are the result of trying to solve numerically a set of non-linear differential equations that describe the evolution (in time and space) of a complete set of climatic elements. These models are constantly evolving for a better representation of a higher number of physical, chemical and biological properties and processes of its components, their interactions, feedback processes and spatial resolution (Fig. 8). However, current coupled Atmosphere-Ocean General Circulation Models (AOGCMs) are already able to provide a comprehensive representation of the climate system that are used to study and simulate the climate, even for operational purposes, which includes monthly, seasonal and interannual climate predictions [40]. A comprehensive assessment of the climate model's types, characteristics and ability, both individually and collectively, to simulate the most important features of the climate is provided in [23].

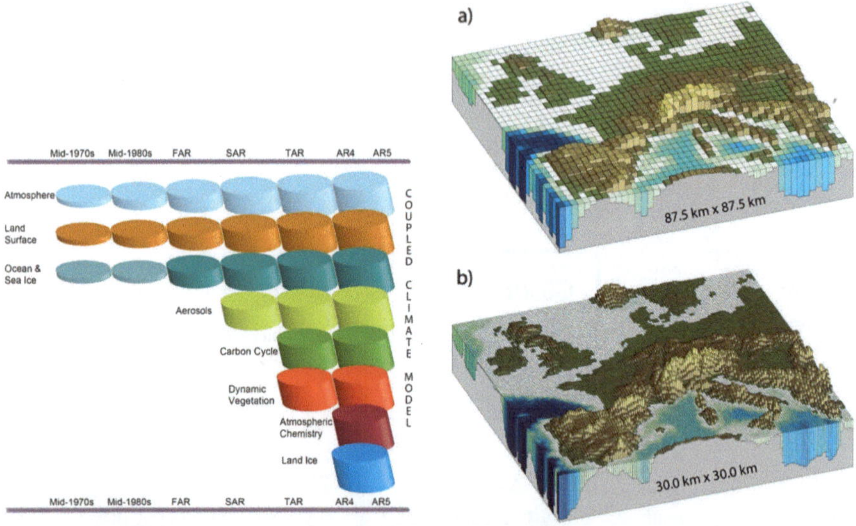

Fig. 8 The evolution of climate models in terms of the different components that were coupled into the climate models with increasing complexity and range of processes has increased over time (illustrated by growing cylinders, in the *left panel*) and the current (**a**) higher resolution models (87.5 × 87.5 km) and (**b**) in the very high resolution models now being tested (30.0 × 30.0 km) (*right panel*). Image from [16]

3 Potential Impacts of Regional Climate Change Projections

This section is devoted to presenting the results of the assessment of regional impacts of climate change in two case studies: the area burned by forest fires, and in the design of storm water runoff systems. In both cases, the study area is the Continental Portugal, the south-westernmost country of Europe located in the SW corner of the Iberian Peninsula (Fig. 9).

3.1 Effects on Wildfire Regime

Three different types of factors influencing wildfires in Portugal: (a) the physiological conditions of the vegetation/forest which includes the type, state and characteristics (e.g., moisture content, fire resistance, organization and cleanliness of the forest); (b) geographical conditions (such as the location and accessibility to the site of the fire, topography, amount and availability of firemen and equipment and other resources for fire combat, the number and intensity of simultaneous fires, amount and type of socioeconomic activities); and, (c) weather and climate.

Weather and climate are among the most important factors of forest fires worldwide. Indeed, the climate defines the existence and type of vegetation in each region and, along with the weather conditions (mainly temperature and

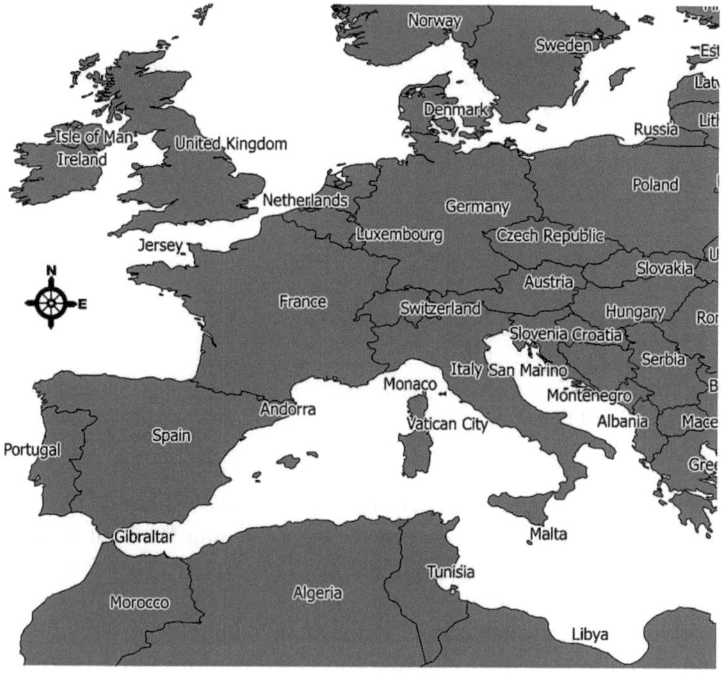

Fig. 9 Geographical location of Continental Portugal in Western Europe

precipitation), are responsible for the physiological state of the vegetation. Weather is also a determinative factor at all stages of the fire: from ignition (lightning), development (wind, humidity, temperature) and extinction (precipitation).

In Portugal, several studies have put in evidence that these factors are responsible for about two-thirds of the variability of the annual total burnt area (hereafter, BA) [61, 63]. Similar findings were found for other regions of Spain [77]. In addition, the role of specific weather parameters in fire activity in Portugal has been demonstrated in several studies, for example, the synoptic weather patterns at different levels of altitude [3, 61], extreme weather conditions [76] and circulation weather types [77].

Meteorological variables are frequently used, alone or in combination with other information, in the development of indices of fire danger/risk for various regions of the world [20, 30, 79]. The Comparative analysis of the performance of several fire danger indices have been performed for southern Europe [8, 81]. In Portugal, the Portuguese Institute of the Sea and the Atmosphere IPMA (Instituto Português do Mar e da Atmosfera) starts to use in 1988 the Portuguese index [25, 35], which is a modified version of the Nesterov index [71] but in 1998 adopted the Canadian Fire Weather Index, also known as FWI, for having greater predictive capacity of fire risk in summer [81].

In fact the FWI is just one of the fire indices that integrate the Canadian Forest Fire Danger Rating System (CFFDRS), which was specifically designed for Canada (Fig. 10). The development of the CFFDRS starts in 1968 and presently comprises

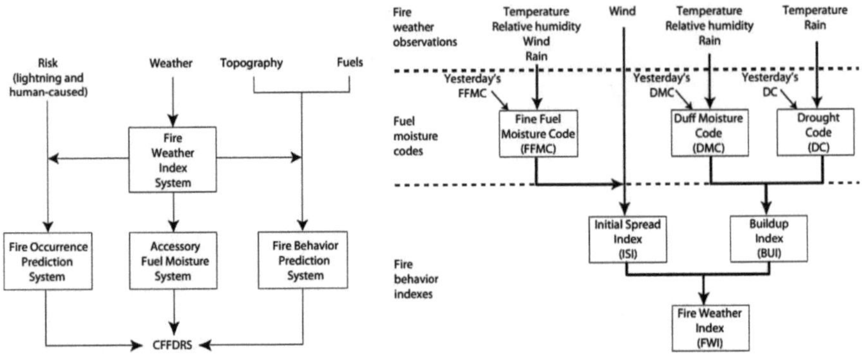

Fig. 10 Structure of the Canadian Forest Fire Danger Rating System, CFFDRS and of the Canadian Forest Fire Weather Index System, CFFWIS. Both adapted/extracted from [48]

two major subsystems: the Canadian Forest Fire Weather Index System (CFFWIS) [79, 80]; and, the Canadian Forest Fire Behaviour Prediction System (CFFBPS). The CFFWIS (Fig. 10) uses exclusively daily weather data to compute a set of fire indices and has shown to be appropriate to rate the risk of forest fires all over the world [86]. The system entails a total of six components: three fuel moisture codes and three fire behaviour indexes. The Fine Fuel Moisture Code (FFMC), the Duff Moisture Code (DMC) and the Drought Code (DC) respectively account for the average moisture content of surface litter and other cured fine fuels, decomposing litter of moderate depth and of deep, compact organic (humus) layers of the soil that presents different drying rates. The Initial Spread Index (ISI) combines the effects of wind and FFMC to estimate the expected rate of fire spread, while the Buildup Index (BUI) combines DMC and DC to account for the total amount of fuel available for combustion. The BUI is finally combined with ISI to produce the FWI and the Daily Severity Rating (DSR). Mathematically, the DSR is defined as a power of the FWI ($DSR = 0.0272FWI^{1.77}$) but rates the difficulty of controlling fires. The FWI is a suitable general index of daily fire danger in forested areas while DSR reflects more accurately the expected efforts required for fire suppression and was specifically designed for averaging either in time or in space [57]. The equations and program code of the CFFWIS as well as a comprehensive description of these indexes may be found in [79, 80], respectively. These moisture codes and fire indexes are numerical ratings and, in this study, were computed on the basis of daily values of air temperature, relative humidity, and wind speed at 12 UTC and 24-h cumulated precipitation for the 1980–2007 period obtained from ERA-40 re-analysis product [78] of the European Centre for Medium-Range Weather Forecasts (ECMWF).

The assessment of the potential impacts of regional climate change projections on wildfire regime was performed following the approach and the methodology described in [63] and is based on the Portuguese Rural Fire Database, PRFD [62]. This dataset provides detailed information for each fire occurred in mainland in the period 1980–2007 which includes: type of land cover affected by the fire (in forest, shrubs and agricultural areas), location of the fire ignition in terms of the name of the administrative regions (district, county and parish) and date and time of ignition and extinction. Monthly and annual cumulated values of burnt area in Portugal were derived from the PRFD for the 28-year period. The annual cycle of monthly burnt area for mainland Portugal reveals that the vast majority of total burnt area (89 %) is due to fires in the summer months (26 % in July, 46 % in August and 17 % in September). The inter-annual variability is also much higher during the summer months and the variability of July and August is about twice the one of September. These results are expected in the Mediterranean region due to the temperate type of climate that induces high levels of water and thermal stress on the vegetation during the hot and dry periods in the late spring and summer [61, 76, 82].

Time series of burnt area in July and August (Fig. 11) account for 72 % of the total burnt area in Portugal and resembles the high inter-annual variability of the annual burnt area time series which suggest that the annual fire regime will be dominated by the events that take place in those two summer months. Therefore the study will be restricted to the burnt areas in the months of July and August.

Annual values of burnt area were clustered in two classes, severe/mild, if the monthly burnt areas of July and August are both greater/lower than the upper /lower terciles of the respective month, i.e. greater than 39,000 ha for July and 44,000 ha for August/lower than 11,000 ha for July and 23,000 ha for August. According to

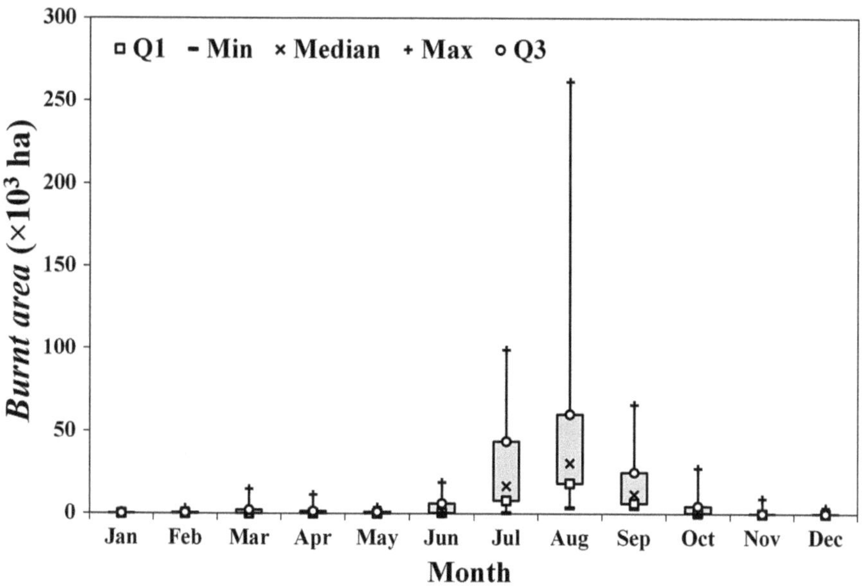

Fig. 11 Box plot of the annual cycle of monthly burnt area in Portugal for the period 1980–2007. *Boxes* indicate monthly values of the lower quartile (Q1), the median and the upper quartile (Q3). Whiskers extend down to the minimum and up to the maximum monthly values

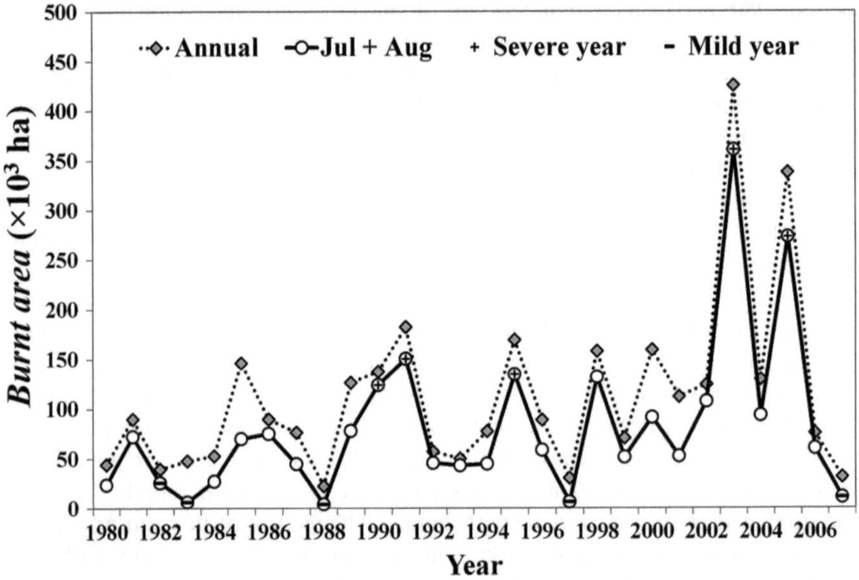

Fig. 12 Inter-annual variability of burnt area in mainland Portugal for yearly (*thin line* with *grey diamonds*) and July plus August (*thick line* with *white circles*) amounts, for the period 1980–2007. A *plus* (+) inside the *white circle* represents a severe summer season, defined as one where the monthly burnt areas of July and August are both greater than the upper tercile of the respective month while a *minus sign* (−) inside the *white circle* represents a mild summer season, defined as one where both the monthly burnt areas of July and August are lower than the respective lower terciles

this criteria, the years of 1990, 1991, 1995, 2003 and 2005 were severe ones whereas the years of 1982, 1983, 1988, 1997 and 2007 were mild ones (Fig. 12).

Composites analysis was used to assess the relationship between of monthly burnt area in July and August and the meteorological and fire risk indexes. Composite analysis includes the composite which is an arithmetic averages for a subsample (or class) and the anomaly which is the departure of composite from the grand average computed for the entire sample. The statistical significance is assessed by estimating percentiles 10 and 90 from a sample of 1,000 composites randomly generated using the bootstrapping technique [21]. As shown in Fig. 13, monthly composites of the air temperature, acumulated precipitation and DSR from January to August present a contrasting behaviour between severe and mild years during the months preceding and during the summer fire season that is worth mentioning and analysing in detail.

In the pre fire season, severe years are associated to positive anomalies of precipitation in the early spring (March), followed by significant negative anomalies of precipitation and air relative humidity (not shown) and positive air temperature anomalies in May and June. This climatic pattern is consistent with the atmospheric circulation from NE, over Portugal during this period (not shown). As expected,

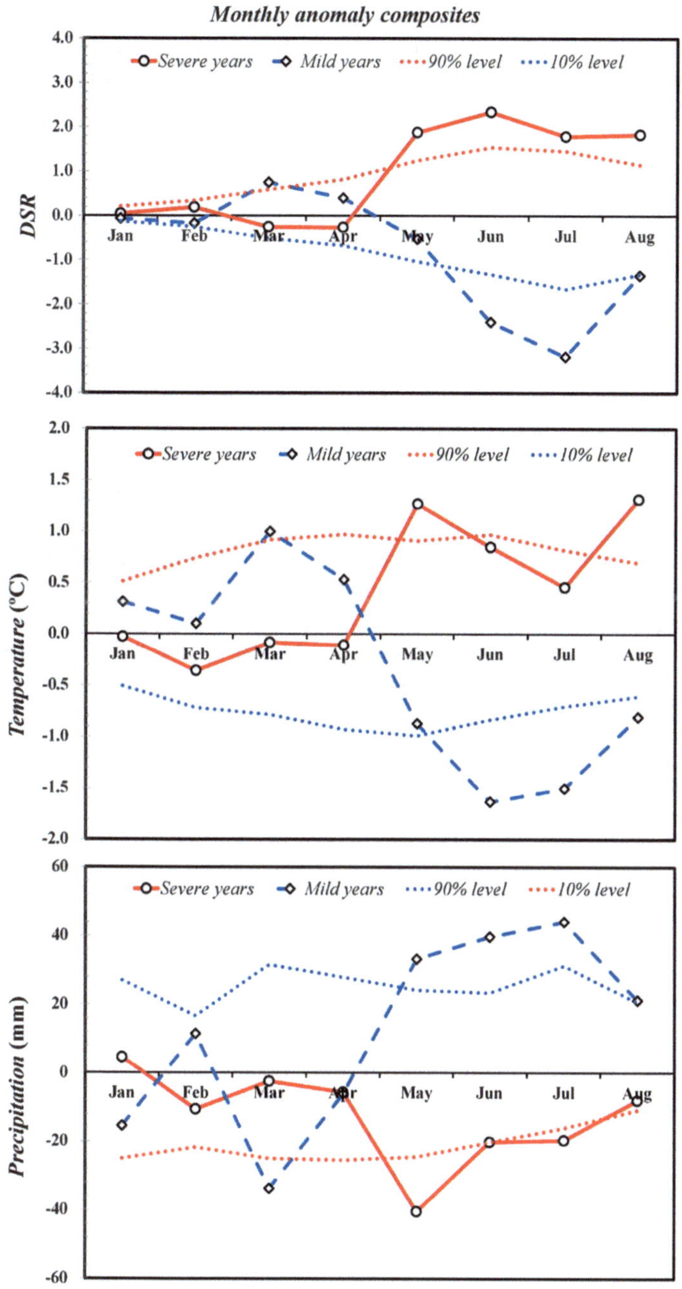

Fig. 13 Monthly anomalies (between January and August) of the daily severity rating (DSR), temperature and cumulated precipitation (*upper*, *middle* and *bottom panels*, respectively), for composites of severe (*solid lines* with *white circles*) and mild (*dashed lines* with *white diamonds*) fire seasons. The 90 and 10 % statistical significant level obtained with bootstrap are also plotted in *dotted lines*

DSR anomalies reflect the above-described cumulative behaviour of temperature, relative humidity (not shown), wind and precipitation, the increasing trend of positive anomalies from April to June. In the case of mild years an opposite behaviour is observed, i.e. there is significant less than usual amount of precipitation and relative humidity (not shown) in March and an abnormal high values of precipitation in the months of May and June, in conjunction with meaningful lower values of temperature, associated with SW surface wind (not shown). During summer major statistically significant differences between severe and mild fire seasons are found in all meteorological variables and, consequently, in DSR. Severe fire seasons are characterised by extreme negative precipitation and humidity anomalies and positive temperature associated to southern leading to utmost DSR anomalies.

Results from composite analysis suggest developing a Burned Area Model (BAM) by means of multiple linear regression analysis of monthly burnt areas in summer using, as predictors, meteorological risk indices (that integrate the effects of the meteorological variables) respecting to the pre-fire and/or fire seasons. Because of the highly asymmetrical character of the monthly means of burnt area in July and August (Fig. 11), the decimal logarithm of monthly burnt area was used as the predictand. It must be emphasized the positive and statistically significant (p-value < 0.001) value of the Pearson Product-Moment correlation coefficient between the decimal logarithm of areas burned in July and in August (0.63) and between the DSR and the decimal logarithm of monthly burnt area for the months of July (0.69) and August (0.71), during the period 1980–2007. However, it is also worth noting the very low correlation ($r = 0.15$, p-value < 0.26) between DSR monthly means of July and August. These results suggest that the BA in July and August are associated with different meteorological fire risk conditions during the summer and the climatological background in the pre-summer season could also condition the fire regimes in those months. Different selection methods (e.g., stepwise, forward, backward and explained variance) were used to select the best and parsimonious BAM which was obtained when using the following equation:

$$\mathrm{Log}_{10}(BA_{J/A}) = 2.6173 + 0.1189 \times DSR_{J/A} + 0.1095 \times DSR_{PF}$$

where Log_{10} is the decimal logarithm, $BA_{J/A}$ is the monthly burnt areas in July or in August, $DSR_{J/A}$ is the monthly mean of DSR in July or in August depending if the predictand is the monthly burnt area in July or August, respectively, DSR_{PF} is the monthly mean of DSR during the pre-fire period (PF), defined as May and June when the predictand is the decimal logarithm of monthly burnt area in July ($\mathrm{Log}_{10}(BA_J)$) and May, June and July when the predictant is the decimal logarithm of monthly burnt area in August ($\mathrm{Log}_{10}(BA_A)$). Both predictors retained in the model are significant at the 5 % level and were selected in the order they appear in the model with an increasing value of the coefficient of determination (r^2) from 0.47, when $BA_{J/A}$ is the only predictor, to 0.61 when both predictors are used in the model. In order to mitigate the effects of over fitting, the performance of the experiments was evaluated using a cross validation procedure that removes 14 year pairs (i.e. July and August for each year) instead of the more usual and less demanding leave-one-out-cross validation scheme [84]. The overall agreement between observed and modelled values using cross-validation of the decimal logarithm of burnt area reveals only a slightly decrease of r^2 to 0.58 (p-value <0.001). This means that the BAM is able to explain, in cross-validation mode, almost 3/5 of the total variance. In addition, the Kolmogorov-Smirnov test confirms that the observed and modelled values of the logarithm of the values of burnt area as well as the residuals of the BAM has a normal distribution. The quality and robustness of these results will be exploited when the developed BAM will be used as a generator of monthly burnt area scenarios in present and future climate conditions.

Values of the meteorological variables needed to compute the DSR for the future were simulated by the Model for Interdisciplinary Research on Climate, MIROC [42] for three grid points of the MIROC 3.2 medium resolution (medres) grid, two located over mainland Portugal and one over Galicia (Spain). Data was extracted for the twentieth century model simulations 20C3M, and for the emission scenario B1 [55] covering the 1951–2000 and 2051–2100 period, respectively. MIROC is a coupled General Circulation Model (GCM) used in the 4th Assessment Report of the Intergovernmental Panel on Climate Change [37] and several model comparison studies indicate MIROC as one of the models with best performance, in particular over the Iberian Peninsula [1, 22, 52, 54, 56, 69, 75]. The B1 scenario corresponds to a rapid economic growth but high level of environmental and social consciousness is accompanied by rapid changes towards a service and information economy and the introduction of cleaning technologies [36]. Averages were made over the three selected grid points for the considered parameters, and monthly means were finally computed.

The BAM was then used with DSR values for the pre-fire and the fire season estimated with GCM outputs respecting to 20C3M and B1 climate scenario to generate time series of burnt areas in July and August for present and future scenarios. Kolmogorov-Smirnov (K-S) tests were then used to check the null hypothesis that the samples of DSR and BA simulated with the BAM has a normal distribution. However, there are changes in both the mean and the variance, even in the case of present climate (20C3M) due to climate change signal as well as to the limitations of BAM and bias of the GCM (noise). In an attempt to remove the noise and for the correct comparison of results, the normal distribution of the BA samples N(5.58, 1.18) obtained with the BAM using data for present (20C3M) scenario was forced to match (i.e. have the same mean and variance of) the BA normal distribution N(4.34, 0.54) obtained with observed data (Table 1). Then the exactly the same correction factors were applied to correct the BA normal distributions, N(6.11, 0.81) and N(6.61, 1.20), obtained with the BAM when using 2051–2078 and 2073–2100 data for B1 of future climate scenario (Table 1).

Descriptive statistics of the normal distributions of the $Log_{10}(BA_{J/A})$ may be found in Table 2. When compared with the present climate scenario (20C3M), there are increases in the means of $Log_{10}(BA)$ with both future climate scenarios periods, respectively of 6 % (4.34–4.58) and 11 % (4.34–4.81) from the 20C3M to the B1 scenario 2051–2078 and 2073–2100 periods. The same does not happen in the case of the standard deviation where a contrast is found for the two periods of the future

Table 1 Means, standard deviations and *p*-values from the one-sample Kolmogorov-Smirnov (K-S) normality test for the following samples of burnt areas in July and August: observed values (1980–2007), simulated values using BAM fed with observed meteorological data (1980–2007) and of simulated values using BAM fed with GCM outputs from the present climate scenario, 20C3M (1973–2000) and from future climate scenario B1 (2051–2078 and 2073–2100)

		Observed	Modelled	20C3M	B1	
		1980–2007	1980–2007	1973–2000	2051–2078	2073–2100
LogBA	Mean	4.34	4.34	5.58	6.11	6.61
	St. deviation	0.54	0.42	1.18	0.81	1.20
	p-value (K-S test)	0.74	0.71	0.72	0.69	0.70

Table 2 Descriptive statistics of the corrected normal distributions of logarithm of monthly burnt area for present (20C3M) and future (B1) climate scenario

	20C3M	B1	
	1973–2000	2051–2078	2073–2100
Mean	4.34	4.58	4.81
St deviation	0.54	0.37	0.55
P5	3.63	3.98	3.95
P10	3.73	4.21	4.13
P25	3.94	4.39	4.45
P50	4.24	4.62	4.79
P75	4.66	4.81	5.11
P90	5.21	5.02	5.55
P95	5.39	5.08	5.77
IQR	0.73	0.43	0.66

climate scenario; the standard deviation remains unchanged from the 20C3M to the last period of B1 scenarios, but presents a decrease of about 30 % (0.54–0.37) from the 20C3M to the first 28-year of B1 scenario. It is also worth noting that differences in percentiles changes with increasing percentiles, e.g. from 0.35 (0.32) in P5 to 0.38 (0.55) in P50 and to −0.31 (0.38) in P95 when going from present climate to first (last) 28-year period of B1 scenario. This is an important aspect, since it reveals that for the 2051–2078 period major increases in burnt area are only expected for values below P75 and the larger increases should be expected for P10 (0.48) values of burnt area while for the 2073–2100 period increases are therefore to be expected for all values of burnt area but larger increases are found for P50 (0.55).

Differences are more impressive when analysing changes in burnt area (and not in the logarithm) from present to future climate scenarios, e.g. by looking at the measures of location and dispersion of the corresponding log-normal distributions (Table 3). Increases of 17,000 ha to 42,000 ha and 61,000 ha may be observed in the median from the 20C3M to the first and last 28-year period of B1 scenario, respectively. On the other hand, the mean changes from 53,000 ha for present climate scenario to 51,000 ha in the 2051–2078 period and increase to 157,000 ha to the 2078–2100 period of B1 scenario. The weight of extremely large values of burnt area is also well apparent given the growing differences between the median values associated to the large positive skewness of the log-normal distributions. Increases

Table 3 Measures of location and dispersion respecting to the lognormal distributions of monthly burnt area for present (20C3M) and future (B1) climate scenarios

	20C3M	B1	
	1973–2000	2051–2078	2073–2100
Mean ($\times 10^3$ ha)	53	51	157
Median ($\times 10^3$ ha)	17	42	61
Interquartile range ($\times 10^3$ ha)	37	41	100
Relative dispersion[a]	2.14	0.98	1.64

[a] Defined as the semi-interquartile range divided by the median

may also be found in dispersion, taking into account that the inter-quartile range increase from 37,000 ha, in the case of the 20C3M scenario, to 41,000 ha and 100,000 ha, in the case of first and last 28-year period of B1 scenario. However, relative dispersion presents a rather different behaviour of the mean and the median, that is, a decrease from 20C3M to the 2051–2078 period of B1 (2.14–0.98) and an increase to the 2073–2100 period (0.98–1.64).

3.2 *Impact of Projected Climate Change in the Design of Storm Water Drainage Infrastructures*

The design of stormwater drainage infrastructure relies on the implicit assumption that the intense precipitation distribution is statistically stationary [68]. However, several European countries have experienced the occurrence of floods in urban areas more frequently in recent years. In addition, latest projections of climate change [39] points to changes in the precipitation regime, in particular an increase frequency and intensity of precipitation extreme events, namely long drought periods and heavy precipitation episodes, even in regions where total precipitation may decrease [17, 39]. All these facts point to: (i) the increased uncertainty about the performance of the storm water drainage infrastructures constructed under current paradigm in the near future; (ii) the need to assess potential changes in the regime of intense rainfall at the regional scale; and, (iii) eventually start to design and construct drainage structures capable of responding to these expectable changes in extreme precipitation regime.

In Portugal, the design of storm water drainage infrastructure is regulated by the Regulatory Decree nr. 23/95 of 23rd August [19], which adopted the Intensity Duration Frequency (IDF) curves developed by Matos and Silva (1986). From the mathematical point of view, the IDF curves reflects the power law behavior dependence of the precipitation intensity (I) with the duration of the precipitation (t) according to

$$I = a \times t^b$$

where a and b are the so called IDF parameters. These curves are of empirical nature and, from the engineering point of view, are of fundamental importance for the design of hydraulic structures, as they provide maximum precipitation intensity related to a given length and a given return period which represent key information for the design of hydraulic structures [10].

The objective of this study is to assess the impacts of potential change in the precipitation regime in the design of drainage systems for rainwater and hence the need to review the legislation that supports the design of these structures. The study relied on a comparative analysis between the IDF parameters provided by the Portuguese legislation and those obtained with precipitation data observed in meteorological stations located and representative of the three precipitation zones defined for Portugal and simulated by COSMO-CLM [COnsortium for Small-scale MOdelling and Climate Limited-area Modelling Community] [67] regional climate model (RCM) for recent past and future climate conditions.

The precipitation data used in this study comprises: (a) hourly time series observed in eight selected weather stations (based on their length and quality), located on the three rainfall regions defined in Portuguese Law [19] provided by the National System of Water Information and Resources, SNIRH (Sistema Nacional de Informação e Recursos Hídricos) (Fig. 14); and (b) and daily time series simulated by the COSMO-CLM with ECHAM5/MPI-OM1 boundary conditions for recent climate conditions (20C, 1961–2000) and for the B1 and A1B (2000–2100) SRES scenarios [55] for the grid cells containing the location of the aforementioned weather stations. The COSMO-CLM model has demonstrated high performance in simulating the precipitation in different regions of Europe and, for that reason, has been used to asses changes in precipitation regime over Europe [28, 44] and specifically over Portugal [15].

The methodology used for the estimation of the IDF curves is based on [10], explained in [64] and embraces: (i) disaggregation (and aggregation) of precipitation from daily to hourly time scales using the method of the fragments [4, 70, 73, 74] and from hourly to sub-hourly scales using the disaggregation coefficients suggested by [10], in order have maximum precipitation for ten different duration times (5, 10, 15, 30 min, 1, 2, 6, 12, 24 and 48 h); (ii) fitting of the Gumbel distribution function to time series of maximum precipitation intensity for each of the ten durations, using likelihood estimation of the location (μ) and scale (σ) parameters in each case (cf. [13]); (iii) use of the Gumbel inverse probability distribution to estimate maximum precipitation intensity values for eight return periods (2, 5, 10, 20, 50, 100, 500 and 1,000 years); (iv) plot of I (mm/h) versus precipitation duration (min) in logarithmic scales (log.log plot) and the subsequent use of regression analysis to estimate the IDF parameters a and b; and, finally (v) correction of the COSMO-CLM model's bias dues to its difficulty in reproducing exactly the observed weather conditions. The goodness of fit of the Gumbel distribution function to the data was assessed with the analysis of Quantile-Quantile plots and the Kolmogorov-Smirnov test (KS test) while the IDF parameters were estimated using robust regression [32, 33, 72] with a statistical significance level of 5 % and the quality of the linear regression

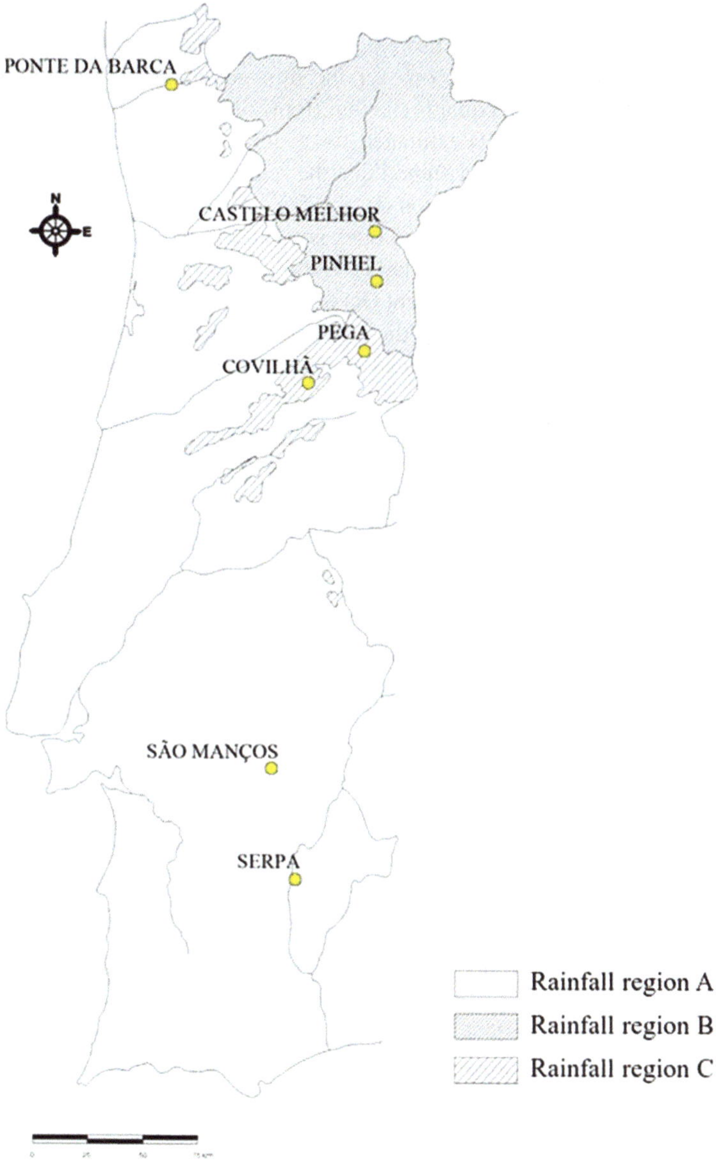

Fig. 14 Rainfall regions defined in the Portuguese Law [19] and the geographical location of the weather stations used in this study

also assessed by the coefficient of determination (r^2), the F-statistic (*p-value*) and the error variance.

The impact of projected climate change in the design of storm water drainage infrastructures was performed directly on their size. For sake of simplicity, from all the different types of these systems, only a specific residential rain gutter and collector will be considered. The flows (Q) were calculated with the rational method using a contribution area of $100\,\text{m}^2$ for the rain gutter and $155\,\text{m}^2$ for the rain collector, a flow coefficient equal to the unit (typically used for building coverings) and precipitation intensity (I) estimated for a duration (t) of 5 min and a return period (T) of 10 years. It was assumed that the rain gutter has a rectangular shape, with a base (B) of 20 cm, inclination (i) of 0.5 % and was dimensioned so that the height of the water depth (h) therein does not exceed 7/10 of the total height of the rain gutter. The Manning-Strickler's formula ($Q = K \times A_f \times R^{2/3} \times i^{1/2}$), was then used with a roughness coefficient (K) of $90\,\text{m}^{1/3}/\text{s}$, which correspond to metal plate. The hydraulic radius (R) and the area occupied by the fluid (A_f), in the case of rectangular sections, are determined respectively by using the following set of equations:

$$Q = K \times A_f \times R^{2/3} \times i^{1/2}$$

$$R = (B \times h)/(B + 2h)$$

$$A_f = B \times h$$

The residential rain collector was also designed using the Manning-Strickler formula for full section, roughness (K) of $120\,\text{m}^{1/3}/\text{s}$ (which correspond to polyvinyl chloride, PVC) and an inclination (i) of 2 %. The hydraulic radius (R), in the case of a filled circular section, is given by $R = D_i/4$, where D_i is the internal diameter of the piping.

The residential rain gutter and collector were designed following the previously described methodology and using IDF curve parameters estimated with observed and simulated data for three time periods (2011–2040, 2041–2070 and 2071–2100) of both future climate scenarios, after correcting the RCM bias. The differences between the drainage structures dimensions estimated for future current and future climate conditions are presented in Table 4.

The first result which should be emphasized is the difference between the dimensions of the organs of collecting rainwater, estimated for different weather stations of the same rainfall region, which may suggest the need to revise the law [19] that determines the same dimensions for all local within the same rainfall region.

In general, the results points to the need of larger rain gutters and collectors in the future. The expected increases tend to be higher for the end of the century and for the conditions of A1B climate scenario. Maximum estimated increases can reach 50 % for the rain gutter and 25 % for the collector but are not identical distributed across the country. Estimated maximum increases in the height of the rain gutter were obtained for the 2071–2100 period of A1B scenario and are of 49 % (in São Manços) for region A, 40 % (in Castelo Melhor) for region B and 19 % (in Pega) for region C. Expected maximum changes in the rain gutter dimensions are also not uniform in all precipitation regions except in region C. In fact, it range between 32 %

Table 4 Projected changes in the dimension of the drainage systems. Dimension of the Gutter (*H*) and Collector (*C*) designed for weather stations located in the three pre-defined rainfall regions, using the precipitation intensity estimated with observed data (Observed) and with data simulated by COSMO-CLM, for three periods of 30 years of the two future scenarios (A1B and B1) as well as the relative differences between these dimensions (ΔH and ΔC)

Station		Scenario	Period	Gutter				Collector			
				H (cm)	ΔH (%)	$\overline{\Delta H_s}$ (%)	$\overline{\Delta H_w}$ (%)	C (cm)	ΔC (%)	$\overline{\Delta C_s}$ (%)	$\overline{\Delta C_w}$ (%)
Region A	Ponte da Barca	Observed		3.71			31	86.51			16
		A1B	2011–2040	4.57	23	19		97.35	13	11	
			2041–2070	4.01	8			90.46	5		
			2071–2100	4.71	27			99.01	14		
		B1	2011–2040	4.79	29	30		99.92	16	16	
			2041–2070	4.91	32			101.4	17		
			2071–2100	4.76	28			99.7	15		
	São Manços	Observed		4.29				93.93			
		A1B	2011–2040	5.9	38	42		112.27	20	22	
			2041–2070	5.96	39			112.91	20		
			2071–2100	6.39	49			117.27	25		
		B1	2011–2040	6.3	47	33		116.35	24	17	
			2041–2070	5.64	31			109.51	17		
			2071–2100	5.21	21			104.83	12		

(continued)

Table 4 (continued)

Station		Scenario	Period	Gutter				Collector			
				H (cm)	ΔH (%)	$\overline{\Delta H_s}$ (%)	$\overline{\Delta H_w}$ (%)	C (cm)	ΔC (%)	$\overline{\Delta C_s}$ (%)	$\overline{\Delta C_w}$ (%)
Region B	Castelo Melhor	Observed		2.93			14	75.54			8
		A1B	2011–2040	3.48	19	21		83.22	10	11	
			2041–2070	3.07	5			77.53	3		
			2071–2100	4.11	40			91.71	21		
		B1	2011–2040	3.49	19	18		83.41	10	10	
			2041–2070	3.19	9			79.3	5		
			2071–2100	3.71	27			86.48	14		
	Pinelo	Observed		3.14				78.49			
		A1B	2011–2040	3.06	−3	9		77.18	−2	5	
			2041–2070a	3.22	3			79.63	1		
			2071–2100	3.96	26			89.75	14		
		B1	2011–2040a	3.33	6	8		81.16	3	5	
			2041–2070a	3.29	5			80.56	3		
			2071–2100	3.6	15			84.88	8		

Region C	Covilhã	Observed		4.27			8	93.64			5
		A1B	2011–2040a	4.43	4	10		95.58	2	5	
			2041–2070	4.55	7			97.08	4		
			2071–2100	5.06	19			103.03	10		
		B1	2011–2040	4.56	7	7		97.2	4	4	
			2041–2070	4.71	10			98.99	6		
			2071–2100a	4.41	3			95.38	2		
	Pega	Observed		3.59				84.8			
		A1B	2011–2040b	3.79	6	12		87.58	3	6	
			2041–2070	3.96	10			89.63	6		
			2071–2100	4.26	19			93.52	10		
		B1	2011–2040a	3.69	3	5		86.18	2	3	
			2041–2070	3.86	8			88.46	4		
			2071–2100	3.77	5			87.42	3		

Projected changes are statistical significant at 99 % level except for the cases identified by superscript lowercase letter (^{a}). Arithmetic averages of ΔH and ΔC for each weather station ($\overline{\Delta H_w}$ and $\overline{\Delta C_w}$) and region and scenario ($\overline{\Delta H_s}$ and $\overline{\Delta H_s}$) are also shown. The former values are calculated over different periods to obtain an average value representative for the mid-twenty-first century, thus sampling decadal variability. The latter are derived to obtain values representative for each region considering scenario uncertainty

(Ponte da Barca) and 49 % (in São Manços) in region A and from 26 % (Pinelo) to 40 % (Castelo Melhor), in region B.

In order to obtain an average value representative for the mid-twenty-first century, averages were built for each station over the three time periods, thus sampling decadal variability. This results on an estimated average increase of 42 % (33 %) in São Manços and 19 % (30 %) in Ponte da Barca for the A1B (B1) scenario in Region A. For region B, changes are larger for Castelo Melhor (about 20 %) than for Pinelo (about 9 %) for both scenarios. On the contrary, little differences were found for region C, with changes ranging between about 11 and 6 %, respectively for scenario A1B and B1 in both weather stations. Furthermore, averages were built over both stations in each region and both scenarios to obtain values representative per region considering scenario uncertainty. Averaged increase in Gutter dimension is likely to be higher in Region A (31 %) than in region B (14 %) and in region C (8 %).

The projected changes for the rain collector size are essentially proportional those of the gutter (Table 4) and for that reason a detailed presentation of the results is omitted. The main results are: (i) changes are typically smaller than for the gutter; (ii) averaged changes in the collector diameter increases from 5 % in region C, to 8 % in region B and 16 % in region A; (iii) different behavior in region A is characterized by higher changes in the weather stations located in the southern (São Manços) than in the northern part (Ponte da Barca) and, in region B, at lower (Castelo Melhor) than at higher altitude (Pinelo); (iii) higher homogeneity in the expected changes in mountainous region C and (iv). Finally, it is important to underline that projected changes in the size of the building drainage systems are statistical significant at the 99 % level in all cases except in the 6 cases (17 %) identified by a dagger in Table 4 and for the station of Pega in the first 30-year period of the A1B scenario. It is important to underline that projected changes in all cases of rainfall region A are statistical significant (99 %) and that statistical significance is higher in the end of the twenty-first century.

4 Conclusions

Climate change was presented as a problem of energy transfer between the Sun and the Earth (e.g. changes in the orbital parameters of our planet) and the radiative balance of the Climate System (i.e. changes in the chemical composition of the atmosphere), adopting a personal presentation of the fundamental concepts and privileging the mathematical description of the processes. However, most of the definitions and concepts used in this chapter may be found in the literature and, specifically, on the glossary of the recent Fifth Assessment Report (AR5) of the Intergovernmental Panel on Climate Change (IPCC) which also provides a comprehensive assessment of the physical science basis of climate change [40].

Results of the assessment of the potential impacts of regional climate change on the area burned by rural fires in Continental Portugal consistently points towards: (i) an increasing risk of fire under future climate conditions; (ii) an increasing possibility of having much larger fire events; (iii) an increasing inter-annual variability of the fire regime, which together with the positive bias will have dramatic consequences at the social, economic and environmental levels. Nevertheless, it is very likely that the simulated amounts of burnt area are overestimated. This is to be attributed at least to three orders of reasons: (i) the use of global (GCM) or regional (RCM) circulation models which are just limited representations of reality; (ii) the use of a linear BAM which is only able to explain a partial amount of the inter-annual variability (even if up to 2/3 of the total variance), prevents the introduction of feedback mechanisms that might reduce the amounts of burnt area and applied to future climate scenarios, i.e. to meteorological conditions beyond the range of the tested domain; and, (iii) not taking into account other important factors for fire occurrence and size such as that are not yet fully understood nor properly modelled, such as those related to changes in fuel structure [58, 59], climate-vegetation dynamics and in conservation planning [45], patterns of lightning strikes [18] and anthropogenic activities and drivers of fire, such as control over ignition, fire management, suppression activities, land use/land cover changes [2, 14, 43, 45, 49].

Finally, the methodology developed to assess the impact of projected climate change in the design of storm water drainage infrastructures ensures robustness, statistical significance and adequate comparative analysis of the results obtained. Differences in the design of storm water drainage systems with observed and simulated data for the future scenarios suggest that the impact of climate change will generally imply: (i) the increase of the dimension of these organs in the future; (ii) this variation is not identical in the three rainfall regions defined for Portugal; (iii) nor between stations within of each of these regions. Moreover, these differences are very similar to those found between the size of the drainage systems designed with the IDF curves stipulated in the Portuguese Law [19] and estimated with the same data simulations [64] which reinforces the need to review the rainfall classification of the territory and update the IDF curves defined in the legislation.

Acknowledgements This was a personal presentation that included results obtained in studies performed in collaboration with other colleagues and, for this reason, the author is grateful to Teresa Calado, Carlos C. DaCamara, and Tomás Calheiros coauthors of [63] as well as Luís Sanches Fernandes, Eduarda Macário, Sónia Gaspar e Joaquim Pinto co-authors of [64]. The author is also thankful to the IPCC (http://www.ipcc.ch/index.htm), the Global Warming Art (http://www.globalwarmingart.com/) and Croatian-English Chemistry Dictionary & Glossary (http://glossary.periodni.com/dictionary.php) for making available and allows the use some of the presented images.

References

1. Ahlfeld, D.P.: Comparison of climate model precipitation forecasts with North American observations. In: Proceedings of the XVI International Conference on Computational Methods in Water Resources (2006)

2. Aldersley, A., Murray, S.J., Cornell, S.E.: Global and regional analysis of climate and human drivers of wildfire. Sci. Total Environ. **409**(18), 3472–3481. (2011)
3. Amraoui, M., Liberato, M.L., Calado, T.J., DaCamara, C.C., Coelho, L.P., Trigo, R.M., Gouveia, C.M.: Fire activity over Mediterranean Europe based on information from Meteosat-8. For. Ecol. Manag. **294**, 62–75 (2013)
4. Arganis-Juarez, M.L., Mora, R.D., Cisneros-Iturbe, H.L., Fuentes-Mariles, G.E.: Génŕation d'échantillons synthétiques des volumes mensuels écoulés vers deux barrages selon la méthode de Svanidze modifiée/Synthetic sample generation of monthly inflows into two dams using the modified Svanidze method. Hydrol. Sci. J. **53**(1), 130–141 (2008)
5. Barry, L., Craig, G.C., Thuburn, J.: Poleward heat transport by the atmospheric heat engine. Nature **415**(6873), 774–777 (2002)
6. Benestad, R.E.: Solar Activity and Earth's Climate. Springer, Berlin (2006)
7. Berger, A.: Milankovitch theory and climate. Rev. Geophys. **26**(4), 624–657 (1988)
8. Bovio, G., Camia, A.: Meteorological indices for large fires danger rating. In: Chuvieco, E. (ed.) A Review of Remote Sensing Methods for the Study of Large Wildland Fires, pp. 73–89. Universidad de Alcalá, Alcalá de Henares Spain (1997)
9. Bradley, R.S.: Quaternary Palaeoclimatology: Methods of Palaeoclimatic Reconstruction. Unwin Hyman, London (1985)
10. Brandão, C., Rodrigues, R., Costa, J.P.: Análise de fenómenos extremos, Precipitações intensas em Portugal Continental http://snirh.pt/snirh/download/relatorios/relatorio_prec_intensa.pdf (2001). Accessed 1 Feb 2013
11. Budyko, M.I.: The effect of solar radiation variations on the climate of the earth. Tellus **21**(5), 611–619 (1969)
12. Campisano, C.J.: Milankovitch cycles, paleoclimatic change, and hominin evolution. Nat. Educ. Knowl. **4**(3), 5 (2012)
13. Coles, S., Bawa, J., Trenner, L., Dorazio, P.: An Introduction to Statistical Modeling of Extreme Values, vol. 208. Springer, New York (2001)
14. Costa, L., Thonicke, K., Poulter, B., Badeck, F.W.: Sensitivity of Portuguese forest fires to climatic, human, and landscape variables: subnational differences between fire drivers in extreme fire years and decadal averages. Reg. Environ. Chang. **11**(3), 543–551 (2011)
15. Costa, A.C., Santos, J.A., Pinto, J.G.: Climate change scenarios for precipitation extremes in Portugal. Theor. Appl. Climatol. **108**(1–2), 217–234 (2012)
16. Cubasch, U., Wuebbles, D., Chen, D., Facchini, M.C., Frame, D., Mahowald, N., Winther, J.-G.: Introduction. In: T.F., Qin, D., Plattner, G.-K., Tignor, M., Allen, S.K., Boschung, J., Nauels, A., Xia, Y., Bex, V., Midgley, P.M. (eds.) Climate Change 2013: The Physical Science Basis. Contribution of Working Group I to the Fifth Assessment Report of the Intergovernmental Panel on Climate Change Stocker. Cambridge University Press, Cambridge (2013)
17. Diodato, N., Bellocchi, G., Romano, N., Chirico, G.B.: How the aggressiveness of rainfalls in the Mediterranean lands is enhanced by climate change. Clim. Chang. **108**(3), 591–599 (2011)
18. Dissing, D., Verbyla, D.L.: Spatial patterns of lightning strikes in interior Alaska and their relations to elevation and vegetation. Can. J. For. Res. **33**, 770–782 (2003)
19. DR. Decreto Regulamentar n. 23/95 (Regulation-decree n. 23/95). Diário da República, 194/95(I-B). www.dre.pt (1995). Accessed 1 Feb 2013
20. Drouet, J.C., Sol, B.: Incendies de forêts: Mise au point d'un indice numérique de risqué météorologique d'incendie. Revue Générale de Sécurité, 92 (1990)
21. Efron, B., Tibshirani, R.J.: An Introduction to the Bootstrap, vol. 57. CRC, Boca Raton (1994)
22. Errasti, I., Ezcurra, A., Sáenz, J., Ibarra-Berastegi, G.: Validation of IPCC AR4 models over the Iberian Peninsula. Theor. Appl. Climatol. **103**(1–2), 61–79 (2011)
23. Flato, G., Marotzke, J., Abiodun, B., Braconnot, P., Chou, S.C., Collins, W., Cox, P., Driouech, F., Emori, S., Eyring, V., Forest, C., Gleckler, P., Guilyardi, E., Jakob, C., Kattsov, V., Reason, C., Rummukainen, M.: Evaluation of climate models. In: T.F., Qin, D., Plattner, G.-K., Tignor, M., Allen, S.K., Boschung, J., Nauels, A., Xia, Y., Bex, V., Midgley, P.M. (eds.) Climate Change 2013: The Physical Science Basis. Contribution of Working Group I to the Fifth

Assessment Report of the Intergovernmental Panel on Climate Change. Cambridge University Press, Cambridge (2013)
24. Friis-Christensen, E., Lassen, K.: Length of the solar cycle—an indicator of solar activity closely associated with climate. Science **254**(5032), 698–700 (1991)
25. Gonçalves, Z.J., Lourenço, L.: Meteorological index of forest fire risk in the Portuguese mainland territory. In: Proceedings of the International Conference on Forest Fire Research. Coimbra B, vol. 7 (1990)
26. Guttman, N.B.: Statistical descriptors of climate. Bull. Am. Meteorol. Soc. **70**(6), 602–607 (1989)
27. Haigh, J.D.: The impact of solar variability on climate. Science **272**(5264), 981–984 (1996)
28. Haslinger, K., Anders, I., Hofstätter, M.: Regional climate modelling over complex terrain: an evaluation study of COSMO–CLM hindcast model runs for the Greater Alpine Region. Clim. Dyn. **40**(1–2), 511–529 (2013)
29. Hays, J.D., Imbrie, J., Shackleton, N.J.: Variations in the Earth's orbit: pacemaker of the ice ages. Science **194**(4270), 1121–1132 (1976)
30. Heikinheimo, M., Venäläinen, A., Tourula, T.: A soil moisture index for the assessment of forest fire risk in the boreal zone. In: COST, vol. 77, No. 79, p. 711 (1998)
31. Herman, J.R., Goldberg, R.A.: Sun, Weather, and Climate. Dover Publications, Inc., New York (1985)
32. Holland, P.W., Welsch, R.E.: Robust regression using iteratively reweighted least-squares. Commun. Stat. Theory Methods **6**(9), 813–827 (1977)
33. Huber, P.J.: Robust Statistics, pp. 1248–1251. Springer, Berlin (2011)
34. Imbrie, J., Boyle, E.A., Clemens, S.C., Duffy, A., Howard, W.R., Kukla, G., et al.: On the structure and origin of major glaciation cycles 1. Linear responses to Milankovitch forcing. Paleoceanography **7**(6), 701–738 (1992)
35. INMG: Nota explicative sobre o Indice de Risco Meteorológico de Incendios Rurais. Divisão de Meteorologia Agrícola, Instituto Nacional de Meteorologia e Geofísica (1988)
36. IPCC SRES SPM: Summary for Policymakers, Emissions Scenarios: A Special Report of IPCC Working Group III (PDF), IPCC (2000) [ISBN 92-9169-113-5]
37. IPCC: Summary for policymakers. In: Solomon, S., Qin, D., Manning, M., Chen, Z., Marquis, M., Averyt, K.B., Tignor, M., Miller, H.L. (eds.) Climate Change 2007: The Physical Science Basis. Contribution of Working Group I to the Fourth Assessment Report of the Intergovernmental Panel on Climate Change. Cambridge University Press, Cambridge (2007)
38. IPCC: In: Solomon, S., Qin, D., Manning, M., Chen, Z., Marquis, M., Averyt, K.B., Tignor, M., Miller, H.L. (eds.) Climate Change 2007: The Physical Science Basis. Contribution of Working Group I to the Fourth Assessment Report of the Intergovernmental Panel on Climate Change. Cambridge University Press, Cambridge (2007)
39. IPCC: In: Field, C.B., Barros, V., Stocker, T.F., Qin, D., Dokken, D.J., Ebi, K.L., Mastrandrea, M.D., Mach, K.J., Plattner, G.-K., Allen, S.K., Tignor, M., Midgley P.M. (eds.) Managing the Risks of Extreme Events and Disasters to Advance Climate Change Adaptation. A Special Report of Working Groups I and II of the Intergovernmental Panel on Climate Change, 582 pp. Cambridge University Press, Cambridge (2012)
40. IPCC: Annex III: Glossary [Planton, S (ed.)]. In: Stocker, T.F., Qin, D., Plattner, G.-K., Tignor, M., Allen, S.K., Boschung, J., Nauels, A., Xia, Y., Bex, V., Midgley, P.M. (eds.) Climate Change 2013: The Physical Science Basis. Contribution of Working Group I to the Fifth Assessment Report of the Intergovernmental Panel on Climate Change. Cambridge University Press, Cambridge (2013)
41. IPCC: Summary for Policymakers. In: Stocker, T.F., Qin, D., Plattner, G.-K., Tignor, M., Allen, S.K., Boschung, J., Nauels, A., Xia, Y., Bex, V., Midgley, P.M. (eds.) Climate Change 2013: The Physical Science Basis. Contribution of Working Group I to the Fifth Assessment Report of the Intergovernmental Panel on Climate Change. Cambridge University Press, Cambridge (2013)
42. K-1 model developers: K-1 coupled model (MIROC) description. K-1 technical report 1. In: Hasumi, H., Emori, S. (eds.) Center for Climate System Research. University of Tokyo (2004)

43. Kloster, S., Mahowald, N.M., Randerson, J.T., Lawrence, P.J.: The impacts of climate, land use, and demography on fires during the 21st century simulated by CLM-CN. Biogeosciences **9**(1), 509–525 (2012)
44. Kotlarski, S., Bosshard, T., Lüthi, D., Pall, P., Schär, C.: Elevation gradients of European climate change in the regional climate model COSMO-CLM. Clim. Chang. **112**(2), 189–215 (2012)
45. Krawchuk, M.A., Moritz, M.A., Parisien, M.A., Van Dorn, J., Hayhoe, K.: Global pyrogeography: the current and future distribution of wildfire. PLoS One **4**(4), e5102 (2009)
46. Kristjánsson, J.E., Kristiansen, J., Kaas, E.: Solar activity, cosmic rays, clouds and climate—an update. Adv. Space Res **34**(2), 407–415 (2004)
47. Laut, P.: Solar activity and terrestrial climate: an analysis of some purported correlations. J. Atmos. Solar Terr. Phys. **65**(7), 801–812 (2003)
48. Lawson, B., Armitage, O.: Weather guide for the Canadian Forest Fire Danger Rating System Natural Resources Canada, Canadian Forest Service, Northern Forestry Centre, Edmonton (2008)
49. Le Page, Y., Oom, D., Silva, J., Jönsson, P., Pereira, J.: Seasonality of vegetation fires as modified by human action: observing the deviation from eco-climatic fire regimes. Glob. Ecol. Biogeogr. **19**(4), 575–588 (2010)
50. Lean, J., Beer, J., Bradley, R.: Reconstruction of solar irradiance since 1610: implications for climate change. Geophys. Res. Lett. **22**(23), 3195–3198 (1995)
51. Lisiecki, L.E., Raymo, M.E.: A Pliocene-Pleistocene stack of 57 globally distributed benthic $\delta^{18}O$ records. Paleoceanography **20**(1), PA1003 (2005)
52. Lucarini, V., Calmanti, S., Dell'Aquila, A., Ruti, P.M., Speranza, A.: Intercomparison of the northern hemisphere winter mid-latitude atmospheric variability of the IPCC models. Clim. Dyn. **28**(7–8), 829–848 (2007)
53. Matos, R., Silva, M.: Estudos de precipitação com aplicação no projeto de sistemas de drenagem pluvial. Curvas Intensidade-Duração-Frequência da precipitação em Portugal. ITH24 LNEC, Lisbon (1986)
54. Maxino, C.C., McAvaney, B.J., Pitman, A.J., Perkins, S.E.: Ranking the AR4 climate models over the Murray-Darling Basin using simulated maximum temperature, minimum temperature and precipitation. Int. J. Climatol. **28**(8), 1097–1112 (2008)
55. Nakicenovic, N., et al.: Special Report on Emissions Scenarios: A Special Report of Working Group III of the Intergovernmental Panel on Climate Change. Cambridge University Press, Cambridge (2000)
56. Nieto, S., Rodríguez-Puebla, C.: Comparison of precipitation from observed data and general circulation models over the Iberian Peninsula. J. Clim. **19**(17), 4254–4275 (2006)
57. NRC (Natural Resources Canada): Canadian Wildland Fire Information System [Online]. http://www.nrcan.gc.ca/home (2011). Accessed Nov 2011
58. Pausas, J.G.: Changes in fire and climate in the eastern Iberian Peninsula (Mediterranean basin). Clim. Chang. **63**(3), 337–350 (2004)
59. Pausas, J.G., Bradstock, R.A.: Fire persistence traits of plants along a productivity and disturbance gradient in mediterranean shrublands of south-east Australia. Glob. Ecol. Biogeogr. **16**(3), 330–340 (2007)
60. Peixoto, J.P., Oort, A.H.: Physics of Climate. American Institute of Physics, New York (1992)
61. Pereira, M.G., Sanches Fernandes, L., Macário, E., Gaspar, S., Pinto, J.G.: Climate change impacts in the design of drainage systems–a case study for Portugal. J. Irrig. Drain. Eng. **141**(2), 05014009 (2015)
62. Pereira, M.G., Malamud, B.D., Trigo, R.M., Alves, P.I., Llasat, M.C.: The history and characteristics of the 1980–2005 Portuguese rural fire database. Nat. Hazards Earth Syst. Sci. **11**(12), 3343–3358 (2011)
63. Pereira, M.G., Calado, T.J., DaCamara, C.C., Calheiros, T.: Effects of regional climate change on rural fires in Portugal. Clim. Res. **57**(3), 187–200 (2013)
64. Pereira M.G., Sanches Fernandes, L., Macário, E., Gaspar, S., Pinto, J.G.: Climate change impacts in the design of drainage systems—a case study for Portugal (2014, accepted)

65. Prévost, P.: Mémoire sur l'équilibre du feu. J. Phys. **38**, 314–322 (1791)
66. Quinn, T.R., Tremaine, S., Duncan, M.: A three million year integration of the Earth's orbit. Astron. J. **101**, 2287–2305 (1991)
67. Rockel, B., Will, A., Hense, A.: The regional climate model COSMO-CLM (CCLM). Meteorol. Z. **17**(4), 347–348 (2008)
68. Rosenberg, E.A., Keys, P.W., Booth, D.B., Hartley, D., Burkey, J., Steinemann, A.C., Lettenmaier, D. P.: Precipitation extremes and the impacts of climate change on stormwater infrastructure in Washington State. Clim. Chang. **102**(1–2), 319–349 (2010)
69. Scherrer, S.C.: Present-day interannual variability of surface climate in CMIP3 models and its relation to future warming. Int. J. Climatol. **31**(10), 1518–1529 (2011)
70. Sharif, M., Burn, D.H.: Improved k-nearest neighbor weather generating model. J. Hydraul. Eng. **12**(1), 42–51 (2007)
71. Shetinsky, E.A.: Protection of forests and forest pyrology. Ecology, Moscow, 209 p. (1994)
72. Street, J.O., Carroll, R.J., Ruppert, D.: A note on computing robust regression estimates via iteratively reweighted least squares. Am. Stat. **42**(2), 152–154 (1988)
73. Svanidse, G.G.: Principles of Estimating River Flow Regulation by the Monte Carlo Method, p. 271. Metsniereba, Tbilisi (1964)
74. Svanidze, G.G.: Mathematical Modeling of Hydrologic Series for Hydroelectric and Water Resources Computations. Water Resources Publications, Fort Collins (1980)
75. Tebaldi, C., Hayhoe, K., Arblaster, J.M., Meehl, G.A.: Going to the extremes. Clim. Chang. **79**(3–4), 185–211 (2006)
76. Trigo, R.M., Sousa, P.M., Pereira, M.G., Rasilla, D., Gouveia, C.M.: Modelling wildfire activity in Iberia with different atmospheric circulation weather types. Int. J. Climatol. (2013). http://onlinelibrary.wiley.com/doi/10.1002/joc.3749/full
77. Van Wagner, C.E., Pickett, T.L.: Equations and FORTRAN Program for the Canadian Forest Fire Weather Index System, vol. 33. Can. For. Serv., Ottawa, Ontario (1985)
78. Uppala, S.M., Kållberg, P.W., Simmons, A.J., Andrae, U., Bechtold, V., Fiorino, M., et al.: The ERA-40 re-analysis. Q. J. R. Meteorol. Soc. **131**(612), 2961–3012 (2005)
79. Van Wagner, C.E.: Development and Structure of the Canadian Forest Fire Weather Index System. Forestry Technical Report 35, Canadian Forestry Service, Ottawa (1987)
80. Van Wagner, C.E., Pickett, T.L.: Equations and FORTRAN Program for the Canadian Forest Fire Weather Index System. Forestry Technical Report 33, Canadian Forestry Service, Ottawa (1985)
81. Viegas, D.X., Bovio, G., Ferreira, A., Nosenzo, A., Sol, B.: Comparative study of various methods of fire danger evaluation in southern Europe. Int. J. Wildland Fire **9**(4), 235–246 (2000)
82. Viegas, D.X., Piñol, J., Viegas, M.T., Ogaya, R.: Estimating live fine fuels moisture content using meteorologically-based indices. Int. J. Wildland Fire **10**(2), 223–240 (2001)
83. Wang, Y., Cheng, H., Edwards, R.L., He, Y., Kong, X., An, Z., et al.: The Holocene Asian monsoon: links to solar changes and North Atlantic climate. Science **308**(5723), 854–857 (2005)
84. Wilks, D.S.: Statistical Methods in the Atmospheric Sciences, vol. 100. Academic, New York (2011)
85. WMO: World Meteorological Organization: Technical Regulations, vol I. WMO-NO. 49. Geneva, Switzerland (1984)
86. Wotton, B.M.: Interpreting and using outputs from the Canadian Forest Fire Danger Rating System in research applications. Environ. Ecol. Stat. **16**(2), 107–131 (2009)

Infinite Horizon Optimal Control for Resources Management in Agriculture

Fernando Lobo Pereira

Abstract This article concerns an optimal control based framework for the optimization of resources in agriculture taking into account the environment sustainability. A decentralized, adaptive, hierarchic architecture to support long term coordinated decision-making strategies is required in order to achieve the common long term desired equilibrium in the environment state, and, at the same time, allow the economic sustainability of a number of distributed farm producers with, possibly conflicting, short term economic goals. The overall coordination is achieved by an adaptive Model Predictive Control structure that, on the one end hand, promotes the long term common good by approximating the solution to an infinite horizon optimal control problem, and, on the other hand, provides agro-chemical indicators to each one of the local farmers. We will emphasize on the importance of optimality results for infinite horizon optimal control problems of the Mayer type depending on the state at the final time while satisfying constraints at both trajectory endpoints.

1 Introduction

A framework based on optimal control is proposed as an advanced tool to support the management and control of resources in agriculture and builds on the issues discussed in [14]. The general motivation relies on the clear perception of the difficulty in meeting the future food needs on earth without destroying the environmental equilibrium, [7].

From the small sample of literature review in [14], it is clear that: (a) the interest in applying control and optimization techniques has been growing substantially, and (b) the overall problem is so complex that, generally speaking, only relatively partial problems have been addressed by these techniques. A cursory review of the literature points out to the extreme wide variety of problems whose complexity range may vary tremendously. This variety arises at two, generally interdependent, levels: (a) specification of the scope of the problem, and (b) the problem specificities once its scope is defined. In the later, the modeling of wealth of dynamics and constraints

F.L. Pereira (✉)
FEUP/SYSTEC and ISR, Rua Dr. Roberto Frias s/n, 4200-465 Porto, Portugal
e-mail: flp@fe.up.pt

J.-P. Bourguignon et al. (eds.), *Mathematics of Energy and Climate Change*,
CIM Series in Mathematical Sciences 2, DOI 10.1007/978-3-319-16121-1_13

so that the essentially relevant features are captured and, at the same time, the tractability constraints are satisfied, and specification of performance functionals and time horizons to be considered are key challenges. In what concerns the former, the extremely vast array of highly intertwined contexts that can be considered—economics, environment, climate, ecology, natural resources (notably, water and soil), spatial (from farm level to a region or country) and time scales (from short horizon of a few production scales to long horizon returns). In order to deal with the complexity for either concentrated or distributed control problem formulations, researchers have resorted to the specification of architectures enabling the organization of complex systems in multiple interacting monolithic simpler subsystems.

The promising perspectives of optimal control were pointed out in the pioneer work of David Zilberman in [24] by outlining a few sketchy statements of optimal control problems in agricultural economics.

In [8, 11, 12, 21], optimal control problems with an increasing degree of sophistication have been considered in order to find the combination of chemical and non-chemical control strategies optimizing the long term economic trade-off between crop yield profits, herbicide costs, and long term adverse effects of weed resistance. In particular, one should point out the effect of weeds resistance to herbicides which, in spite of the modeling difficulties, proved to be an important factor in the problem formulation. Various techniques, such as nonlinear programming, Pontryagin Maximum Principle and dynamic programming, have been used to solve the formulated optimal control problems. The general conclusion of these studies is that a better performance is achieved if a wider range of control methods are available and that, even without explicitly considering environmental factors in the problem, the optimal solutions are environmentally friendlier than the conventional ones.

The optimal control of pests has been investigated in a number of articles, notably [5, 18, 23]. Besides the economic issues addressed in the optimal weed control problems, now, issues concerning long term ecosystem equilibrium have been considered. Moreover, optimal strategies seeking long term environmental equilibria have been designed for problems with multiple pest species and by combining the use of fertilizers and pesticides. The Pontryagin Maximum Principle and dynamic programming techniques, [1, 4, 16] have been used to solve these problems.

Other classes of optimal control problems for resources management in agriculture considering more global contexts have also been considered. While soil erosion and lake restoration from phosphorus runoff were taken into account in [9], the problem of optimal carbon sequestration in agricultural soils (either by reduced tillage, organic farming and other carbon input to the soil techniques, switch to perennial crops, etc.) that takes into account the fact that carbon de-sequestration is far faster than sequestration is formulated in [6, 19].

It is important to observe that the optimal control problems considered so far were monolithic due to the fact that its formulation did not require the consideration of multiple, distributed, independent, albeit interacting and, possibly, conflicting, subsystems. However, when that is the case, a meaningful problem formulation requires an appropriate architectural arrangement in order to optimize resources for complex problems for which, usually, decisions affect multiple conflicting interests, different sets of stakeholders, and impacts in different time horizons.

In [17], a systemic approach is adopted in order to formulate optimization and optimal control problems to optimize the economic valuable botanical yield components based on a functional-structural plant growth model in terms of the source-sink dynamics. This model encompasses all pertinent ingredients which encompass both botanical, and ecological yield components, and all other environmental factors. A key challenge here is to ensure the compatibility of this model with the plant model in terms of spatial and temporal scales. In [10], optimal control modeling was used to analyze how public resources should be allocated to small-scale water protection efforts in agriculture or, alternatively, to investments in large-scale waste water treatment plants to control point source loads. In [20], an analysis is performed in the context of the Australian agriculture to show the need of increasingly adaptive policies to take into account the evolution of perceptions of the state of the system and of the intervening processes in the multiple spatial and temporal scales, as well as the increased role of environment changes for which climate variability plays a prominent role. Finally, a much more general context is considered in [7] where it is argued that the optimal development path in the sense of the max-min criterion of intergenerational justice is too demanding to be practical and too costly for the economically less competitive. This calls for a development policy following an optimal growth approach while encompassing measures to mitigate the intergenerational and intra-generational welfare inequalities.

The crescendo of complexity that emerges from this small sample of publications points out to a number of research directions that go well beyond the conventional ones naturally associated with the usual "monolithic" optimal control problem formulation, such as: (i) modeling of system's dynamics, constraints and performance criteria; (ii) approaches to solve it, which may draw from optimization and control theory, and computational procedures; and (iii) framework to integrate the generated output in appropriate decision-making and control support systems. Another class of challenges that have not been addressed to the same extent, concerns the formulation and solution of decision and control problems which encompass the following issues:

- Spatial heterogeneity due to soil composition, groundwater distribution, solar and wind exposure, etc.
- Ecosystem interactions to account for multiple crops, species of weeds, and pests.
- Environmental effects such as soil, groundwater, air, and carbon emission/sequestration.
- Geographic boundary effects, notably with distinct agricultural production facilities.
- Multiple goals, possibly conflicting and manifesting in different time horizons.
- Meteorological variability, as well as, unanticipated climate change trends.

This requires a very ambitious research agenda to be addressed by a strongly interdisciplinary research program involving key stakeholders.

A general abstract control architecture satisfying some key identified requirements is discussed in the next section. Then, in Sect. 3 some key results in Optimal

Control and Model Predictive Control are considered as providing the basis for decision and control synthesis in the context of the control architecture. Finally, some brief conclusions are given in Sect. 4.

2 A Control Architecture

A careful analysis of some key references, e.g., [7, 9, 10, 17, 19, 20, 23, 24], led to the distillation of the following general requirements to be satisfied by a comprehensive dynamic optimization based framework to support decision-making and control:

- Long (say, infinite) time and short time horizon optimal control strategies should be articulated in spite of the, possibly conflicting, goals to be considered in the different time horizons.
- Scalability in time and space to deal with complexity and heterogeneity.
- Coordinated decentralization of the decision and control system composed by multiple independent decision makers.
- Adaptivity to take into account climate change trends, other environment changes, economic and social trends, and, possibly disruptive, technological developments.
- Robustness of the solution with respect to modeling uncertainties and perturbations.

These requirements are described here in general and relatively abstract terms. Obviously, they have to be further detailed in the specific context in which an optimal control based management and control support tool would be designed. This task will be part of the tool in question design activity.

The layered control architecture depicted in Fig. 1 was proposed in [14].

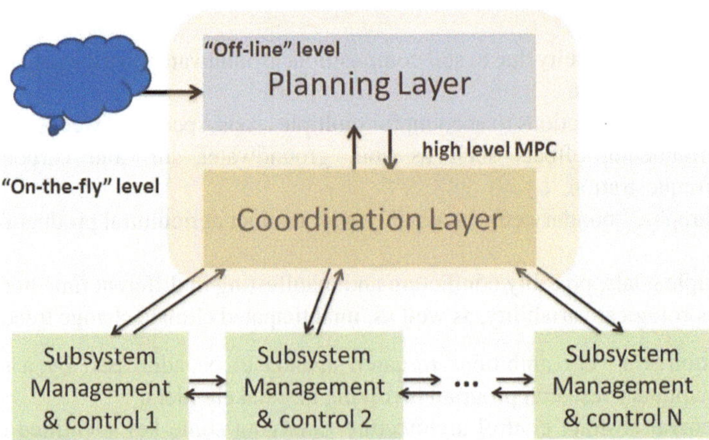

Fig. 1 Layered control architecture

The layers composing the system are: Planning, Coordination and Execution. The Planning layer considers the global issues and takes aggregated information of the system from the coordination layer and external sources to generate long-term planning targets which are passed to the Coordination layer. If significant inconsistencies between the current and the executed plans are detected the global plan is updated. Moreover, detected significant trends can be incorporated into the models so that plans reflect the overall system's evolution. The Coordination layer receives the planned targets and generates shorter term targets for each one of the subsystems to achieve their coordination. It aggregates status data from the subsystems, to provides "feedback" data to the planning layer. The Execution layer of each subsystem computes its control strategy by taking into account the local goals, constraints, and the coordinating targets provided by the Coordination layer.

In all these layers, optimization, and specially optimal control, plays a key role, even when consensus might have to be generated in order to coordinate subsystems with conflicting goals. However, the performance criteria and targets at the overall planning level are different than those at the level of each one of the subsystems. Clearly, this modular structure accommodates spatial heterogeneity, decentralization, adaptivity, and concurs to the subordination of strategies optimizing shorter term local goals subject to the local constraints (e.g., to ensure economic sustainability) to global, longer term, goals (e.g., environment sustainability) via the coordination layer that bridges both perspectives.

3 Model Predictive Control Based Framework

In spite of its already long tradition, Model Predictive Control (MPC) is still a growing topic due to the fact that its wide range of schemes yield control strategies combining "optimality" with adaptivity. Here, we are particularly interested in control designs that ensure control strategies that are: (a) robust to episodic perturbations, (b) adaptive to sustained changing trends, and (c) "optimal" in the sense of the trade-off between conflicting short term economic local goals with long term global environment targets. In this context, we envisage MPC schemes generating control strategies that approximate solutions to infinite horizon optimal control problems, [15].

Essentially, the variant of the general MPC scheme, [13], that we are considering here is as follows:

1. Initialization: $t = t_0, x(t_0)$.
2. Solve (P_T) over $[t_0, t_0 + T]$ to obtain an optimal reference trajectory x^*.
3. Compute an optimal feedback control u^* during $[t_0, t_0 + \Delta]$ to track x^* restricted to this time interval.
4. Sample the state variable x at $t_0 + \Delta$ to obtain $\bar{x} = x(t_0 + \Delta)$.
5. Slide the time origin by Δ time units, let $x(t_0) = \bar{x}$, and go to step (2).

Here, T and Δ are the durations of the optimization (or prediction) and of the control horizons, respectively. Notice that feedback control is considered both during the intervals between consecutive sampling times, and when solving the open loop optimal control problem (P_T) which, by using the sampled state variable, takes into account perturbations that might affect the state trajectory. Remark also, that identification methods can be used to adapt models in (P_T) if warranted by the observed deviations in the values of the sampled state variable. The optimal control problem (P_T) can be stated as follows:

$$
\begin{aligned}
(P_T)\ \text{Minimize}\ \ & g(x(T)) + \int_{t_0}^{t_0+T} l(s,x(s),u(s))ds \\
\text{subject to}\ \ & \dot{x} = f(t,x,u),\ \ \mathcal{L}\text{-a.e.} \\
& x(t) \in X_t,\ \ u \in \mathcal{U},\ \ x(t_0) \in C_0.
\end{aligned}
$$

Here, the functions $(f,l) : [t_0, t_0+T] \times \mathbb{R}^n \times \mathbb{R}^m \to \mathbb{R}^n \times \mathbb{R}$ represent the controlled dynamics and the running cost of the system, $g : \mathbb{R}^n \to \mathbb{R}$ is the endpoint cost functional, and $(X_t, U_t) \subset \mathbb{R}^n \times \mathbb{R}^m$ are, respectively, the pointwise state and control constraint sets. A wide range of optimality conditions are currently available to support solution methods to this problem, [1–4, 16].

Since one key objective of the proposed resources optimization framework is to reconcile long term goals driving the system to an environmental equilibrium with short term goals ensuring economic competitiveness, the MPC scheme running at the Planning and Coordination layers should generate strategies that asymptotically approximate the solution to an "infinite-horizon" optimal control problem, that is

$$
\begin{aligned}
\text{Minimize}\ \ & g_\infty(\xi) + \int_{t_0}^{\infty} l(t,x(t),u(t))dt \\
\text{subject to}\ \ & \dot{x} = f(t,x,u)\ \ \mathcal{L}\text{-a.e.},\ \ x(t_0) \in C_0, \\
& \xi \in C_\infty,\ \ \lim_{t\to\infty} x(t) = \xi,\ \ u \in \mathcal{U},
\end{aligned}
$$

where the function $g_\infty(\cdot)$ is the performance functional that forces the system to be driven to the desired long term equilibrium. Following the arguments in [15], the optimal control problem (P_T) to be considered in the MPC scheme, so that its solutions approximate the infinite horizon ones, is:

$$
\begin{aligned}
(P_T)\ \text{Minimize}\ \ & V(t_0+T, x(t_0+T)) + \int_{t_0}^{t_0+T} l(s,x(s),u(s))ds \\
\text{subject to}\ \ & \dot{x} = f(t,x,u),\ \ u \in \mathcal{U},\ \ x(t_0) \in C_0,
\end{aligned}
$$

where the value function $V(\cdot,\cdot)$ defined by

$$
V(\tau,z) := \min_{(x,u)\in ACP(\tau,z)} \left\{ g(\xi) + \int_{\tau}^{\infty} l(t,x(t),u(t))dt \right\},
$$

being $ACP(\tau, z)$ the set of all feasible asymptotically convergent control processes (x, u) satisfying $x(\tau) = z$.

Under appropriate assumptions, the value function can be obtained by solving an Hamilton-Jacobi partial differential equation

$$\begin{cases} \frac{\partial}{\partial t} V(t,x) + \min_{u \in \Omega} \langle \frac{\partial}{\partial x} V(t,x), f(t,x,u) \rangle = 0 \\ V(T, x(T)) = g(x(T)), \end{cases}$$

for which the existence of solution is ensured by the assumptions and by adopting compatible notions of derivatives and of solution concept, [22]. The huge difficulties associated with the computational complexity of this equation are well known even for finite horizon problems. Thus, in [15], an alternative approach to this problem is considered by examining necessary conditions of optimality for infinite horizon optimal control problems particularly well suited for this class of applications. In this article, we consider the problem on $[0, \infty)$,

$$\begin{aligned} (P_\infty) \text{ Minimize } \; & h(x(0), \xi) \\ \text{such that } \; & \dot{x}(t) = f(t, x(t), u(t)) \; \mathcal{L} - a.e. \\ & x(0) \in C_0, \; \lim_{t \to \infty} x(t) = \xi \in C_\infty \\ & u(t) \in \Omega \subset \mathbb{R}^m, \end{aligned}$$

where C_0 and C_∞ are compact sets and the remaining ingredients are as above. In spite of the significant body of literature on this class of problems, the degenerative effect of the infinite horizon still constitutes a huge challenge. The maximum principle proposed in [15], features a new transversality condition at infinity—based on the concept of directional inclusion at infinity—that, for asymptotically convergent control processes, yield an interesting trade-off between the applicability breadth (dictated by the required assumptions) and the information provided by the optimality conditions.

4 Conclusions

The decision support system based on optimal control discussed in this article satisfies some important requirements arising in resources management and control in agriculture. These requirements were extracted from a significant review of pertinent literature. The planning and coordinated control layers of the proposed multi-layer control structure relies on an “infinite horizon” Model Predictive Control scheme. Some theoretical challenges inherent to optimality conditions for this class of problems were examined. The complexity inherent to the formulation of the optimization problems at various levels of the control architecture is huge, and, thus, this article can be regarded as a roadmap pointing out to a number of research issues which may lead to tools to design decision-making, management and control support systems.

Acknowledgements The author thanks the support of grants FCT n° PEST-OE/EEI/UI0147/2014, and Project USP-UP "Controle para a Optimização de Recursos na Exploração Agrícola".

References

1. Arutyunov, A.: Optimality Conditions: Abnormal and Degenerate Problems. Springer, Dordrecht (2000)
2. Arutyunov, A., Karamzin, D., Lobo Pereira, F.: The maximum principle for optimal control problems with state constraints by R.V. Gamkrelidze: revisited. J. Optim. Theory Appl. **149**, 474–493 (2011)
3. Clarke, F.: The maximum principle in optimal control, then and now. J. Control Cybern. **34**, 709–722 (2005)
4. Clarke, F., Ledyaev, Y., Stern, R., Wolenski, P.: Nonsmooth Analysis and Control Theory. Springer, New York (1998)
5. Christiaans, T., Eichner, T., Pethig, R.: Optimal pest control in agriculture. J. Econ. Dyn. Control **31**, 3965–3985 (2007)
6. Djezou, W.: Agriculture and deforestation: What optimal conversion of forest land to agriculture in Côte d'Ivoire? Technical Report, University of Cocody-Abidjan (2009)
7. Farzin, Y.: Sustainability, optimality, and development policy. Technical Report, Dept. Agricultural & Resource Econom, UCB, Davis (2008)
8. Gorddard, R., Pannell, D., Hertzler, G.: An optimal control model for integrated weed management under herbicide resistance. Aust. J. Agric. Econ. **39**(1), 71–87 (1995)
9. Hediger, W.: Optimal control of soil erosion and phosphorous runoff from agricultural land. Agricultural Economics, Swiss Federal Inst of Technology (2000)
10. Huhtala, A., Laukkanen, M.: Optimal control of dynamic point and non-point pollution in a coastal ecosystem: agricultural abatement versus investment in waste water treatment plants. MTT Agrifood Research Finland (2011)
11. Jones, R., Cacho, O.: A dynamic optimisation model of weed control. Working Paper Ser. Agricultural & Resource Economics, New England University (2000)
12. Manalil, S., Busi, R., Renton, M., Powles, S.: Rapid evolution of herbicide resistance by low herbicide dosages. Weed Sci. **59**(2), 210–217 (2011)
13. Mayne, D., Rawlings, J., Rao, C., Scokaert, P.: Constrained model predictive control: stability and optimality. Automatica **36**, 789–814 (2000)
14. Pereira, F.L., Fontes, F., Ferreira, M., Pinho, M., Oliveira, V., Costa, E., Silva, G.: An optimal control framework for resources management in agriculture. J. Conf. Pap. Math. **2013**, Art. ID 769598 (2013)
15. Pereira, F., Silva, G.: Necessary Conditions of optimality for state constrained infinite horizon differential inclusions. In: Proceedings of the 50th IEEE Conference on Decision and Control & European Control, pp. 6717–6722, Orlando (2011)
16. Pontryagin, L., Boltyanskiy, V., Gamkrelidze, R., Mishchenko, E.: Mathematical Theory of Optimal Processes Interscience, New York (1962)
17. Qi, R.: Optimization and optimal control of plant growth: application of greenlab model for decision aid in agriculture. Ph.D. thesis, École Centrale des Arts et Manufactures, École Centrale Paris (2004)
18. Rafikov, M., Balthazar, J.M.: Optimal pest control problem in population dynamics. Comput. Appl. Math. **24**(1), 65–81 (2005)

19. Ragot, L., Schuber, K.: The optimal carbon sequestration in agricultural soils: does the dynamics of the physical process matter? Technical Report 2006.40, Centre d'Economie de la Sorbonne UMR 8174 (2006)
20. Risbey, J., Kandlikar, M., Dowlatabadi, H., Graetz, D.: Scale, context, and decision making in agricultural adaptation to climate variability and change. In: Mitigation and Adaptation Strategies for Global Change, pp. 137–165. Springer, Netherlands (1999)
21. Stiegelmeier, E., Oliveira, V., Silva, G., Karam, D., Furlan, M., Kajino, H.: Herbicide application optimization model for weed control using the resistance dynamics. Technical Report, São Paulo University (2012)
22. Vinter, R.: Optimal Control. Birkhauser, Boston, Basel, Berlin (2000)
23. Wetzstein, M., Szmedra, P., Musser, W., Chou, C.: Optimal agricultural pest management with multiple species. Northeast. J. Agric. Resour. Econ. **14**(1), 71–77 (1985)
24. Zilberman, D.: The use and potential of optimal control models in agricultural economics. West. J. Agric. Econ. **7**, 395–406 (1982)

Distributed Reasoning

Pedro Rodrigues and João Gama

Abstract This paper discusses the problem of learning a global model from local information. We consider ubiquitous streaming data sources, such as sensor networks, and discuss efficient learning distributed algorithms. We present the generic framework of distributed sources of data, an illustrative algorithm to monitor the global state of the network using limited communication between peers, and an efficient distributed clustering algorithm.

1 Introduction

Data are distributed in nature. Nowadays, detailed data for almost any task are collected over a broad area, and streams in at a much greater rate than ever before. In particular, advances in miniaturization, the advent of widely available and cheap computer power, and the explosion of networks of all kinds gave life to inanimate things. Simple objects that surround us are gaining sensors, computational power, and actuators, and are changing from static, inanimate objects into adaptive, reactive systems. Sensor networks and digital social networks are present everywhere [7].

Examples of network data include smart grids consisting of millions of automated electronic meters. The meters would generate an overwhelming amount of distributed data that can be handled with emergent techniques: data streams management and processing approaches. A key characteristic of smart grids is the *intelligent layer* that analysis the data produced by these meters allowing companies to develop powerful new capabilities in terms of grid management, planning and customer services for energy efficiency. The development of the market with a growing share of load management incentives and the increasing number of local generators will bring new difficulties to grid management and exploitation. Present monitoring systems suffer for the lack of machine learning technologies that can

P. Rodrigues
LIAAD-INESC TEC, University of Porto, Porto, Portugal
e-mail: pprodrigues@med.up.pt

J. Gama (✉)
LIAAD-INESC TEC and Faculty of Economics, University of Porto, Porto, Portugal
e-mail: jgama@fep.up.pt

J.-P. Bourguignon et al. (eds.), *Mathematics of Energy and Climate Change*,
CIM Series in Mathematical Sciences 2, DOI 10.1007/978-3-319-16121-1_14

modify the behavior of monitoring systems based on the sequence patterns arriving over time. From a data mining point of view, a smart grid forms a network (eventually decomposable) of distributed sources of high-speed data streams. The dynamics of data are unknown; the topology of network changes over time, the number of meters tends to increase and the context where the meter acts evolves over time. This way, several data mining tasks are involved: prediction, cluster analysis (profiling), event and anomaly detection, correlation analysis, etc. All these characteristics constitute challenges and opportunities for applied research in distributed data mining. The requirements of near real-time analysis for multiple time horizons and multiple space aggregations make these analyses an even harder research challenge.

In this work we focus on *clustering*, one of the most used data mining techniques. The goal in cluster analysis is the assignment of a set of observations (or objects) into groups so that observations in the same group are similar in some sense.

The paper is organized as follow. In Sect. 2 we present the distributed network framework and an illustrative example about distributed reasoning. In Sect. 3, we present a distributed clustering algorithm for sensor networks. In the context of this work, a cluster is defined to be a set of sensors. The key characteristic of the proposed algorithm is that each sensor processes locally their own data, and communicate with neighbours in order to learn a global view of the network. The last section concludes the paper by presenting the lessons learned.

2 Network Data Model

The goal of our study are networks of interconnected nodes. Nodes, or sensors or peers, are sensing the environment measuring some quantity of interest. Individually, each peer has a local and limited information about the environment. If sensors communicate, the network might have a global perspective of the environment. Figure 1 illustrates this context.

2.1 The Framework

Network topology is the organizational hierarchy of the interconnected nodes. Different network topologies can affect throughput, but reliability is often more critical.

A common structure is the *star network*, where all nodes are connected to a special central node, the coordinator. This is the typical layout found in a wireless sensor networks. Another popular layout is the *mesh network*, where each node is connected to an arbitrary number of neighbours in such a way that there is at least one traversal from any node to any other. The main purpose of a *mesh network* is fault tolerance.

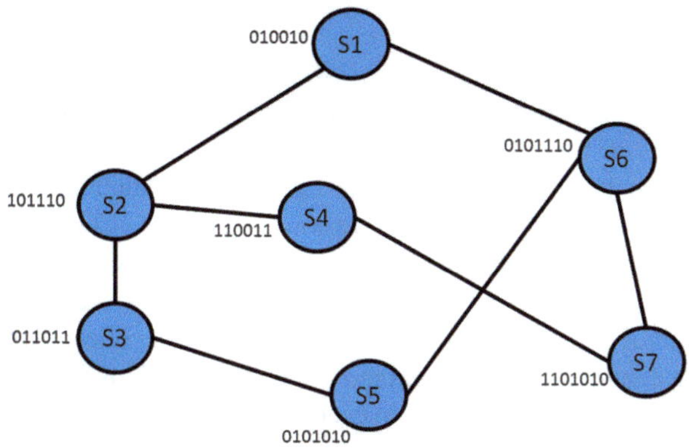

Fig. 1 A network of interconnected nodes. *Circles* represent sensors, *edges* represent communication paths

Routing is the process of selecting network paths to carry network traffic. Some popular routing schemes are: *unicast*: delivers a message to a single specific node; *broadcast*: delivers a message to all nodes in the network; *anycast*: delivers a message to a group of nodes, typically the ones nearest to the source.

In data-mining problems, a user runs queries over the data produced by the sensors. A query is defined over the data produced by all the sensors:

$$Query = Q(\bigcup_{i=0}^{n} S_i)$$

We can consider two types of queries:

1. One-shot queries: What is the current state of the network?
2. Continuous queries: Track and monitor the state of network at any time.

Continuous queries are of particular interest because they are used for monitoring purposes, understanding dynamics, detect anomalies and changes.

In the network data model, data is vertically distributed. Answering continuous queries, requires specific characteristics of the algorithms. Following [2, 13], the requirements for processing continuous queries are:

- Single pass: process each observation once;
- Small space: constant space;
- Small processing time;
- Reduced communications.

Local approaches are the most efficient ones [5]. They preserve privacy and security issues but require some sort of synchronization between peers [8].

2.2 An Illustrative Example

In this section, we present an illustrative application of ubiquitous reasoning. The problem consists of monitoring data produced in a sensor network. The sensors monitor the concentration of air pollutants. Each sensor maintains a data vector with measurements of the concentration of various pollutants (CO_2, SO_2, O_3, etc.). A function on the average of the data vectors determines the Air Quality Index (AQI). The goal consists of trigger an alert whenever the AQI exceeds a given threshold. The problem involves computing a function over the data collected in all sensors. A trivial solution consists of sending data to a central node. This might be problematic due to huge volume of data collected in each sensor and the large number of sensors.

Sharfman et al. [12] present a distributed algorithm to solve this type of problems. They present a geometric interpretation of the problem. Figure 2 illustrate the instance space. Each axis corresponds to one pollutant. For visualization purposes, we represent only two pollutants. The gray dots corresponds to the sensor's measurements, and the black dot to the aggregation vector, the AQI index. The gray region corresponds to the alarm region. The goal is detect whenever the AQI index is inside the gray region. In Fig. 2 we present three examples. The first one, all sensors and the AQI index are outside the alarm region. In the second plot, the AQI index is outside the alarm region, although one of the sensors is inside the alarm region. The third plot, illustrate the case where the AQI index is inside the alarm region, although all sensors are outside the alarm region. These examples illustrate that information of individual sensors is not enough to make a decision about the global state of the network. Sensors need to share information to reach a correct decision.

The method is based on local computations with reduced communications between sensors. The base idea is that the aggregated function is always inside the convex-hull of the vectors space (see Fig. 3a, b). Suppose that all points share a reference point. Each sensor can compute a sphere with diameter the current measurement and the reference point. If all spheres are in the normal region, the

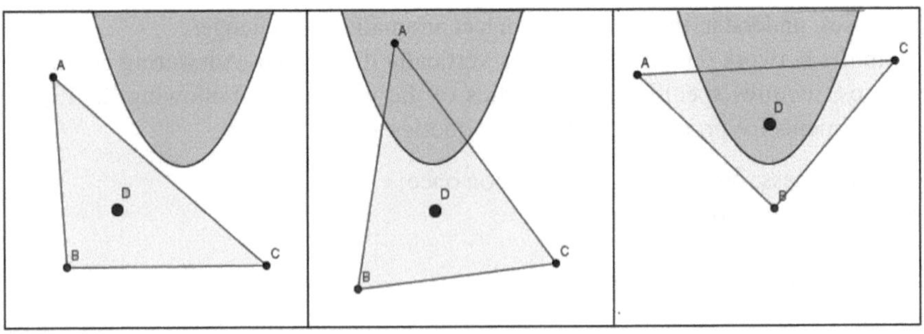

Fig. 2 The vector space: the *gray dots* (*A, B, C*) corresponds to the sensor's measurements; and the *black dot* (*D*) to the aggregation vector. The *gray region* corresponds to the alarm region. The *left* and *central figures* illustrates a normal air condition. The *right figure* presents an alarm condition, where none of the sensors is inside the alarm region

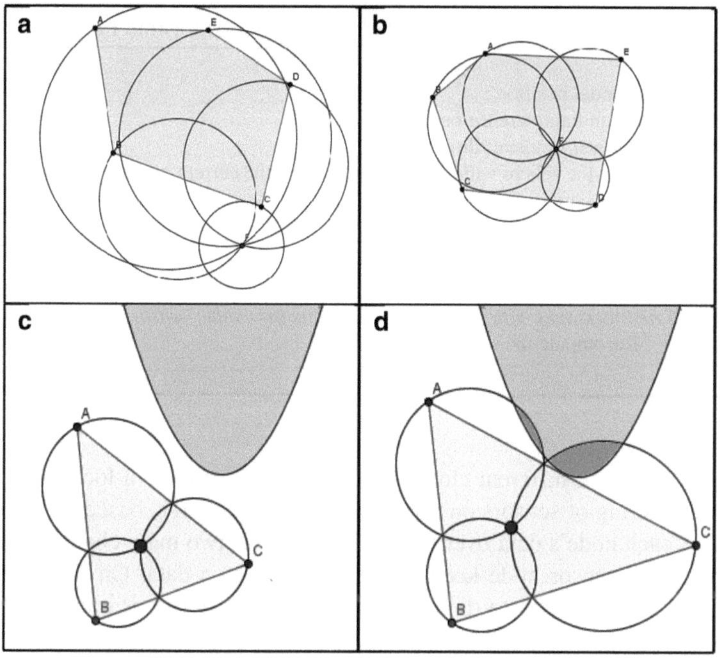

Fig. 3 The bounding theorem: the convex-hull of sensors is bounded by the union of spheres. Sensors only need to communicate their measurements when the spheres are non-monochromatic

aggregated value is also in the normal region. This holds, because the convex-hull of all vertex is bounded by the union of the spheres (see Fig. 3c, d). In the case that a sphere is not monochromatic, the node triggers the re-calculation of the aggregated function. Sensors broadcast their current measurements, and a new common point is computed.

The algorithm guarantees that any alarm is detected and no false alarms are signalled. The algorithm only uses local constraints. Mostly only local computations are required and this minimizes the communications between sensors.

3 Clustering Distributed Data Sources

Clustering is the most popular technique for data understanding. The basic idea behind clustering streaming data sources is to find groups of sources that behave similarly through time, which is usually measured in terms of the distance between the data series or the data distribution. Let X be a sensor node producing observations x_i at each time step i. The goal of an incremental clustering system for streaming data sources is to find (and make available at any time i) a partition $C(i)$ of data sources, where data sources in the same cluster tend to be more alike

Algorithm 1: The Monitoring Threshold Functions Algorithm (sensor node).

```
begin
    Broadcast Initial position ;
    Compute an initial reference point ;
    foreach new measurement do
        Compute the sphere with diameter defined by the current measurements and the
        reference point and check its colour;
        if sphere non monochromatic then
            Broadcast the actual measurement;
            Recompute a new reference point;
        if new messages with current measurements from other sensors received then
            Recompute the reference point;
```

than data sources in different clusters [3, 9, 11]. We propose a local algorithm to perform clustering of sensors on ubiquitous sensor networks, based on the moving average of each node's data over time. *L2GClust* has two main characteristics. On one hand, each sensor node keeps a sketch of its own data. On the other hand, communication is limited to direct neighbours, so clustering is computed at each node. The moving average of each node is approximated using memoryless fading average, while clustering is based on the furthest point algorithm applied to the centroids computed by the node's direct neighbours. This way, each sensor acts as data stream source but also as a processing node, keeping a sketch of its own data, and a definition of the clustering structure of the entire network of data sources.

In this work we search for a definition of k clusters of sensor nodes, with k previously known by the system. Although this simple example lacks some of the common characteristics of real-world scenarios (e.g. unknown number or clusters or unbalanced data), its extension is straightforward. If the number of clusters to find is unknown, each node could search for a clustering with different number of clusters. As only centroids are transmitted and used as single points (as if operating with ensembles of clusters), there's no need to know how many points come from each node; all centroids that are received are included in the clustering as single points. For unbalanced data (in terms of the assignment of nodes to clusters) we believe that the convergence would take longer but deeper analysis is required in future work.

As previously stated, we consider that each sensor produces a univariate stream of data, and we want to define a clustering structure for the sensors, where sensors producing streams, which are alike, are clustered together. Hence, we should consider techniques that project each sensor's data stream into a reduced set of dimensions that suffice to extract similarity with other sensors. These estimates can be seen as the sensor's current view of its own data, giving a sign of where in the data-space this sensor is included [10]. One-way to summarize a data stream x is by computing its sample mean $\hat{\mu}_x$ and standard deviation $\hat{\sigma}_x$. Our approach is to keep track of the moving average of each sensor, as an estimate of the sample mean of most recent data.

Each sensor produces data continuously. Given this, each sensor s is responsible of keeping its own estimate of the sample mean ($\hat{\mu}_s$) in a online fashion. Moving averages are usually easy to compute, if we can keep a small buffer of data points [10]. However, in such resource-demanding scenarios, this is seldom the case. Nonetheless, sum-based statistics computed on sliding windows can be approximated by weighting the sums using fading statistics [4]. The *α-fading sum* $S_{x,\alpha}(i)$ of observations from a stream x is computed at time $\forall i > 0$, as: $S_{x,\alpha}(i) = x_i + \alpha \times S_{x,\alpha}(i-1)$, where $S_{x,\alpha}(0) = 0$. In the computation, α ($0 \ll \alpha < 1$) is a constant determining the forgetting factor of the sum. This way, the *α-fading average* at observation $\forall i > 0$ is then computed as: $M_{x,\alpha}(i) = \frac{S_{x,\alpha}(i)}{N_\alpha(i)}$, where $N_\alpha(i) = 1 + \alpha \times N_\alpha(i-1)$ is the corresponding *α-fading increment*, with $N_\alpha(0) = 0$. An important feature of the α-fading increment is that: $\lim_{i \to +\infty} N_{\alpha<1}(i) = \frac{1}{1-\alpha}$. Each value of α, which should be close to 1 (e.g. 0.999), will converge to sliding windows of different sizes. This way, at each observation i, $N_\alpha(i)$ gives an approximated value for the weight given to recent observations used in the α-fading sum.

3.1 Local Clustering of Stream Sources

The goal is to have at each local site an approximation of the global clustering structure of the entire sensor network. Each sensor should include incremental clustering techniques which operate with distance metrics developed for the dimensionally-reduced sketches of the data streams. Also, and although in several real-world scenarios this is not true, we should not assume the sample mean of each sensor to be correlated with its physical location and connectivity, as the matching between data clusters and physical clusters is a promising strategy for sensor network comprehension, so we should not bias the clustering solution [10]. Given the simple sketch definition, the dissimilarity between two sensors x and y is the absolute distance between their sample means, $d(x, y) = |\hat{\mu}_x - \hat{\mu}_y|$.

3.1.1 Neighbourhood Interaction

Each sensor x is not only able to sketch its own data in a dimensionally-reduced definition (the fading average $M_{x,\alpha}$), but it is also able to interact with its neighbouring nodes η_x. The main characteristic of our approach is that, at each new observation i produced by sensor x, instead of sending its own sketch $M_{x,\alpha}$ to its neighbours η_x, the node sends its own estimate of the global clustering $C_x(i)$. Note that, with this approach, each node needs to keep an estimate of the global cluster centers $C_x(i) \approx C_g(i)$. This estimate can be seen as the sensor's current view of the entire network which, together with its own sketch, gives a sign of where in the entire network data-space this sensor is included.

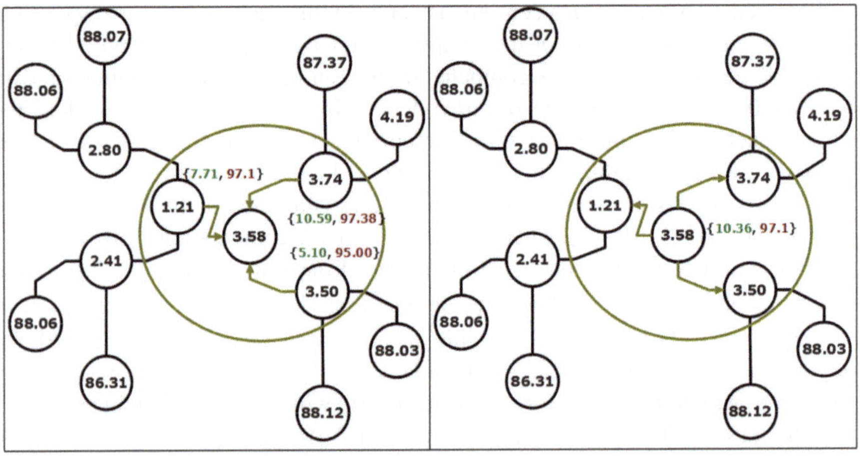

Fig. 4 The two main local steps in L2GClust. In the *left figure*, each node receives data from direct neighbours. Each node recomputes their centroids and send the new centroids to the neighbour nodes (*right figure*)

At first observations, each sensor node x has only access to its own sketch $M_{x,\alpha}(i)$. However, with neighbour nodes broadcasting their approximations of the global clustering structure $C_y(i), \forall y \in \eta_x$, node x suddenly has access to several data points which are believed by other nodes to be the real cluster centers. Let $P_x(i)$ be the complete set of clustering definitions $\{C_j(i) \mid j \in \eta_x\}$ received by node x between observations x_{i-1} and x_i. The set of points used in the clustering step includes: $\hat{\mu}_x$, the node's own sketch; $C_x(i-1)$, the node's approximation of global cluster centers (computed before observation x_i); and $P_x(i)$, the centroids sent by node's direct neighbours. Therefore, $C_x(i)$ is computed by clustering the set of points $\{M_{x,\alpha}(i)\} \cup C_x(i-1) \cup P_x(i)$.

The idea behind this step is to aggregate all the locally defined centers and apply a clustering procedure on these centers, considering them as points for the clustering. This way, next time this sensor uses or transmits its estimate $C_x(i)$ of the global clustering structure, it is already updated with its most recent sketch and neighbours' information (Fig. 4).

3.1.2 Furthest-Point Clustering

In the general task of finding k centers given m points, there are two major objectives: minimize the *radius* (maximum distance between a point and its closest cluster center) or minimize the *diameter* (maximum distance between two points assigned to the same cluster) [1]. The *Furthest Point* algorithm [6] gives a guaranteed 2-approximation for both the *radius* and *diameter* measures. It begins by picking an arbitrary point as the first center, c_1, then finding the remainder centers c_i iteratively as the point that maximizes its distance from the previously chosen centers $\{c_1, \ldots, c_{i-1}\}$. After k iterations, one can show that the chosen centers $\{c_1, c_2, \ldots, c_k\}$ represent a factor 2 approximation to the optimal clustering [1].

This strategy gives a guaranteed definition of the cluster centers, computed by finding the center k_i of each cluster after attracting remainder points to the closest center c_i. Since we are applying clustering to cluster centroids, we are in fact merging clustering definitions, a known technique which has been argued to give good results [1].

4 Conclusions

In this paper, we have discussed the problem of learning global models from distributed local information. We have presented a clustering algorithm for data streams generated on wide sensor networks producing high speed data, from a dynamic (time-changing) environment. The algorithms run locally in each node of the network, processing their own data and communicating aggregated data to its neighbours. This is an important characteristic in several applications, because it preserves user's privacy. A good characteristic of the proposed systems is the ability to adapt to resource-restricted environments: system granularity can be defined given the resources available in the network's processing sites. The proposed algorithms reduce both the dimensionality and the communication burdens, by exploiting limited computational resources at each local sensor.

Acknowledgements This work was supported by Sibila and Smartgrids research projects (NORTE-07-0124-FEDER-000056/59), financed by North Portugal Regional Operational Programme (ON.2 O Novo Norte), under the National Strategic Reference Framework (NSRF), through the Development Fund (ERDF), and by national funds, through the Portuguese funding agency, Fundação para a Ciência e a Tecnologia (FCT), and by European Commission through the project MAESTRA (Grant number ICT-2013-612944).

References

1. Cormode, G., Muthukrishnan, S., Zhuang, W.: Conquering the divide: continuous clustering of distributed data streams. In: ICDE: Proceedings of the International Conference on Data Engineering, Istanbul, pp. 1036–1045 (2007)
2. Du, W., Deng, J., Han, Y., Varshney, P., Katz, J., Khalili, A.: A pairwise key predistribution scheme for wireless sensor networks. ACM Trans. Inf. Syst. Secur. **8**(2), 228–258 (2005)
3. Gama, J.: Knowledge Discovery from Data Streams. Data Mining and Knowledge Discovery. Chapman & Hall/CRC Press, Atlanta (2010)
4. Gama, J., Sebastião, R., Rodrigues, P.P.: On evaluating stream learning algorithms. Mach. Learn. **90**(3), 317–346 (2013)
5. Giannella, C., Liu, K., Olsen, T., Kargupta, H.: Communication efficient construction of decision trees over heterogeneously distributed data. In: Proceedings of the Fourth IEEE International Conference on Data Mining, pp. 67–74. IEEE, Washington (2004)

6. Gonzalez, T.F.: Clustering to minimize the maximum intercluster distance. Theor. Comput. Sci. **38**(2/3), 293–306 (1985)
7. Kargupta, H., Joshi, A., Sivakumar, K., Yesha, Y.: Data Mining: Next Generation Challenges and Future Directions. AAAI Press/MIT Press, Menlo Park (2004)
8. May, M., Saitta, L. (eds.): Ubiquitous Knowledge Discovery. Lecture Notes in Artificial Intelligence, vol. 6202. Springer, Heidelberg (2010)
9. Rodrigues, P.P., Gama, J., Lopes, L.M.B. Clustering distributed sensor data streams. In: European Conference on Machine Learning and Knowledge Discovery in Databases. Lecture Notes in Computer Science, Antwerp, vol. 5212, pp. 282–297. Springer, Heidelberg (2008)
10. Rodrigues, P.P., Gama, J., Lopes, L.: Knowledge discovery for sensor network comprehension. In: Cuzzocrea, A. (ed.) Intelligent Techniques for Warehousing and Mining Sensor Network Data, pp. 118–134. Information Science, Hershey (2010)
11. Rodrigues, P.P., Gama, J., Araújo, J., Lopes, L.M.B.: L2gclust: local-to-global clustering of stream sources. In: Chu, W.C., Wong, W.E., Palakal, M.J., Hung, C.-C. (eds.) SAC, pp. 1006–1011. ACM, New York (2011)
12. Sharfman, I., Schuster, A., Keren, D.: A geometric approach to monitoring threshold functions over distributed data streams. ACM Trans. Database Syst. **32**(4), 301–312 (2007)
13. Zhu, S., Setia, S., Jajodia, S.: LEAP: efficient security mechanisms for large-scale distributed sensor networks. In: CCS '03, pp. 62–72. ACM Press, New York

Multiscale Internet Statistics: Unveiling the Hidden Behavior

Paulo Salvador, António Nogueira, and Eduardo Rocha

Abstract Being able to characterize and predict the behavior of Internet users based only on layer 2 statistics can be very important for network managers and/or network operators. Operators can perform a low level monitoring of the communications at the network entry points, independently of the data encryption level and even without being associated with the network itself. Based on this low level data, it is possible to optimize the access service, offer new security threats detection services and infer the users behavior, which consists of identifying the underlying web application that is responsible by the layer 2 traffic at different time instants and characterize the usage dynamics of the different web applications. Several identification methodologies have been proposed over the years to classify and identify IP applications, each one having its own advantages and drawbacks: port-based analysis, deep packet inspection, behavior-based approaches, learning theory, among others. Although some of them are very efficient when applied to specific scenarios, all approaches fail when only low level statistics are available or under data encryption restrictions. In this work, we propose the use of multiscaling traffic characteristics to differentiate web applications and the use of a Markovian model to characterize the dynamics of user actions over time. By applying the proposed methodology to Wi-Fi layer 2 traffic generated by users accessing different common web services/contents through HTTP (namely social networking, web news and web-mail applications), it was possible to achieve a good prediction of the different users behaviors. The classification results obtained show that the developed multiscaling traffic Markovian model has the potential to efficiently identify, model and predict Internet users behaviors based only on layer 2 traffic statistics.

P. Salvador (✉) • A. Nogueira
Instituto de Telecomunicações/DETI, University of Aveiro, Aveiro, Portugal
e-mail: salvador@ua.pt; nogueira@ua.pt

E. Rocha
Instituto de Telecomunicacções, Aveiro, Portugal

Leipzig University of Applied Sciences, Leipzig, Germany
e-mail: rocha.eduardo@external.hft-leipzig.de

J.-P. Bourguignon et al. (eds.), *Mathematics of Energy and Climate Change*,
CIM Series in Mathematical Sciences 2, DOI 10.1007/978-3-319-16121-1_15

1 Introduction

Nowadays, Internet can be seen as an ever-changing platform where new different types of services and applications are constantly emerging. Consequently, novel and more complex communications paradigms are continuously appearing, creating network traffic that results from multiple simultaneous interactions. This growing traffic complexity, together with the emergence of highly stealth security attacks, create huge challenges for network operators to improve resources utilization, network performance, service personalization and security. Moreover, network operators are limited by legal and technical constrains when they need to analyze confidential/protected traffic data. These constrains, and the need to optimize the users' quality of experience, lead to an increasing need for new ways of unveiling the hidden behaviors of users, applications, services and networks, based only on low level traffic statistics.

In this way, a new network analysis paradigm that captures, analyzes, characterizes, models and (if possible) predicts the multiscale traffic dynamics, must be applied. The concept behind this approach is the fact that Internet traffic is generated and shaped by several events and mechanisms occurring in different time scales. High timescale events are associated with human behaviors and actions. Service and high level network mechanisms events, such as the establishment of traffic sessions and the corresponding control mechanisms (traffic shaping), originate mid-range timescale components. Protocol and low level network mechanisms, such as packets arrivals and queuing, are mapped to very low timescale events. All these mechanisms and events are correlated since the traffic of any Internet application is generated by user requests and controlled by service daemons and traffic control mechanisms, influencing how packet arrivals will occur. For example, when a user performs a request using an Internet application, such as clicking on a link in a web site or requesting an on-line video, several processes are created by the operating system. Each one of these processes creates a set of Internet sessions, each generating a traffic flow and a sequence of packets.

This chapter presents some results of applying this multiscale analysis to Internet traffic generated by different users accessing distinct services. These results include a multiscale analysis based on wavelet transforms applied to several low level traffic metrics, such as the number of transmitted bytes and packets. The analysis of the energy variation at different scales allows the creation of bi-dimensional behavior descriptors, in terms of the energy variation at a specific timescale. These descriptors will allow the differentiation and identification of users, applications and services behaviors.

The proposed methodology can be applied to scenarios where existing identification approaches are not applicable at all or have limited efficiency, like low level monitoring and service optimization at Wi-Fi [16] or WiMax [12] access points and Universal Mobile Telecommunications System (UMTS) [8, 31] or Long Term Evolution (LTE) [1] base stations.

Besides this identification effort, this chapter also proposes the use of a Markovian model [25, 29] to characterize the various dynamics of the user actions over time. This methodology will be able to identify and predict the different user behaviors, even if this information is somehow hidden when performing a classical statistical analysis of the generated traffic.

The results obtained by applying the proposed classification methodology to layer 2 traffic promiscuously captured in the vicinity of a Wi-Fi network access point (without authenticating) show that it is able to achieve a good identification accuracy. It was possible to identify, model and predict the behavior of users accessing three common web applications: social networking (without chatting and game interactions), news web journals and web-mail.

For validation purposes, the ground-truth of the data was assured by asking a pre-determined set of users to replicate their traditional Internet behavior using a controlled environment (user terminals and network).

The remaining part of this paper is organized as follows: Sect. 2 presents some of the most relevant related work on statistical classification of web applications and user behavior modeling; Sect. 3 presents some important background on traffic dynamics and multiscaling analysis; Sect. 4 presents the details of the proposed identification methodology and user behavior model; Sect. 5 presents the results of a proof-of-concept of the methodology and, finally, Sect. 6 presents some brief conclusions about the presented identification methodology and user behavior model.

2 Related Work

Identifying different behaviors of Internet users by analyzing the application types they are running is the key issue of many crucial network monitoring and management tasks, such as quality of service improvement, network equipment optimization or detection of security threats. Most existing approaches are based on static information about the applications (such as the name and type of the application, its owner, the execution time, or the host on which the application was executed). However, such approaches are not applicable to scenarios involving low level monitoring, traffic encryption or under stringent confidentiality requirements, since they rely on analyzing specific fields of the packet header.

One of the first and most common forms of traffic classification is port-based classification, which relies on the port numbers employed by the application at the transport layer. However, since many modern applications use dynamic ports negotiation, port-based classification became ineffective [18, 30], with accuracy ranges between 30 and 70 %. Chronologically, the next proposed classification technique was deep packet inspection (DPI) or payload-based classification, which requires the inspection of the packets' payload: this classifier extracts the application payload from the layer 4 data unit and searches for a signature that can identify the flow type. Although DPI is widely used by today's traffic classifier vendors, being very accurate [20, 30] for some scenarios, it is unable to deal with low level or encrypted data.

Since different applications typically generate different traffic patterns, the study of the statistical properties of the traffic flows can be a very efficient identification methodology. The statistical approach to classification is based on collecting statistical data of the network flow, such as the mean packet size, flow duration, number of bytes per time interval, number of packets per time interval, etc., and has been the subject of intensive research in recent years.

Paxson et al. [26] established a relationship between flow application type and flow properties (such as the number of bytes and the flow duration). In [9], the authors proposed a methodology for separating chat traffic from other Internet traffic using statistical properties such as packet sizes, number of bytes, duration and packets inter arrival times. In [21], Mcgregor et al. explored the possibility of forming clusters of flows based on flow properties such as packet size statistics (e.g., minimum and maximum), byte count, idle times, etc., using an expectation maximization (EM) algorithm to find the clusters' distribution density functions. A study focusing on identifying flow application categories rather than specific individual applications was presented in [28]. Although it was limited by a small dataset, the authors have been able to show that the k-nearest neighbor algorithm and other techniques can achieve good results, correctly identifying around 95 % of the flows. In reference [32], the authors were able to obtain an average success rate of 87 % in the separation of individual applications using an EM based clustering algorithm. In [23], Moore et al. studied the basic Navie Bayes algorithm, enhanced by certain refinements, showing that it is able to achieve an accuracy level of 95 %.

In [3], realtime classification was addressed by studying the feasibility of application identification at the beginning of a TCP connection: based on an analysis of packet traces collected on eight different networks, the authors found that it is possible to distinguish the behavior of an application from the observation of the size and the direction of the first few packets of the TCP connection. Three techniques were applied to cluster TCP connections: K-Means, Gaussian Mixture Model and spectral clustering. Crotti et al. [6] presented a realtime classification mechanism based on three simple properties of the captured IP packets: their size, inter-arrival time and arrival order. Based on new structures called protocol fingerprints, which express these quantities in a compact way, and on a simple classification algorithm based on normalized thresholds, the proposed technique showed promising results on classifying of a reduced set of protocols. In [11], a traffic classification approach based on Support Vector Machines (SVM) was proposed: using a simple optimization algorithm, a statistical traffic classifier was able to perform correctly with only a few hundred samples for training. Note that these algorithms were tested only against basic application protocols. Encrypted applications communications add additional constraints to the detection problem by making the traffic packet headers and data inaccessible to network based monitoring systems. Therefore, the detection methods that rely on packets headers/data information are completely inappropriate in encrypted communications scenarios [18, 24].

Bar-Yanai et al. [2] introduces a hybrid statistical algorithm that integrates the k-nearest neighbors and k-means machine learning algorithms. The proposed algorithm is fast, accurate and is insensitive to encrypted traffic, overcoming several weaknesses of the DPI approach (like asymmetric routing and packet ordering). The strength of the algorithm was demonstrated on encrypted BitTorrent, which is known to use packet encryption, port alternation and packet padding (on initial flow packets) to avoid detection.

The BLINC [17] approach is based on observing and identifying patterns of host behavior at the transport layer, analyzing the social, functional and application level patterns. The fact that this approach relies on layer 3 and layer 4 traffic statistics makes it impossible to be used by an operator in certain entry points of the network where only low level data is available.

The work published in [10] demonstrated that cluster analysis can be effectively used to identify similar groups of traffic using only transport layer statistics. The K-Means and DBSCAN (Density Based Spatial Clustering of Applications with Noise) unsupervised clustering techniques and the AutoClass algorithm, which is a probabilistic model-based clustering technique that allows the automatic selection of the number of clusters, were used to achieve an accurate traffic identification. The accuracy of the clustering techniques was compared using the HTTP, P2P, POP3 and SMTP protocols. The connections that DBSCAN labeled as noise reduced the overall accuracy of this algorithm, since they are considered as misclassification mistakes. However, DBSCAN presented the highest accuracy when classifying three of the studied protocols, while K-Means was the fastest approach.

Instead of classifying traffic based on statistics of individual flows, the authors in [14] focused on building behavioral profiles describing the dominant patterns of a target application. A two-level matching mechanism is then used to classify captured traffic, where the first determines if a host participates in the application by comparing its behavior with the profiles. Subsequently, each flow of the host is compared to the profiles in order to identify the ones that were generated by the studied application. The selected target application was P2P and several rules were obtained for TCP and UDP connections, which are merged from different training traces. Then, the authors looked back at the behavior of each host to construct application profiles. The classification results proved that their approach could accurately identify BitTorrent traffic. However, the number of rules required for classification was very high, which raised doubts about the scalability of the approach and its ability to classify traffic on-the-fly.

In a recent work [15], the authors propose a "two-way" application of *k*-means clustering techniques that consists in analyzing a bidirectional flow as two unidirectional flows. The authors argue that, in this way, they are able to increase the classification accuracy by as much as 18 % when compared to other similar approaches. In addition, they state that their approach generates fewer clusters, which implies that fewer calculations have to be performed to classify traffic. Several discriminators were proposed and the authors used their own version of the Sequential Forward Selection (SFS) algorithm to choose the best discriminators. It starts by clustering the training data according to each of the several discriminators separately and then determines the best ones by evaluating how many flows were assigned to the correct cluster. In the following iterations, the previously selected discriminators are combined with all the others individually to cluster the data. The best combination is selected until no improvement is made. The k-means clustering technique was used due to its fast training times and ease of implementation. Their results showed indeed an increase in the accuracy when compared to some other works.

Rocha et al. [27] presented a methodology for the detection of security attacks and the classification of Internet flows that relies on multidimensional Gaussian distributions [22]. In this way, it is possible to account for the correlation between the values that are obtained for the different dimensions, allowing to infer even more accurate probability distributions. The proposed approach starts by performing a multiscale analysis to the sampled IP data-streams, obtaining multiscale estimators for all streams; the estimators are subsequently processed by mapping a dimension to each timescale, so that the multivariate distributions (for each protocol) can be inferred; an algorithm will then find the dimensions where the separation between the several distributions is most noticeable and each of the traffic streams is then classified according to the probability of belonging to each one of the inferred distributions.

3 Traffic Dynamics and Multiscaling Analysis

The classification approaches that will be proposed are based on the decomposition and analysis of the network traffic at several time scales, i.e. different aggregation levels, in order to identify and model the different characterizing frequency spectrum components. In this manner, a *Multi-Scale Signature* is obtained for each application class, allowing the construction of accurate traffic and user profiles.

Figure 1 shows three different frequency spectrum regions, together with their corresponding events (*Power* is related to the energy of the different types of events). Human events, which are associated in the Internet world to human/user behaviors and actions, can be mapped to low frequency components. Network events, such as the establishment of traffic sessions and the corresponding control mechanisms (traffic shaping), originate mid-range frequency components, while protocol and Internet events, such as packets arrivals, create components in the high-frequency spectrum region. All these mechanisms and events are correlated since traffic belonging to any Internet application is generated by user requests (which are low-frequency events) and controlled by Internet sessions and traffic control mechanisms (which are mid-range frequency events), creating events such as Internet packet arrivals that originate high frequency components. For instance, when a user performs a request using an Internet application, such as clicking on a link in a web site or requesting an on-line video, several processes are created by the operating system. Each one of these processes creates a set of Internet sessions,

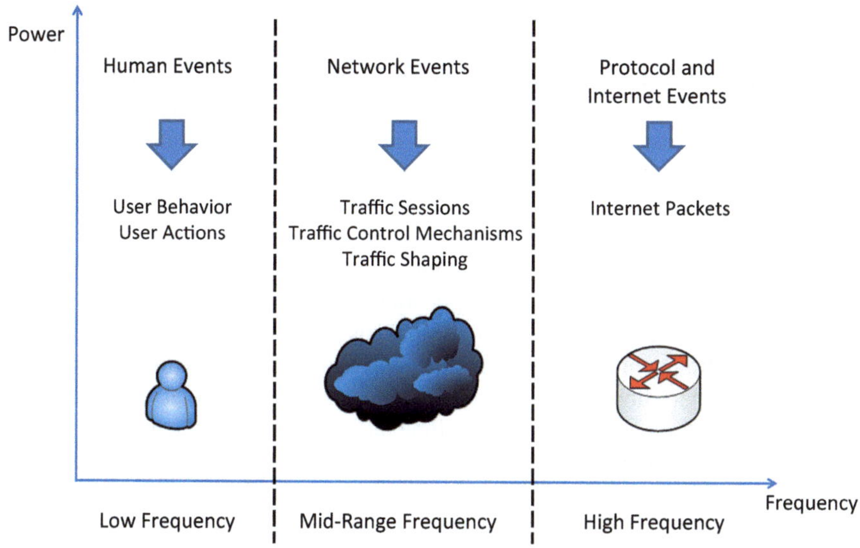

Fig. 1 Frequency regions mapping into network and users mechanisms

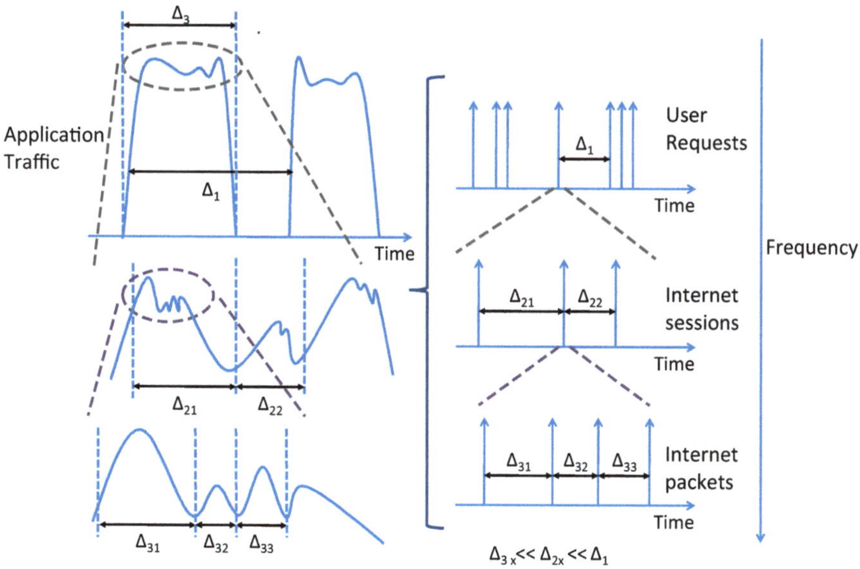

Fig. 2 Multi-scale traffic dynamics

each generating a traffic flow. At the network layer, each one of these connections will transmit and receive the requested data in several packets.

The analysis of each mechanism can then be performed by using the appropriate aggregation scale or frequency, as shown in Fig. 2, which illustrates how the mechanisms of the different scales are connected and how they shape Internet traffic. The graph on the left side represents the traffic generated by an Internet application, from which we are able to infer all the mechanisms present at the several scales of analysis and their corresponding traffic patterns. The time interval between user requests is represented by Δ_1, while Δ_{2x} represents the time intervals between the starting instants of the Internet sessions and Δ_{3x} the time intervals between the different Internet packets. The analysis of such events, which are characteristic of each application class, and of the corresponding frequency components allow a simple but effective traffic assignment. This analysis allows us to associate captured traffic to the corresponding Internet application class.

The inability of conventional Fourier analysis to preserve the time dependence and describe the evolutionary spectral characteristics of non-stationary processes requires tools that allow time and frequency localization. Wavelet transforms can provide information concerning both time and frequency, which allows local, transient or intermittent components to be elucidated [5]. Such components are often obscured due to the averaging inherent within spectral only methods, like Fast Fourier Transform (FFT) [4], for example.

Wavelets are mathematical functions that are used to divide a given signal into its different frequency components. They consist of a short duration wave that has limited energy. Wavelets enable the analysis of each one of the signal components in an appropriate scale. Starting with a mother wavelet $\psi(t)$, a family $\psi_{\tau,s}(t)$ of "wavelet daughters" can be obtained by simply scaling and translating $\psi(t)$:

$$\psi_{\tau,s}(t) = \frac{1}{\sqrt{|s|}}\psi(\frac{t-\tau}{s}) \tag{1}$$

where s is a scaling or dilation factor that controls the width of the wavelet (the factor $\frac{1}{\sqrt{|s|}}$ being introduced to guarantee preservation of the energy, $\|\psi_{\tau,s}\| = |\psi|$) and τ is a translation parameter controlling the location of the wavelet. Scaling a wavelet simply means stretching it (if $|s| > 1$) or compressing it (if $|s| < 1$), while translating it simply means shifting its position in time.

Given a signal $x(t) \in L^2(\Re)$ (the set of square integrable functions), its Continuous Wavelet Transform (CWT) with respect to the wavelet ψ is a function of time (τ) and scale (s), $W_{x;\psi}(\tau, s)$, obtained by projecting $x(t)$ onto the wavelet family $\{\psi_{\tau,s}\}$:

$$W_{x;\psi}(\tau, s) = \int_{+\infty}^{-\infty} x(t)\frac{1}{\sqrt{|s|}}\psi(\frac{t-\tau}{s})dt \tag{2}$$

By analogy with the terminology used in the Fourier case, the energy components of the signal are given by the square of the CWT components of the signal

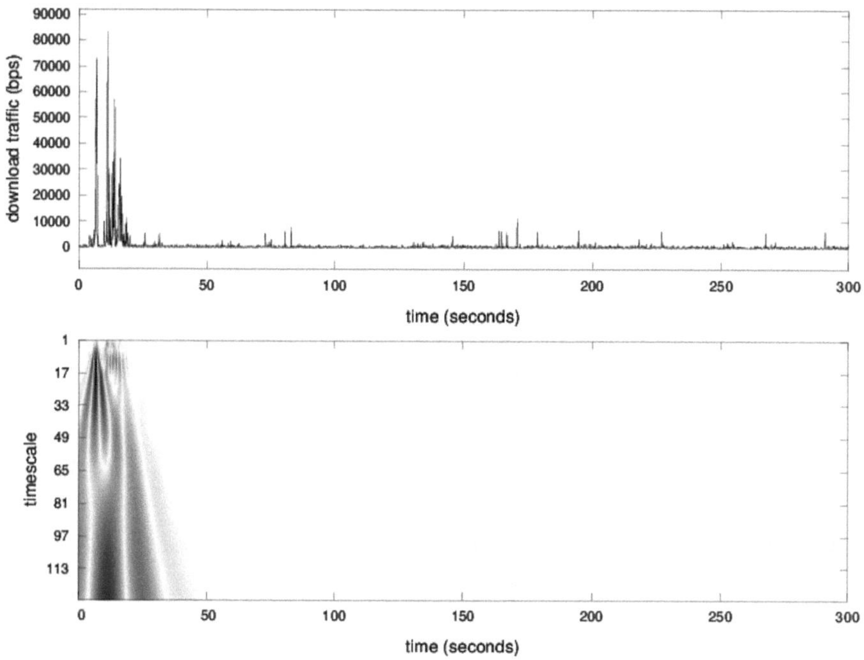

Fig. 3 Each application type will have a distinct Multi-Scale Signature: example of the Web-Mail application

and the (local) Wavelet Power Spectrum (sometimes called Scalogram or Wavelet Periodogram) is defined as the normalized energy over time and scales:

$$E_x(\tau, s) = 100\frac{\left|W_{x;\psi}(\tau, s)\right|^2}{\sum_{\tau'}\sum_{s'}\left|W_{x;\psi}(\tau', s')\right|^2} \tag{3}$$

Figures 3, 4 and 5 show examples of scalograms. Scalograms reveal much information about the nature of non-stationary processes that was previously hidden, so they are applied to a lot of different scientific areas: diagnosis of special events in structural behavior during earthquake excitation, ground motion analysis, transient building response to wind storms, analysis of bridge response due to vortex shedding, among others [13].

Using this mathematical tool, the previously mentioned *Multi-Scale Signature* of each application type can be easily obtained, facilitating the construction of accurate traffic and user profiles. Figures 3, 4 and 5 show examples of these signatures for three particular services: Web-Mail, Facebook and Web-News, respectively.

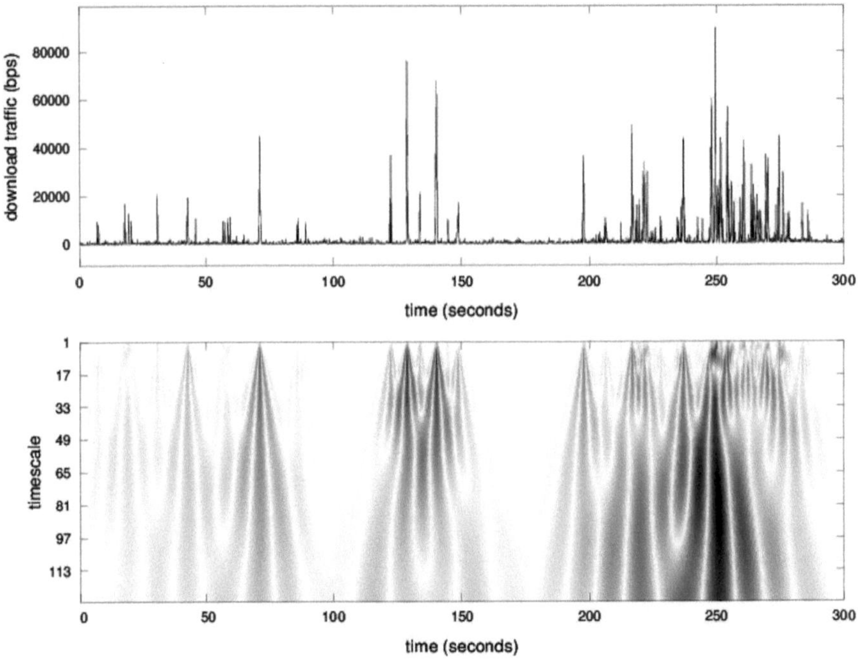

Fig. 4 Each application type will have a distinct Multi-Scale Signature: example of the Facebook application

4 Multiscaling Behavior Modeling

4.1 Multiscale Traffic Data

Let us assume that process $x(t)$ represents a counting statistic of a layer 2 traffic trace to and from a specific user terminal (e.g., number of frames on the upload direction, number of bytes in the download direction, etc.). The user is identified by a layer 2 address depending on the underlying communications technology. It is possible to apply a multiscaling analysis to process $x(t)$ by calculating the scalogram using Eq. (3). We characterize the multiscale user behavior by the estimator of the standard deviation of that user's traffic energy within a time window for a set of timescales. Therefore, a traffic process energy standard deviation at time interval k and time scale s using a sliding time window of width W can be defined as:

$$\hat{D}_x(k,s) = \sqrt{\frac{1}{W-1} \sum_{\tau \in [k-W,k]} \left(E_x(\tau,s) - \overline{E_x(k,s)}\right)^2} \tag{4}$$

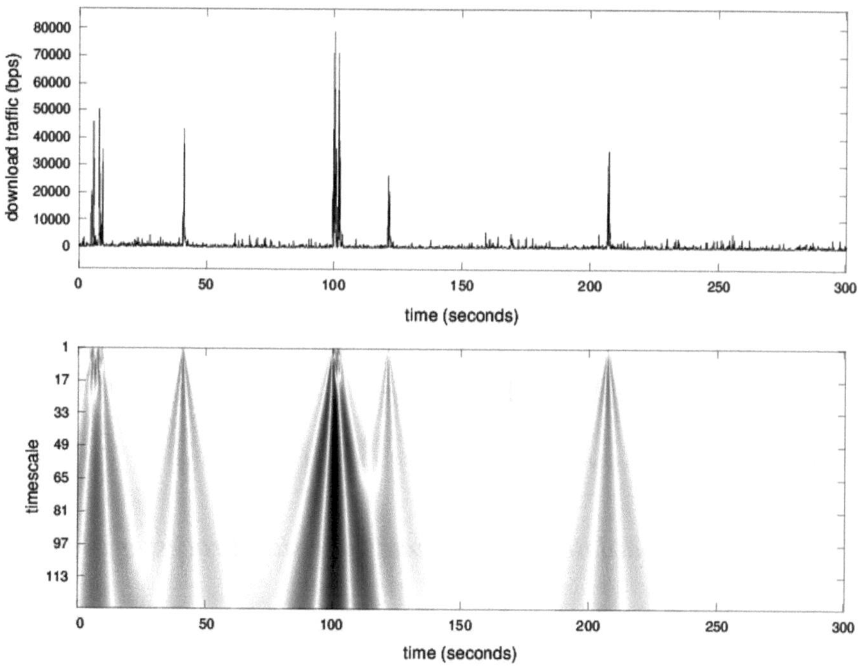

Fig. 5 Each application type will have a distinct Multi-Scale Signature: example of the Web-News application

with $k = \{W, W+1, W+2, \ldots\}$ and

$$\overline{E_x(k,s)} = \frac{1}{W} \sum_{\tau' \in [k-W,k]} E_x(\tau', s) \tag{5}$$

Choosing J timescales ($\{s_1, s_2, \ldots, s_J\}$) of interest, it is possible to define a vector $B_{x,k}$ that describes the inferred localized multiscaling characteristics (at time interval k) of the traffic process x:

$$B_{x,k} = \{\hat{D}_x(k, s_j), j = 1, \ldots, J\} \tag{6}$$

Since each application type is characterized by a distinct multi-scale signature, the most important time-scales of activity can be identified by inferring the mean of the activity energy for each traffic process, while the constancy of the pseudo-periodicity can be quantified by inferring the standard deviation of the activity energy. This is illustrated in Fig. 6.

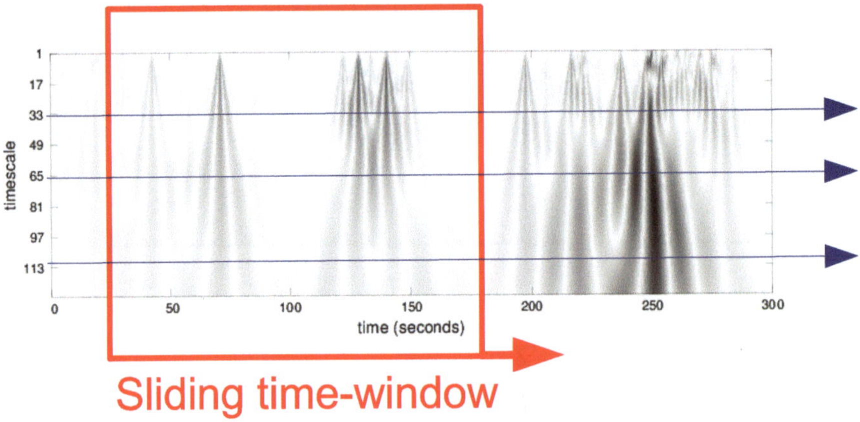

Fig. 6 Mean and standard deviation of the energy over a set of chosen timescales

4.2 *Markov Modulated Multivariate Gaussian Processes Model*

The discrete time Markov Modulated multivariate Gaussian Process (dMMGP) model that will be proposed characterizes position and mobility of a subject based on the following assumptions: (a) the multiscaling behavioral metrics for the use of a specific web application can be described by a multivariate Gaussian distribution, (b) the time scales of importance can be pre-determined, (c) a ground truth for the multiscaling characteristics of web applications usage can be pre-established and (d) the transition between applications can be described by an underlying (homogeneous) Markov chain where each state maps the multiscaling behavior characteristics of a specific web application usage, as illustrated in Fig. 7.

The dMMGP can then be described as a J-dimensional random process (B) with a multivariate Gaussian distribution that characterizes the behavior of a user in an universe of A possible applications in a J-dimensional environment (for J time scales of importance), whose parameters are a function of the state (S) of the modulator Markov chain (B, S) with A states. The dMMGP model states will map the applications multiscale characteristics and the dMMGP model transitions will define the user behavior/dynamics on the usage of the different applications. The former will be inferred based on pre-established ground truth (set of known flows) for the web applications multiscaling characteristics and the later will be inferred based on the dynamics of the mapping of a set of flows of specific users to the application multiscale characteristics (i.e. model states).

More precisely, the (homogeneous) Markov chain

$$(B, S) = \{(B_k, S_k),\ k = 0, 1, \ldots\}$$

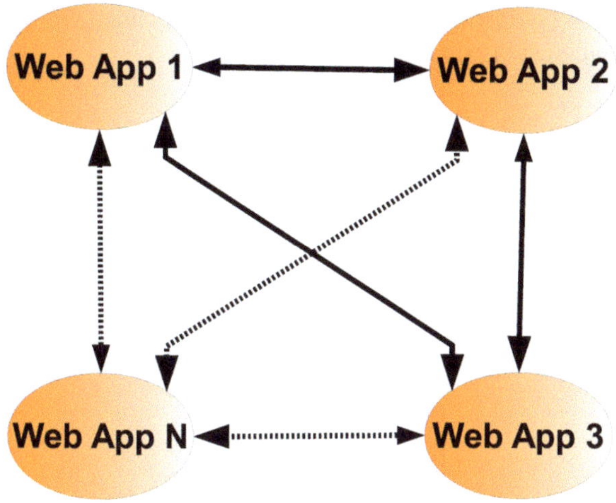

Fig. 7 Markov chain that describes the multiscaling behavior characteristics of the different web application usage profiles

with state space $\mathbb{R}^J \times U$, with $U = \{1, 2, \ldots, A+1\}$, is a dMMGP if and only if for $k = 0, 1, \ldots,$

$$P(B_{k+1} = \mathbf{b}, S_{k+1} = n | S_k = m) = p_{mn} \Gamma_n(\mathbf{b}) \tag{7}$$

where $\mathbf{b} \in \mathbb{R}^J$ is a generic multiscale component in a J-dimensional environment, p_{mn} represents the probability of a transition from state m to state n of the underlying Markov chain in time interval $[k, k+1]$, and

$$\Gamma_a(\mathbf{b}) = (2\pi)^{-\frac{J}{2}} \mathbf{\Sigma}_a^{-\frac{1}{2}} e^{-\frac{1}{2}(\mathbf{b}-\mathbf{m}_a)^T \Sigma_a^{-1}(\mathbf{b}-\mathbf{m}_a)} \tag{8}$$

is the multivariate Gaussian distribution of the multiscaling characteristics of application a flows, it is centered in m_a and has covariance matrix Σ_a.

Whenever (7) holds, we say that (B, S) is a dMMGP with a set of modulating states with size A and parameter matrices **P**, **M** and **S**. Matrix **P** is the transition probability matrix of the modulating Markov chain S,

$$\mathbf{P} = \begin{bmatrix} p_{11} & p_{12} & \cdots & p_{1A} \\ p_{21} & p_{22} & \cdots & p_{2A} \\ \cdots & \cdots & \cdots & \cdots \\ p_{A1} & p_{A2} & \cdots & p_{AA} \end{bmatrix} \tag{9}$$

while matrix **M** defines the mean values of each multiscaling Gaussian distribution:

$$\mathbf{M} = \left[\mathbf{m}_1 \; \mathbf{m}_2 \; \ldots \; \mathbf{m}_A\right] \tag{10}$$

where $\mathbf{m}_a$ is a $J \times 1$ vector. Matrix $\mathbf{S}$ contains the covariance (sub-)matrices of each multiscaling Gaussian distribution:

$$\mathbf{S} = \left[\boldsymbol{\Sigma}_1\ \boldsymbol{\Sigma}_2 \ldots \boldsymbol{\Sigma}_A\right] \tag{11}$$

where $\boldsymbol{\Sigma}_a$ is a $J \times J$ matrix. Moreover, we denote by $\boldsymbol{\Pi} = [\pi_1, \pi_2, \ldots, \pi_A]$ the stationary distribution of the underlying Markov chain.

Matrix $\mathbf{P}$ will be unique for each user, and will characterize his/her behavior on the usage of the applications characterized by matrices $\mathbf{M}$ and $\mathbf{S}$. The overall multiscaling behavior of a user can be statistically described by a stationary probability density defined by a weighted sum of A multivariate Gaussian distributions:

$$f(\mathbf{b}) = \sum_{a=1}^{A} \pi_a \Gamma_a(\mathbf{b}), \mathbf{b} \in \mathbb{R}^J \tag{12}$$

where $\mathbf{b}$ is a multiscale component that belongs to the J-dimensional domain of chosen timescales.

4.3 Model Inference Procedure

Assuming that we have a ground-truth for a set of A web applications, analyzed over F flows, over K time windows in J timescales of interest, we can define the multiscale profile of an application $a(a = 1, \ldots, A)$ as $G_{a,f,k}$, inferred using Eq. (6) considering that process $x(t)$ is the f-th flow of application a, with $a = 1, \ldots, A$, $f = 1, \ldots, F$ and $k = 1, \ldots, K$, i.e.:

$$G_{a,f,k} = B_{x,k}, x \leftrightarrow \text{flow}\, f \text{of application } a \tag{13}$$

The $\mathbf{M}$ and $\mathbf{S}$ matrices of the dMMPGP model can then be inferred as

$$\mathbf{m}_a = \frac{1}{KF} \sum_{f=1}^{F} \sum_{k=1}^{K} G_{a,f,k} \tag{14}$$

$$\boldsymbol{\Sigma}_a = \frac{1}{KF-1} \sum_{f=1}^{F} \sum_{k=1}^{K} \left(\left(G_{a,f,k} - \mathbf{m}_a\right) \left(G_{a,f,k} - \mathbf{m}_a\right)^T \right) \tag{15}$$

The final step of the inference procedure is to infer matrix $\mathbf{P}$, i.e. the transition probabilities between the states defined in the first step. This task is achieved by probabilistically mapping each multiscaling behavior of each unknown flow trace

$x(t)$ $B_{x,k}, k = 1, \ldots, K$ to one state/application and then averaging the probabilistic transitions between states, according to a probability vector:

$$\mathbf{q}_k = \{\Gamma_1(B_{x,k}), \ldots, \Gamma_A(B_{x,k})\}, k = 0, 1, \ldots, K \tag{16}$$

4.4 Behavior Prediction

Defining $\mathbf{c}_k = \{c_{k,a} : a = 0, 1, \ldots, A\}, k = 0, 1, \ldots, K$, where $\mathbf{c}_k$ is the probability vector defining that within time-window $[k - W, k]$ the user is using application a, and based on Eq. (12) we can define the multivariate distribution of the predicted multiscaling behavior of the user in a future time-window (z observations in the future) as:

$$\sum_{a=1}^{A} \mathbf{c}_{k+z}\Gamma_a \tag{17}$$

with

$$\mathbf{c}_{k+z} = \mathbf{c}_k \mathbf{P}^z \tag{18}$$

where $\mathbf{c}_{k+z}$ represents the probabilistic vector that quantifies the probability of a web application to be in use k time windows in the future.

So, after inferring a behavior model for each user, the multiscale characteristics of the current user's data flow are probabilistically mapped to a specific state and, based on the inferred underlying Markov chain transitions, it is possible to predict which application the user will be "using" in the future. This is illustrated in Fig. 8.

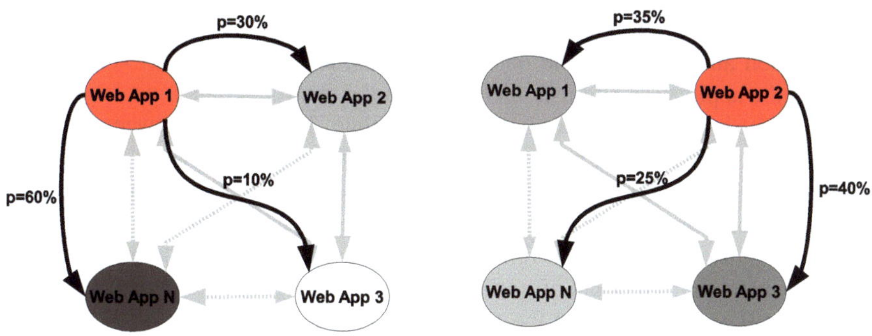

Fig. 8 Predicting which application a user will be using in the future

5 Proof of Concept

5.1 Data-Set

The test data-set was obtained by capturing, in promiscuous mode, the layer 2 traffic having as source or destination a specific Wi-Fi network access point. The traffic capture was performed without authenticating to the network and consisted only of 802.11 frames. In a controlled environment, where all terminals were using a bare installation of Linux with a daemon that recorded all browse requests, a set of invited users were asked to access and use their usual web applications, maintaining their typical behavior. This approach allowed us to create the ground-truth of a mapping between layer 2 data traces and their originating users and web applications. Within the context of this proof of concept, we only used the data traces that were created by users accessing three general web applications: social networking, namely Facebook (without chatting and game interactions), news web journals and web-mail access. The total number of data sets was divided in two: the first half was used to infer the underlying dMMGP model of the behavior of each application and user, while the second half of the data sets was used to validate the inferred models by comparing the predicted multiscale behavior (and associated web application usage sequence) of each user. The raw statistical process used was the amount of bytes transmitted from the Wi-Fi access point to each user, sampled every 0.1 s. Sampling the raw statistics in 0.1 s allows our method to measure and incorporate some of the most characteristic multiscale dynamics of an application: (a) the lower timescales that are strictly related with the way that specific application handles the multiple data sessions, (b) the medium timescales that are related with the application algorithmic dynamics and (c) the higher timescales that reflect mainly the user interactions dynamics [7]. For the purpose of the model inference, we use time windows with a width of 120 s ($W = 1200$) and considered time windows in 20 s interval. The choice of these values is a tradeoff between the amount of (past) data necessary to fully characterize the traffic dynamics and the amount of data that can be process and analyzed in pseudo-real time. The heavier computational tasks, which are the construction and update of the behavior models, are made off-line. However, to perform the application and user identification the measured data must be matched with previously inferred models in pseudo real-time. The interval between windows of classification was chosen in order to minimize the delay between the moment of an user application change and its effective detection by our methodology. With an appropriate choice of parameters, namely window size and interval of processing, this methodology is fully scalable since the computation power required is proportional to the amount of traffic (number of users) under analysis.

Figures 9 and 10 depicted the 80 and 90 % quantile frontiers of the inferred multivariate Gaussian distributions of the multidimensional characteristics of each application (using just 3 timescales) for all users. These distributions reveal that the multiscale characteristics of the three web applications are distinct and have a small overlap in the universe of the three dimensions/scales considered.

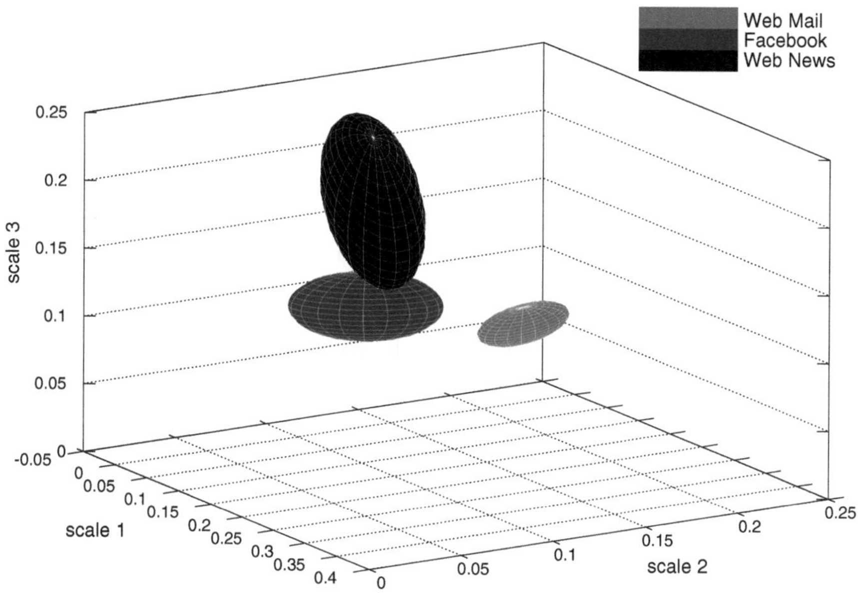

Fig. 9 80 % quantile frontiers of the inferred multivariate Gaussian distributions of the multidimensional characteristics of each application

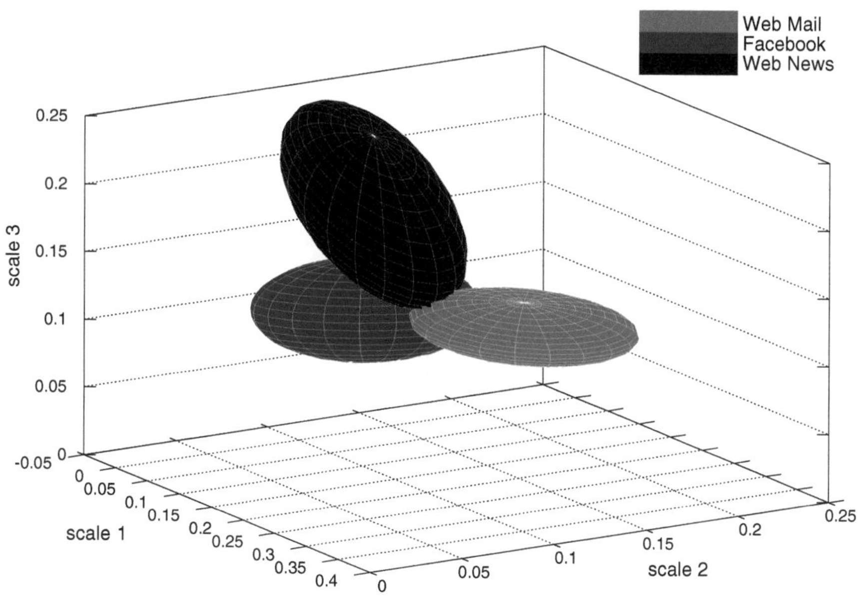

Fig. 10 90 % quantile frontiers of the inferred multivariate Gaussian distributions of the multidimensional characteristics of each application

Table 1 Identification of the current web application results

	Web-mail (%)	Facebook (%)	Web-news (%)
Web-mail	69.95	7.84	22.20
Facebook	4.12	83.13	12.74
Web-news	7.08	24.35	68.55

After inferring the underlying dMMGP model, we use the test data traces to test the precision of the model in identifying the current web application of an user every 20 s. In this test, we were able to obtain a precision of 72.4 % of correctly classified windows and the identification results presented in Table 1. The results show a very good agreement between the identified web application and the real application, considering the reduced amount of information (in terms of raw data and time span of the observation) used for the identification.

Using the test data traces to test the precision of the model in identifying the web applications that are in use 60 s in the future we obtained a precision of 55.3 % of correctly classified windows. The results show that the identification/predicting results are still significantly above the pure random guess.

From these results we can conclude that this methodology was able to obtain very good classification and prediction results considering the reduced amount of information (only network layer 2 sampled statistics) and that the web applications under consideration may, in some particular cases, be very similar. Most of the errors can be explained by the fact that some Web-news pages are very similar to social networking applications pages and even incorporate social network features within its own Web-pages. Also, when the Web-news web pages have less content the user dynamics may get similar to Web-mail or Facebook interactions (i.e., small data chunks exchanged at small intervals).

6 Conclusions and Future Work

This paper presented an approach that uses multiscaling traffic characteristics to differentiate between different web applications and a Markovian model that is able to characterize the dynamics of user actions over time. By applying this methodology to Wi-Fi layer 2 traffic generated by users accessing different common web services/contents through HTTP (namely, social networking, web news and web-mail applications), it was possible to achieve a good matching and prediction of the users behaviors. The proposed methodology may be applied to preallocate resources in network access points based on past user behavior and pseudo real-time predictions of short term requirements.

As future work, we plan to test our methodology incorporating more applications with completely different behavior (such as video streaming, P2P file transferring, online games, etc.). This will require the improvement of the inner algorithms of the methodology to accommodate multiple and dynamic timescale ranges.

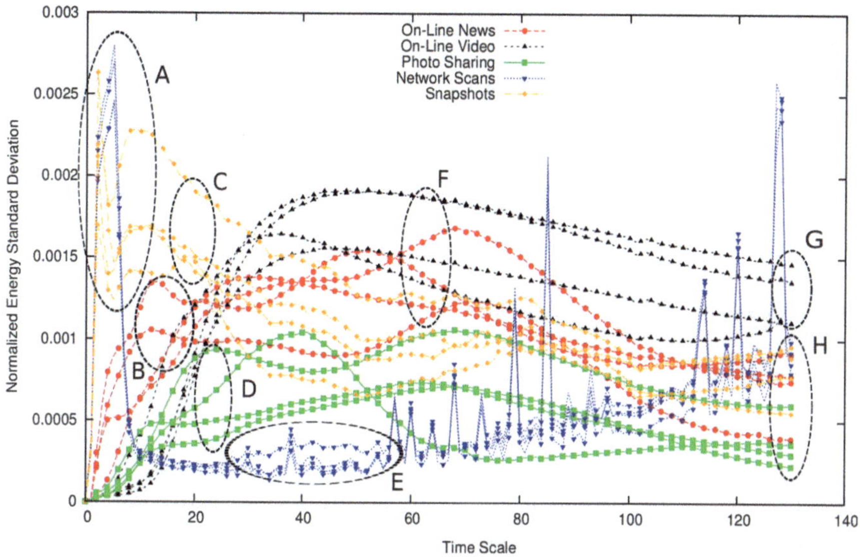

Fig. 11 Internet applications and corresponding frequency mapping regions

We have already started a multi-scale decomposition of known traffic from several classes of web applications (online news, on-line video services and photo-sharing) and, in order to assess the ability of identifying compromised hosts, traffic from two widely deployed illicit applications: network scans and snapshots. Network scans were simulated using a known scanning tool [19] to replicate the behavior of a compromised host scanning for available services, and corresponding vulnerabilities, in other connected hosts. The second illicit application consisted of taking snapshots of the users' desktops and uploading them every time the user performed a click. The purpose of this security attack is stealling confidential information.

Several differentiating regions, shown in Fig. 11 and mapped into the corresponding application in Table 2, emerged in the frequency spectrum. Region A encompasses very low frequency events which can be associated to commands sent to compromised hosts in order to perform a scan or an upload of stolen confidential informations. This region can, thus, be associated to stealth attacks. Network scans also map to another differentiating region (region E), since these scans do not generate substantial mid range frequency components due to the low variance between scanning probes and to the low number and variation of the created traffic flows. Snapshots can be identified by analyzing other frequency spectrum components in which differentiating regions, such as region C, emerge due to the high energy associated to the creation of file transfer sessions, which are automatic mechanisms associated to user clicks on web-pages. Region B encompasses other types of low-frequency events corresponding to user requests typically associated to online news applications. Region D includes low-frequency events occurring

Table 2 Internet applications with their corresponding frequency mapping regions and classification results

Internet applications	Regions	Classification accuracy (%)
On-line news	B and F and H	90.00 [89.00–91.00]
On-line video	F and G	88.90 [88.10–89.70]
Photo-sharing	D and H	85.75 [84.70–86.80]
Network scans	A and E	99.00 [98.00–100.00]
Information theft	A and C and H	90.00 [89.20–90.80]

periodically with low variation and, consequently, can be associated to photo-sharing applications where users perform periodic requests for downloading images shared by other users. For analyzing the creation of traffic sessions and the presence of traffic control mechanisms, region F was created and accounts for applications presenting significant mid-range frequency components. This region can be associated to on-line news and video applications, which are frequently mixed since many news pages present embedded videos that create a significant amount of events in this frequency range. Finally, regions G and H differentiate applications presenting significant and low high-frequency events, respectively. Region G region includes applications generating a high amount of network traffic, which can be associated to on-line video services. On the other hand, region H includes applications that do not present significant high-frequency components, indicating that they generate a small amount of network traffic. This set includes on-line news, photo-sharing, network scans and snapshots.

This classification procedure allow us to achieve an accurate classification of the traffic generated by the different applications and an accurate identification of some of the most used Internet attacks, as shown in Table 2. Nevertheless, some classification overlaps can occur due to similarities between the different classes: On-Line News and Video classes are intrinsically connected since many journal web-pages contain embedded videos; on the other hand, some Photo-Sharing traffic was assigned to the News class due to irregular user interactions.

Finally, our short term plans include the developing of a prototype and test in a 3G/4G network base station for optimal dynamic allocation of resources.

Acknowledgements This work was supported by the Fundação para a Ciência e a Tecnologia (FCT) through PEst-OE/EEI/LA0008/2013 and PTDC/EEI-TEL/2016/2012.

References

1. Astely, D., Dahlman, E., Furuskar, A., Jading, Y., Lindstrom, M., Parkvall, S.: Lte: the evolution of mobile broadband. IEEE Commun. Mag. **47**(4), 44–51 (2009). doi:10.1109/MCOM.2009.4907406
2. Bar-Yanai, R., Langberg, M., Peleg, D., Roditty, L.: Realtime classification for encrypted traffic. In: SEA'10, pp. 373–385 (2010)
3. Bernaille, L., Teixeira, R., Salamatian, K.: Early application identification. In: Proceedings of the 2006 ACM CoNEXT Conference (CoNEXT '06), pp. 6:1–6:12. ACM, New York (2006)
4. Brigham, E.: Fast Fourier Transform and Its Applications. Prentice Hall, Englewood Cliffs (1988)
5. Byrnes, J., Hargreaves, K.A., Berry, K.: Wavelets and Their Applications. Springer, Heidelberg (1994)
6. Crotti, M., Dusi, M., Gringoli, F., Salgarelli, L.: Traffic classification through simple statistical fingerprinting. SIGCOMM Comput. Commun. Rev. **37**, 5–16 (2007)
7. Crovella, M., Krishnamurthy, B.: Internet Measurement: Infrastructure, Traffic and Applications. Wiley, New York (2006)
8. Dahlman, E., Gudmundson, B., Nilsson, M., Skold, A.: UMTS/IMT-2000 based on wideband CDMA. IEEE Commun. Mag. **36**(9), 70–80 (1998). doi:10.1109/35.714620
9. Dewes, C., Wichmann, A., Feldmann, A.: An analysis of internet chat systems. In: Proceedings of the 3rd ACM SIGCOMM Conference on Internet Measurement (IMC '03), pp. 51–64. ACM, New York (2003)
10. Erman, J., Arlitt, M., Mahanti, A.: Traffic classification using clustering algorithms. In: 2006 SIGCOMM Workshop on Mining Network Data (MineNet '06), pp. 281–286. ACM, New York (2006). doi:http://doi.acm.org/10.1145/1162678.1162679
11. Este, A., Gringoli, F., Salgarelli, L.: Support vector machines for tcp traffic classification. Comput. Netw. **53**, 2476–2490 (2009). doi:10.1016/j.comnet.2009.05.003
12. Ghosh, A., Wolter, D., Andrews, J., Chen, R.: Broadband wireless access with WiMax/802.16: current performance benchmarks and future potential. IEEE Commun. Mag. **43**(2), 129–136 (2005). doi:10.1109/MCOM.2005.1391513
13. Gurley, K., Kareem, A.: Applications of wavelet transforms in earthquake, wind, and ocean engineering. Eng. Struct. **21**, 149–167 (1999)
14. Hu, Y., Chiu, D.M., Lui, J.C.: Profiling and identification of p2p traffic. Comput. Netw. **53**(6), 849–863 (2009). doi:10.1016/j.comnet.2008.11.005 [Traffic Classification and Its Applications to Modern Networks]
15. Hurley, J., Garcia-Palacios, E., Sezer, S.: Classifying network protocols: a 'two-way' flow approach. IET Commun. **5**(1), 79–89 (2011). doi:10.1049/iet-com.2009.0776
16. IEEE Standard for Information Technology: Telecommunications and Information Exchange Between Systems - Local and Metropolitan Area Networks - Specific Requirements - Part 11: Wireless LAN Medium Access Control (MAC) and Physical Layer (PHY) Specifications. IEEE Std 802.11-2007 (Revision of IEEE Std 802.11-1999), pp. 1–1076 (2007). doi:10.1109/IEEESTD.2007.373646
17. Karagiannis, T., Papagiannaki, K., Faloutsos, M.: BLINC: multilevel traffic classification in the dark. In: Proceedings of the 2005 Conference on Applications, Technologies, Architectures, and Protocols for Computer Communications (SIGCOMM '05), pp. 229–240. ACM, New York (2005). doi:10.1145/1080091.1080119
18. Kim, H., Claffy, K., Fomenkov, M., Barman, D., Faloutsos, M., Lee, K.: Internet traffic classification demystified: myths, caveats, and the best practices. In: Proceedings of the 2008 ACM CoNEXT Conference (CoNEXT '08), pp. 11:1–11:12. ACM, New York (2008)
19. Lyon, G.F.: Nmap Network Scanning: The Official Nmap Project Guide to Network Discovery and Security Scanning, Insecure (2009)
20. Madhukar, A., Williamson, C.L.: A longitudinal study of p2p traffic classification. In: Proceedings of the IEEE MASCOTS, pp. 179–188 (2006)

21. McGregor, A., Hall, M., Lorier, P., Brunskill, J.: Flow clustering using machine learning techniques. Lecture Notes in Computer Science **3015**, 205–214 (2004)
22. Miller, K.S.: Multidimensional Gaussian Distributions. Wiley, New York (1964)
23. Moore, A., Zuev, D.: Internet traffic classification using bayesian analysis techniques. In: ACM SIGMETRICS, pp. 50–60 (2005)
24. Nguyen, T.T.T., Armitage, G.J.: A survey of techniques for internet traffic classification using machine learning. IEEE Commun. Surv. Tutorials **10**(4), 56–76 (2008)
25. Pacheco, A., Tang, L.C., Prabhu, N.U.: Markov-modulated processes & semiregenerative phenomena. World Scientific, Hackensack (2009)
26. Paxson, V.: Empirically derived analytic models of wide-area tcp connections. IEEE/ACM Trans. Netw. **2**, 316–336 (1994)
27. Rocha, E., Salvador, P., Nogueira, A.: Detection of illicit network activities based on multivariate gaussian fitting of multi-scale traffic characteristics. In: 2011 IEEE International Conference on Communications (ICC 2011), pp. 1–6 (2011). doi:10.1109/icc.2011.5962651
28. Roughan, M., Sen, S., Spatscheck, O., Duffield, N.G.: Class-of-service mapping for qos: a statistical signature-based approach to ip traffic classification. In: Internet Measurement Conference '04, pp. 135–148 (2004)
29. Salvador, P., Valadas, R., Pacheco, A.: Multiscale fitting procedure using Markov modulated poisson processes. Telecommun. Syst. J. **23**(1–2), 123–148 (2003)
30. Sen, S., Spatscheck, O., Wang, D.: Accurate, scalable in-network identification of p2p traffic using application signatures. In: Proceedings of the 13th International Conference on World Wide Web (WWW '04), pp. 512–521. ACM, New York (2004)
31. van Nielen, M.: UMTS: a third generation mobile system. In: Proceedings Third IEEE International Symposium on Personal, Indoor and Mobile Radio Communications 1992 (PIMRC '92), pp. 17–21 (1992). doi:10.1109/PIMRC.1992.279971
32. Zander, S., Nguyen, T.T.T., Armitage, G.J.: Automated traffic classification and application identification using machine learning. In: LCN '05, pp. 250–257 (2005)

The Role of Clouds, Aerosols and Galactic Cosmic Rays in Climate Change

Filipe Duarte Santos

Abstract A review of the role played by clouds, by natural and anthropogenic aerosols and by their interaction, on climate, is presented. The suggestion that galactic cosmic rays may affect the interaction between clouds/aerosols and climate is here discussed in the context of the CLOUD (Cosmics Leaving Outdoor Droplets) experiment at CERN. The experiment has shown that cosmic rays enhance aerosol nucleation and cloud condensation but the effect is too weak to have an impact on climate during a solar cycle or over the last century. The CLOUD experiment has also revealed a nucleation mechanism involving the formation of clusters containing sulphuric acid and oxidized organic molecules.

1 Clouds

The Earth system is continually seeking to establish equilibrium between energy it receives from the Sun and energy it emits back out to space. Clouds contribute to this radiative balance because they reflect, absorb and radiate energy. They can warm or cool the Earth depending on their altitude, composition, optical depth and size. About 20 % of the incoming shortwave solar radiation is reflected by clouds and about 4 % is absorbed [10]. As regards the outgoing infrared radiation clouds are responsible for about 26 % of the emissions. To understand the role of clouds in climate it is therefore essential to know how they absorb and emit shortwave solar radiation and infrared radiation. Clouds tend to cool and warm the atmosphere by reflecting the incoming solar radiation and by inhibiting the emission of infrared radiation from the Earth's surface to space. The cooling effect depends on the difference between the cloud and surface albedos and on the amount of incident solar radiation. The smaller the surface albedo below a cloud and the larger the incident solar radiation the greater is its cooling effect. Clouds can also exert a warming effect by absorbing part of the infrared radiation emitted by the surface and by re-emitting a smaller amount of infrared radiation because the top of the

F.D. Santos (✉)
CCIAM Group, Center for Ecology, Evolution and Environmental Changes – cE3c, Faculdade de Ciências, University of Lisbon, Lisboa, Portugal
e-mail: fdsantos@fc.ul.pt

J.-P. Bourguignon et al. (eds.), *Mathematics of Energy and Climate Change*,
CIM Series in Mathematical Sciences 2, DOI 10.1007/978-3-319-16121-1_16

cloud is at a lower temperature compared to the surface. This effect is particularly strong in high altitude clouds. In general high clouds tend to have a warming effect, while low clouds tend to have a cooling effect. Clouds also produce precipitation from water vapor releasing heat to the atmosphere in the process. Thus clouds play a very important role since they can modify the radiative energy balance and water exchanges that contribute to determine the climate.

Anthropogenic climate change, which results from emissions of greenhouse gases into the atmosphere associated with some human activities, has a variety of impacts on cloud processes that represent cloud-climate feedbacks. These occur through changes in cloud cover, cloud-top height and cloud optical properties. The identification of the cloud-climate feedbacks and the determination of their sign and magnitude can only be addressed through the use of global climate models (GCM) that simulate the Earth climate system. The problem however is that cloud processes span scales from the submicron scale of cloud condensation nuclei to cloud system scales of up to thousands of kilometres. It is impossible at present to make numerical simulations for this range of spatial scales in the GCMs due to limitations in computing power. The representation of microphysical processes in clouds, such as cloud droplets and ice crystals formation, turbulence, cumulus convection, and aerosol and chemical transport is made through parameterizations in the GCMs. Recently global cloud-resolving models have been run with grid spacing as small as 3.5 km [11]. However, such models can only be used for relatively short simulations of a few months to a year or two on the fastest supercomputers. It is likely that in the future they may also provide long term climate projections. The limitations in resolution of the GCMs are one of the main sources of uncertainties in the simulations of cloud-climate feedbacks. Realistic simulation of cloud processes and the response of clouds to climate change is one of the greatest challenges of climate modelling. The recent Intergovernmental Panel on Climate Change (IPCC) Fifth Assessment Report [3] concludes that the net radiative feedback due to all cloud types in the atmosphere is likely positive, although a small negative feedback is not excluded. The likely range of the cloud feedback parameter is −0.2–2.0 $W\,m^{-2}\,K^{-1}$. Most climate models simulate a decrease in low-cloud amounts, which increases global warming, but the deficiencies in the model representation of low clouds at the global scale and the uncertainties in the mechanisms of cloud formation gives a low confidence to this result. In conclusion the most uncertain radiative feedbacks in GCMs continue to be the cloud feedbacks.

2 Aerosols

Another important source of uncertainty in our simulations of the climate system and the future climate is the role played by atmospheric aerosols. These are small particles that come from natural and anthropogenic sources and have a major impact on climate and also on our health. The most important groups of aerosols are sulphates, nitrates, organic carbon, black carbon, mineral dust and sea salt. Often aerosols clump together to form hybrid particles such as those that result from clumping carbon from soot or smoke with sulphates and nitrates. Atmospheric aerosols have a relatively short lifetime of about 1 day to a few weeks in the troposphere and about 1 year in the stratosphere.

Aerosols affect the climate through aerosol-radiation and aerosol-cloud interactions. As regards the former interaction aerosols scatter and absorb the incoming solar radiation, thereby modifying the Earth's radiative balance. Some aerosols, such as pure sulphate aerosols, scatter most of the solar radiation and are very weakly absorbing. Others, such as soot, are strongly absorbing. Aerosol scattering tends to cool the climate, while aerosol absorption has the opposite effect of warming the climate. Anthropogenic emissions of aerosols and their gaseous precursors have increased substantially since the beginning of the industrial revolution inducing a net cooling effect on climate that compensated part of the warming effect of anthropogenic greenhouse gas emissions. To project the global impact of the aerosol-radiation interaction on the future climate requires the use of data from ground-based networks and from satellite-based sensors and the use of climate models. There are still many uncertainties on aerosol monitoring at the regional and global levels, in particular as regards to soot and the role it plays in the atmosphere.

The aerosol-cloud interaction arises because aerosols serve as cloud condensation nuclei and ice nuclei upon which liquid droplets and ice crystals can form. In general, a large concentration of aerosols tends to produce more low clouds and also liquid clouds which are brighter and have longer lifetime because of the larger number of smaller cloud droplets. Both these effects tend to cool the climate. However, it is very difficult to model the overall impacts of aerosols on cloud amounts and cloud properties because of the complexity of the aerosol-cloud interactions and the need to represent them through parameterizations in the climate models.

Although it is likely that aerosol anthropogenic emissions had an overall cooling effect on climate in the twentieth century there is limited evidence for a rather weak aerosol-climate feedback during the twenty-first century [3]. Over the last two decades, anthropogenic aerosol emissions have decreased in industrialised countries, but increased in developing countries, particularly in China and India. In the medium and long term it is likely that the aerosol anthropogenic emissions will tend to decrease due to their damaging impact on health, thereby increasing the global warming induced by greenhouse gases.

3 Galactic Cosmic Rays, Climate Change and the CLOUD Experiment

There is another mechanism that may affect the interaction between clouds/aerosols and climate. Galactic cosmic rays (GCR) are the primary source of ionization in the atmosphere above 1 km altitude. Since high solar activity reduces the flux of GCR reaching the Earth atmosphere, through deflection of the low energy GCR, it was suggested in 1959 [8] that GCR may act to amplify a presumed impact of variations in solar activity on climate through an enhanced production of charged aerosols that may grow to become cloud condensation nuclei (CCN). If cosmic rays do affect cloudiness, they would provide a link through which solar activity could affect climate [17].

There have been many studies that seek to establish correlations between the flux of cosmic rays reaching the Earth and the properties of aerosols and clouds. These correlations would imply an interaction between GCR and climate, a field that adopted the name of cosmoclimatology [9, 15, 16]. However, correlations between GCR and low-level cloud cover data obtained using satellite remote sensing observations over periods of one decade or less have not proved robust when extending the time period under consideration [1]. Some studies found small but significant positive correlations between GCR and high- and mid-altitude clouds [7, 13] but these variations were very weak and dependent on how Forbush events of rapid decrease in GCR intensity were selected [3].

Other studies have addressed the investigation of the physical mechanisms linking cosmic rays to cloudiness. The most studied is the ion-aerosol clean air mechanism in which the atmospheric ions produced by the GCR have an impact on CCN concentrations and cloud properties through aerosol nucleation and growth. However aerosol nucleation rates resulting from variations in ionization rates induced by CGR flux changes are poorly known [4].

The CLOUD (Cosmics Leaving OUtdoor Droplets) experiment at CERN has recently addressed these questions [6]. It consists of a chamber that simulates the atmosphere. Due to its cleanliness and to controlled amounts of trace gases, the CLOUD chamber allows the measurement of nucleation and of the molecular makeup and growth of newly-formed molecular clusters from single molecules up to stable aerosol particles. Furthermore the chamber has the capability to measure nucleation enhancement by cosmic rays that are simulated using the CERN pion beam. The chamber is exposed to a 3.5 GeV/c secondary charged pion beam from the CERN Proton Synchrotron, spanning GCR intensity range from ground level to the stratosphere. The experiment has shown that GCR-induced ionization enhances water sulphuric acid nucleation in the middle and upper troposphere, but is very unlikely to give a significant contribution to nucleation taking place in the continental boundary layer. It is now clear that cosmic rays enhance aerosol nucleation and cloud condensation nuclei production in the free troposphere, but the effect is too weak to have an impact on climate during a solar cycle or over the last century. It is too weak because the radiative forcing produced by cosmic rays is very small compared with the total global average radiative forcing in the present day atmosphere. The main components of this total radiative forcing result from changes in the atmospheric concentrations of well-mixed greenhouse gases (CO_2, CH_4, N_2O, among others), ozone, stratospheric water vapour, aerosols, and from changes in surface albedo and in the solar irradiance. The first five components have an anthropogenic origin and only the last one is natural. Volcanic radiative forcing produced by major eruptions is also a natural radiative forcing with a very irregular temporal pattern. The globally averaged solar cycle modulation of the Earth's radiative forcing arising from the increase in atmospheric ionization by GCR, from solar maximum to minimum, through charged nucleation of aerosol, the direct aerosol effect and the cloud albedo effect amounts to $0.05\,\mathrm{W\,m^{-2}}$ [5]. This value is considerably smaller than the change in radiative forcing caused by the changes in total solar irradiance during a solar cycle, which is $-0.24\,\mathrm{W\,m^{-2}}$, from solar maximum to minimum. Both these forcings are smaller than the present day total global average anthropogenic radiative forcing, which is estimated at $+2.3\,\mathrm{W\,m^{-2}}$ [3]. The natural radiative forcing due to changes in average solar irradiance is estimated at $0.05\,\mathrm{W\,m^{-2}}$, and therefore much smaller than the anthropogenic radiative forcing. It should also be noticed that the total anthropogenic radiative forcing has been increasing from $0.57\,\mathrm{W\,m^{-2}}$ in 1950, to $1.25\,\mathrm{W\,m^{-2}}$ in 1980 to $2.3\,\mathrm{W\,m^{-2}}$ presently. In conclusion there is at present no evidence for a causal connection between variations in cosmic ray intensity and the observed climate change [3].

The CLOUD chamber allows the study of particle aerosol nucleation in the atmosphere knowing the participating molecules, which is a necessary step to understand the processes of aerosol formation and their effects on clouds and climate. While sulphuric acid is the main driver of nucleation, the nucleation rate is also affected by ammonia, amines and volatile organic vapours. The CLOUD experiment has shown that dimethylamine above three parts per trillion by volume can enhance particle formation rates more than 1000-fold compared with ammonia, which is sufficient to account for the particle formation rates observed in the atmosphere [2, 14]. The nucleation rate measurements made in the CLOUD chamber are well reproduced by a dynamical model that simulates cluster collision and coagulation rates which are computed from kinetic gas theory. Equilibrium constants are computed from quantum chemical calculations of binding energies of molecular clusters, and evaporation. Cluster fission rates are then obtained from detailed balance. All possible cluster-cluster processes have been included. The electrostatic enhancement of ion-molecule collisions is calculated by using dipole moments and polarizabilities obtained from quantum chemistry. The model has no fitted parameters. The experiments also reveal a nucleation mechanism involving the formation of clusters containing sulfuric acid and oxidized organic molecules from the very first step. Inclusion of this mechanism in a global aerosol model yields a photochemically and biologically driven seasonal cycle of particle concentrations in the continental boundary layer, in good agreement with observations [12]. The newly discovered role played by amines in cloud formation, suggests that natural and anthropogenic sources of amines can have an influence on climate. Anthropogenic amine emissions could be considered as a possible geoengineering technology. Furthermore it is important to note that CO_2 scrubbing using amines is the most frequent method for post-combustion capture. Amine technology has already been used for decades to capture CO_2 from both flue gas and natural gas. The widespread use of this technology for carbon capture from fossil fuel power plants would tend to increase the anthropogenic amine emissions and their possible cooling effect on climate.

References

1. Agee, E.M., Kiefer, K. Cornett, E.: Relationship of lower troposphere cloud cover and cosmic rays: an updated perspective. J. Clim. **25**, 1057–1060 (2012)
2. Almeida J., Amorim, A., Santos, F.D., et al.: Molecular understanding of sulphuric acidamine particle nucleation in the atmosphere. Nature **502**(7471), 359–363 (2013)
3. IPCC: IPCC Fifth Assessment Report (2014)
4. Kazil, J., Harrison, R.G., Lovejoy, E.R.: Tropospheric new particle formation and the role of ions. Space Sci. Rev. **137**, 241–255 (2008)
5. Kazil, J., Zhang, K., Stier, P., Feichter, J., Lohmann, U., O'Brien, K.: The present-day decadal solar cycle modulation of Earth's radiative forcing via charged H2SO4/H2O aerosol nucleation. Geophys. Res. Lett. **39**, L02805 (2012)
6. Kirkby J., Amorim, A., et al.: Role of sulphuric acid, ammonia and galactic cosmic rays in atmospheric aerosol nucleation. Nature **476**(7361), 429–433 (2011)
7. Laken, B.A., Kniveton, D.R., Frogley, M.R.: Cosmic rays linked to rapid mid-latitude cloud changes. Atmos. Chem. Phys. **10**, 10941–10948 (2010)
8. Ney, E.: Cosmic radiation and the weather. Nature **183**, 451–452 (1959)
9. Palle Bago, E., Butler, C.J.: The influence of cosmic rays on terrestrial clouds and global warming. Astron. Geophys. **41**, 4.18 (2000)
10. Peixoto, J.P., Oort, A.: Physics of Climate. American Institute of Physics, New York (1992)
11. Putman, W.M., Suarez, M.: Cloud-system resolving simulations with the NASA Goddard Earth Observing System global atmospheric model (GEOS-5). Geophys. Res. Lett. **38**, L16809 (2011)
12. Riccobono, F., Amorim, A., Santos, F.D., et al.: Oxidation products of biogenic emissions contribute to nucleation of atmospheric particles. Science **344**, 717–721 (2014)
13. Rohs, S., Spang, R., Rohrer, F., Schiller, C., Vos, H.: A correlation study of high-altitude and midaltitude clouds and galactic cosmic rays by MIPAS-Envisat. J. Geophys. Res. **115**, D14212 (2010)
14. Schobesberger, A.A., Santos, F.D., et al.: Molecular understanding of atmospheric particle formation from sulphuric acid and large oxidized organic molecules. Proc. Natl. Acad. Sci. USA **110**, 17223–17228 (2013)
15. Svensmark, H.: Cosmoclimatology: a new theory emerges. News Rev. Astron. Geophys. **48**, 18 (2007)
16. Svensmark, H., Friis-Christensen, E.: Variation in cosmic ray flux and global cloud coverage a missing link in solar-climate relationship. J. Atmos. Sol. Terr. Phys. **59**, 1225 (1997)
17. Tinsley, B.A., Fangqun, Y.: In: Pap, J.M., Fox, P., Frohlich, C., Hudson, H.S., Kuhn, J., McCormack, J., North, G., Sprigg, W., Wu, S.T. (eds.) Atmospheric Ionization and Clouds as Links Between Solar Activity and Climate, in Solar Variability and Its Effects on Climate. Geophysical Monograph, vol. 141, p. 321. American Geophysical Union, Washington, DC (2004) [ISBN 0-87590-406-8]

Long Time Behaviour and Self-similarity in an Addition Model with Slow Input of Monomers

Rafael Sasportes

Abstract We consider a coagulation equation with constant coefficients and a time dependent power law input of monomers. We discuss the asymptotic behaviour of solutions as $t \to \infty$, and we prove solutions converge to a similarity profile along the non-characteristic direction.

1 Introduction

We study some aspects of the long time behaviour of a system with an infinite number of ordinary differential equations modelling the kinetics of particle coagulation; we consider a mean-field point island deposition growth process, with Becker-Döring type kinetics with critical island size $i = 1$. In [4] a different island growth model is considered, for which clusters of size j $(1 < j \leq i)$ do not arise.

The system we consider is composed of a large number of particles, each particle consisting of an integer number of monomers with mass 1, so that a j-cluster (a particle formed by j monomers) will have mass j. We assume these clusters can bind together to form larger clusters, and that we only have *binary* reactions, in the sense that we only consider aggregation of two clusters at a time, one of them being a monomer; we do not consider, for example, simultaneous aggregation of three clusters. The cluster interaction is assumed to follow the mass action law of chemical kinetics. Let $(c_j(t))_{j=1}^{\infty}$ be the sequence whose elements are the concentration of clusters of mass j at some time t, and we want to study the evolution of $c_j(t)$ as $t \to +\infty$, either pointwise in j (i.e., for each fixed j), or when j also converges to $+\infty$ in some way related to the convergence of t. The evolution of the cluster

R. Sasportes (✉)
DCeT, Universidade Aberta, Lisboa, Portugal

CAMGSD, Instituto Superior Técnico, Universidade de Lisboa, Lisboa, Portugal
e-mail: rafael.sasportes@uab.pt

J.-P. Bourguignon et al. (eds.), *Mathematics of Energy and Climate Change*,
CIM Series in Mathematical Sciences 2, DOI 10.1007/978-3-319-16121-1_17

population can be described by the following coagulation kinetic equations

$$\begin{aligned} \dot{c}_1 &= -c_1^2 - c_1 \sum_{j=1}^{\infty} c_j \\ \dot{c}_j &= c_1(c_{j-1} - c_j), \quad j \geq 2. \end{aligned} \tag{1}$$

From the first equation in (1) it is clear that the number of monomers is decreasing; as described in more detail in [6], Eq. (1) are a special case of the Becker-Döring coagulation equations, corresponding to a situation where the only effective reactions are the ones involving monomers. Thus the special role played by monomers is expected to freeze the dynamics when we run out of monomers. In the context of aggregation models of cluster growth [3] we consider an "addition" model [7] where cluster growth can only occur by the addition of movable monomers to the immovable clusters [3]. We provide a source of monomers by adding a source term $J_1(t)$ to the right hand side of the c_1-equation in (1). One way to externally supply monomers is to define the input term $J_1(t)$ independently of the state of the system. This is a reasonable assumption in a number of applications, including in simple models of polymerization and of epitaxial growth [2]. The easiest hypothesis about $J_1(t)$, which turns out to be very useful in applications, is to make it a time independent constant. Another possible choice, quite interesting from a mathematical viewpoint, is to consider for J_1 a power law $J_1(t) = \alpha t^{\omega}$, with $\alpha > 0$ and $\omega \in \mathbb{R}$. The constant case was considered in [6], using an approach based on methods (Poincaré compactification and center manifold) that are not available for the general power law case; the case $\omega > -1/2$ was considered in [5]. For $\omega \leq -1/2$ partial results were obtained in [8]. In this paper we restrict ourselves to $\omega = -1/2$. A formal analysis was presented in [9], and we use the ansatz provided by Wattis [9] to rigorously analyse the addition model with a power law input of monomers $J_1(t) = \alpha t^{-1/2}$, namely

$$\begin{aligned} \dot{c}_1 &= \alpha t^{-1/2} - c_1^2 - c_1 \sum_{j=1}^{\infty} c_j \\ \dot{c}_j &= c_1(c_{j-1} - c_j)\, j \geq 2. \end{aligned} \tag{2}$$

We study two aspects of the dynamical behaviour of solutions to (2). First, we want to establish the componentwise behaviour of the solution as time $t \to +\infty$ and the behaviour of the total amount of clusters. The second aspect of the dynamics we are interested in is the occurrence of similarity behaviour. Our first step consists in transforming the infinite dimensional system (2) into a problem that is almost exactly solvable. Introducing the total number of clusters as a new macroscopic variable $c_0(t)$ defined by $c_0(t) = \sum_{j=1}^{\infty} c_j(t)$, and formally differentiating termwise, we conclude that c_0 satisfies the evolution equation

$$\dot{c}_0 = \alpha t^{-1/2} - c_0 c_1.$$

Using c_0, we can write system (2) as

$$\begin{aligned} \dot{c}_0 &= \alpha t^{-1/2} - c_0 c_1, \\ \dot{c}_1 &= \alpha t^{-1/2} - c_0 c_1 - c_1^2, \\ \dot{c}_j &= c_1(c_{j-1} - c_j), \quad j \geq 2. \end{aligned} \tag{3}$$

If $\sum_{j=1}^{\infty} c_j(0) < \infty$ then if (c_0, c_1, c_j), $j \geq 2$ is a solution of system (3) then (c_1, c_j), $j \geq 2$ is a solution of system (2). The proof can be done as in [6, Theorem 2.1].

The equations governing the dynamics of $c_0(t)$ and $c_1(t)$ actually define a nonautonomous bidimensional system

$$\begin{aligned} \dot{c}_0 &= \alpha t^{-1/2} - c_0 c_1 \\ \dot{c}_1 &= \alpha t^{-1/2} - c_0 c_1 - c_1^2, \end{aligned} \tag{4}$$

and we can now study the dynamics of (4) in a way totally independent of the remaining components of the infinite dimensional system. In order to solve this system we use an ansatz for a convenient change of variables suggested by Wattis [9, Table 2] and obtained via formal asymptotics. Based on [9, Table 2] we expect solutions (c_0, c_1) of system (4) to have the following behaviour as $t \to +\infty$

$$c_0(t) \sim \left(3\alpha^2\right)^{1/3} (\log t)^{1/3} \text{ and } c_1(t) \sim (\alpha/3)^{1/3} t^{-1/2} (\log t)^{-1/3}, \tag{5}$$

in the following sense

$$\lim_{t\to+\infty} c_0(t) \left(3\alpha^2 \log t\right)^{-1/3} = 1 \text{ and } \lim_{t\to+\infty} c_1(t)(\alpha/3)^{-1/3} t^{1/2} (\log t)^{1/3} = 1.$$

This suggests that defining functions $C_0(t)$ and $C_1(t)$ by

$$C_0(t) := \left(3\alpha^2\right)^{-1/3} (\log t)^{-1/3} c_0(t) \text{ and } C_1(t) := (\alpha/3)^{-1/3} t^{1/2} (\log t)^{1/3} c_1(t), \tag{6}$$

they might both be expected to converge to 1 as $t \to +\infty$, and reciprocally, if this happens then c_0 and c_1 will behave as stated in (5). To prove this convergence behaviour we need an equation for the evolution of (C_0, C_1). We begin by differentiating (6), and then replacing it into system (4). We then change the time scale $t \mapsto \tau$ by letting

$$\frac{d\tau}{dt} = \left(3\alpha^2\right)^{1/3} (\log t)^{1/3}. \tag{7}$$

Considering $t > 1$ we have a well defined change of variables, and defining

$$x(\tau) := C_1(t(\tau)) \text{ and } y(\tau) := C_0(t(\tau)),$$

and denoting $\frac{\mathrm{d}}{\mathrm{d}\tau}(\cdot)$ by $(\cdot)'$ we finally obtain an equation for (x, y):

$$\begin{aligned} x' &= 1 - xy - \hat{c}(\tau)x^2 + \hat{d}(\tau)x \\ y' &= \hat{c}(\tau)(1 - xy - \hat{c}(\tau)y), \end{aligned} \tag{8}$$

where

$$\hat{c}(\tau) = c\,(t(\tau)) := (9\alpha)^{-1/3}\,(t(\tau))^{-1/2}\,(\log t(\tau))^{-2/3}\,,$$

and

$$\hat{d}(\tau) = d\,(t(\tau)) := \hat{c}^2(\tau)(3/2\log t(\tau) + 1).$$

In [5] we have seen that for $\omega > -1/2$, the change of variables corresponding to (7) can be explicitly solved; for $\omega = -1/2$ we do not have an explicit expression for t as a function of τ, and we will use some preliminary results to obtain what we need: the asymptotic relationship between the two time scales.

For $t \in [1, +\infty[$ we have $\mathrm{d}\tau/\mathrm{d}t = \left(3\alpha^2\right)^{1/3}(\log t)^{1/3} > 0$; since $\lim_{t\to\infty} \mathrm{d}\tau/\mathrm{d}t = +\infty$, we can conclude that $\tau(t)$ (resp. $t(\tau)$) is a strictly increasing function of t (resp. τ). This allows us to conclude that $\tau \to +\infty$ (resp. $t \to +\infty$) as $t \to +\infty$ (resp. $\tau \to +\infty$). To get a better estimate on the asymptotic behaviour of $\tau(t)$, using integration by parts, we obtain from (7)

$$\tau(t) = t(3\alpha^2\log t)^{1/3}\,(1 + \mathrm{o}(1)) \;\text{ as } t \to +\infty.$$

This allows us to write $t(\tau) = \tau(3\alpha^2\log\tau)^{-1/3}\,(1 + \mathrm{o}(1))$ as $\tau \to +\infty$.

We also have as $\tau, t \to +\infty$ that

$$\tau(t) = \mathrm{O}\left(t(\log t)^{1/3}\right) \text{ and } t(\tau) = \mathrm{O}\left(\tau(\log\tau)^{-1/3}\right),$$

and also

$$\hat{c}(\tau) = \mathrm{O}\left((\tau\log\tau)^{-1/2}\right) \text{ and } \hat{d}(\tau) = \mathrm{O}\left(\tau^{-1}\right).$$

In the next section, we will study the bidimensional system (4); then in Sect. 3 we will study the long time behaviour of solutions, and in Sect. 4 we will study the existence of self-similar behaviour.

2 The Bidimensional System

Since we are only interested in non-negative solutions to (4), by *solution* we shall mean *non-negative solution*. The main result of this section concerns the asymptotic behaviour of c_0 and c_1.

Theorem 1 *Let* $\alpha > 0$, *and* (c_0, c_1) *be any solution of (4). Then*

1. $(3\alpha^2)^{-1/3} (\log t)^{-1/3} c_0(t) \to 1$ *as* $t \to +\infty$,
2. $(\alpha/3)^{-1/3} t^{1/2} (\log t)^{1/3} c_1(t) \to 1$ *as* $t \to +\infty$,
3. $(3/\alpha \log t)^{2/3} t \left(\alpha t^{-1/2} - c_0 c_1\right) \to 1$ *as* $t \to +\infty$.

To prove this theorem we use two propositions. These propositions follow closely what was done in a series of lemmas in [5, 6], and the proofs differ mainly because now we have a log term and also, as mentioned already, because we do not have an explicit expression for the change of variables defined by (7). We start by showing that non-negative solutions to (8) remain non-negative as $\tau \to +\infty$, then we show how the x and y boundedness are closely related, and finally we show that every solution to (8) with positive initial data is bounded.

Proposition 1 *For the system of equations (4) the following holds*

1. *The first quadrant* $\{x \geq 0,\ y \geq 0\}$ *is positively invariant for (8).*
2. y *(resp.* x*) is bounded* $\Longleftrightarrow$ x *(resp.* y*) is bounded away from zero.*
3. *Every solution to (8) with positive initial data is bounded.*

An immediate consequence of Proposition 1 is that solutions to (8), with positive initial data, are bounded and bounded away from zero; we also have that the conclusions of Proposition 1 still hold if the initial condition is nonnegative. Proposition 1 also implies that every orbit of (8) is bounded and bounded away from zero. We are now ready to study the ω-limit set of (8). We start by showing that the ω-limit set of every orbit is contained in the hyperbola $\{xy = 1\}$, then we fully identify it by showing that both x and y converge to 1, and finally we establish the convergence rate of $x(\tau)y(\tau)$ as $\tau \to +\infty$.

Proposition 2 *For the system of Eq. (8) the following holds*

1. *Let* (x, y) *be any solution to (8) then* $x(\tau)y(\tau) \to 1^-$ *as* $\tau \to +\infty$.
2. $\lim_{\tau\to+\infty} x(\tau) = 1$ *and* $\lim_{\tau\to+\infty} y(\tau) = 1$.
3. *Let* (x, y) *be any solution to (8) then we have* $\lim_{\tau\to+\infty} \frac{1-x(\tau)y(\tau)}{\hat{c}(\tau)} = 1$.

Recalling the definition of x, y and $\hat{c}$, Theorem 1 follows from the last two statements in Proposition 2.

3 Long Time Behaviour of the System

Given a solution of (4), we introduce a new time scale

$$\varsigma(t) := \varsigma_0 + \int_{t_0}^{t} c_1(s)\mathrm{d}s, \tag{9}$$

where ς_0 is a positive constant, and we consider the new phase variables

$$\tilde{c}_j(\varsigma) := c_j(t(\varsigma)), \tag{10}$$

where $t(\varsigma)$ is the inverse function of $\varsigma(t)$. When $c_1(t) > 0$, these are well defined and ς is an increasing function of t. In these new variables, the equations for c_j in (3) now become

$$\tilde{c}_j{}' = \tilde{c}_{j-1} - \tilde{c}_j, \qquad j \geq 2,$$

where $(\cdot)' = \frac{\mathrm{d}}{\mathrm{d}\varsigma}(\cdot)$. This system of differential equations is a lower triangular linear system and thus can be explicitly solved in terms of the function $\tilde{c}_1(\varsigma)$ starting from the equation for $j = 2$ and applying the variation of constants formula recursively:

$$\tilde{c}_j(\varsigma) = \mathrm{e}^{-\varsigma} \sum_{k=2}^{j} \frac{\varsigma^{j-k}}{(j-k)!} c_k(0) + \frac{1}{(j-2)!} \int_0^{\varsigma} \tilde{c}_1(\varsigma - s) s^{j-2} \mathrm{e}^{-s} \mathrm{d}s. \tag{11}$$

From now on we will only consider the new time scale defined by (9).

We now establish convergence results similar to those of Theorem 1 but for all values of j, and in both time scales.

Proposition 3 *With c_j, ς and $\tilde{c}_j(\varsigma)$ as given by (9) and (10)*

1. $\varsigma(t) = (8\alpha/3)^{1/3}\, t^{1/2} (\log t)^{-1/3} (1 + \mathrm{o}(1))$, *as* $t \to +\infty$,
2. $(1/2)^{1/3}(3/\alpha)^{2/3}\varsigma(\log \varsigma)^{2/3}\tilde{c}_j(\varsigma) = 1 + \mathrm{o}(1), \forall j \geq 1$, *as* $\varsigma \to +\infty$,
3. $(\alpha/3)^{-1/3}\, t^{1/2} (\log t)^{1/3}\, c_j(t) \longrightarrow 1, \forall j \geq 1$, *as* $t \to +\infty$.

By definition we have $\mathrm{d}\varsigma/\mathrm{d}t = c_1(t)$ and we already know from Theorem 1 the asymptotic behaviour of c_1, hence we have the following estimates

$$\forall \varepsilon > 0, \exists T_\varepsilon \colon \forall t > T_\varepsilon, 1 - \varepsilon < t^{1/2} (3 \log t/\alpha)^{1/3} c_1(t) < 1 + \varepsilon$$
$$\Longrightarrow t^{-1/2} (3 \log t/\alpha)^{-1/3} (1 - \varepsilon) < c_1(t) < t^{-1/2} (3 \log t/\alpha)^{-1/3} (1 + \varepsilon).$$

We are thus naturally led to estimate the integral $\int_{t_0}^{t} s^{-1/2} (\log s)^{-1/3}\, \mathrm{d}s$, as $t \to +\infty$, to obtain, as $t \to +\infty$

$$\varsigma(t) = (8\alpha/3)^{\frac{1}{3}}\, t^{1/2} \,(\log t)^{-\frac{1}{3}}\, (1 + \mathrm{o}(1)). \tag{12}$$

Using Eq. (12) we obtain the following relation between the logarithms of $\varsigma(t)$ and $t(\varsigma)$

$$\log \varsigma(t) = \frac{1}{2} \log t(\varsigma)(1 + \mathrm{o}(1)),$$

and using this last equation,

$$t(\varsigma) = (4\alpha/3)^{-2/3}\, \varsigma^2\, (\log \varsigma)^{2/3}\, (1 + \mathrm{o}(1)), \tag{13}$$

as $\varsigma \to \infty$. Using (13) we obtain the asymptotic behaviour of $\tilde{c}_1(\varsigma)$

$$\lim_{\varsigma\to+\infty} (1/2)^{1/3}\, (3/\alpha)^{2/3}\, \varsigma\, (\log \varsigma)^{2/3}\, \tilde{c}_1(\varsigma) = 1. \tag{14}$$

Using (14) and the representation of the $\tilde{c}_j$ given by (11) we can establish the behaviour of c_j in terms of the original t variable. To this end, letting

$$g(\varsigma) := (1/2)^{1/3}(3/\alpha)^{2/3}\varsigma(\log \varsigma)^{2/3}, \tag{15}$$

we can write $g(\varsigma)\tilde{c}_1(\varsigma) = 1 + \mathrm{o}(1)$, as $\varsigma \to +\infty$. Multiplying (11) by $g(\varsigma)$ we obtain

$$g(\varsigma)\tilde{c}_j(\varsigma) = g(\varsigma)\mathrm{e}^{-\varsigma} \sum_{k=2}^{j} \frac{\varsigma^{j-k}}{(j-k)!} c_k(0) + \frac{g(\varsigma)}{(j-2)!} \int_0^{\varsigma} \tilde{c}_1(\varsigma - s) s^{j-2} \mathrm{e}^{-s} \mathrm{d}s. \tag{16}$$

The first term on the right hand side of (16), corresponding to the non-monomeric initial data contribution, can be written as

$$\varsigma(\log \varsigma)^{2/3}\mathrm{e}^{-\varsigma} \sum_{k=2}^{j} \frac{\varsigma^{j-k}}{(j-k)!} c_k(0) = \mathrm{O}\left((\log \varsigma)^{2/3}\varsigma^{j-1}\mathrm{e}^{-\varsigma}\right) = \mathrm{o}\left(\mathrm{e}^{-\lambda\varsigma}\right) \text{ as } \varsigma \to +\infty,$$

for every $\lambda < 1$ and fixed j.

For the second term in the right hand side of (16) we start by changing integration variables $s \mapsto y = s/\varsigma$, which allows us to write the integral term as an integral over the fixed interval $[0, 1]$. Defining the function

$$\psi(\cdot) := g(\cdot)\tilde{c}_1(\cdot), \tag{17}$$

we obtain

$$\frac{g(\varsigma)}{(j-2)!}\int_0^{\varsigma}\tilde{c}_1(\varsigma-s)s^{j-2}\mathrm{e}^{-s}\mathrm{d}s$$
$$=\frac{\varsigma^{j-1}(\log\varsigma)^{2/3}}{(j-2)!}\int_0^1\frac{\psi(\varsigma(1-y))y^{j-2}}{(1-y)(\log\varsigma(1-y))^{2/3}}\mathrm{e}^{-\varsigma y}\mathrm{d}y. \quad (18)$$

In order to evaluate the integral in (18) we split the interval of integration at the $y=1$ singularity as $(0,1-\varepsilon)$ and $(1-\varepsilon,1)$, for a fixed $\varepsilon\in(0,1)$. For the integral over $(1-\varepsilon,1)$ we know that since $\tilde{c}_1$ is continuous and goes to zero at infinity, by (14), there exists a positive constant M satisfying $0\le\tilde{c}_1(x)\le M$ for $x\in[0,+\infty[$ and hence

$$\varsigma^j(\log\varsigma)^{2/3}\int_{1-\varepsilon}^1\tilde{c}_1(\varsigma(1-y))y^{j-2}\mathrm{e}^{-\varsigma y}\mathrm{d}y\le\varsigma^j(\log\varsigma)^{2/3}M\int_{1-\varepsilon}^1\mathrm{e}^{-\varsigma y}\mathrm{d}y \quad (19)$$
$$=M\varsigma^{j-1}(\log\varsigma)^{2/3}\mathrm{e}^{-\varsigma}\left(\exp(1-\varepsilon)-1\right),$$

and this term is exponentially small when $\varsigma\to+\infty$.

For the integral over $(0,1-\varepsilon)$, we use the fact that $y<1-\varepsilon\Rightarrow\varsigma(1-y)>\varsigma\varepsilon\to+\infty$ as $\varsigma\to+\infty$, then for ς sufficiently large, we can use (14) and (17) to conclude that $\psi=1+\mathrm{o}(1)$ in the interval we are considering, and thus

$$\forall\delta_1>0,\exists T_1(\delta_1)\colon\forall\varsigma>T_1(\delta_1),\psi(\varsigma(1-y))\in[1-\delta_1,1+\delta_1],$$

and hence as $\varsigma\to+\infty$ we have

$$(1-\delta_1)I_j(\varsigma)\le\int_0^{1-\varepsilon}\frac{(\log\varsigma)^{2/3}\psi(\varsigma(1-y))y^{j-2}}{(1-y)(\log\varsigma(1-y))^{2/3}}\mathrm{e}^{-\varsigma y}\mathrm{d}y\le(1+\delta_1)I_j(\varsigma), \quad (20)$$

where

$$I_j(\varsigma):=\int_0^{1-\varepsilon}\left(1+\frac{\log(1-y)}{\log\varsigma}\right)^{-2/3}\frac{y^{j-2}}{1-y}\mathrm{e}^{-\varsigma y}\mathrm{d}y.$$

For $y\in[0,1-\varepsilon[$, we now have that

$$\forall\delta_2>0,\exists T_2(\delta_2)\colon\forall\varsigma>T_2(\delta_2),\left(1+\frac{\log(1-y)}{\log\varsigma}\right)^{-2/3}\in[1-\delta_2,1+\delta_2].$$

Hence for ς sufficiently large, it is enough to estimate the integral $\int_0^{1-\varepsilon} \frac{y^{j-2}}{1-y} e^{-\varsigma y} dy$, which, using Watson's lemma, is equal to

$$\int_0^{1-\varepsilon} \frac{y^{j-2}}{1-y} e^{-\varsigma y} dy = \frac{\Gamma(j-1)}{\varsigma^{j-1}} + O\left(\frac{1}{\varsigma^j}\right), \text{ as } \varsigma \to +\infty.$$

Putting this last expression in (20) results in

$$\frac{\varsigma^{j-1}}{(j-2)!} \int_0^{1-\varepsilon} \frac{(\log \varsigma)^{2/3} \psi(\varsigma(1-y)) y^{j-2}}{(1-y)(\log \varsigma(1-y))^{2/3}} e^{-\varsigma y} dy = 1 + O(\varsigma^{-1}) \text{ as } \varsigma \to +\infty. \tag{21}$$

Gathering (19) and (21), we have the following generalization of (14), as $\varsigma \to +\infty$

$$(1/2)^{1/3} (3/\alpha)^{2/3} \varsigma (\log \varsigma)^{2/3} \tilde{c}_j(\varsigma) = 1 + o(1), \forall j \geq 1,$$

or in the original t variable [using (12)], as $t \to +\infty$

$$t^{1/2} \left(3 \log t/\alpha\right)^{1/3} c_j(t) \to 1, \forall j \geq 1.$$

This concludes the proof of Proposition 3.

4 Self-similar Behaviour

We can now turn to the results concerning convergence of solutions to self-similar profiles.

Da Costa and Sasportes [5] showed that when the input of monomers is given by $J_1(t) = \alpha t^{\omega}$, with $\omega > -1/2$ we have a similarity profile $\Phi_{1,\omega} : \mathbb{R}^+ \setminus \{1\} \to \mathbb{R}$ defined by

$$\Phi_{1,\omega}(\eta) := \begin{cases} (1-\eta)^{\frac{\omega-1}{\omega+2}} & \text{if } 0 < \eta < 1 \\ 0 & \text{if } \eta > 1. \end{cases}$$

The following result states that choosing $\omega = -1/2$, the function $\Phi_{1,-1/2}$ is still a similarity profile for the solutions to (2) along non-characteristic directions.

Theorem 2 *Let (c_j) be any non-negative solution of (2) with initial data satisfying $\exists \rho > 0, \mu > 1 : \forall j, c_j(0) \leq \rho/j^{\mu}$. Let $\varsigma(t)$ and $\tilde{c}_j(\varsigma)$ be as in (9) and (10), respectively. Then for $\eta = j/\varsigma$ fixed and $0 < \eta \neq 1$, we have*

$$\lim_{j,\varsigma \to +\infty} (1/2)^{1/3} (3/\alpha)^{2/3} \varsigma (\log \varsigma)^{2/3} \tilde{c}_j(\varsigma) = \Phi_{1,-1/2}(\eta).$$

4.1 Monomeric Initial Data

For monomeric initial data, the representation formula for $\tilde{c}_j$ [given by (11)] shows that we only have the integral term, and multiplying (11) by $g(\varsigma)$ we have

$$g(\varsigma)\tilde{c}_j(\varsigma) = \frac{g(\varsigma)}{(j-2)!}\int_0^{\varsigma} \tilde{c}_1(\varsigma - s)s^{j-2}e^{-s}ds. \tag{22}$$

In order to evaluate the right hand side of (22) we replace the discrete variable j by a continuous one x, allowing us to use Stirling's asymptotic formula for the Γ function. Let φ_1 on $[2,\infty)\times[0,\infty)$ be defined by

$$\varphi_1(x,\varsigma) := \frac{g(\varsigma)}{\Gamma(x-1)}\int_0^{\varsigma} \tilde{c}_1(\varsigma - s)s^{x-2}e^{-s}ds.$$

When $x \geq 2$ is an integer, the function φ_1 clearly satisfies $\varphi_1(x,\varsigma) = g(\varsigma)\tilde{c}_x(\varsigma)$, and we shall use φ_1 instead of the definition of $\tilde{c}_j$. Using Stirling's asymptotic formula $\Gamma(x) = e^{-x}x^{x-1/2}\sqrt{2\pi}\left(1+O(x^{-1})\right)$ as $x\to\infty$, the recursive relation $\Gamma(x-1) = \Gamma(x)/(x-1)$, letting $\eta := x/\varsigma$, and changing variable $s \mapsto y = s/\varsigma$, we can write,

$$\varphi_1(\varsigma\eta,\varsigma) = \frac{1}{\sqrt{2\pi}}\left(\frac{9}{2\alpha^2}\right)^{1/3}\eta^{3/2-\eta\varsigma}\varsigma^{1/2}\varsigma(\log\varsigma)^{2/3}\times$$
$$\times\left(1+O\left(\varsigma^{-1}\right)\right)\int_0^1 \tilde{c}_1(\varsigma(1-y))\frac{\exp(\varsigma(\eta\log y - y + \eta))}{y^2}dy. \tag{23}$$

In order to make clear the asymptotic behaviour of $\tilde{c}_1(\varsigma)$ we multiply (and divide) inside the previous integral by $g(\varsigma(1-y))$, as defined in (15) and (17), and we obtain

$$\varphi_1(\varsigma\eta,\varsigma) = \frac{1}{\sqrt{2\pi}}\left(\frac{9}{2\alpha^2}\right)^{1/3}\eta^{\frac{3}{2}-\eta\varsigma}\varsigma^{1/2}\varsigma(\log\varsigma)^{2/3}\left(\frac{9}{2\alpha^2}\right)^{-1/3}\varsigma^{-1}\times$$
$$\times\left(1+O\left(\varsigma^{-1}\right)\right)\int_0^1 \psi(\varsigma(1-y))\frac{\exp(\varsigma(\eta\log y - y + \eta))}{(\log(\varsigma(1-y)))^{2/3}y^2(1-y)}dy.$$

Simplifying and grouping the logarithmic terms we obtain

$$\varphi_1(\varsigma\eta,\varsigma) = \frac{1}{\sqrt{2\pi}}\eta^{\frac{3}{2}-\eta\varsigma}\varsigma^{1/2}\left(1+O\left(\varsigma^{-1}\right)\right)\times$$
$$\times\int_0^1 \psi(\varsigma(1-y))\left(1+\frac{\log(1-y)}{\log\varsigma}\right)^{-2/3}\frac{\exp(\varsigma(\eta\log y - y + \eta))}{y^2(1-y)}dy. \tag{24}$$

Rearranging the last expression, the proof reduces to the asymptotic evaluation as $\varsigma \to +\infty$ of the function $I(\eta, \varsigma)$ defined by

$$I(\eta,\varsigma) := \varsigma^{1/2}\eta^{-\eta\varsigma}e^{\varsigma\eta}\times$$
$$\times \int_0^1 \psi(\varsigma(1-y))\left(1+\frac{\log(1-y)}{\log\varsigma}\right)^{-2/3}\frac{\exp(\varsigma(\eta\log y - y))}{y^2(1-y)}dy. \quad (25)$$

We start by showing that for $\eta > 1$ we have $I(\eta,\varsigma) \to 0$, as $\varsigma \to +\infty$. In order to study the behaviour of φ_1 we first split the interval of integration as $(0, 1-\varepsilon)$ and $(1-\varepsilon, 1)$, for a fixed $\varepsilon \in (0,1)$.

In $(0, 1-\varepsilon)$ both $\psi(\varsigma(1-y))$ and $\left(1+\frac{\log(1-y)}{\log\varsigma}\right)^{-1}$ are $1+o(1)$ for large values of ς, and hence to evaluate (25) it is enough to estimate

$$\begin{aligned}
&\varsigma^{1/2}\eta^{-\eta\varsigma}e^{\varsigma\eta}\int_0^{1-\varepsilon}\frac{\exp(\varsigma(\eta\log y - y))}{y^2(1-y)}dy\\
&\quad = \varsigma^{1/2}\eta^{-\eta\varsigma}\exp(\varsigma\eta)\int_0^{1-\varepsilon}\frac{y^{-2}\exp(\varsigma(\eta\log y - y))}{(1-y)}dy\\
&\quad \le \varsigma^{1/2}\eta^{-\eta\varsigma}\exp(\varsigma\eta)\exp\left(\max_{t\in[0,1-\varepsilon]} g_1(t)\right)\int_0^{1-\varepsilon}\frac{1}{1-y}dy\\
&\quad = \varsigma^{1/2}\eta^{-\eta\varsigma}\exp(\varsigma\eta)\exp\left(\max_{t\in[0,1-\varepsilon]} g_1(t)\right)\log\varepsilon^{-1}, \qquad (26)
\end{aligned}$$

where $g_1(t) := (\varsigma\eta - 2)\log t - \varsigma t$. For $\varsigma > 2/(\eta-1)$ and $t \le 1$, the function g_1 satisfies $g_1'(t) = (\varsigma\eta-2)/t - \varsigma \ge (\varsigma\eta - 2) - \varsigma = \varsigma(\eta-1) - 2 > 0$, and hence $g_1(t) \le g_1(1-\varepsilon) = -\varsigma(1-\varepsilon-\eta\log(1-\varepsilon)) - 2\log(1-\varepsilon)$. Plugging this result back in (26) we have

$$\begin{aligned}
&\varsigma^{1/2}\eta^{-\eta\varsigma}e^{\varsigma\eta}\ \exp\left(\max_{t\in[0,1-\varepsilon]} g_1(t)\right)\log\varepsilon^{-1}\\
&\qquad = \frac{\varsigma^{1/2}\log\varepsilon^{-1}}{(1-\varepsilon)^2}\exp\left(-\varsigma\left(\eta\log\eta - \eta + (1-\varepsilon) - \eta\log(1-\varepsilon)\right)\right),
\end{aligned}$$

and so it is enough to check that we have $\eta\log\eta - \eta + (1-\varepsilon) - \eta\log(1-\varepsilon) > 0$ for $\varsigma > 2/(\eta-1)$ and $\eta > 1$. But

$$\begin{aligned}
\eta\log\eta - \eta + (1-\varepsilon) - \eta\log(1-\varepsilon) > 0 &\Leftrightarrow (1-\varepsilon) - \eta > \eta\log\frac{1-\varepsilon}{\eta}\\
&\Leftrightarrow \frac{1-\varepsilon}{\eta} - 1 > \log\frac{1-\varepsilon}{\eta},
\end{aligned}$$

and, letting $z = (1-\varepsilon)/\eta$, this last inequality amounts to $z > \log z + 1$ which holds for all $z \neq 1$, and that is the case since $\eta > 1 \Rightarrow \eta \neq 1-\varepsilon$. This concludes the proof in the interval $(0, 1-\varepsilon)$.

We now show that the integral over $(1-\varepsilon, 1)$ also goes to 0 as $\varsigma \to +\infty$. Since $\tilde{c}_1$ is continuous and goes to 0 as $\varsigma \to +\infty$ it is bounded in $[1-\varepsilon, 1]$, and so there is a constant $M > 0$ such that $\tilde{c}_1(\varsigma(1-y)) < M, \forall y \in [1-\varepsilon, 1]$. Now we have to estimate

$$\begin{aligned}
\eta^{-\eta\varsigma}\varsigma^{3/2}(\log\varsigma)^{2/3}&\int_{1-\varepsilon}^{1}\frac{\exp\left(\varsigma(\eta\log y - y + \eta)\right)}{y^2}\mathrm{d}y\\
&= \varsigma^{3/2}(\log\varsigma)^{2/3}\int_{1-\varepsilon}^{1}\frac{\exp\left(-\varsigma(\eta\log\eta - \eta\log y + y - \eta)\right)}{y^2}\mathrm{d}y\\
&= \varsigma^{3/2}(\log\varsigma)^{2/3}\int_{1-\varepsilon}^{1}\frac{\exp(-\varsigma h(y))}{y^2}\mathrm{d}y\\
&< \varsigma^{3/2}(\log\varsigma)^{2/3}\exp\left(-\varsigma\min_{t\in[1-\varepsilon,1]}h(t)\right)\int_{1-\varepsilon}^{1}\frac{1}{y^2}\mathrm{d}y\\
&= \varsigma^{3/2}(\log\varsigma)^{2/3}\frac{\varepsilon}{1-\varepsilon}\exp\left(-\varsigma h(1)\right),
\end{aligned} \tag{27}$$

where $h(t) := \eta\log\eta - \eta\log t + t - \eta$ has a unique minimum at $t = 1$, and since $h(1) = \eta\log\eta + 1 - \eta$, and $\eta\log\eta + 1 - \eta > 0$ for $\eta \neq 1$ the expression in (27) is exponentially small as $\varsigma \to +\infty$. This concludes the proof for $\eta > 1$.

For $\eta < 1$ we use a similar approach, but the situation being slightly more delicate, since now the (unique) maximum of $\eta\log y - y$ is attained at an interior point ($1 > \eta \in (0,1)$), we need to split the integral by writing it as a sum of integrals over $(0,\varepsilon)$, $(\varepsilon, 1-\varepsilon)$ and $(1-\varepsilon, 1)$. With g and ψ defined as above, for every $\varepsilon > 0$ we split the integral over $[0,1]$ as the sum of three integrals: I_1 over $(0,\varepsilon)$, I_2 over $(\varepsilon, 1-\varepsilon)$ and I_3 over $(1-\varepsilon, 1)$. We will show that both I_1 and I_3 go to zero, and that the only non zero contribution comes from the integral over $(\varepsilon, 1-\varepsilon)$. Given $\eta < 1$, we choose $\varepsilon > 0$ in such a way that $\eta \in (\varepsilon, 1-\varepsilon)$, for instance $\varepsilon < \min\{\eta/a, 1-\eta\}$, with $a > 1$.

For the integral over I_1, we now have that both $\psi(\varsigma(1-y))$ and $\left(1 + \frac{\log(1-y)}{\log\varsigma}\right)^{-1}$ are $1 + o(1)$ when estimating the integral for large values of ς; and hence to evaluate the integral over I_1 we can use an argument similar to the one we already used in the $\eta > 1$ case. To evaluate I_1 it is then enough to estimate, as $\varsigma \to +\infty$, the value of

$$\varsigma^{1/2}\eta^{-\eta\varsigma}e^{\varsigma\eta}\int_0^{\varepsilon}\frac{\exp(\varsigma(\eta\log y - y))}{y^2(1-y)}\mathrm{d}y.$$

As in $(0, 1-\varepsilon)$, using $g_1(t) = (\varsigma\eta - 2)\log t - \varsigma t$, we have $0 < t < \varepsilon < \eta/a < \eta$ and hence for $\varsigma > 2/(1-1/a)\eta$, we can conclude that g_1' satisfies $tg_1'(t) = \varsigma(\eta - t) - 2 > 0$, since $\varsigma > 2/(\eta - \eta/a) > 2/(\eta - t)$ and hence $g_1(t) \le g_1(\varepsilon) = -\varsigma(\varepsilon - \eta\log\varepsilon) - 2\log\varepsilon$. We then have the following estimates

$$\begin{aligned}
\varsigma^{1/2}\eta^{-\eta\varsigma}\mathrm{e}^{\varsigma\eta}\int_0^{\varepsilon}\frac{\exp(\varsigma(\eta\log y - y))}{y^2(1-y)}\mathrm{d}y &= \varsigma^{1/2}\eta^{-\eta\varsigma}\mathrm{e}^{\varsigma\eta}\int_0^{\varepsilon}\frac{\exp(g_1(y))}{1-y}\mathrm{d}y \\
&\le \varsigma^{1/2}\eta^{-\eta\varsigma}\mathrm{e}^{\varsigma\eta}\exp\big(\max_{t\in[0,\varepsilon]} g_1(t)\big)\int_0^{\varepsilon}\frac{1}{1-y}\mathrm{d}y \\
&= \varsigma^{1/2}\eta^{-\eta\varsigma}\exp(\varsigma\eta + g_1(\varepsilon)\log(1-\varepsilon)^{-1} \\
&= \varsigma^{1/2}\varepsilon^{-2}\log(1-\varepsilon)^{-1}\exp(-\varsigma(\eta\log\eta - \eta + \varepsilon - \eta\log\varepsilon)).
\end{aligned}$$

And so we only need to check that $\eta\log\eta - \eta + \varepsilon - \eta\log\varepsilon > 0$, which is true since this last expression is always positive, except for $\eta = \varepsilon$ where it is zero, and we chose $\varepsilon < \eta$. Hence $I_1 \to 0$, as $\varsigma \to +\infty$.

For I_3, the integral over $[1-\varepsilon, 1]$, we have $0 \le \varsigma(1-y) \le \varsigma\varepsilon$, and we use Eq. (23), which involves $\tilde{c}_1$, and we have to evaluate, as $\varsigma \to +\infty$,

$$\eta^{-\eta\varsigma}\varsigma^{3/2}(\log\varsigma)^{2/3}\int_{1-\varepsilon}^{1}\frac{\exp(\varsigma(\eta\log y - y + \eta))}{y^2}\mathrm{d}y.$$

This can be done as before, by showing that the function $h(y) := \eta\log\eta - \eta - \eta\log y + y$ is always positive for $y \in [1-\varepsilon, 1]$, remembering that we picked $\varepsilon < 1-\eta$, and hence $y > 1-\varepsilon > \eta$, when evaluating I_3. And so recalling that $h(y) \ge 0$, and $h(y) > 0$ for $y \ne \eta$, we conclude that I_3 is also exponentially small as $\varsigma \to +\infty$.

For the integral I_2, we use again the fact that for $y \in (\varepsilon, 1-\varepsilon)$, we have $\left(1 + \frac{\log(1-y)}{\log\varsigma}\right)^{-2/3} \to 1$ as $\varsigma \to +\infty$, and so we rewrite (23) as

$$\begin{aligned}
&\sqrt{2\pi}\,\varphi_1(\varsigma\eta,\varsigma) = \left(2\alpha^2/9\right)^{-1/3}\eta^{3/2-\eta\varsigma}\varsigma^{1/2}\varsigma(\log\varsigma)^{2/3}\times \\
&\times\left(1 + \mathrm{O}\left(\varsigma^{-1}\right)\right)\int_{\varepsilon}^{1-\varepsilon}\tilde{c}_1(\varsigma(1-y))\frac{\exp(\varsigma(\eta\log y - y + \eta))}{y^2}\mathrm{d}y \\
&= \eta^{3/2-\eta\varsigma}\varsigma^{1/2}\times \\
&\times\left(1 + \mathrm{O}\left(\varsigma^{-1}\right)\right)\int_{\varepsilon}^{1-\varepsilon}\psi(\varsigma(1-y))\frac{(\log\varsigma)^{2/3}\exp(\varsigma(\eta\log y - y + \eta))}{y^2(1-y)(\log\varsigma(1-y))^{2/3}}\mathrm{d}y \\
&= \eta^{3/2-\eta\varsigma}\varsigma^{1/2}\left(1 + \mathrm{O}\left(\varsigma^{-1}\right)\right)\times \\
&\times\int_{\varepsilon}^{1-\varepsilon}\psi(\varsigma(1-y))\left(\frac{\log\varsigma}{\log\varsigma(1-y)}\right)^{2/3}\frac{\exp(\varsigma(\eta\log y - y + \eta))}{y^2(1-y)}\mathrm{d}y.
\end{aligned}$$

Since in this case $\psi(\varsigma(1-y))$ and $\left(\frac{\log\varsigma}{\log\varsigma(1-y)}\right)^{2/3}$ are $1+\mathrm{o}(1)$ when $\varsigma\to\infty$, it holds that

$$\forall\delta>0,\exists T(\delta)\colon\forall\varsigma>T(\delta),\psi(\varsigma(1-y))\left(\frac{\log\varsigma}{\log\varsigma(1-y)}\right)^{2/3}\in[1-\delta,1+\delta].$$

It is then enough to study the limit, as $\varsigma\to+\infty$, of the function

$$J(\eta,\varsigma):=\eta^{3/2-\eta\varsigma}\varsigma^{1/2}\int_{\varepsilon}^{1-\varepsilon}\frac{\exp(\varsigma(\eta\log y-y+\eta))}{y^2(1-y)}\mathrm{d}y,$$

since we can write $(1-\delta)J(\eta,\varsigma)\le I_2\le(1+\delta)J(\eta,\varsigma)$, for ς sufficiently large.

Applying Laplace's method for the asymptotic evaluation of integrals [1, p. 431] to the integral

$$\int_{\varepsilon}^{1-\varepsilon}\frac{\exp(-\varsigma(y-\eta\log y-\eta))}{y^2(1-y)}\mathrm{d}y=\int_{\varepsilon}^{1-\varepsilon}\frac{\exp(-\varsigma\phi(y))}{y^2(1-y)}\mathrm{d}y,$$

where $\phi:(0,1)\to\mathbb{R}$ defined by $\phi(y):=y-\eta\log y-\eta$, is smooth and has a unique minimum, attained at $y=\eta\in(\varepsilon,1-\varepsilon)$ with value $\phi(\eta)=-\eta\log\eta$ and $\phi''(\eta)=\eta^{-1}$, we obtain, as $\varsigma\to+\infty$,

$$\begin{aligned}\int_{\varepsilon}^{1-\varepsilon}&\frac{\exp(-\varsigma(y-\eta\log y-\eta))}{y^2(1-y)}\mathrm{d}y=\\&=\frac{\sqrt{2\pi\eta/\varsigma}\exp(\varsigma\eta\log\eta)}{\eta^2(1-\eta)}+\mathrm{O}\Big(\frac{\exp(\varsigma\eta\log\eta)}{\varsigma^{3/2}}\Big).\end{aligned}\qquad(28)$$

Now from (23) and (28), we obtain for $\eta<1$, as $\varsigma\to+\infty$

$$\begin{aligned}\varphi_1(\varsigma\eta,\varsigma)&=\frac{1}{\sqrt{2\pi}}\eta^{3/2-\eta\varsigma}\varsigma^{1/2}\exp(\varsigma\eta\log\eta)\frac{1}{\eta^2(1-\eta)}\sqrt{2\pi\eta/\varsigma}+\mathrm{O}\left(\varsigma^{-1}\right)\\&=\frac{1}{1-\eta}\left(1+\mathrm{o}(1)\right).\end{aligned}$$

This concludes the proof in the monomeric case.

4.2 Non monomeric Initial Data

If the initial condition is not monomeric we have the contribution from the sum term in the right hand side of (11). Multiplying it by $g(\varsigma)$ we now have to prove that

$$\lim_{j,\varsigma\to+\infty} g(\varsigma)\mathrm{e}^{-\varsigma}\sum_{k=2}^{j}\frac{\varsigma^{j-k}}{(j-k)!}c_k(0)=0,\, \eta=j/\varsigma \text{ fixed, and } \eta\neq 1.$$

Since we want to show the limit is zero, we will drop the constants in the definition of g, and so we only consider the terms $\varsigma(\log\varsigma)^{2/3}$. The proof is based on the same argument used in [6, Sect. 5.2]. Defining $\nu := \eta^{-1}$, letting $\varsigma = j\nu$, and using the assumption on the initial condition in Theorem 2, namely $c_j(0) \leq \rho/j^{\mu}$, we then have

$$\begin{aligned}\varsigma(\log\varsigma)^{2/3}\mathrm{e}^{-\varsigma}\sum_{k=2}^{j}\frac{\varsigma^{j-k}}{(j-k)!}c_k(0) &\leq \rho j\nu(\log j\nu)^{2/3}\exp(-j\nu)\sum_{k=2}^{j}\frac{(j\nu)^{j-k}}{(j-k)!k^{\mu}}\\ &:= \rho\varphi_2(\nu,j).\end{aligned}$$

Our goal is to prove that $\varphi_2(\nu,j)\to 0$ as $j\to\infty$, for all positive $\nu\neq 1$. We can adapt the results in the study of φ_2 presented in [6, Sect. 5.2], noticing that we only need to multiply all the estimates in [6, Sect. 5.2] by $\sqrt{j\nu}(\log j\nu)^{2/3}$. The estimates show that now in order for φ_2 to converge to zero we need to consider initial data satisfying $c_j(0)\leq\rho/j^{\mu}$, but in this case with $\mu>1$. The $\log j$ term growing much slower than $\sqrt{j}$ has no influence on the convergence of φ_2 to zero. This completes the proof of the theorem.

4.3 On the Self-similar Behaviour Along the Characteristic Direction

In the case with input αt^{ω} with $\omega > -1/2$, we have seen [5] that for values of $\omega < 1$ the singularity of the self-similar solution $\Phi_{1,\omega}$ can be dealt with by considering a different similarity variable and a different time-scaling, allowing us a sort of inner expansion for the characteristic direction $\eta = 1$, and we obtained a function $\Phi_{2,\omega}$ satisfying

$$\tilde{c}_j(\varsigma)\sim\varsigma^{(\omega-1)/(2\omega+4)}\Phi_{2,\omega}\left((j-\varsigma)/\sqrt{\varsigma}\right).$$

It is worth noticing that for $\omega > -1/2$ the similarity variable was independent of ω, and the exponent of the time scaling variable, although ω-dependent was always *half* the exponent used for $\Phi_{1,\omega}$. Now we also have a singularity at $\eta = 1$ and so it is natural to check if this similarity variable also gives rise to a solution, and if that is the case, one for which $\eta = 1$ is no longer a singularity.

Choosing the similarity variable $(j - \varsigma)/\sqrt{\varsigma}$ and replacing ς by $\varsigma^{1/2}$ in the expression in the limit in Theorem 2

$$\varsigma^{1/2}\left(\log(\varsigma)^{1/2}\right)^{2/3} = (1/2)^{2/3}\varsigma^{1/2}(\log\varsigma)^{2/3},$$

and letting $\Phi_{2,-1/2} : \mathbb{R} \to \mathbb{R}$ be defined by

$$\Phi_{2,-1/2}(\xi) := \mathrm{e}^{-\xi^2/2}\int_0^{+\infty} y^{-1}\mathrm{e}^{-\xi y^2 - y^4/2}\mathrm{d}y,$$

we hope it is equal to the limit for $\xi = (j - \varsigma)/\sqrt{\varsigma}$ fixed and $\xi \in \mathbb{R}$, of

$$\lim_{j,\varsigma\to+\infty} C_\alpha \varsigma^{1/2}(\log\varsigma)^{2/3}\tilde{c}_j(\varsigma), \tag{29}$$

where $C_\alpha > 0$ is a constant that only depends on α. We now show that this limit does not exist.

Following a strategy similar to the one we used in [6] for $\omega > -1/2$, for monomeric initial data we have to estimate

$$(\log\varsigma)^{2/3}\varsigma^{1/2}\tilde{c}_j(\varsigma) = \frac{(\log\varsigma)^{2/3}\varsigma^{1/2}}{\Gamma(j-1)}\int_0^{\varsigma}\tilde{c}_1(\varsigma - s)s^{j-2}\mathrm{e}^{-s}\mathrm{d}s,$$

as $\varsigma \to +\infty, j \to +\infty, (j - \varsigma)/\sqrt{\varsigma}$ fixed.

We define the function φ_3 in $[2,\infty) \times [0,\infty)$ by

$$\varphi_3(x,\varsigma) := \frac{(\log\varsigma)^{2/3}\varsigma^{1/2}}{\Gamma(x-1)}\int_0^{\varsigma}\tilde{c}_1(\varsigma - s)s^{j-2}\mathrm{e}^{-s}\mathrm{d}s,$$

and using the similarity variable $\xi = (j-\varsigma)/\sqrt{\varsigma}(= (x-\varsigma)/\sqrt{\varsigma})$ we rewrite φ_3 as

$$\varphi_3(\varsigma + \xi\sqrt{\varsigma},\varsigma) = \frac{(\log\varsigma)^{2/3}\varsigma^{1/2}}{\Gamma(\varsigma + \xi\sqrt{\varsigma} - 1)}\int_0^{\varsigma}\tilde{c}_1(\varsigma - s)s^{\varsigma+\xi\sqrt{\varsigma}-2}\mathrm{e}^{-s}\mathrm{d}s.$$

If $2 \leq x = j \in \mathbb{N}$, we have $\varphi_3(j,\varsigma) = (\log\varsigma)^{2/3}\varsigma^{1/2}\tilde{c}_j(\varsigma)$, and hence we need to evaluate the limit

$$\lim_{\varsigma\to+\infty}\varphi_3(\varsigma + \xi\sqrt{\varsigma},\varsigma). \tag{30}$$

Changing variables $s \mapsto w := \sqrt{\sqrt{\varsigma} - s/\sqrt{\varsigma}}$, in such a way that $\varsigma - s = \sqrt{\varsigma}w^2$ and $\mathrm{d}s = -2\sqrt{\varsigma}w\mathrm{d}w$, we obtain

$$\varphi_3(\varsigma + \xi\sqrt{\varsigma}, \varsigma) = \frac{(\log \varsigma)^{2/3}\, \varsigma^{1/2}}{\Gamma(\varsigma + \xi\sqrt{\varsigma} - 1)} \times$$
$$\times \int_0^{\varsigma^{1/4}} \tilde{c}_1(\sqrt{\varsigma}w^2)(\varsigma - \sqrt{\varsigma}w^2)^{\varsigma + \xi\sqrt{\varsigma} - 2} \exp(-\varsigma + \sqrt{\varsigma}w^2) 2\sqrt{\varsigma}w\mathrm{d}w. \qquad (31)$$

Using (17), and letting $D = 2(\alpha/3)^{2/3}$, we rewrite (31) as

$$\begin{aligned}\varphi_3(\varsigma + \xi\sqrt{\varsigma}, \varsigma) &= D\frac{(\log \varsigma)^{2/3}\, \varsigma^{1/2}}{\Gamma(\varsigma + \xi\sqrt{\varsigma} - 1)} \int_0^{\varsigma^{1/4}} \left(\log(\sqrt{\varsigma}w^2)\right)^{-2/3} (\sqrt{\varsigma}w^2)^{-1} \times \\ &\quad \times \psi(\sqrt{\varsigma}w^2)(\varsigma - \sqrt{\varsigma}w^2)^{\varsigma + \xi\sqrt{\varsigma} - 2} \exp(-\varsigma + \sqrt{\varsigma}w^2)\sqrt{\varsigma}w\mathrm{d}w \\ &= D\frac{\varsigma^{\varsigma + \xi\sqrt{\varsigma} - 3/2} \mathrm{e}^{-\varsigma}}{\Gamma(\varsigma + \xi\sqrt{\varsigma} - 1)} \int_0^{\varsigma^{1/4}} \left(1 + \frac{4\log w}{\log \varsigma}\right)^{-2/3} \psi(\sqrt{\varsigma}w^2) \times \\ &\quad \times \left(1 - \frac{w^2}{\sqrt{\varsigma}}\right)^{\varsigma + \xi\sqrt{\varsigma} - 2} \exp(\sqrt{\varsigma}w^2)\frac{1}{w}\mathrm{d}w.\end{aligned}$$

Using Stirling's asymptotic expansion for the Gamma function we can write

$$\frac{1}{\Gamma(\varsigma + \xi\sqrt{\varsigma} - 1)} = \frac{1}{\sqrt{2\pi}} \frac{\exp(\varsigma + \xi\sqrt{\varsigma})}{(\varsigma + \xi\sqrt{\varsigma})^{\varsigma + \xi\sqrt{\varsigma} - 3/2}} (1 + \mathrm{o}(1)),$$

as $\varsigma \to +\infty$, and hence φ_3 can be written, as $\varsigma \to +\infty$,

$$\varphi_3(\varsigma + \xi\sqrt{\varsigma}, \varsigma) = \frac{D}{\sqrt{2\pi}} \frac{\varsigma^{\varsigma + \xi\sqrt{\varsigma} - 3/2} \exp(\xi\sqrt{\varsigma})}{(\varsigma + \xi\sqrt{\varsigma})^{\varsigma + \xi\sqrt{\varsigma} - 3/2}} (1 + \mathrm{o}(1)) \times \qquad (32)$$
$$\times \int_0^{\varsigma^{1/4}} \left(1 + \frac{4\log w}{\log \varsigma}\right)^{-2/3} \psi(\sqrt{\varsigma}w^2)\left(1 - \frac{w^2}{\sqrt{\varsigma}}\right)^{\varsigma + \xi\sqrt{\varsigma} - 2} \exp(\sqrt{\varsigma}w^2)\frac{1}{w}\mathrm{d}w.$$

To estimate the multiplicative prefactor in (32) as $\varsigma \to +\infty$ we can write it as

$$\frac{\varsigma^{\varsigma + \xi\sqrt{\varsigma} - 3/2} \exp(\xi\sqrt{\varsigma})}{(\varsigma + \xi\sqrt{\varsigma})^{\varsigma + \xi\sqrt{\varsigma} - 3/2}} = \exp\left(-\frac{\xi^2}{2}\right)(1 + \mathrm{o}(1)), \qquad (33)$$

where in (33) to compute the limit as $\varsigma \to +\infty$ we use the change of variables $\varsigma \mapsto x := \xi/\sqrt{\varsigma}$ to obtain $\left(\mathrm{e}(1+x)^{-1/x}\right)^{\xi^2/x}$ and then we apply L'Hôpital's rule twice. Using this last expression we can write (32) as $\varsigma \to +\infty$ in the following way

$$\varphi_3(\varsigma + \xi\sqrt{\varsigma}, \varsigma) = \frac{D}{\sqrt{2\pi}} \exp(-\xi^2/2)(1+\mathrm{o}(1)) \times \tag{34}$$
$$\times \int_0^{\varsigma^{1/4}} \left(1 + \frac{4\log w}{\log \varsigma}\right)^{-2/3} \psi(\sqrt{\varsigma}w^2)\left(1 - \frac{w^2}{\sqrt{\varsigma}}\right)^{\varsigma+\xi\sqrt{\varsigma}-2} \exp(\sqrt{\varsigma}w^2)\frac{1}{w}\mathrm{d}w.$$

In the case where $\omega > -1/2$ in [6] the proof continued with a study of the integral term in (34) by considering first w (and also $\sqrt{\varsigma}w^2$) close to zero, and then w away from zero, and showing that the integral, for small values of w, could be made arbitrarily small, while the remaining integral, for w away from zero converged as $\varsigma \to +\infty$. In the case at hand we no longer have convergence, essentially due to the singularity arising from $1/w$ in the integrand of (34). We now consider $\varepsilon > 0$ arbitrarily small and $1 < T < \varsigma^{1/4}$ and we show that the integral over $[\varepsilon, T]$ can be made arbitrarily large. We start by splitting the integral over $[0, \varsigma^{1/4}]$ in (34) as a sum over 3 intervals $[0, \varepsilon]$, $[\varepsilon, T]$ and $[T, \varsigma^{1/4}]$. The integrals over $[0, \varepsilon]$ and $[T, \varsigma^{1/4}]$ are both non negative and for $w \in [\varepsilon, T]$ we have $\sqrt{\varsigma}w^2 \geq \sqrt{\varsigma}\varepsilon^2 \to +\infty$ as $\varsigma \to +\infty$, and so as by (14), (15), and (17), it follows that $\left(1 + \frac{4\log w}{\log \varsigma}\right)^{-2/3} \psi(\sqrt{\varsigma}w^2) = 1 + \mathrm{o}(1)$ as $\varsigma \to +\infty$.

This means that for $w \in [\varepsilon, T]$ the integral we want to evaluate is asymptotically equal to

$$(1+\mathrm{o}(1)) \int_\varepsilon^T \left(1 - \frac{w^2}{\sqrt{\varsigma}}\right)^{\varsigma+\xi\sqrt{\varsigma}-2} \exp(\sqrt{\varsigma}w^2)\frac{1}{w}\mathrm{d}w.$$

To estimate this last integral we have as $\varsigma \to +\infty$

$$\left(1 - \frac{w^2}{\sqrt{\varsigma}}\right)^{\varsigma+\xi\sqrt{\varsigma}-2} \exp(\sqrt{\varsigma}w^2) = \exp\left(-\frac{w^4}{2} - \xi w^2\right)\left(1+\mathrm{o}(1)\right), \tag{35}$$

where (35) is obtained by first changing variables $\varsigma \mapsto x = 1/\sqrt{\varsigma}$ and then applying L'Hôpital's rule. From (35) we conclude that there exists a continuous function $g(w,\varsigma)$ defined for $\varsigma^{1/4} > \varepsilon$ and $w \in [\varepsilon, T]$ and satisfying $1 + g(w,\varsigma) \geq 0$ and $g(w,\varsigma) \to 0$ as $\varsigma \to +\infty$ for each fixed w, such that

$$\left(1 - \frac{w^2}{\sqrt{\varsigma}}\right)^{\varsigma+\xi\sqrt{\varsigma}-2} \exp(\sqrt{\varsigma}w^2) = \exp(-w^4/2 - \xi w^2)\left(1 + g(w,\varsigma)\right). \tag{36}$$

We now estimate the integral

$$\int_{\varepsilon}^{T} \exp(-w^4/2 - \xi w^2)\big(1 + g(w,\varsigma)\big)\frac{1}{w}\mathrm{d}w. \tag{37}$$

Using (36)

$$1 + g(w,\varsigma) = \Big(1 - \frac{w^2}{\sqrt{\varsigma}}\Big)^{\varsigma+\xi\sqrt{\varsigma}-2} \exp(\sqrt{\varsigma}w^2 + w^4/2 + \xi w^2),$$

which implies, as $\varsigma \to +\infty$, $\Big(1 - \frac{w^2}{\sqrt{\varsigma}}\Big)^{\xi\sqrt{\varsigma}} \geq \frac{1}{2}\exp(-w^2\xi)$ and

$$\left(e^{w^2}\Big(1 - \frac{w^2}{\sqrt{\varsigma}}\Big)^{\sqrt{\varsigma}}\right)^{\sqrt{\varsigma}} \geq \frac{1}{2}\exp(-w^4/2).$$

Since $\Big(1 - \frac{T^2}{\sqrt{\varsigma}}\Big)^{-2} > 1$ we have

$$1 + g(w,\varsigma) = \Big(1 - \frac{w^2}{\sqrt{\varsigma}}\Big)^{\varsigma+\xi\sqrt{\varsigma}-2} \exp(\sqrt{\varsigma}w^2 + (w^4/2) + \xi w^2) \geq \frac{1}{4}\exp(-w^4/2),$$

and hence the integral in (37) can be estimated as

$$\begin{aligned}\int_{\varepsilon}^{T} \exp(-w^4/2 - \xi w^2)\big(1 + g(w,\varsigma)\big)\frac{1}{w}\mathrm{d}w &\geq \frac{1}{4}\int_{\varepsilon}^{T} \exp(-w^4 - \xi w^2)\frac{1}{w}\mathrm{d}w \\ &> L_1 \int_{\varepsilon}^{T} \frac{1}{w}\mathrm{d}w, \end{aligned} \tag{38}$$

where $L_1 = L_1(\xi, T) := \exp(-T^4 - \xi T^2)/4$. The integral in (38) can be made arbitrarily large, since we can choose $\varepsilon > 0$ suitably small, and hence since the integral in (34) is (strictly) larger than the integral in (37), this concludes the proof that the limit in (30) [and hence in (29)] does not exist.

5 Concluding Remarks

We studied the addition model with input $J_1 = \alpha t^{-1/2}$ and showed the existence of self-similar behaviour along non-characteristic directions.

Along the characteristic direction we considered a different similarity variable, $\xi = (j - \varsigma)/\sqrt{\varsigma}$. This new "layer" variable of width $\sqrt{\varsigma}$ around $j = \varsigma$ provides a kind of expansion of the singularity of the scaling transformation $\Phi_{1,-1/2}$. For this similarity variable, we concluded that $\Phi_{2,-1/2}$ does not scale like $\varsigma^{1/2}(\log \varsigma)^{2/3}$.

Whether there is some similarity variable and some constants a and b such that the similarity function scales like $\varsigma^a(\log\varsigma)^b$ remains an open question.

Acknowledgements The author thanks an anonymous referee for providing insightful comments and valuable suggestions. Discussions with F.P. da Costa and J.T. Pinto are gratefully acknowledged. This work was partially supported by FCT through CAMGSD.

References

1. Ablowitz, M.J., Fokas, A.S.: Complex Variables, 2nd edn. Cambridge Texts in Applied Mathematics. Cambridge University Press, Cambridge (2003)
2. Bartelt, M.C., Evans, J.W.: Exact island-size distributions for submonolayer deposition: influence of correlations between island size and separation. Phys. Rev. B **54**, R17359–R17362 (1996)
3. Brilliantov, N.V., Krapivsky, P.L.: Non-scaling and source-induced scaling behaviour in aggregation models of movable monomers and immovable clusters. J. Phys. A Math. Gen. **24**, 4787–4803 (1991)
4. Costin, O., Grinfeld, M., O'Neill, K.P., Park, H.: Long-time behaviour of point islands under fixed rate deposition. Commun. Inf. Syst. **13**, 183–200 (2013)
5. da Costa, F.P., Sasportes, R.: Dynamics of a non-autonomous ODE system occurring in coagulation theory. J. Dyn. Differ. Equat. **20**, 55–85 (2008)
6. da Costa, F.P., van Roessel, H., Wattis, J.A.D.: Long-time behaviour and self-similarity in a coagulation equation with input of monomers. Markov Process. Relat. Fields **12**, 367–398 (2006)
7. Hendriks, E.M., Ernst, M.H.: Exactly soluble addition and condensation models in coagulation kinetics. J. Colloid Interface Sci. **97**, 176–194 (1984)
8. Sasportes, R.: Dynamical problems in coagulation equations. Ph.D. Thesis, Universidade Aberta, Lisboa (2007). http://hdl.handle.net/10400.2/1909 Accessed 13 Jun 2014
9. Wattis, J.A.D.: Similarity solutions of a Becker-Döring system with time-dependent monomer input. J. Phys. A Math. Gen. **37**, 7823–7841 (2004)

Modelling the Fixed Bed Adsorption Dynamics of CO_2 / CH_4 in 13X Zeolite for Biogas Upgrading and CO_2 Sequestration

José A.C. Silva and Alírio E. Rodrigues

Abstract The sorption of CO_2 and CH_4 in binderless beads of 13X zeolite has been investigated between 313 and 423 K and total pressure up to 0.5 MPa through fixed bed adsorption experiments. Experimental selectivities CO_2 / CH_4 range from 37 at a low pressure of 0.0667 MPa to approximately 5 at the high temperature of 423 K. The breakthrough curves measured show a plateau of pure CH_4 of approximately 6 min depending of the operating conditions chosen. A mathematical model was developed and tested predicting with good accuracy the behaviour of the fixed bed adsorption experiments being a valuable tool for the design of cyclic adsorption processes for biogas upgrading and CO_2 capture using 13X zeolite.

1 Introduction

The reduction of CO_2 and CH_4 emissions to atmosphere is a matter of great concern nowadays since both gases can contribute significantly to the so-called greenhouse effect that describes the trapping of heat near earth's surface by gases in the atmosphere. At the same time CO_2 / CH_4 separations are of interest in treating gas streams like landfill gas, biogas and coal-bed methane. Accordingly, there is a need to investigate on this topic and that can be done with improved efficient technologies to separate or remove CO_2 and CH_4 from exhaust gases. Two recent reviews discuss this matter with great detail concerning the use of adsorbents (porous solids)

J.A.C. Silva (✉)
Escola Superior de Tecnologia e Gestão, Instituto Politécnico de Bragança, Apartado 134, 5301-857 Bragança, Portugal
e-mail: jsilva@ipb.pt

A.E. Rodrigues
Faculdade de Engenharia, Departamento de Engenharia Química, LSRE – Laboratory of Separation and Reaction Engineering, Universidade do Porto, Rua do Dr. Roberto Frias, 4200-465 Porto, Portugal
e-mail: arodrig@fe.up.pt

J.-P. Bourguignon et al. (eds.), *Mathematics of Energy and Climate Change*, CIM Series in Mathematical Sciences 2, DOI 10.1007/978-3-319-16121-1_18

based technologies to handle CO_2 capture and CO_2 / CH_4 separations (see [1, 2]). Biogas is mainly composed by CH_4 (60–70 %) and CO_2 (30–40 %) and to obtain a high energy content CO_2 needs to be separated from CH_4. For this purpose a variety of solid physical adsorbents have been considered including molecular sieve zeolites and a new class of adsorbents named Metal-Organic Frameworks (MOFs). The technology for biogas upgrading using adsorbents is called Pressure Swing Adsorption (PSA). With this technique, carbon dioxide is separated from the biogas by adsorption under elevated pressure. The adsorbing material, is regenerated by a sequential decrease in pressure before the column is reloaded again, hence the name of the technique. A review about the use of PSA technology for Biogas Upgrading is described by Grande [3]. The modelling of fixed bed adsorption dynamics is of fundamental importance for the design of industrial adsorbers due to the complexity of these systems, that involve several mechanisms for mass and heat transfer coupled with thermodynamic models that describe the equilibrium between gas and solid phases (see [4]).

In this work, we will present fixed bed adsorption data of CO_2 and CH_4 on zeolite 13X at 313, 373 and 423 K and pressures up to 0.5 MPa. A mathematical model is developed and validated to describe the fixed bed experimental data which could be used in the implementation (simulation) of cyclic adsorption processes (PSA, TSA) for the purification of biogas or CO_2 sequestration.

2 Mathematical Model for Simulation of Fixed Bed Adsorption Dynamics

In practice the separation of gaseous mixtures by porous solids adsorbents are performed in a fixed bed. The perspective of the design engineer is to predict the response of the column at the outlet (breakthrough curve) after a step change in the concentration at inlet. The transient dynamic response must take into account distinct levels of porosity: bed, particles and crystals, each one corresponding to bulk porosity, macropores and micropores, respectively (Fig. 1). Each level presents different resistances to mass transfer; some of these resistances are placed in series and can be grouped into a single parameter (e.g., film, macropore, and micropore resistances) in order to simplify the numerical procedure (Linear-Driving-Force (LDF) model). At the same time since adsorption is an exothermic phenomena the importance of heat effects should also be considered in the design of such adsorbers.

Let us consider that at time zero a mixture of known composition and an inert gas is introduced at the inlet of the column.

The following additional assumptions are made:

1. Ideal gas;
2. There is no pressure drop in the column;
3. The flow pattern is described by the axial dispersed plug flow model;

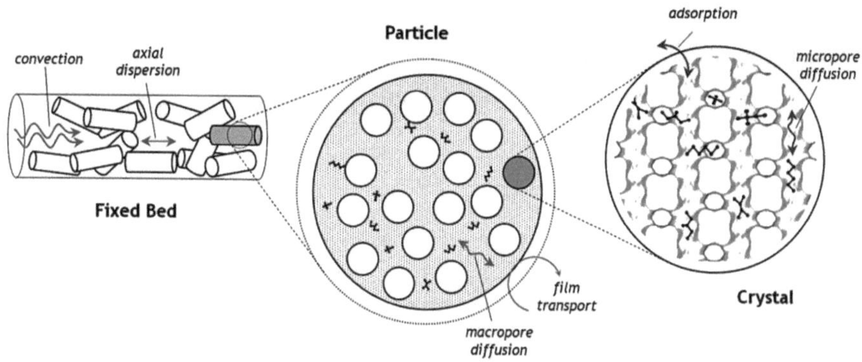

Fig. 1 Schematic representation of a fixed bed adsorption system showing distinct resistances to mass transfer at different scales

4. The mass transfer between bulk gas phase and adsorbent particle is accounted by a Linear-Driving-Force (LDF) model (see [4]).
5. A resistance to heat transfer exists in the external fluid film around the solid.
6. The adsorption equilibrium isotherm is described by the Fowler Model described below in the text (see also [5]).

Table 1 shows the mathematical model equations that describe this system and Table 2 the nomenclature of the variables used.

2.1 Numerical Solution of Model Equations

The set of coupled partial differential equations was reduced first to a set of ordinary differential/algebraic equations (DAE's) applying orthogonal collocation technique to the spatial coordinate (see [6]), where the first and second order differential terms were substituted by collocation matrices $A(i,j)$ and $B(i,j)$, respectively. The collocation points were given by the zeros of Jacobi polynomials $P_N^{(\alpha,\beta)}(x)$, with $\alpha = \beta = 0$ calculated by subroutine JCOBI and the collocation matrices $A(i,j)$ and $B(i,j)$ found by subroutine DFOPR. Both subroutines can be found in a FORTRAN code in Villadsen and Michelsen [6]. The number of interior interpolation points N was chosen to give stability to the numerical solution of discretized equations. The resulting system was solved using a fifth order Runge-Kutta code (ODE's) in conjunction with a Gauss elimination (Algebraic equations). Sixteen collocation points appeared to give satisfactory accuracy for all calculations performed. This gives for two adsorbable species 128 (64×2) ODE's being integrated at the same time: 32 (16×2) from the Mass balance to adsorbable species; 32 (16×2) from the equation representing the Mass transfer rate, 32 (16×2) from the Energy balance in the gas phase and 32 (16×2) from the energy balance for the solid phase. At

Table 1 Mathematical model for the study of fixed bed adsorption dynamics

Overall mass balance:

$$\frac{\partial F}{\partial z} + \varepsilon_b \frac{\partial C}{\partial t} + (1 - \varepsilon_b) \sum_{\iota=1}^{ncp} \frac{\partial \bar{q}_\iota}{\partial t} = 0.$$

Boundary condition:

$$z = 0; \quad t > 0 \quad F = F_f.$$

Mass balance to sorbate species i:

$$-\varepsilon_b D_{ax} \frac{\partial}{\partial z}\left(C \frac{\partial y_i}{\partial z} \right) + \frac{\partial (F y_i)}{\partial z} + \varepsilon_b \frac{\partial (C y_i)}{\partial t} + (1 - \varepsilon_b) \frac{\partial \bar{q}_\iota}{\partial t} = 0.$$

Boundary conditions

$$z = 0; \quad t > 0, \quad \varepsilon_b D_{ax} C \frac{\partial y_i}{\partial z} = F(y_i - y_{if}),$$

$$z = L; \quad t > 0, \quad \frac{\partial y_i}{\partial z} = 0.$$

Mass transfer rate to solid:

$$\frac{\partial \bar{q}_\iota}{\partial t} = k_{LDF} C (y_i - \bar{y}_\iota).$$

Energy balance (gas phase):

$$-K_{ax} \frac{\partial^2 T}{\partial z^2} + F c_{pg} \frac{\partial T}{\partial z} + \varepsilon_b C c_{pg} \frac{\partial T}{\partial t} + (1 - \varepsilon_b) a_p h_p (T - T_s) + a_c h_w (T - T_w) = 0.$$

Boundary conditions

$$z = 0; \quad t > 0, \quad K_{ax} C \frac{\partial T}{\partial z} = F c_{pg} (T - T_f),$$

$$z = L; \quad t > 0, \quad \frac{\partial T}{\partial z} = 0.$$

Energy balance (solid phase):

$$c_{ps} \frac{\partial T_s}{\partial t} = a_p h_p (T - T_s) + \sum_{\iota=1}^{ncp} (-\Delta H_i) \frac{\partial \bar{q}_\iota}{\partial t}.$$

Initial conditions

$$t = 0; \ \forall z; \ y_i = \bar{q}_\iota > 0, \quad K_{ax} C \frac{\partial T}{\partial z} = F c_{pg} (T - T_f),$$

$$z = L; \quad t > 0, \quad \frac{\partial T}{\partial z} = 0.$$

Table 2 Nomenclature for the mathematical model

a_c	Specific area of column (m^{-1})
a_p	Specific area of particle (m^{-1})
c_{pg}	Heat capacity of gas (J/mol K)
c_{ps}	Heat capacity of solid (J/m^3 K)
C	Total gas phase concentration (mol/m^3)
C_f	Total gas phase concentration at inlet of the column (mol/m^3)
d_p	Particle diameter (m)
d_c	Column diameter (m)
D_{ax}	Axial mass dispersion coefficient (m^2/s)
F	Total molar flux (mol/m^2 s)
F_f	Total molar flux at inlet of the column (mol/m^2 s)
h_p	Film heat transfer coefficient (W/m^2 K)
h_w	Wall heat transfer coefficient (W/m^2 K)
K_{ax}	Axial heat dispersion coefficient (W/m K)
$-\Delta H_i$	Heat of adsorption of species i (J/mol)
k_{LDF}	Linear Driving Force (LDF) mass transfer coefficient (s^{-1})
L	Column length (m)
m_a	Mass of adsorbent (g)
ncp	Number of components (–)
P_c	Total gas pressure in the column (kPa)
$\bar{q}_\iota$	Average adsorbed phase concentration of species i in the pores of the adsorbent in equilibrium with the gas phase $\bar{y}_\iota$ (mol/m^3)
t	Time (s)
T	Temperature in bulk gas phase (K)
T_f	Temperature in bulk gas phase at inlet of column (K)
T_s	Temperature in solid phase (K)
T_w	Temperature at the wall of the column (K)
v	Interstitial velocity (m/s)
y_i	Mole fraction of sorbate species i in bulk phase
$\bar{y}_\iota$	Average gas phase concentration of species i in the pores of the adsorbent (mol/m^3)
y_{if}	Mole fraction of sorbate species i at inlet of column
z	Axial coordinate in bed (m)
Greek letters	
ε_b	Bed porosity

the same time there are 32 (16 × 2) equations being solved by Gaussian elimination from the equation representing the Overall mass balance.

2.2 *Binary Isotherms*

The binary adsorption equilibrium of CO_2 and CH_4 in 13X zeolite is described by the Fowler isotherm [5],

$$\frac{1}{p_1}\frac{\theta_1}{(1-\Theta)} = b_1 \exp\left(-\frac{w_{11}\theta_1}{RT} - \frac{w_{12}\theta_2}{RT}\right), \tag{1}$$

$$\frac{1}{p_2}\frac{\theta_2}{(1-\Theta)} = b_2 \exp\left(-\frac{w_{22}\theta_2}{RT} - \frac{w_{21}\theta_1}{RT}\right), \tag{2}$$

where $\theta = q/q_m$ is the degree of filling of sites, b is an equilibrium constant, p the pressure, q the amount adsorbed and q_m is the amount adsorbed at the saturation of the adsorbent, w is the extra energy when sorbate molecules occupy adjacent sites, R the ideal gas constant and T the temperature, Θ is the total fractional loading (species 1 and 2) and the subscripts 1 and 2 refer to the two species of the binary system. Table 3 shows the model parameters.

Table 3 Isotherm model parameters for single and binary sorption of CO_2 and CH_4 in binderless beads of 13X zeolite

		CO_2(1)	CH_4(2)
q_m	(mmol/g_{ads})	7.4	7.4
$-\Delta H$	(kJ/mol)	43.1	8.9
w_{11}	(kJ/mol)	12.3	–
w_{22}		–	–
		313 K	
b	(atm^{-1})	21.3	0.0643
$-w_{12}/RT$	(–)	1.39	
		373 K	
b	(atm^{-1})	1.49	0.0374
$-w_{12}/RT$	(–)	1.25	
		423 K	
b	(atm^{-1})	0.286	0.0256
$-w_{12}/RT$	(–)	1.10	

2.3 *Adsorbent and Sorbates*

The powder of 13X from which the binderless beads were formed is from Chemiewerk Bas Kostritz GmbH (Germany) with a Si/Al ratio of 1.18. Metakaolin is used to manufacture the beads. The synthesis and characterization procedure is described in detail elsewhere (see [7]). Briefly, the beads formed consist in spherical particles with a diameter ranging from 1.2 to 2.0 mm. The size of the zeolite crystals are around 2 m. The sorbate and inert gases were furnished by Air Liquid with the following purities: methane N35 (99.95 %), carbon dioxide N48 (99.998 %), and helium ALPHAGAZ 2 (99.9998 %).

2.4 *Multicomponent Fixed Bed Experiments*

The adsorption column is a 4.6 mm i.d. stainless steel column with 80 mm length placed inside a chromatographic oven. A typical experiment consists in measuring continuously the concentration histories at the outlet of the column using a thermal conductivity detector (TCD) and a mass spectrometer (MS) after feeding the column with a mixture of CO_2 and CH_4 of known composition. When the saturation is reached, the column is regenerated. Details of the apparatus and experimental procedure can be found elsewhere (see [8]).

3 Results and Discussion

3.1 *Binary Breakthrough Curves*

In practice we wish to separate the CO_2 from CH_4 by feeding the fixed bed containing adsorbent particles with mixtures of known composition. When in contact with the adsorbent the mixture is selectively adsorbed giving rise to a breakthrough curve at the outlet with a different composition of the one in bed inlet until the fixed bed is completely saturated. In this section we give an overview of two typical binary breakthrough curves obtained in the 13X zeolite from where the amounts adsorbed and the selectivities of CO_2 and CH_4 were calculated. Figure 2 shows the experimental breakthrough curve after feeding the column with a 50/50 CO_2 / CH_4 mixture diluted with helium (inert) at the temperature of 313 K at a total pressure in the column of 0.5 MPa. We plot the breakthrough curve in terms of the normalized mole flow of the adsorptive species F_i/F_0, as a function of time. Figure 2 shows that at this temperature the separation between CO_2 and CH_4 at the end of the column is significant given rise to a long plateau of pure CH_4 of almost 4 min. It can be also seen that CH_4 breaks the column practically at 1 min due to the very low affinity of this compound with the adsorbent. Another interesting feature

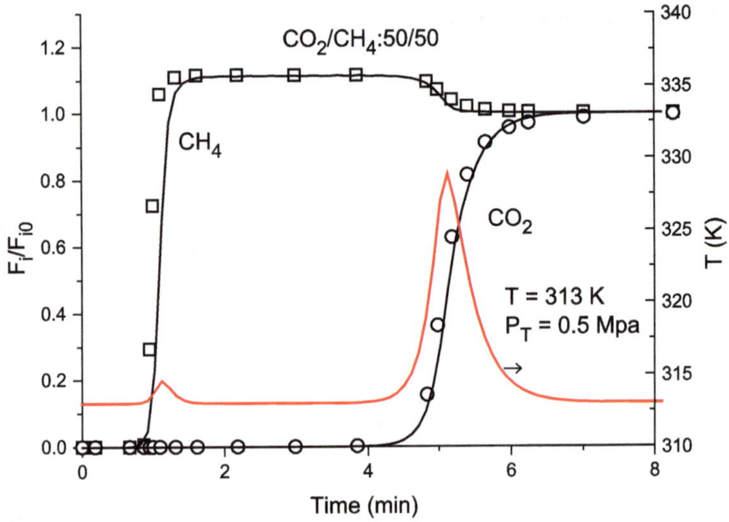

Fig. 2 Binary 50/50 breakthrough curve of CO_2 / CH_4 in 13X zeolite at the temperature of 313 K and total pressure in the column of 0.5 MPa. *Points* are experimental data and *lines* represent model predictions (*black* for fluxes and *red* for temperature). The experimental conditions and model parameters are specified in Table 4. Feed conditions: $y_{CO_2 f} = 0.333$; $y_{CH_4 f} = 0.333$; $P_f = 0.5\,\text{MPa}$; $T_f = 313\,\text{K}$; $F_f = 2.21\,\text{mol/m}^2\,\text{s}$; $C_f = 194\,\text{mol/m}^3$

observed in the figure is that the mass transfer zone for CH_4 is very steep being much more dispersive for CO_2. The model described in Table 1 was used to simulate the binary breakthrough shown in Fig. 2. The model parameters are described in Table 4. To capture the profile of the breakthrough curve it is necessary to use a non-isothermal model. The black lines in the figure show that the model describes with good accuracy the concentration profiles of both components. The figure also shows the temperature at the outlet of the bed (red line) where it can be seen a temperature wave accompanying the breakthrough of each compound. The peak rise is small (only 2 K) when CH_4 breaks but is significant in the case of the CO_2 (20 K). This is due to the low adsorption of CH_4 when compared to CO_2 and also the difference in the heats of adsorption of both compounds (8.9 kJ/mol for CH_4 and 43.1 kJ/mol for CO_2, see Table 3). It should be expected much higher peak rises in large industrial columns since they operate in adiabatic conditions. Figure 3 shows a breakthrough curve for a different ratio of CO_2(25)/CH_4(75) in the feed. It can be seen that in this case the plateau of pure CH_4 increases to around 6 min. The model predicts again with good accuracy the concentration profiles of both components. The peak rise of CO_2 also increases to around 30 K. Both breakthrough curves represented in Figs. 2 and 3 indicate that the binderless 13X zeolite is efficient for the removal of CO_2 from its binary CO_2 / CH_4 mixtures. At the same time the mathematical model is capable to capture with good accuracy the concentration profiles of both compounds as well as the plateau of pure CH_4 observed experimentally.

Table 4 Model parameters for the simulation of the breakthrough experiments

C_{pg}	29.1 J/mol K	K_{LDF}	7.2 s^{-1}
C_{ps}	0.8 J/g K	K_{ax}	0.465 W/m K
d_p	1.6×10^{-3} m	L	0.08 m
d_c	4.6×10^{-3} m	m_a	0.817 g
D_{ax}	8.25×10^{-5} m^2/s (see note 1)	Greek letters	
h_p	319 W/m^2 K (see note 2)	ε_b	0.38
h_w	250 W/m^2 K (see note 3)		

Note 1: Calculated by the Correlation $D_{ax} = 0.7D_m + 0.5d_p v$ taken from Ruthven [4]. The axial mass Peclet number is 162
Note 2: This value was estimated from the limit of $Nu = 2$ and it can be considered a very high value, which means that the temperature between solid and bulk gas phase is in equilibrium
Note 3: This parameter was obtained through the fitting of the experimental breakthrough curve
Note 4: The isotherm model parameters are the ones shown for the Fowler isotherm (see Table 3)

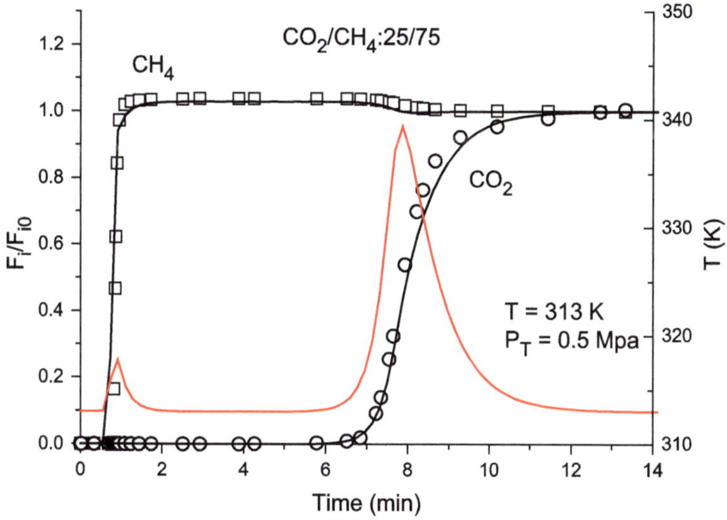

Fig. 3 Binary 25/75 breakthrough curve of CO_2 / CH_4 in 13X zeolite at the temperature of 313 K and total pressure in the column of 0.5 MPa. *Points* are experimental data and *lines* represent model predictions (*black* for fluxes and *red* for temperature). The experimental conditions and model parameters are specified in Table 4. Feed conditions: $y_{CO_2 f} = 0.166$; $y_{CH_4 f} = 0.486$; $P_f = 0.5$ MPa; $T_f = 313$ K; $F_f = 2.47$ mol/m^2 s; $C_f = 194$ mol/m^3

3.2 *Sorption Selectivities*

The primary requirement for an economic separation is an adsorbent with sufficiently high adsorption selectivity, S, defined on a molar basis by $S = (q_1/p_1)/(q_2/p_2)$, where in this case component 1 (CO_2) is the more adsorbed component. Figure 4 shows the temperature dependent sorption selectivity obtained for several 50/50 and 25/75 CO_2 / CH_4 mixture experiments as a function of total

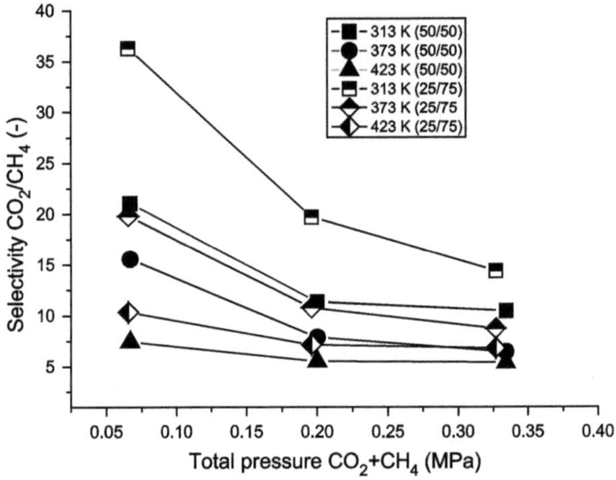

Fig. 4 Selectivities as function of temperature and total mixture pressure of adsorbable species for several binary 50/50 and 25/75 mixture ratio experiments

pressure of the sorbate species. One point in the figure means one experimental binary breakthrough curve. The SCO_2/CH_4 is very high at the low temperature of 313 K and total pressure of sorbates of 0.0666 MPa being 36.3 and 21.1 for the 25/75 and 50/50 mixtures, respectively. As the pressure increases the selectivities decreases but the values are still considerable high at 313 K and total pressure of 0.334 MPa ranging from 14.4 to 10.4 for the 25/75 and 50/50 mixtures, respectively. When temperature increases the selectivities decreases being the lowest value observed 5.4 at 423 K, partial pressure, 0.334 MPa in a 50/50 mixture. These results show that the binderless beads o 13X zeolite can be considered an excellent separator of mixtures CO_2 / CH_4 when appropriate operating conditions are chosen.

4 Conclusions

We performed a study of the sorption of binary mixtures of CO_2 and CH_4 in binderless beads of 13X zeolite between 313 and 423 K and total pressure up to 0.5 MPa. Experimental selectivities CO_2 / CH_4 range from 37 at a low pressure of 0.0667 MPa and temperature of 313 K to approximately 5 at the high temperature of 423 K. The efficiency of the separation of mixtures CO_2 / CH_4 in the binderless beads of 13X zeolite starting from a fresh column is illustrated through two breakthrough curves where pure plateaus of CH_4 are observed (6 and 4 min for mixtures 25(CO_2)/75(CH_4) and 50(CO_2)/50(CH_4) respectively) at 313 K and total pressure 0.5 MPa. The mathematical fixed bed adsorption dynamic model developed taking into account several resistances to mass and heat transfer coupled to the thermodynamic model of adsorption of Fowler was validated through the experimental data proving to be a valuable tool for the design of cyclic adsorption processes for biogas upgrading and CO_2 capture.

Acknowledgements This work was co-financed by FCT/MEC, FEDER under Programe PT2020 (Project UID/EQU/50020/2013).

References

1. D'Alessandro, D.M., Smit, B., Long, J.R.: Carbon dioxide capture: prospects for new materials. Angew. Chem. Int. Ed. **49**(35), 6058–6082 (2010)
2. Férey, G., Serre, C., Devic, T., Maurin, G., Jobic, H., Llewellin, P.L., Weireld, G., Vimont, A., Daturi, M., Chang, J.S.: Why hybrid porous solids capture greenhouse gases? Chem. Soc. Rev. **40**, 550–562 (2011)
3. Grande, C.A.: Biogas upgrading by pressure swing adsorption. In: Bernardes, M.A.S. (ed.) Biofuel's Engineering Process Technology, Chap. 3, pp. 65–84. InTech, Rijeka (2011)
4. Ruthven, D.M.: Principles of Adsorption and Adsorption Processes. Wiley, New York (1984)
5. Silva, J.A.C., Schumann, K., Rodrigues, A.E.: Sorption and kinetics of CO_2 and CH_4 in binderless beads of 13X zeolite. Microporous Mesoporous Mater. **158**, 219–228 (2012)
6. Villadsen, J.V., Michelsen, M.L.: Solution of differential equation models by polynomial approximation. Prentice-Hall, Englewood Cliffs (1978)
7. Schumann, K., Unger, B., Brandt, A., Scheffler, F.: Investigation on the pore structure of binderless zeolite 13X shapes. Microporous Mesoporous Mater. **154**, 119–123 (2012)
8. Silva, J.A.C., Cunha, A.F., Schumann, K., Rodrigues, A.E.: Binary adsorption of CO_2 / CH_4 in binderless beads of 13X zeolite. Microporous Mesoporous Mater. **187**, 100–107 (2014)

Detection of Additive Outliers in Poisson INAR(1) Time Series

Maria Eduarda Silva and Isabel Pereira

Abstract Outlying observations are commonly encountered in the analysis of time series. In this paper a Bayesian approach is employed to detect additive outliers in order one Poisson integer-valued autoregressive time series. The methodology is informative and allows the identification of the observations which require further inspection. The procedure is illustrated with simulated and observed data sets.

1 Introduction

It is well known that unusual observations and intervention effects often occur in data sets and can have adverse effects on model identification and parameter estimation. Time series of counts are no exception. In the last decades time series of counts have become available in a wide variety of fields including: actuarial science, computer science, economics, epidemiology, finance, hydrology, meteorology and environmental studies. These data are naturally non-normal and present non linear characteristics. The need to analyse such data adequately led to a multiplicity of approaches and a diversification of models that explicitly account for the discreteness of the data, see [10] for a recent review. In this paper we focus on the class of Poisson integer valued autoregressive models of order 1, INAR(1). This model, first proposed by [1], has been extensively studied in the literature and applied to many real-world problems because of its simplicity and easiness of interpretation. In fact, any data series that may be thought of as the number of members (e.g. people, firms or orders) of a queue, the number of units in a stock or inventory, or the outcome of a birth-and-death process, or a branching process with immigration may be modelled by the INAR class. The point is that the INAR class has found applications across many disciplines. Hence, it is timely to study the problem of outlier detection given its relevance for inference and diagnostics.

M.E. Silva (✉)
CIDMA & Faculdade de Economia, Universidade do Porto, Porto, Portugal
e-mail: mesilva@fep.up.pt

I. Pereira
CIDMA & Departamento de Matemática, Universidade de Aveiro, Aveiro, Portugal
e-mail: isabel.pereira@ua.pt

J.-P. Bourguignon et al. (eds.), *Mathematics of Energy and Climate Change*,
CIM Series in Mathematical Sciences 2, DOI 10.1007/978-3-319-16121-1_19

In the framework of Gaussian linear time series the problem of detecting and estimating outliers and other intervention effects has been investigated by several authors including [4, 6, 11, 15]. However, the problem of modelling outliers and other intervention effects in the context of time series of counts has, as yet, received little attention in the literature albeit its relevance for inference and diagnostics. Moreover, in this context additional motivation stems from the fact that the usual techniques for outlier removal are not adequate since often lead to non integer values. In the context of time series of counts, [7] investigate the problem of modelling intervention effects in INGARCH models and [2, 3] consider Conditional Least Squares (CLS) estimation of the parameters of an INAR(1) model contaminated, at known time periods, with innovational and additive outliers, respectively. Here the problem of detecting outliers is considered under a Bayesian perspective. Bayesian approaches have been used to detect outliers in ARMA models by [11] and in bilinear models by [5]. This approach has the advantage of not requiring beforehand knowledge on the number and location of outliers in the series and of treating equally all the observations (with and without outliers). In fact, all the observations have the same prior probability of being an outlier. Then, at each time point we estimate the posterior probability of occurrence of an outlier via Gibbs sampling. The Gibbs sampler [8] is a Markovian updating scheme enabling the obtention of samples from a joint distribution via iterated sampling from full conditional distributions. The method may be briefly described considering the case of three parameters $(\theta_1, \theta_2, \theta_3)$ with a complex (posterior) joint and marginal distributions. Let $\mathbf{y}$ be the observed time series and $f_i(\theta_i|\theta_k, \theta_l, \mathbf{y})$ be the conditional distribution of θ_i given the remainder parameters θ_k, θ_l and data, $\mathbf{y}$. The Gibbs sampler employed in this paper then works as follows: given initial values $(\theta_1^{(0)}, \theta_2^{(0)}, \theta_3^{(0)})$, draw $\theta_1^{(1)}$ from $f_1(\theta_1|\theta_2^{(0)}, \theta_3^{(0)}, \mathbf{y})$, then draw $\theta_2^{(1)}$ from $f_2(\theta_2|\theta_3^{(0)}, \theta_1^{(1)}, \mathbf{y})$ and finally, complete the first iteration by drawing $\theta_3^{(1)}$ from $f_3(\theta_3|\theta_1^{(1)}, \theta_2^{(1)}, \mathbf{y})$. After a large number of iterations, say $M+N$ we obtain a sample $(\theta_1^{(j)}, \theta_2^{(j)}, \theta_3^{(j)}), j = M+1, M+2, \ldots, M+N$ whose empirical distribution can approximate the desired posterior marginals. This methodology provides estimates for the probability of outlier occurrence at each time point leading to an effective outlier detection.

To motivate our approach, we represent in Fig. 1a a data set studied by [16] concerning the number of different IP addresses (approximately equivalent to the number of different users) accessing the server of the pages of the Department of Statistics of the University of Würzburg in 2-min periods from 10 am to 6 pm on the 29th November 2005, in a total of 241 observations. Henceforth, this data set will be denoted as IP data and is analysed in detail in Sect. 4.1. Figure 1b represents the posterior probability of outlier occurrence at time t and clearly indicates $t = 224$ as an outlying observation. This result agrees with that of [16] who uses statistical process control techniques.

The paper is organized as follows. Section 2 introduces the first order Poisson integer-valued autoregressive model contaminated with outliers. Section 3 explains the procedure for outlier detection and discusses several computational issues.

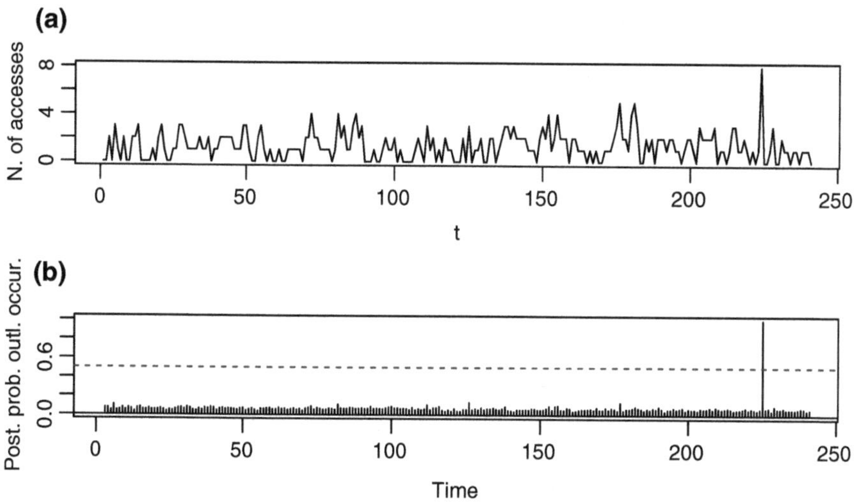

Fig. 1 Number of different IP addresses accessing the server of the pages of the Department of Statistics of the University of Würzburg between 10 am and 6 pm on 29 November 2005 (**a**); posterior probability of outlier occurrence (**b**)

Section 4 illustrates the methodology on several sets of simulated data as well as on the IP data set. Section 5 concludes the paper.

2 INAR(1) Models with Additive Outliers

The Poisson INAR(1), PoINAR(1), model, first introduced by [1] and [12] is defined by the recursive equation

$$X_t = \alpha \circ X_{t-1} + e_t, \quad t \in \mathbb{N}_0, \tag{1}$$

where $\circ$ denotes the binomial thinning operator and $\{e_t\}$, the arrival process, is a sequence of independent and identically distributed Poisson variables, $e_t \sim \text{Po}(\lambda)$, independent of the thinning operations. The binomial thinning is defined as $\alpha \circ X_{t-1} \overset{\mathcal{D}}{=} \sum_{j=1}^{X_{t-1}} \xi_{t,j}$, with $\xi_{t,j}, j = 1, \ldots, X_{t-1}$ a sequence of independent Bernoulli random variables (r.v.'s) with probability of success $\text{P}(\xi_{tj} = 1) = \alpha$. Thus $\alpha \circ X_{t-1} | X_{t-1} \sim Bi(X_{t-1}, \alpha)$, ensuring the discreteness of the process. In fact, the thinning operator $\circ$ acts as the analogue of the usual multiplication used in the continuous-valued autoregressive, AR(1), processes. This concept of thinning is well known in classical probability theory and has been used in the Bienaymé-Galton-Watson branching processes literature as well as in the theory of stopped-sum distributions. Under the above conditions if $X_0 \sim \text{Po}(\lambda/(1-\alpha))$, then the process is strictly stationary and $X_t \sim \text{Po}(\lambda/(1-\alpha))$, yielding a Poisson marginal. X_t behaves like a queue, with arrivals at time t represented by e_t and survivors remaining in the queue, from $t-1$ to t, by $\alpha \circ X_{t-1}$. Alternatively the model may be thought of as a birth-and-death, or stock, process, with additions (births) being generated by e_t and losses (deaths) by $X_{t-1} - \alpha \circ X_{t-1}$.

Note that the Poisson INAR(1) process is a Markov process and that the distribution of X_t given X_{t-1}, $\mathrm{p}(X_t|X_{t-1})$ is the convolution of the two components, binomial and Poisson, as follows:

$$\mathrm{p}(x_t|x_{t-1}) = \sum_{i=0}^{M_t} \binom{x_{t-1}}{i} \alpha^i \, (1-\alpha)^{x_{t-1}-i} \, \frac{\mathrm{e}^{-\lambda} \, \lambda^{x_t - i}}{(x_t - i)!} \tag{2}$$

where $M_t = \min(x_{t-1}, x_t)$ and $\binom{\cdot}{\cdot}$ is the standard combinatorial symbol.

Assume now that the observed time series of counts $y_1, \ldots, y_n$ may be contaminated with one or more additive outliers at unknown time points. Roughly speaking an additive outlier can be interpreted as a measurement error or as an impulse due to some unspecified exogenous source at time τ_i, $i = 1, \ldots, k$. When outliers are present, X_t is unobservable. Then the proposed model for Y_t is the following

$$Y_t = X_t + \eta_t \delta_t, \quad 1 \le t \le n \tag{3}$$

where X_t is a PoINAR(1) process satisfying (1), $\delta_1, \ldots, \delta_n$ are Bernoulli variables with $\mathrm{P}(\delta_t = 1) = \epsilon$, independent of X_t and $\eta_1, \ldots, \eta_n$ are integer valued independent random variables, also independent of X_t and of δ_t. This means that if $\delta_t = 1$ the observed value Y_t is contaminated with an additive outlier (AO) of magnitude η_t. Henceforth, model (3) will be called a Poisson INAR(1) contaminated with outliers.

To obtain the likelihood of the data let $\mathbf{y} = (y_1, \ldots, y_n)$, $\boldsymbol{\Theta} = (\alpha, \lambda)$, $\boldsymbol{\delta} = (\delta_1, \ldots, \delta_n)$, $\boldsymbol{\eta} = (\eta_1, \ldots, \eta_n)$ and assume that there is no outlier in the first observation, that is $y_1 = x_1$. Moreover, under (3) $X_t = Y_t - \eta_t \delta_t$ is a PoINAR(1). Then conditioning on the first observation the likelihood of $\mathbf{y}$ is given by

$$L(\boldsymbol{\Theta}, \boldsymbol{\delta}, \boldsymbol{\eta}, \epsilon, \mathbf{y}) = \prod_{t=2}^{n} \epsilon^{\delta_t} (1-\epsilon)^{1-\delta_t}$$

$$\times \sum_{i=0}^{M_t} \frac{\lambda^{y_t - \delta_t \eta_t - i}}{(y_t - \delta_t \eta_t - i)!} \binom{y_{t-1} - \delta_{t-1}\eta_{t-1}}{i} \mathrm{e}^{-\lambda} \, \alpha^i \, (1-\alpha)^{y_{t-1} - \delta_{t-1}\eta_{t-1} - i}$$

where now $M_t = \min(y_{t-1} - \eta_{t-1}\delta_{t-1}, y_t - \eta_t \delta_t)$.

3 Bayesian Outlier Detection in PoINAR(1) Models

In this section we describe the Bayesian approach via Gibbs sampling to estimate model (3).

In addition to the data and the likelihood, the Bayesian model specification also requires a prior distribution on the parameters. The prior distribution for the contamination parameter ϵ is $\epsilon \sim \text{Be}(h, g)$, with expectation $E(\epsilon) = h/(h+g)$.The prior distribution for η_t is $Po(\beta)$. Regarding the PoINAR(1) parameters α and λ we choose for prior distributions the conjugate of Binomial and Poisson, respectively and thus $\alpha \sim \text{Be}(a, b)$, $\lambda \sim \text{Ga}(c, d)$ [14]. The choice of the set of hyperparameters a, b, c, d, β, h, g is discussed in Sect. 3.2.

Under the above assumptions the prior distribution for $(\boldsymbol{\Theta}, \boldsymbol{\delta}, \boldsymbol{\eta}, \epsilon)$, denoted $\pi(\boldsymbol{\Theta}, \boldsymbol{\delta}, \boldsymbol{\eta}, \epsilon)$ is given by

$$\pi(\boldsymbol{\Theta}, \boldsymbol{\delta}, \boldsymbol{\eta}, \epsilon) \propto \mathrm{e}^{-d\lambda}\, \lambda^{c-1}\, \alpha^{a-1}\, (1-\alpha)^{b-1}\, \epsilon^{h-1}\, (1-\epsilon)^{g-1} \prod_{t=2}^{n} \mathrm{e}^{-\beta} \frac{\beta^{\eta_t}}{\eta_t!}. \tag{4}$$

The posterior distribution for $(\boldsymbol{\Theta}, \boldsymbol{\delta}, \boldsymbol{\eta}, \epsilon)$ is then given by

$$\begin{aligned}
\pi(\boldsymbol{\Theta}, \boldsymbol{\delta}, \boldsymbol{\eta}, \epsilon | \mathbf{y}) &\propto \pi(\boldsymbol{\Theta}, \boldsymbol{\delta}, \boldsymbol{\eta}, \epsilon)\, L(\boldsymbol{\Theta}, \boldsymbol{\delta}, \boldsymbol{\eta}, \epsilon, \mathbf{y}) \\
&\propto \mathrm{e}^{-[d\lambda + n\beta]}\, \lambda^{c-1}\, \alpha^{a-1}\, (1-\alpha)^{b-1}\, \epsilon^{h-1}\, (1-\epsilon)^{g-1} \\
&\quad \frac{\beta^{\sum_{t=2}^{n} \eta_t}}{\prod_{t=2}^{n} \eta_t!}\, L(\boldsymbol{\Theta}, \boldsymbol{\delta}, \boldsymbol{\eta}, \epsilon, \mathbf{y})
\end{aligned} \tag{5}$$

with $0 < \alpha < 1, \lambda > 0, 0 < \epsilon < 1, \delta_t = 0, 1$ and $\eta_t = 0, 1, \ldots, t = 2, 3, \ldots, n$.

This approach is attractive since it enables to measure the likelihood that at each time point the observed value Y_t is affected by an outlier as well as describe the distribution of the outlier size. However, the complexity of the posterior distribution (5) (and consequently of the marginals) makes them analytically intractable. Hence MCMC (Marlov Chain Monte Carlo) methods are required. The model parameters are then estimated by sampling from the complete conditional distribution of each parameter, conditional on the previous sampled values of the other parameters.

3.1 Full Posterior Distributions

The full conditional posterior distributions for α and λ are given by [14]

$$\begin{aligned}
\pi(\alpha | \lambda, \boldsymbol{\delta}, \boldsymbol{\eta}, \epsilon, \mathbf{y}) \propto \quad & \alpha^{a-1}\, (1-\alpha)^{b-1} \\
\prod_{t=2}^{n} \epsilon^{\delta_t} (1-\epsilon)^{1-\delta_t} & \sum_{i=0}^{M_t} \frac{\lambda^{y_t - \eta_t \delta_t - i}}{(y_t - \eta_t \delta_t - i)!} \binom{y_{t-1} - \eta_{t-1}\delta_{t-1}}{i} \\
& \alpha^{i}\, (1-\alpha)^{y_{t-1} - \eta_{t-1}\delta_{t-1} - i}
\end{aligned} \tag{6}$$

and

$$\pi(\lambda|\alpha,\boldsymbol{\delta},\boldsymbol{\eta},\epsilon,\mathbf{y}) \propto \quad \lambda^{c-1}\mathrm{e}^{-(d+(n-1))\lambda}$$
$$\prod_{t=2}^{n}\epsilon^{\delta_t}(1-\epsilon)^{1-\delta_t}\sum_{i=0}^{M_t}\frac{\lambda^{y_t-\eta_t\delta_t-i}}{(y_t-\eta_t\delta_t-i)!}\binom{y_{t-1}-\eta_{t-1}\delta_{t-1}}{i}$$
$$\alpha^i\,(1-\alpha)^{y_{t-1}-\eta_{t-1}\delta_{t-1}-i}. \tag{7}$$

Now, with respect to the full conditional distribution of $\boldsymbol{\delta}$ and $\boldsymbol{\eta}$ we reason as follows. Let $j=2,\ldots,n$, and for each j define $\boldsymbol{\Upsilon}_{\boldsymbol{\delta}}=(\alpha,\lambda,\boldsymbol{\eta},\epsilon,\boldsymbol{\delta}_{(-j)})$ and $\boldsymbol{\Upsilon}_{\boldsymbol{\eta}}=(\alpha,\lambda,\boldsymbol{\delta}_{(-j)},\epsilon,\boldsymbol{\eta}_{(-j)})$ where $\boldsymbol{\delta}_{(-j)}$ and $\boldsymbol{\eta}_{(-j)}$ denote the vectors $\boldsymbol{\delta}$ and $\boldsymbol{\eta}$, respectively, each with the jth component deleted. To derive the full conditional distribution of $\boldsymbol{\delta}$ first note that $\delta_j|(y,\boldsymbol{\Upsilon}_{\boldsymbol{\delta}})\sim Ber(p_j)$. Accordingly, we can write

$$p_j=\mathrm{P}(\delta_j=1|\mathbf{y},\boldsymbol{\Upsilon}_{\boldsymbol{\delta}})=\frac{\mathrm{P}(\delta_j=1,\mathbf{y}|\boldsymbol{\Upsilon}_{\boldsymbol{\delta}})}{f(\mathbf{y}|\boldsymbol{\Upsilon}_{\boldsymbol{\delta}})}. \tag{8}$$

But

$$f(\mathbf{y}|\boldsymbol{\Upsilon}_{\boldsymbol{\delta}})=f(\mathbf{y}|\delta_j=1,\boldsymbol{\Upsilon}_{\boldsymbol{\delta}})\mathrm{P}(\delta_j=1|\boldsymbol{\Upsilon}_{\boldsymbol{\delta}})+f(\mathbf{y}|\delta_j=0,\boldsymbol{\Upsilon}_{\boldsymbol{\delta}})\mathrm{P}(\delta_j=0|\boldsymbol{\Upsilon}_{\boldsymbol{\delta}})$$

with $\mathrm{P}(\delta_j=1|\boldsymbol{\Upsilon}_{\boldsymbol{\delta}})=\epsilon$.

Therefore

$$p_j=\frac{\epsilon f(\mathbf{y}|\delta_j=1,\boldsymbol{\Upsilon}_{\boldsymbol{\delta}})}{\epsilon f(\mathbf{y}|\delta_j=1,\boldsymbol{\Upsilon}_{\boldsymbol{\delta}})+(1-\epsilon)f(\mathbf{y}|\delta_j=0,\boldsymbol{\Upsilon}_{\boldsymbol{\delta}})}. \tag{9}$$

To compute $f(\mathbf{y}|\delta_j=1,\boldsymbol{\Upsilon}_{\boldsymbol{\delta}})$ first note that Y_t inherits the Markovian property of X_t and consequently the outlier at time j affects the model for $t=j$ and $t=j+1$. Therefore

$$\begin{aligned} f(\mathbf{y}|\delta_j=1,\boldsymbol{\Upsilon}_{\boldsymbol{\delta}}) =& f(y_j,y_{j+1}|y_{j-1},\delta_j=1,\boldsymbol{\Upsilon}_{\boldsymbol{\delta}})\\ =& f(y_j,y_{j+1}|y_{j-1},\delta_j=1,\alpha,\lambda,\eta_j)\\ =& f(y_j|y_{j-1},\delta_j=1,\alpha,\lambda,\eta_j)\\ &\times f(y_{j+1}|y_j,\delta_j=1,\alpha,\lambda,\eta_j). \end{aligned} \tag{10}$$

Moreover, assuming that $Y_{j-1}=X_{j-1}$ and $Y_{j+1}=X_{j+1}$ meaning that there are no patches of outliers we have

$$f(y_j|y_{j-1},\delta_j=1,\alpha,\lambda,\eta_j)=\mathrm{e}^{-\lambda}\sum_{i=0}^{M_j^{**}}\binom{y_{j-1}}{i}\alpha^i\,(1-\alpha)^{y_{j-1}-i}\,\frac{\lambda^{y_j-\eta_j-i}}{(y_j-\eta_j-i)!} \tag{11}$$

and

$$f(y_{j+1}|y_j, \delta_j = 1, \alpha, \lambda, \eta_j)$$
$$= e^{-\lambda} \sum_{i=0}^{M_j^*} \binom{y_j - \eta_j}{i} \alpha^i (1-\alpha)^{y_j - \eta_j - i} \frac{\lambda^{y_{j+1}-i}}{(y_{j+1}-i)!} \tag{12}$$

with $M_t^{**} = \min(y_{t-1}, y_t - \eta_t)$ and $M_t^* = \min(y_t - \eta_t,\ y_{t+1})$.

Similarly, if $\delta_j = 0$ then $X_j = Y_j$ and therefore

$$f(\mathbf{y}|\delta_j = 0, \boldsymbol{\Upsilon}_{\boldsymbol{\delta}}) = f(\mathbf{y}|\delta_j = 0, \alpha, \lambda, \eta_j)$$
$$= \prod_{t=j}^{j+1} \sum_{i=0}^{M_t} \binom{y_{t-1}}{i} \alpha^i (1-\alpha)^{y_{t-1}-i} e^{-\lambda} \frac{\lambda^{y_t - i}}{y_t - i!} \tag{13}$$

Now, to derive the conditional posterior distribution of $\boldsymbol{\eta}$ note that if $\delta_j = 0$, no outlier at $t = j$, there is no information about η_j except the prior. Then $\eta_j|(\mathbf{y}, \delta_j = 0, \boldsymbol{\Upsilon}_{\boldsymbol{\eta}}) \sim Po(\beta)$. However, if $\delta_j = 1$, $\mathbf{y}$ contains information about η_j. Therefore we have

$$\pi(\eta_j\, |\mathbf{y}, \delta_j = 1, \boldsymbol{\Upsilon}_{\boldsymbol{\eta}}) =$$
$$\frac{\pi(\eta_j|\delta_j = 1, \boldsymbol{\Upsilon}_{\boldsymbol{\eta}}) f(\mathbf{y}|\delta_j = 1, \eta_j, \boldsymbol{\Upsilon}_{\boldsymbol{\eta}})}{\sum_{\eta_j=0}^{\infty} \pi(\eta_j|\delta_j = 1, \boldsymbol{\Upsilon}_{\boldsymbol{\eta}}) f(\mathbf{y}|\delta_j = 1, \eta_j, \boldsymbol{\Upsilon}_{\boldsymbol{\eta}})}$$

$$\propto e^{-\beta} \beta^{\eta_j}/(\eta_j!)\, f(y_j, y_{j+1}\, |\eta_j, \delta_j = 1, \alpha, \lambda, \epsilon),$$
$$\eta_j = 0, 1, 2, \ldots \tag{14}$$

with $f(y_j, y_{j+1}\, |\eta_j, \delta_j = 1, \alpha, \lambda, \epsilon)$ as given in (10).

Finally, the conditional posterior distribution for ϵ depends only on $\boldsymbol{\delta}$. Since the prior distribution of ϵ is $Be(h, g)$ the conditional posterior is given by

$$\epsilon|\mathbf{y}, \lambda, \boldsymbol{\eta}, \boldsymbol{\delta} \equiv \epsilon|\boldsymbol{\delta} \sim Be(h + k, g + n - 1 - k) \tag{15}$$

where k is the number of outliers (number of δ_j's equal to 1).

3.2 Computational Issues

The full conditional distributions of $\alpha, \lambda, \boldsymbol{\delta} = (\delta_2, \ldots, \delta_n)$, $\boldsymbol{\eta} = (\eta_2, \ldots, \eta_n)$ and ϵ do not have standard forms, therefore we have to use the Metropolis-Hastings algorithm to draw a sample of a Markov chain which converges to the joint posterior distribution of the parameters. Since they are not log-concave densities we use the Gibbs methodology within the Metropolis step. In particular Adaptive Rejection Metropolis sampling—ARMS [9]—is used inside the Gibbs sampler. When the number of iterations is sufficiently large, the Gibbs draw can be regarded as a sample from the joint posterior distribution. Thus, complete distributions for the estimated parameters are obtained.

Two key issues in the successful implementation of this methodology are: deciding the length of the chain and the burn-in period and establishing the convergence of the chain. We use a burn-in period of M iterations and then iterate the Gibbs sampler for a further N iterations, but retain only each Lth element in the sample. This thinning strategy reduces the autocorrelation within the chain.

We now discuss the other relevant issue in the proposed Bayesian approach: the choice of the hyperparameters for prior distributions. Recall from Sect. 2 that the prior distributions for α and λ are $\mathrm{Be}(a, b)$ and $\mathrm{Ga}(c, d)$, respectively. In the absence of further or inside information we set $a = b = c = d = 0.001$ to use non informative prior distributions (Beta and Gamma distributions with large variability). For the prior distribution for the size of the outlier at time t, $\eta_t \sim Po(\beta)$ two approaches are pursued: an informative setup in which β_{info} is set equal to three times the standard deviation of the 1-step-ahead prediction error and also a non-informative setup with $\beta_{ninfo} = 30$ to reflect large variability. Finally, regarding the prior distribution for the probability of outliers occurrence, $\epsilon \sim Be(h, g)$, we choose $h = 5$, $g = 95$ to express the view that outliers occur occasionally. The posterior probability of outlier occurrence is then estimated and inspected to identify potential outliers. In an automated procedure a cut-off value, typically $c = 0.2$, may be used.

4 Illustration

In this section we document the performance of the above procedure with simulated data sets of 100 observations. In all the examples the Gibbs sampler is iterated $M + N = 5005$ times and the $L = 5$th value of the last $N = 2505$ iterations is kept, providing sample sizes of 501 values from which the probability of outlier occurrence at each time point as well as all the other parameter estimates are computed. The parameters α and λ are computed as the posterior mean. To ensure an integer value, the size of the outlier η is computed as the posterior median. The results are reported for $\beta_{ninfo} = 30$ since they do not differ from those obtained with β_{info}. We simulate time series from several PoINAR(1) processes without and with outliers of different sizes introduced at different times. The range of values considered for α and λ allow to illustrate the performance of the methodology for time series with small and large variability.

Table 1 reports the results from the application of the methodology to time series simulated from INAR(1) models with $\alpha = 0.15, 0.5, 0.85$ and $\lambda = 1, 3, 5$, all contaminated with three outliers of different sizes. The data generating model is identified in the column headed *Model* by the parameters of the contaminated PoINAR(1) model: α, λ and η_S, which indicates contamination with an outlier of size η at time S. Finally, all the outliers detected by the algorithm, that is all the time points for which the posterior probability of outlier occurrence is over the threshold $\hat{p} = 0.2$, are indicated by the time of occurrence, estimated size, and associated posterior probability.

The results indicate that the procedure detects all the additive outliers in PoINAR(1) time series with high (near 1) estimated probabilities of occurrence. Moreover, the occurrence of false detections was null in the contaminated time series as well as in outlier free time series whose results are not reported here.

Note that the convergence of the MCMC algorithm was duly analysed with the usual diagnostic tests available in [13].

4.1 IP Data Example

Let us consider once again the motivating example of Sect. 1, regarding the number of different IP addresses accessing the server of the Department of Statistics of the University of Würzburg on November 29th, 2005, between 10 am and 6 pm, represented in Fig. 1 [16]. The sample mean and variance of the series are $\bar{x} = 1.32, \hat{\sigma}^2 = 1.39$. The autocorrelation and partial autocorrelation functions indicate that a model of order one is appropriate. CLS estimates for α and λ are $\hat{\alpha} = 0.22$ and $\hat{\lambda} = 1.03$, respectively. The result of applying the proposed methodology is represented in Fig. 1b indicating the possible occurrence of an outlier at time $t = 224$. The estimated size of the outlier is $\hat{\eta} = 7$. It is interesting to note that setting the time of the outlier to $t = 224$ and using the results from [2] the CLS estimate for η is $\hat{\eta}_{CLS} = 6.73$. Removing the effect of the outlier at $t = 224$ the mean and variance of the resulting series are 1.29 and 1.2, respectively. The autocorrelation and partial autocorrelation functions still indicate that a model of order one is appropriate. CLS estimates for the parameters are now $\hat{\alpha}_{CLS} = 0.29$ and $\hat{\lambda}_{CLS} = 0.91$ in accordance with the estimates obtained from the Gibbs sampling, $\hat{\alpha}_{Bayes} = 0.27$ and $\hat{\lambda}_{Bayes} = 0.89$, whose posterior distributions are represented in Fig. 2.

Table 1 Results from Gibbs sampling in simulated INAR(1) time series with parameters α, λ, contaminated at time S with an outlier of size η_S

Model		Estimates	Outliers detected Time	Size	Probability	Model		Estimates	Outliers detected Time	Size	Probability
α	0.15	0.07				α	0.85	0.90			
λ	1	1.27				λ	1	0.83			
η_{34}	7		34	8	0.89	η_9	7		9	7	0.87
η_{50}	5		50	8	0.89	η_{29}	13		29	12	0.99
η_{63}	9		63	9	1.00	η_{75}	18		75	19	0.99
α	0.15	0.01				α	0.85	0.86			
λ	3	3.60				λ	3	2.62			
η_{24}	9		24	11	0.99	η_9	31		9	29	0.92
η_{28}	13		28	13	0.99	η_{29}	13		29	10	0.99
η_{65}	6		65	7	0.99	η_{75}	22		75	22	0.99
α	0.15	0.01				α	0.85	0.85			
λ	5	5.3				λ	5	4.60			
η_{33}	7		33	11	0.99	η_{38}	40		38	37	0.92
η_{70}	12		70	13	1.00	η_{41}	28		41	27	0.99
η_{10}	16		10	18	0.98	η_{78}	17		78	20	0.99
α	0.5	0.41				α	0.85	0.90			
λ	1	0.94				λ	1	0.83			
η_9	10		9	11	0.90	η_9	7		9	7	0.87
η_{27}	4		27	7	0.85	η_{29}	13		29	12	0.99
η_{97}	7		97	8	0.81	η_{75}	18		75	19	0.99
α	0.5	0.59				α	0.85	0.86			
λ	3	2.28				λ	3	2.62			
η_{99}	17		99	17	0.99	η_9	31		9	29	0.92
η_{17}	12		17	16	0.99	η_{29}	13		29	10	0.99
η_7	7		7	7	0.97	η_{75}	22		75	22	0.99
α	0.5	0.51				α	0.85	0.85			
λ	5	4.30				λ	5	4.60			
η_{29}	10		29	14	0.91	η_{38}	40		38	37	0.92
η_{22}	21		22	22	0.99	η_{41}	28		41	27	0.99
η_{19}	15		19	17	0.99	η_{78}	17		78	20	0.99

5 Concluding Remarks

In this paper, a retrospective analysis of the Poisson INAR(1) model for time series of counts under a Bayesian approach is carried out. The Bayesian framework is more flexible than a classical likelihood approach leading to the identification of observations that may require further scrutinizing. In fact, by estimating the probability that each observation is affected by an outlier under a certain model, the procedure is useful for detecting suspicious observations but also possible model

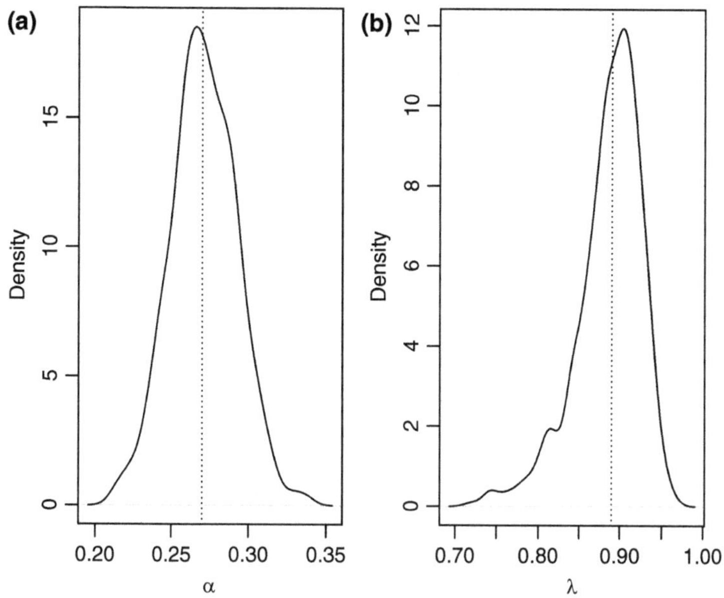

Fig. 2 Posterior distribution of α and λ. The *dotted lines* represent the estimates $\hat{\alpha}_{Bayes} = 0.27$ and $\hat{\lambda}_{Bayes} = 0.89$

inadequacies since the presence of many outliers may indicate the wrong choice of model. There are, thus, several extensions to this work that are being investigated, namely: the detection of patches of outliers that may cause masking and swamping effects; development of strategies for including different outliers effects and other interventions; other distributional assumptions; higher-order models.

Acknowledgements This work was supported by Portuguese funds through the CIDMA—Center for Research and Development in Mathematics and Applications, and the Portuguese Foundation for Science and Technology (FCT-Fundação para a Ciência e a Tecnologia), within project PEst-OE/MAT/UI4106/2014.

References

1. Al-Osh, M.A., Alzaid, A.A.: First-order integer-valued autoregressive (INAR(1)) process. J. Time Ser. Anal. **8**, 261–275 (1987)
2. Barczy, M., Ispány, M., Pap, G., Scotto, M., Silva, M.E.: Additive outliers in INAR(1) models. Stat. Pap. **53**, 935–949 (2011)
3. Barczy, M., Ispány, M., Pap, G., Scotto, M., Silva, M.E.: Innovational outliers in INAR(1) models. Commun. Stat. Theory Methods **39**, 3343–3362 (2010)
4. Chang, I., Tiao, G.C., Chen, C.: Estimation of time series parameters in the presence of outliers. Technometrics **30**, 193–204 (1988)

5. Chen, C.W.: Detection of additive outliers in bilinear time series. Comput. Stat. Data Anal. **24**, 283–294 (1997)
6. Chen, C., Liu, L.M.: Joint estimation of model parameters and outlier effects in time series. J. Am. Stat. Assoc. **88**, 284–297 (1993)
7. Fokianos, K., Fried, R.: Interventions in ingarch processes. J. Time Ser. Anal. **31**, 210–225 (2010)
8. Gelfand, A.E., Smith, A.F.M.: Sampling-based approaches to calculating marginal densities. J. Am. Stat. Assoc. **85**, 398–409 (1990)
9. Gilks, W.R., Best, N.G., Tan, K.K.C.: Adaptive rejection metropolis sampling within Gibbs sampling. J. R. Stat. Soc. Ser. C (Appl. Stat.) **44**, 455–472 (1995)
10. Jung, R., Tremayne, A.: Useful models for time series of counts or simply wrong ones? Adv. Stat. Anal. **55**, 59–91 (2011)
11. Justel, A., Peña, D., Tsay, R.: Detection of outlier patches in autoregressive time series. Stat. Sin. **11**, 651–673 (2001)
12. McKenzie, E.: Some simple models for discrete variate time series. J. Am. Water Resour. Assoc. **21**, 645–650 (1985)
13. Plummer, M., Best, N., Cowles, K., Vines, K.: CODA: convergence diagnosis and output analysis for MCMC. R News **6**, 7–11 (2006)
14. Silva, I., Silva, M., Pereira, I., Silva, N.: Replicated INAR(1) processes. Methodol. Comput. Appl. Probab. **7**, 517–542 (2005)
15. Tsay, R.S.: Time series model specification in the presence of outliers. J. Am. Stat. Assoc. **81**, 132–141 (1986)
16. Weiß, C.H.: Controlling correlated processes of poisson counts. Qual. Reliab. Eng. Int. **23**, 741–754 (2007)

From Ice to Penguins: The Role of Mathematics in Antarctic Research

José C. Xavier, S.L. Hill, M. Belchier, T.J. Bracegirdle, E.J. Murphy, and J. Lopes Dias

Abstract Mathematics underpins all modern Antarctic science as illustrated by numerous activities carried out during the international year "Mathematics for Planet Earth". Here, we provide examples of some ongoing applications of mathematics in a wide range of Antarctic science disciplines: (1) Feeding and foraging of marine predators; (2) Fisheries management and ecosystem modelling; and (3) Climate change research. Mathematics has allowed the development of diverse models of physical and ecological processes in the Antarctic. It has provided insights into the past dynamics of these systems and allows projections of potential future conditions, which are essential for understanding and managing the effects of fishing and climate change. Highly specific methods and models have been developed to address particular questions in each discipline, from the detailed analyses of remote-sensed predator tracking data to the assessment of the outputs from multiple global climate models. A key issue, that is common to all disciplines, is how to deal with the inherent uncertainty that arises from limited data availability and the assumptions or simplifications that are necessary in the analysis and modeling of interacting processes. With the continued rapid development of satellite-based and remote observation systems (e.g. ocean drifters and automatic weather stations), and of new methods for genetic analyses of biological systems, a step-change is occurring in the magnitude of data available on all components of Antarctic systems. These changes in data availability have already led to the development of new methods and algorithms for their efficient collection, validation, storage

J.C. Xavier (✉)
Department of Life Sciences, Marine and Environmental Research Centre (MARE), University of Coimbra, 3001–401 Coimbra, Portugal

British Antarctic Survey (BAS), Natural Environment Research Council, High Cross, Madingley Road, CB3 0ET Cambridge, UK
e-mail: jxavier@zoo.uc.pt

S.L. Hill, • M. Belchier • T.J. Bracegirdle • E.J. Murphy
British Antarctic Survey (BAS), Natural Environment Research Council, High Cross, Madingley Road, CB3 0ET Cambridge, UK

J.L. Dias
Lisbon School of Economics and Management (ISEG), Technical University of Lisbon, Rua do Quelhas, 6, 1200-781 Lisbon, Portugal
e-mail: jldias@iseg.utl.pt

J.-P. Bourguignon et al. (eds.), *Mathematics of Energy and Climate Change*,
CIM Series in Mathematical Sciences 2, DOI 10.1007/978-3-319-16121-1_20

and analysis. Further progress will require the development of a wide range of new and innovative mathematical approaches, continuing the trend of world science becoming increasingly international and interdisciplinary.

1 Introduction

The Polar Regions are the cornerstones of the global ecosystem, barometers of the health of the planet, and messengers of global processes [49, 56]. Because it strongly influences the global climate and harbours unique and diverse biological communities, the Antarctic plays a distinct and critical role in both the physical Earth system and the ecosystem that it supports. Antarctica is renowned as being the highest, driest, windiest and coldest continent, boasting the lowest recorded temperature on Earth, −93.2 °C, on the East Antarctic Plateau (http://www.nasa.gov/press/2013/december/nasa-usgs-landsat-8-satellite-pinpoints-coldest-spots-on-earth/#.UqndqvvwoYt, accessed 12/12/13), but it is surrounded by the Southern Ocean which, in contrast, is very thermally stable (with some locations varying as little as 0.2 °C over a year) [25]. Some of the key science on globally important issues is conducted in the Antarctic, often coordinated by the Scientific Committee on Antarctic Research and various the bodies that administer the Antarctic Treaty System. These issues include sea level rise, climate change, ocean acidification, biodiversity change, the ozone hole and global ocean circulation [3, 25, 56, 72, 89, 91, 92, 97, 98]. Furthermore, the Antarctic continues to spark the curiosity and imagination of people around the world. It appeals to the sense of adventure and fear of the unknown. These are perfect ingredients for education and outreach [52, 102, 103, 117], providing an excellent way to transmit basic concepts about a wide range of Science, Technology, Engineering and Mathematics (STEM) disciplines. During the international year "Mathematics for Planet Earth", numerous activities related to mathematics were carried out throughout the world, including the International Conference and Advanced School Planet Earth, Mathematics of Energy and Climate Change, held in Lisbon (Portugal), in 21–28 March 2013. Here, we follow discussions at that conference with a selective review of how mathematics is applied in a wide range of Antarctic science disciplines

2 The Scope of Mathematical Analyses in Antarctic Science

Mathematics underpins all modern Antarctic science. It is central to of the data collection process, for example in generating efficient algorithms to allow data storage and transfer, and for the calibration and validation of data from in-situ and remote instrumentation (e.g. automatic weather stations and satellite-based instruments). Mathematics is used in analyses and modeling of all aspects of Antarctic science including weather and climate, ice sheet and sea ice dynamics, ocean circulation, biogeochemical cycles and ecosystem processes. The range and complexity of applications is wide, from simple analyses of small scale experiments to high resolution satellite-based studies that provide circumpolar views, and from simple theoretical to fully coupled atmosphere-ocean-ice models. Here, we illustrate the development and application of mathematical analyses by considering three major areas of Antarctic science: (1) Feeding and foraging of marine top predators; (2) Fisheries management and ecosystem modeling; and (3) Climate change research. These sections provide illustrations in three distinct types of scientific activity. Studies of the feeding and foraging of marine top predators are strongly field based and require extensive sample collection and analysis and careful design of sampling methods. With the advent of satellite instrumentation, remote tracking now provides high resolution information on position and movement, which has revolutionized analyses of predator foraging. This has been associated with the development of a range of other remote devices, and has generated a step-change in the size of datasets available. The second area, fisheries management and ecosystem modeling, provides an illustration of the use of mathematical methods in an applied arena to generate robust policy advice. The final area, climate change, illustrates the challenge of developing projections of the impacts of future change that can only be addressed through mathematical analysis and modelling.

3 The Application of Mathematics in the Study of Feeding and Foraging Ecology of Marine Top Predators

In order to understand how the Southern Ocean food web operates, it is essential to understand what animals eat and where they feed. To obtain reliable estimates from the available data requires a wide range of mathematical tools (particularly statistical and modeling tools). The most common source of feeding data is the stomach contents of sampled predators. To characterize the diet of top predators, such as penguins (Fig. 1) or albatrosses, prey in these stomach contents are generally quantified by their frequency of occurrence, or the number or mass of each prey species [87, 112]. To identify the prey (generally fish, cephalopods such as squid and octopods, and crustaceans), scientists often have to use hard structures that are not destroyed by digestion. These structures include the sagittae otoliths (colloquially the "ear bones") of fish. Otoliths are calcium carbonate structures located directly behind the brain of teleost (bony) fish. For crustaceans, scientists use their carapaces and for cephalopods their beaks. Cephalopod beaks are chitinous structures, whose function is similar to that of teeth in carnivorous mammals: to grasp, kill and dismember their prey.

Allometric regression equations can be derived to describe the relationship between the mass and length of complete individuals of known species and the size of their otoliths, carapaces or beaks. When applied to hard structures found in stomach contents these equations provide a valuable mathematical tool for

Fig. 1 Penguins, such as macaroni penguins, are important components of the Southern Ocean ecosystem and are difficult to directly observe because of the remote and hostile conditions in which they live and the considerable distances that they travel. Nonetheless Antarctic scientists have used mathematical tools to develop ways to study their behavior and ecology in the wild, including their feeding and foraging ecology (photo by José Xavier)

reconstructing what a predator has been consuming [14, 37, 88, 110]. However, numerous prey species do not have allometric equations because they are still poorly known (therefore scientists must rely on equations from closely related species) and many of the available allometric equations are based on a limited number and size range of specimens. Therefore, future work must focus on obtaining more complete fish, crustaceans and cephalopods to improve allometric equations, and on characterizing the uncertainty that is inherent in the application of such methods [112, 113, 115, 116, 118, 119].

One of the key issues that marine ecologists need to address, when assessing the diet and feeding ecology of a predator, is how many samples are needed. This is essential to characterize their diet correctly, and so to provide fundamental data for food web studies, particularly modeling the present ecosystem status and predicting future changes. Mathematically, this is an interesting challenge. A randomization technique was used to estimate the number of stomach samples from albatrosses needed to reach two saturation points: (1) the maximum cumulative number of species; and (2) where each of the five most important species (i.e. >5 % of the diet, by mass) was present in at least one sample [113]. For each sampling event, the program randomly selected one of the samples and checked the species present. If one or more of the required species were absent, the program randomly selected another sample that had not yet been selected, and the process was repeated until one of the two saturation points was reached. The entire process was repeated 100 times. This study also compared different ways of collecting samples (i.e. using stomach contents or voluntary regurgitations, named boluses) which permitted the investigation of biases associated with each sampling method [113].

Other techniques for analyzing diet use tissues from stomach samples to identify prey species, or from predators (e.g. flesh, feathers, blood) to identify their habitat or trophic level (e.g. DNA analyses, stable isotopes, fatty acids, trace elements, chemical pollutants) with considerable success [21, 28, 28, 53, 86, 96]. Each of these techniques involves its own unique analytical challenges. Mixed data sources (e.g. diet data obtained from stomach contents and data on the stable isotope signatures from predators and prey) can be compared to calibrate different methods. A Bayesian multisource stable isotope mixing model (SIAR: Stable Isotope Analyses in the statistical package R) has been used to estimate the probable contributions of each prey to the diet of each individual and hence the predator's level of specialization on particular prey items [74]. This method indicated that for wandering albatrosses *Diomedea exulans* fish was the main component (56.4 %) of the diet, followed by cephalopods (43.6 %). These proportions were similar to those from analysis of stomach contents, showing the usefulness of these models for future research [20].

Advances in micro-technology (and the decreasing of the size of tracking devices) in the last two decades have revolutionized our understanding of the foraging behavior of predators [12, 12, 36, 77, 79, 79]. Seabirds can travel great distances (hundreds to thousands of kilometres) and exhibit a number of unique physiological adaptations for such highly pelagic lifestyles [29]. Albatrosses and petrels spend the great majority of their lives at sea, and the use of tracking technology is the most effective and, in many respects the only, means for gaining detailed insights into their foraging behaviour [51, 73, 83].

Satellite sensors, combined with "ground truth" data from in situ surveys, are contributing to a better understanding of ocean systems by providing large scale and long-term data on biological bulk parameters such as chlorophyll, and on ecologically relevant physical parameters, such as sea surface temperature or ice cover [97]. These data, combined with tracking technology can be used to answer scientific questions about foraging behavior and how animals use their ocean habitat. For example, satellite-tracking on animals in the late 1990s involved the deployment of a Platform Terminal Transmitter (PTT) that sends a short radio signal typically every 90 s to polar-orbiting NOAA satellites. The precision of the location estimate can vary from meters to hundred of meters, depending on the number of satellites in view at that place and time, the design and power of the transmitter, and the speed of the animal [108]. More recently, Global Positioning System (GPS) loggers have been widely used, mostly because of their higher precision (within 10 m) [107] and ability to record positions at various time intervals, from minutes to days (depending on the amount of time the device is on) [70]. However, if scientists are more interested in knowing where animals are for a longer period of time (e.g. large migration studies), geolocators or Global Location Sensing (GLS) loggers are extremely useful [27, 78]. GLS loggers record ambient light. This allows the estimation of sunset and sunrise times from curve thresholds. These times in turn allow the estimation of latitude from day length, following standard astronomical algorithms, and longitude from the time of local mid-day with respect to GMT and Julian day. The disadvantages are that the animal must be recaptured (as with most GPS loggers), only two locations can be calculated per day, latitude estimation is impossible for variable periods around the equinoxes, and the precision is relatively low, with an average error of 186 km estimated for free-ranging albatrosses [78].

These examples demonstrate that the reliable estimation of animal location, and its associated error, is a fundamental part of modern animal ecology. There are many existing techniques for handling location error, but these are often *ad hoc* or are used in isolation from each other. There is a Bayesian framework for determining location that uses all the data available, is flexible enough to be used with all tagging techniques, and provides location estimates with built-in measures of uncertainty [95]. Bayesian methods allow the contributions of multiple data sources to be decomposed into manageable components. Sumner et al. [95] showed that many of the problems with uncertainty in archival tag and satellite tracking data can be reduced and quantified using readily available tools.

With these mathematical tools applied to the feeding and foraging of top predators, it has been possible to model potential areas where poorly known organisms may be distributed. Indeed, the distribution of many cephalopod, crustacean and fish species in the Southern Ocean, and adjacent waters, is poorly known, particularly during times of the year when research surveys are rare [16, 35, 90, 111]. Analysing the stomach samples of satellite-tracked higher predators has been advocated as a potential method by which such gaps in knowledge can be filled. This approach showed that wandering albatrosses, *Diomedea exulans*, foraged in up to three different water-masses, the Antarctic zone (AZ), the sub-Antarctic zone (SAZ) and the sub-Tropical zone (STZ) [114]. A probabilistic mathematical model was applied to the tracking and diet data collected from wandering albatrosses to construct a large scale map of where various prey were captured. Furthermore, robustness and sensitivity analyses were used to test model assumptions about the time spent foraging and relative catch efficiencies and to evaluate potential biases associated with the model. The analysts were able to predict the distributions of a multiple cephalopod, crustacean and fish species [114]. This method is likely to be used in the future to predict the distributions of poorly known species, such as large oceanic cephalopods, that are not effectively sampled using nets [109, 115, 119].

In summary, mathematical methods are critically important to studies in marine ecology, including those related to the feeding and foraging ecology of top predators. The many examples range from producing mathematical mixed models to quantify the consumption of prey to providing the algorithms that allow the tracking of top predators in the Southern Ocean.

4 The Application of Mathematics in Fisheries Management and Ecosystem Modelling

4.1 Management of Southern Ocean Fisheries

Fishing is one of the main economic activities in the Southern Ocean, alongside science and tourism [34]. The responsible management of these fisheries is therefore an important applied ecology problem, which has led to innovative approaches that make extensive use of mathematics and modelling. Fishing removes animals from an ecosystem which would otherwise continue feeding, growing, reproducing and being fed upon. Such removals can reduce the ability of fished populations to replace themselves and they can have wider impacts on other populations by changing the balance of predators and prey. The Southern Ocean ecosystem has already experienced considerable perturbation as a result of past harvesting which started in the late 1770s and led to localised extinctions of Antarctic fur seals *Arctocephalus gazella* and the commercial extinction of many baleen whale species which had previously consumed an estimated 175 Mt yr^{-1} of Antarctic krill *Euphausia superba* [57].

Antarctic krill is a swarming shrimp-like animal that grows to a maximum of about 6 cm, and is now the target of an expanding fishery [39, 71]. There is also commercial harvesting of various fish species including the high-value Antarctic toothfish, *Dissostichus mawsoni*, and Patagonian toothfish, *Dissostichus eleginoides* [34]. Fish products are generally sold for direct human consumption while krill is usually processed to produce fishmeal for aquaculture, and oil which is sold as a health supplement [39, 71]. These fisheries are managed by the Commission for the Conservation of Antarctic Marine Living Resources (CCAMLR), which was established in 1982. CCAMLR is responsible for ensuring that fisheries do not cause long-term damage to fished populations or the wider ecosystem. Consequently fisheries management draws on a much broader research effort which aims to understand the dynamics and structure of Southern Ocean ecosystems.

One of the key challenges faced by the scientists who advise CCAMLR is uncertainty. Assessments of the state of fished populations are affected by sometimes considerable estimation error and there are no failsafe models to indicate how these populations change in response to fishing, environmental variability and changes in other populations. The uncertainty about how fishing will affect complex ecosystems is even greater. Thus many of the major uses of mathematics in Southern Ocean fisheries management and related research address uncertainty in some form. These uses include producing useful estimates of the state of fished stocks from limited observations, identifying safe catch levels, understanding ecosystem structure and dynamics and evaluating potential risks to the wider ecosystem.

CCAMLR uses the precautionary approach to identify catch levels for Southern Ocean fisheries. Hill [39], paraphrasing Garcia [32], states that the precautionary approach aims to "reduce the probability of occurrence of bad events within acceptable limits when the potential for these events is plausible, but not necessarily demonstrated, and the potential costs are significant." Hill [39] also suggests that the precautionary approach should reduce the risk of harm to the ecosystem by setting low catch limits and protecting areas from fishing until there is evidence that the risks associated with more intensive fishing are acceptable.

4.2 Analyses and Models for the Management of Fin Fish Stocks

In order to provide robust management advice to CCAMLR on sustainable catch limits that are consistent with the precautionary approach, fisheries scientists undertake regular (annual or biennial) assessments of exploited fin fish stocks. In common with fisheries management elsewhere, these stock assessments use a vast array of fishery-dependent (e.g. catch rates) and fishery-independent (e.g. local biomass estimates from scientific fishing) data to describe the past and current status of a stock and to project the potential response of the stock to current and future management options (e.g. catch limits). Mathematical techniques lie at the heart of all stock assessments and are used in the construction of assessment models for each fished stock. These models are generally based on population dynamics models that can have varying degrees of complexity. The choice of stock assessment model will depend both on the quality and availability of data on catch and fishing effort and knowledge and availability of information on stock size, geographical stock boundaries, and species-specific life history traits such as growth, natural mortality and sexual maturity. Our understanding of these processes is usually summarized in formal mathematical models (e.g. the von Bertallanfy growth equation) [19].

The two species of toothfish mentioned above are exploited by deepwater demersal longline fisheries in various locations throughout the Southern Ocean [33]. Several of these fisheries have taken place for over two decades and consequently assessments can draw on a large amount of fishery-dependent and ecological data. These "established" fisheries, that include those carried out within the Ross Sea and at a number of sub-Antarctic islands (i.e. South Georgia [48] and the Heard and McDonald islands [18]) are assessed using age—structured, Bayesian "integrated" stock assessment methods. The input data for these assessments include trawl survey estimates of recruitment, commercial catch at length or age data, standardised catch rate data, mark-recapture data from multi-year tagging programmes, and estimates of natural mortality, growth, the length-weight relationship and maturity data. Given the integrated nature of these assessments in which many datasets are used concurrently to estimate parameters, much attention is given to the statistical weighting of each dataset. Bayesian methods are frequently used in the estimation procedure and uncertainty in the dynamics is evaluated using Markov Chain Monte Carlo (MCMC) methods [60, 61].

A number of smaller fisheries for the two toothfish species exist in the high seas areas of the Southern Ocean, in particular on the seamounts of the Indian Ocean sector for which data on stock size and biological parameters is far more sparse [2]. In these new, exploratory and research fishery areas, biomass estimates of the local toothfish population are usually derived from mark recapture data and calculated using the Lincoln -Peterson equation which estimates population size as the product of the numbers of animals captured in each of two events divided by the number that were captured twice (i.e. in both events) [58]. This biomass estimate allows suitable catch limits to be obtained by the application of a conservative exploitation rate. The scientific purpose of these fisheries, which are considered "data poor", is the collection of high quality data on abundance and the biological characteristics of the stock with the aim of developing fully integrated stock assessments in the near future. As more abundant and robust data become available for these fisheries, more complex population dynamics models are developed and tested in the transition towards a fully integrated assessment.

As with all biological systems there are varying degrees of uncertainty associated with the data used within the stock assessment models. A suite of mathematical procedures has been developed to address this uncertainty in order to improve model fits within the stock assessments of Southern Ocean fin fish populations. These procedures are part of an integrated approach which aims to reduce the uncertainty in the projections used to evaluate management options. The areas of greatest uncertainty have included the estimation of levels of illegal fishing [1], tagging [120], cetacean depredation [22], unaccounted fishing mortality [106], appropriate model weighting for catch-at-age data, catch [17] and natural mortality [19], among others.

4.3 *Analyses and Models for the Management of Antarctic Krill Stocks*

Antarctic krill is a highly abundant species. Atkinson et al. [6] used various statistical models to estimate the gross growth potential, the amount of new biomass that would be produced by growth each year if all animals survived. These estimates ranged from 342 to 536 Mt yr^{-1} depending on the model used. For comparison, total global marine fisheries landings are approximately 80 Mt yr^{-1} [30]. Of course, Antarctic krill do not achieve their full growth potential because many of them do not survive the year. The vast population of Antarctic krill is continually grazed by an array of predators including pelagic and demersal fish, penguins and other seabirds, whales, seals and even benthic invertebrates. Many of these predators rely on Antarctic krill as their main source of food [57, 104, 112]. For this reason the precautionary approach for Antarctic krill has to consider the indirect effects of fishing on predators since it effectively removes part of their food supply [34]. Management which includes such considerations is sometimes known as Ecosystem Based Management [63].

In a logistic biomass growth model, the per-capita rate of increase is highest at half of the asymptotic biomass. This leads to the hypothesis that fished populations are most productive if reduced to half of their pre-fishing biomass [84]. However, the requirement to explicitly manage potential impacts on Antarctic krill predators led to a more precautionary objective: to ensure that, in the long-term, fishing does not reduce the Antarctic krill population by more than 25 % on average [23, 42]. Scientists use a stochastic population projection model to identify catch levels that meet this criterion. The model runs multiple simulations with random deviates in various population parameters (e.g. recruitment, natural mortality, age at maturity) and a range of different catch levels, until it finds the correct level. During this process catch levels are also assessed against another criterion: that the risk of the breeding population falling below 20 % of its initial biomass is no more than 10 %. The highest catch level that meets both criteria is selected to manage the fishery [23].

Smith et al. [93] used nine ecosystem dynamics models (that is models of the interacting dynamics of multiple species) to assess the potential impacts of fishing on the rest of the ecosystem. This study considered fisheries for lower trophic level species, such as capelin (*Mallotus villosus*), herring (*Clupea* spp.) and anchovies (e.g. *Engraulis* spp.) in other oceans and did not directly consider the Antarctic krill fishery. Nonetheless it found that allowing a fishery to deplete the biomass of the fished population by no more than 25 % provided reasonable catch levels while achieving "much lower impacts on marine ecosystems" than the higher depletion rates allowed by many fisheries management regimes. This suggests that the general approach used for setting Antarctic krill catch limits might be appropriately precautionary. However, a recent study shows that the catch limits selected using this approach are sensitive to assumed levels of recruitment variability and that recruitment variability in real Antarctic krill stocks might be higher than that assumed in model projections [55]. The ecosystem impacts of fishing depend not just on how much biomass of the fished species it removes, but also where it removes biomass from. Although Antarctic krill is widely distributed throughout the Southern Ocean, the vast majority of the catch (83 % of all reported catch to date) [39] is taken from the Scotia Sea and southern Drake Passage (Fig. 2) and is in fact concentrated in just 26 % of this area [33]. Specifically, fishing occurs in and close to the shallow waters that surround the many islands in this area. Fishing does not generally occur in the more hostile waters of the open ocean where Antarctic krill is still abundant but much less likely to occur in the dense aggregations that the fishery targets [45]. Scientists have used ecosystem dynamics models to assess the risk that such spatially restricted fishing poses to Antarctic krill predators [81, 105]. These models are spatially resolved to distinguish the various shallow water and open ocean areas and they represent the interactions between Antarctic krill, the fishery, and several groups of competing predators. The exact nature of these interactions is uncertain and there is very little information about past dynamics from which to infer the interactions. Consequently, the modelers did not attempt to devise a single best model to project the consequences of future fishing. Instead they attempted to evaluate the uncertainty in such projections and they translated this uncertainty into estimates of the risks associated with candidate management options (Fig. 3).

The approach to this uncertainty about the true nature of the modeled interactions was to use multiple plausible "scenarios" or plausible representations of the system [44]. The word "scenario" here means a model and its data (sensu Rademeyer et al. [85]). The scenarios were based on two different model structures, described in Plagányi and Butterworth [81] and Watters et al. [105] and several alternative parameterizations of each model structure. These alternative parameterizations were chosen specifically to bracket key uncertainties. For example, the speed at which

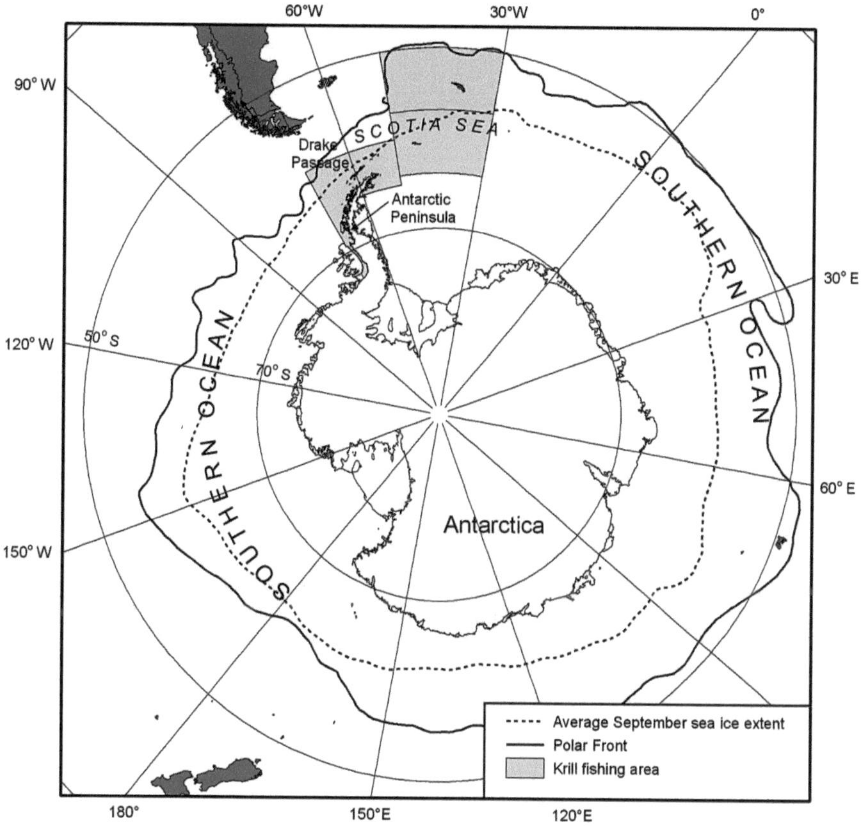

Fig. 2 The Antarctic continent and the Southern Ocean which surrounds it. The Polar Front is the approximate northern limit of the Southern Ocean ecosystem. Antarctic krill is a major component of this ecosystem, the main prey item for a diverse suite of predators and the focus of a developing fishery. Although Antarctic krill fishing is permitted in much of the Southern Ocean, the vast majority of the catch to date has been taken in the Scotia Sea and southern Drake Passage region [39, 71]. Antarctic krill abundance in this region is correlated with September sea ice extent [4], the average position of which (1979–2004) is shown

Antarctic krill are transported on ocean currents is not known, but the actual speed is likely to lie between a minimum of zero and a maximum of the speed of passive particles drifting with the currents, which can be deduced from ocean circulation models [43]. Watters et al. [105] developed four parameterizations, each of which combined one of these extreme values for plausible transport speeds with an extreme plausible value for a second key uncertainty affecting the functional relationship between prey availability and the proportion of the predator population that is able to breed.

Another important innovation recognized the impossibility of predicting with accuracy the future state of the system when it is influenced by multiple interacting

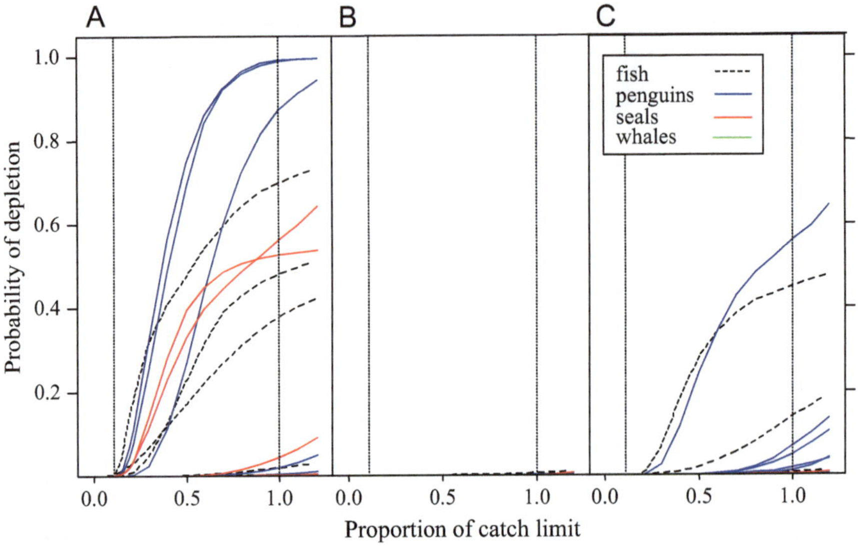

Fig. 3 The format used to present the results of model projections with intentionally high levels of uncertainty [38, 105]. The varied results of multiple simulations were used to calculate the probability of each modelled predator population falling below a threshold level. The results show how this risk increases with catch level ("proportion of catch limit") and varies between three alternative spatial distributions of the catch limit (labelled *A* to *C*)

drivers including climate variability and change, and recovery from historic over-harvesting of whales. The modelers therefore assessed the marginal effects of Antarctic krill fishing in their projections by comparing them to otherwise identical projections without fishing.

The modelers performed 1001 stochastic projections with each scenario for each evaluated management option (consisting of a catch limit and its spatial distribution amongst modeled spatial units). One of the risks evaluated was the probability of each modeled predator population falling below 75 % of its size in comparable projections without fishing. The models generated several thousand projections per management option with which to calculate this probability. The analysts presented results in the format shown in Fig. 3, which shows the coherent accumulation of risk with increasing catch limit, and identifies the least risky spatial distribution (where the catch limit in each spatial unit is proportional to the total predator demand for Antarctic krill in the same unit, labeled "B" in the figure). Hill [38] demonstrated that this distribution remains the least risky even if a different reference level (other than 75 %) or scheme for aggregating modeled predator populations is used.

Because managers need to consider the implications of management options for the fished stock, the fishery and predators in multiple areas, the models assessed each of these risks. One important consideration is that various model outputs (e.g. the biomass of the fished stock versus the biomass of one of its predators) have

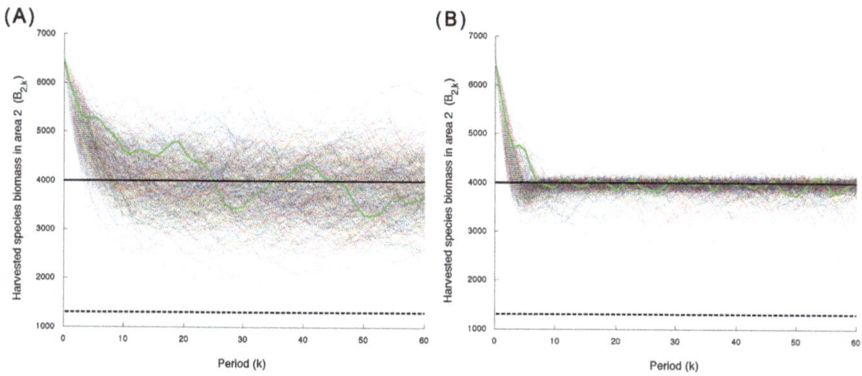

Fig. 4 The biomass of a modelled krill-like species in 500 stochastic simulations using two different harvest control strategies: (**a**) a fixed catch limit, as is currently used to manage the Antarctic krill fishery, and (**b**) a feedback method which uses model predictive control (MPC) to adjust catch limits in response to information about the harvested stock and its predators [40]

different levels of sensitivity to perturbations in model parameters, suggesting that uncertainties in these parameters could bias comparisons of different risks [41].

Scientists advising CCAMLR are attempting to develop a feedback management approach for the Antarctic krill fishery that will modify spatially resolved catch limits in response to information about the local and larger-scale state of the ecosystem. Such an approach is difficult to design and implement when there are multiple objectives for multiple connected areas, and when the system's dynamics are complex and uncertain. Hill and Cannon [40] used a branch of control theory called model predictive control (MPC) to show that such an approach is feasible in principle and more likely than the current fixed catch limit to simultaneously achieve objectives for the state of the Antarctic krill stock, the state of multiple predator populations and the state of fishery catches (Fig. 4). Their study applied MPC to a relatively simple ecosystem dynamics model consisting of two connected areas, each containing a single prey population and a single predator population. Their study also clarified the information requirements of such an approach, which include regular estimates of each of the relevant state variables or, at least, reliable ways of inferring these from the other state estimates and, critically, a clear set of quantitative objectives for each relevant state. Defining such objectives is a major challenge facing CCAMLR and other organizations around the world which seek to implement Ecosystem Based Management [38, 59].

4.4 Ecosystem Modelling

Models exploring the interactions between different populations in the ecosystem are useful for devising and assessing Ecosystem Based Management approaches.

Some pioneering models of this type were developed for the Southern Ocean ecosystem in the 1980s [10, 11] and the modeling effort that continued to develop since then was the subject of a detailed review by Hill et al. [42]. More recent developments include the ecosystem dynamics models described above and a suite of Ecopath-type food web models [7, 26, 46, 80]. Ecopath-type models compile available data on the diet, biomass, and production and consumption rates of the numerous organisms in a particular food web [82]. Modelers generally aggregate these organisms in so-called functional groups to reduce the number of model parameters. The modelers then adjust the parameters to satisfy the "mass balance" constraint that the rate of biomass production by any prey group cannot exceed the rate of consumption of that prey biomass by its predators.

One use of Ecopath-type food web models is to identify which functional groups are likely to be strongly affected by changes in the abundance of fished species [101]. A related use is to explore the potential responses to a plausible change in one part of the food web. For example, Ballerini et al. (in press) converted their model of the winter food web in Marguerite Bay into a bottom-up model, in which consumer biomass increases or decreases with the availability of prey. They increased the modeled biomass of small phytoplankton relative to large phytoplankton while maintaining a constant total phytoplankton biomass. This change is a consistent with recently observed effects [65]. The model predicted reduced production of Antarctic krill and its predators as a result of this change. Hill et al. [46] reduced Antarctic krill biomass by 80 % in their model of the South Georgia shelf food web and readjusted the parameters to achieve mass balance. They found that, without compensating effects, this produced a similar decline in the biomass of Antarctic krill predators (fish, seals, penguins and other seabirds). However, a combination of compensating effects (an increase in grazing zooplankton called copepods and a shift in predator diets to take advantage of this increased copepod biomass) could minimize the impacts on Antarctic krill predators. This illustrates the wide range of outcomes that are possible within the current uncertainties on ecological knowledge. Future model development and data collection should aim to better characterize these uncertainties so that it is possible to assess which outcomes are most likely.

The existing suite of food web models for the Southern Ocean provide a valuable resource for comparing the structure and operation of the different regional food-webs [68, 69]. However, each of the existing regional models was developed by a different modeling team, using patchy and uncertain data, and each model was designed and analyzed to address a unique set of research questions. The differences between the models therefore include real underlying ecological differences, differences due to sampling error in the available data, and differences in the assumptions and subjective decisions made by the various modeling groups. The challenge of distinguishing real ecological differences from these sources of uncertainty is likely to be a major theme in future food web modeling.

In summary, mathematics and modeling are critical to understanding ecosystem structure and dynamics, assessing potential responses to change and developing appropriate fisheries management approaches. CCAMLR's commitment to Ecosystem Based Management and the relative paucity of ecological data for the Southern

Ocean produce some interesting challenges that have led to innovative ecological modeling and analysis. It is practically impossible to identify definitive models of ecosystem structure or dynamics and consequently much of this innovation and many of the ongoing challenges concern the appropriate treatment of uncertainty.

5 The Application of Mathematics in Antarctic Climate Change Research

The analysis and projection of climate change using mathematical modeling currently receives much attention from scientists, politicians, the media and the general public. Due to the increased rates of environmental change in the Antarctic, considerable research effort has been devoted to modeling the Antarctic atmosphere and the Southern Ocean, and to quantifying physical and biological aspects of change. Most global climate models suggest that regional temperature increases will be greatest and most rapid at higher latitudes [49, 98]. Rapid increases are already evident over the Antarctic Peninsula where, in the last half century, air temperatures have risen by 2–3 °C. To the west of the Antarctic Peninsula sea ice has also declined and ocean temperatures have increased by 1 °C over five decades [64, 99]. Although climate models have successfully helped to build a broad picture of the causes of recent regional change, there are still many gaps in knowledge which affect the ability of climate models to reliably represent the Antarctic climate and the behavior of its ice caps and sea ice. A recent study by Turner et al. [100] examined the annual cycle and trends in Antarctic sea ice extent (SIE) for 18 climate models. Many of the models have an annual SIE cycle that differs markedly from that observed over the last 30 years. In contrast to the satellite data, which exhibit a slight increase in SIE, the mean SIE of the models over 1979–2005 shows a decrease in each month [100]. The models have very large differences in SIE over 1860–2005. The negative SIE trends in most of the model runs over 1979–2005 are a continuation of an earlier decline. There are two major gaps in knowledge that hamper the understanding of the observed increase. Possibly the most important is the limited observational record, in which reliable Antarctic-wide estimates of SIE are only available after approximately 1979. It is therefore very difficult to estimate the size of natural fluctuations in ice extent, which may have contributed to the recent changes. Related to this, the other major gap is in understanding the processes for change that need to be mathematically represented in climate models. At present is seems that the processes responsible for the observed SIE increase over the last 30 years are not being simulated correctly [100].

Another important prediction of climate change models is changing patterns of precipitation, altering the water input to terrestrial ecosystems. Spatially detailed predictions are not yet available for Antarctica, although water is possibly the single most important factor limiting the distribution of Antarctic terrestrial biota [25]. In some Antarctic terrestrial systems local environmental changes result in greater energy input and warming, which may be accompanied by a lengthening of the season in which liquid water is available. There is evidence that such changes might increase production, biomass, population size, community complexity and the rate of colonization by previously absent organisms [13, 25]. However, at fine scales a decrease or total loss of water input could lead to local extinctions and drastic changes in local ecosystem structure [24].

Increases in primary biological productivity are already being seen at the margins of the Antarctic continent. These occur in areas of sea-ice loss where recent ice shelf retreat has occurred [76]. However, Antarctic marine species are generally amongst the least capable of adapting to environmental change. There are three main reasons [75]: (1) The geographical range over which they can live or disperse is restricted; (2) they have evolved to live in a very specific environment and tolerate only a narrow range of environmental conditions; and (3) they have long life histories and consequently slow rates of adaptation. Statistical analysis of experimental research provides evidence that the shallow mega- and macrobenthos are also very sensitive to temperature change (stenothermal). Being warmed by about 5 °C over periods greater than one month kills most species tested to date, but even smaller temperature rises (2 or 3 °C above normal) drastically hinder their ability to perform critical functions, such as avoiding predators [9]. In pelagic waters, changes to key pelagic species have also been notable. Regression analysis indicates a statistically significant relationship between Antarctic krill abundance and winter SIE in the western Scotia Sea, and there were apparent declines in both between the 1970s and 1980s [4, 99]. Further SEI declines would likely lead to more changes in the distribution and abundance of Antarctic krill.

Some studies have combined climate projections from global climate models with statistical models linking ecological processes to environmental variables (e.g. [40, 54, 54]). For example, Hill et al. [40] used sea surface temperature projections from 16 climate models with a statistical model linking Antarctic krill growth [5] to sea surface temperature. They found that plausible future warming is likely to lead to substantial reductions in the ability of Antarctic krill to produce new biomass throughout the northern Scotia Sea (Fig. 5). This is where large populations of Antarctic fur seals, *Arctocephalus gazella*, penguins and flying seabirds feed on Antarctic krill during the summer breeding season. A reduction in Antarctic krill biomass could result in greater predation on alternative prey and therefore negative impacts on some fish species [3]. As mentioned in the previous section, food web models have been used to assess how such changes could propagate through the food web [7, 46]. A key issue taken into account by Hill et al. [47] is that different climate models give different projections, contributing to uncertainty in estimates of future change. Assessing and quantifying this uncertainty is an important mathematical challenge in itself and an active area of research is in developing statistical approaches to combining information from different climate models [15, 94].

The effects of climate change in populations of top predators, such as penguins, have also been considered [8]. Even apparently straightforward tasks, like obtaining an estimate of the total number of penguins, are not easy and require plenty of

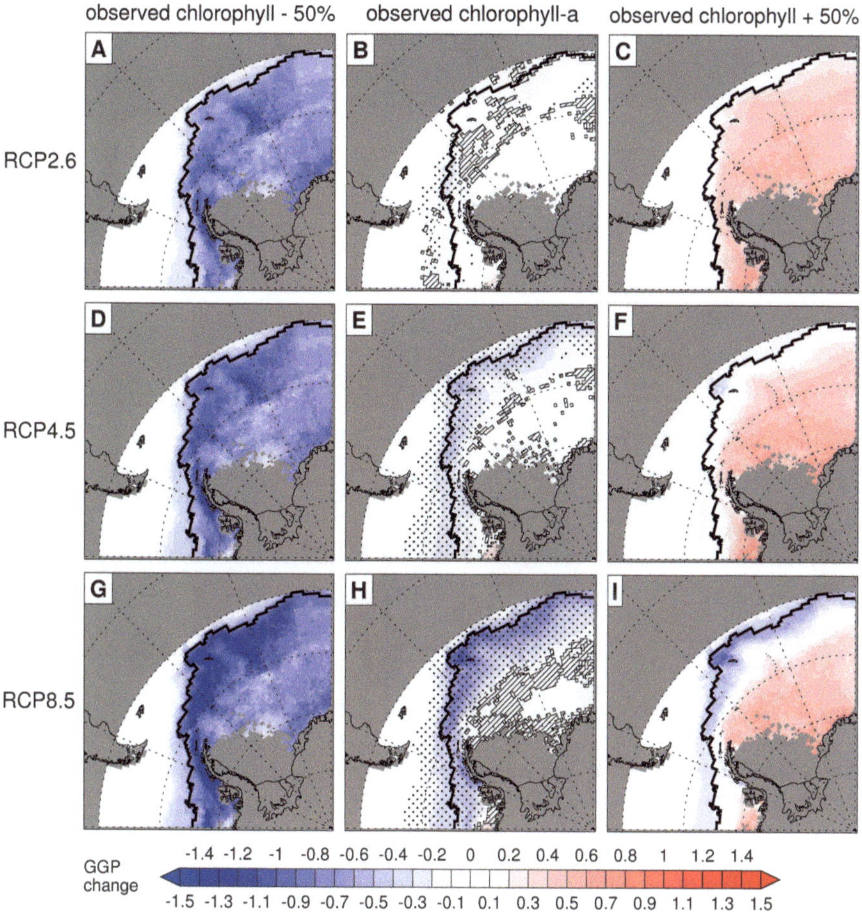

Fig. 5 The results of a study which used projections from multiple climate models to drive a statistical model of Antarctic krill growth [40]. The figure shows the spatial pattern of projected change in gross growth potential (GGP, an indicator of new biomass production) from 1997–2011 to 2070–2099. The growth model represents the influence of both temperature and food availability. The panels are arranged from *top to bottom* in order of increasing projected warming from three different representative control pathways (RCPs, which control the radiative forcing and hence warming in climate models). The figures are arranged from *left to right* in order of increasing final food availability indicated by chlorophyll concentration, including a 50 % decrease and a 50 % increase from current (observed) concentrations. Additionally, the central column shows the degree of agreement between climate models: Cells where 50 % or more of the models project significant GGP change are highlighted with stippling if 90 % or more of models agree on the sign of the change, and are *highlighted* with *hatched lines* if fewer than 90 % agree

mathematical tools. As an example, a recent study aimed to estimate the population of emperor penguins, *Aptenodytes fosteri*, using a single synoptic survey in 2009 [31]. The analysts examined the whole continental coastline of Antarctica using

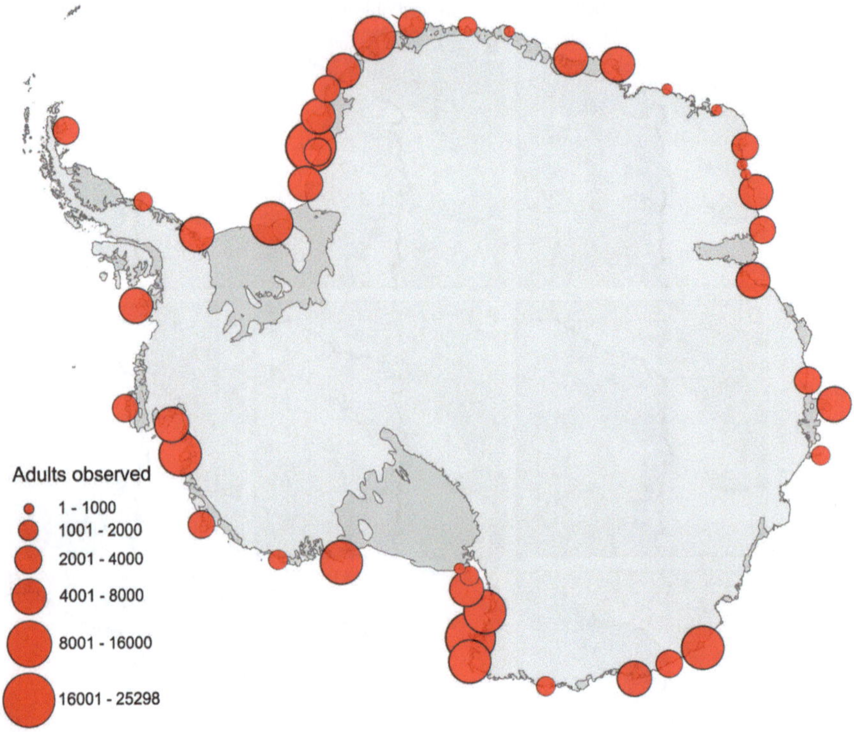

Fig. 6 The location of all known emperor penguin colonies in Antarctica and the estimated number of adults present in each at the time of a 2009 satellite survey [31]

a combination of medium resolution and very high resolution satellite imagery to identify emperor penguin colonies and the area occupied by penguins in each. They obtained actual counts of penguins from eleven ground truthing sites and used robust regression to model the relationship between the number of adult penguins and the area they occupy. They then used the model to estimate of the number of adult penguins at every colony. Finally they scaled this number up to estimate the total population of adults, including those that were absent at the time of the survey, using information about rates of participation in breeding and breeding success (Fig. 6). The final estimate of 238,000 adults present, out of a total population of 595,000 compares with the previously published estimate of 135,000–175,000 breeding pairs [62]. The revised, comprehensive estimate of the total breeding population can be used in population models and will provide a baseline for long-term research [31] which is necessary because global and regional emperor penguin populations are likely to be strongly affected by climate change [8, 50, 98].

In summary, climate research has always depended upon mathematics to build models and implement analyses. These models and analyses have increased our understanding of the past, and are now being used to project future climate conditions. This is particularly important in the Antarctic where recent changes in some areas are amongst the most extreme on earth. Ecological models are now being linked to climate projections from global climate models, providing a new era of research and bringing disciplines together. Future Antarctic climate research will include foci on characterizing and reducing the uncertainty in model outputs (e.g. by collecting further data and by improving the precision on the variables collected), on improving understanding and representation of climate processes (to improve model performance and the reliability of projections) and on working together with other science disciplines to provide robust evidence on the range of climate impacts, from sea level changes to biodiversity effects, that will inform policy decisions.

6 Final Considerations

Mathematical analyses are crucial in all areas of Antarctic science and central to addressing issues of global importance. In each scientific area highly specific methods and models have been developed to address particular questions, from the detailed analyses of remotely sensed predator tracking data to the assessment of the outputs from multiple climate models to determine the potential impacts of future global climate change. A key issue, that is common to all scientific disciplines, is how to deal with the inherent uncertainty associated with the analysis of process interactions in Antarctic systems. Major uncertainties are often the result of limited data availability, due to the difficulties of operating in remote Antarctic systems. However, over the last decade a series of long-term sampling programmes and large-scale international integrated projects (such as Census of Antarctic Marine Life (http://www.caml.aq/), ANDRILL (http://www.andrill.org), GLOBEC (http://www.globec.org/)), and a rapid increase in the volume of remotely sensed information available, have changed the scale of the data available for analysing these systems. This increase in data availability has led to the development of new methods and algorithms for their efficient collection, validation, storage and analyses. With the continued rapid development of satellite-based and remote observation systems (e.g. ocean drifters and automatic weather stations), and of new methods for genetic analyses of biological systems, a step-change is occurring in the magnitude of data available on all components of Antarctic systems. Dealing with these data will require a similar step-change in the use of mathematics in all aspects of Antarctic science.

Many of the issues of global importance in Antarctic science are at the interfaces between traditional disciplines (e.g. biology and physics or oceans and the cryosphere). In many of these areas new methodological and analytical approaches and models are required. For example, addressing questions about how climate change and direct human impacts (such as fishing) will affect ecosystems requires integrated studies that link knowledge of biogeochemical cycles, species and food webs [68, 69]. This requires integrated whole ecosystem (also termed "end-to-end") analyses at local (10 s to 100 km), regional (100 s to 100 km) and circumpolar scales (10,000 s km) [66, 68]. Such whole system integration has become a central focus of international activities in many areas of Antarctic science, and particularly in Southern Ocean studies aimed at linking climate and ecosystem processes [66]. There are major theoretical and analytical challenges in developing such integrated analyses and models. These include questions about how different physical, chemical and biological processes link across a range of scales [67], how different model structures can be coupled together to ensure appropriate feedbacks and system behaviour [68] and how to control and characterize the uncertainty that often multiplies as models integrate more processes [44]. This will require the development of a wide range of new and innovative mathematical approaches.

Acknowledgements We thank Alberto Pinto and his team for organizing such an excellent conference in bringing scientists from a wide range of research disciplines during the year on "Mathematics for Planet Earth". We also thank Janet Silk and Peter Fretwell for preparing Figs. 2 and 6 respectively. This work has the support of the Ministry of Science and Higher Education in Portugal (Fundação para a Ciência e a Tecnologia) and the research programs CEPH, SCAR AnT-ERA, PROPOLAR and ICED. It is a contribution to the British Antarctic Survey programmes Ecosystems and Climate which are funded by the Natural Environment Research Council.

References

1. Agnew, D.J., Kirkwood, G.P.: A statistical method for estimating the level of IUU fishing: application to CCAMLR subarea 48.3. CCAMLR Sci. **12**, 119–141 (2005)
2. Agnew, D.J., Edwards, C., Hillary, R., Mitchell, R., Lopez Abellan, L.J.: Status of the coastal stocks of *Dissostichus* spp. In east Antarctica (Divisions 58.4.1 and 58.4.2). CCAMLR Sci. **16**, 71–100 (2009)
3. Anisimov, O.A., Vaughan, D.G., Callaghan, T.V., Furgal, C., Marchant, H., Prowse, T.D., Vilhjálmsson, H., Walsh, J.E.: Polar regions (Arctic and Antarctic). In: Parry, M.L., Canziani, O.F., Palutikof, J.P., van der Linden, P.J., Hanson, C.E. (eds.) Climate Change 2007: Impacts, Adaptation and Vulnerability. Contribution of Working Group II to the Fourth Assessment Report of the Intergovernmental Panel on Climate Change, pp. 653–685. Cambridge University Press, Cambridge (2007)
4. Atkinson, A., Siegel, V., Pakhomov, E.A., Rothery, P.: Long-term decline in krill stock and increase in salps within the Southern Ocean. Nature **432**, 100–103 (2004)
5. Atkinson, A., Shreeve, R.S., Hirst, A.G., Rothery, P., Tarling, G.A., Pond, D.W., Korb, R., Murphy, E.J., Watkins, J.L.: Natural growth rates of Antarctic krill (*Euphausia superba*): II. Predictive models based on food, temperature, body length, sex, and maturity stage. Limnol. Oceanogr. **51**, 973–987 (2006)
6. Atkinson, A., Siegel, V., Pakhomov, E.A., Jessopp, M.J., Loeb, V.: A re-appraisal of the total biomass and annual production of Antarctic krill. Deep Sea Res. I Oceanogr. Res. Pap. **56**, 727–740 (2009)
7. Ballerini, T., Hofmann, E.E., Ainley, D.G., Daly, K., Marrari, M., Ribic, C.A., Smith, W.O., Steele, J.H.: Productivity and linkages of the food web of the southern region of the Western Antarctic Peninsula continental shelf. Prog. Oceanogr. **122**, 10–29 (2014)
8. Barbraud, C., Heimerskirch, H.: Emperor penguins and climate change. Nature **411**, 183–186 (2001)
9. Barnes, D.K.A., Peck, L.S.: Vulnerability of Antarctic shelf biodiversity to predicted regional warming. Clim. Res. **37**, 149–163 (2008)

10. Beddington, J.R., May, R.M.: Maximum sustainable yields in systems subject to harvesting at more than one trophic level. Math. Biosci. **51**, 261–281 (1980)
11. Beddington, J.R., May, R.M.: The harvesting of interacting species in a natural ecosystem. Sci. Am. **247**, 42–50 (1982)
12. Block, B.A., Jonsen, I.D., Jorgensen, S.J., Winship, A.J., Shaffer, S.A., Bograd, S.J., Hazen, E.L., Foley, D.G., Breed, G.A., Harrison, A.L., Ganong, J.E., Swithenbank, A., Castleton, M., Dewar, H., Mate, B.R., Shillinger, G.L., Schaefer, K.M., Benson, S.R., Weise, M.J., Henry, R.W., Costa, D.P.: Tracking apex marine predator movements in a dynamic ocean. Nature **475**(7354), 86–90 (2011). doi:10.1038/nature10082
13. Bokhorst, S., Huiskes, A., Convey, P., Sinclair, B.J., Lebouvier, M., Van de Vijver, B., Wall, D.H.: Microclimate impacts of passive warming methods in Antarctica: implications for climate change studies. Polar Biol. **34**, 1421–1435 (2011)
14. Boltovskoy, D.: South Atlantic Zooplankton. Backhuys Publishers, Leiden (1999)
15. Bracegirdle, T.J., Stephenson, D.B.: Higher precision estimates of regional polar warming by ensemble regression of climate model projections. Clim. Dyn. **39**, 2805–2821 (2012)
16. Brandt, A., Gooday, A.J., Brandao, S.N., Brix, S., Brokeland, W., Cedhagen, T., Choudhury, M., Cornelius, N., Danis, B., De Mesel, I., Diaz, R.J., Gillan, D.C., Ebbe, B., Howe, J.A., Janussen, D., Kaiser, S., Linse, K., Malyutina, M., Pawlowski, J., Raupach, M., Vanreusel, A.: First insights into the biodiversity and biogeography of the Southern Ocean deep sea. Nature **447**(7142), 307–311 (2007)
17. Candy, S.G.: Modelling catch and effort data using generalised linear models, the tweedie distribution, random vessel effects and random stratum-by-year effects. CCAMLR Sci. **11**, 59–80 (2004)
18. Candy, S.G., Constable, A.J.: An integrated stock assessment for the Patagonian toothfish (*Dissostichus eleginoides*) for the Heard and McDonald islands using CASAL. CCAMLR Sci. **15**, 1–34 (2008)
19. Candy, S.G., Constable, A.J., Lamb, T., Williams, R.: A von Bertalanffy growth model for toothfish at heard island fitted to length-at-age data and compared to observed growth from mark-recapture studies. CCAMLR Sci. **14**, 43–66 (2007)
20. Ceia, F.R., Phillips, R.A., Ramos, J.A., Cherel, Y., Vieira, R.P., Richard, P., Xavier, J.C.: Short- and long-term consistency in the foraging niche of wandering albatrosses. Mar. Biol. **159**, 1581–1591 (2012)
21. Cherel, Y., Hobson, K.: Stable isotopes, beaks and predators: a new tool to study the trophic ecology of cephalopods, including giant and colossal squids. Proc. R. Soc. Lond. B **272**, 1601–1607 (2005)
22. Clark, J.M., Agnew, D.J.: Estimating the impact of depredation by killer whales and sperm whales on longline fishing for toothfish (*Dissostichus eleginoides*) around South Georgia. CCAMLR Sci. **17**, 163–178 (2010)
23. Constable, A.J., De la mare, W.K., Agnew, D.J., Everson, I., Miller, D.: Managing fisheries to conserve the Antarctic marine ecosystem: practical implementation of the Convention on the Conservation of Antarctic Marine Living Resources (CCAMLR). ICES J. Mar. Sci. **57**, 778–791 (2000)
24. Convey, P.: Antarctic ecosystems. In: Levin, S.A. (ed.) Encyclopedia of Biodiversity, pp. 179–188. Academic, Waltham (2013)
25. Convey, P., Aitken, S., di Prisco, G., Gill, M.J., Coulson, S.J., Barry, T., Jónsdóttir, I.S., Dang, P.T., Hik, D., Kulkarni, T., Lewis, G.: The impacts of climate change on circumpolar biodiversity. Biodiversity **13**, 134–143 (2012)
26. Cornejo-Donoso, J., Antezana, T.: Preliminary trophic model of the Antarctic Peninsula Ecosystem (Sub-area CCAMLR 48.1). Ecol. Model. **218**, 1–17 (2008)
27. Croxall, J.P., Silk, J.R.D., Phillips, R.A., Afanasyev, V., Briggs, D.R.: Global circumnavigations: tracking year-round ranges of nonbreeding albatrosses. Science **307**, 249–250 (2005)
28. Deagle, B.E., Gales, N.J., Evans, K., Jarman, S.N., Robinson, S., Trebilco, R., Hindell, M.A.: Studying seabird diet through genetic analysis of faeces: a case study on macaroni penguins (*Eudyptes chrysolophus*). PLoS One **2**, e831 (2007)

29. Egevang, C., Stenhouse, I.J., Phillips, R.A., Petersen, A., Fox, J.W., Silk, J.R.D.: Tracking of Arctic terns *Sterna paradisaea* reveals longest animal migration. In: Proceedings of the National Academy of Sciences (2010)
30. FAO: The state of world fisheries and aquaculture. In: Part 1. World Review of Fisheries and Aquaculture. Food and Agriculture Organization of the United Nations, Rome (2012)
31. Fretwell, P.T., LaRue, M.A., Morin, P., Kooyman, G.L., Wienecke, B., Ratcliffe, N., Fox, A.J., Fleming, A.H., Porter, C., Trathan, P.N.: An emperor penguin population estimate: the first global, synoptic survey of a species from space. PLoS One **7**, e33751 (2012)
32. Garcia, S.M.: The precautionary approach to fisheries and its implications for fishery research, technology and management: an updated review.. In: FAO. Precautionary Approach to Fisheries. Part 2: Scientific Papers, vol. 350 Part 2. p. 210. FAO Fisheries Technical Paper, Rome (1996)
33. Grant, S.M., Hill, S.L., Fretwell, P.: Spatial distribution of management measures, Antarctic krill catch and Southern Ocean bioregions: implications for conservation planning. CCAMLR Sci. **20**, 1–20 (2013)
34. Grant, S.M., Hill, S.L., Trathan, P.N., Murphy, E.J.: Ecosystem services of the Southern Ocean: trade-offs in decision-making. Antarct. Sci. **25**, 603–617 (2013)
35. Griffiths, H.J.: Antarctic Marine Biodiversity: What Do We Know About the Distribution of Life in the Southern Ocean? PLoS One **5**(8), e11683 (2010)
36. Hazen, E.L., Jorgensen, S., Rykaczewski, R.R., Bograd, S.J., Foley, D.G., Jonsen, I.D., Shaffer, S.A., Dunne, J.P., Costa, D.P., Crowder, L.B., Block, B.A.: Predicted habitat shifts of Pacific top predators in a changing climate. Nat. Clim. Chang. **3**(3), 234–238 (2013)
37. Hecht, T.: A guide to the otoliths of Southern Ocean fishes. S. Afr. J. Antarct. Res. **17**, 2–87 (1987)
38. Hill, S.L.: From strategic ambiguity to technical reference points in the Antarctic krill fishery: the worst journey in the world? Environ. Conserv. **40**, 394–405 (2013)
39. Hill, S.L.: Prospects for a sustainable increase in the availability of long chain omega 3s: lessons from the Antarctic krill fishery. In: De Meester, F., Watson, R.R., Zibadi, S. (eds.) Omega 6/3 Fatty Acids Functions, Sustainability Strategies and Perspectives, pp. 267–296. Humana Press, New York (2013)
40. Hill, S.L., Cannon, M.: A potential feedback approach to ecosystem based management: model predictive control of the Antarctic krill fishery. CCAMLR Sci. **20**, 119–138 (2013)
41. Hill, S.L., Matthews, J.: The sensitivity of multiple output statistics to input parameters in a krill-predator- fishery ecosystem dynamics model. CCAMLR Sci. **20**, 97–118 (2013)
42. Hill, S.L., Murphy, E.J., Reid, K., Trathan, P.N., Constable, A.: Modelling Southern Ocean ecosystems: krill, the food-web, and the impacts of fishing. Biol. Rev. **81**, 581–608 (2006)
43. Hill, S.L., Reid, K., Thorpe, S.E., Hinke, J., Watters, G.M.: A compilation of parameters for ecosystem dynamics models of the Scotia Sea- Antarctic Peninsula region. CCAMLR Sci. **14**, 1–25 (2007)
44. Hill, S.L., Watters, G.M., Punt, A.E., McAllister, M.K., Le Quere, C., Turner, J.: Model uncertainty in the ecosystem approach to fisheries. Fish Fish. **8**, 315–333 (2007)
45. Hill, S.L., Trathan, P.N., Agnew, D.J.: The risk to fishery performance associated with spatially resolved management of Antarctic krill (*Euphausia superba*) harvesting. ICES J. Mar. Sci. **66**, 2148–2154 (2009)
46. Hill, S.L., Keeble, K., Atkinson, A., Murphy, E.J.: A food web model to explore uncertainties in the South Georgia shelf pelagic ecosystem. Deep Sea Res. II **59–60**, 237–252 (2012)
47. Hill, S.L., Phillips, T., Atkinson, A.: Potential climate change effects on the habitat of Antarctic krill in the Weddell Quadrant of the Southern Ocean. PLoS One **8**, e72246 (2013)
48. Hillary, R.M., Kirkwood, G.P., Agnew, D.J.: An assessment of toothfish in Subarea 48.3 using CASAL. CCAMLR Sci. **13**, 65–95 (2006)
49. IPCC: Climate Change 2007: Synthesis Report. Contribution of Working Groups I, II and III to the Fourth Assessment Report of the Intergovernmental Panel on Climate Change. IPCC, Geneva (2007)

50. Jenouvrier, S., Holland, M., Stroeve, J., Barbraud, C., Weimerskirch, H., Serreze, M., Caswell, H.: Effects of climate change on an emperor penguin population: analysis of coupled demographic and climate models. Glob. Chang. Biol. **18**(9), 2756–2770 (2012)
51. Jouventin, P., Weimerskirch, H.: Satellite tracking of wandering albatrosses. Nature **343**, 746–748 (1990)
52. Kaiser, B., Zicus, S., Allen, B.: Polar Science and Global Climate: An International Resource for Education & Outreach. Pearson, Essex (2010)
53. Karnovsky, N.J., Hobson, K.A., Iverson, S.J.: From lavage to lipids: estimating diets of seabirds. Mar. Ecol. Prog. Ser. **451**, 263–284 (2012)
54. Kawaguchi, S., Ishida, A., King, R., Raymond, B., Waller, N., Constable, A., Nicol, S., Wakita, M., Ishimatsu, A.: Risk maps for Antarctic krill under projected Southern Ocean acidification. Nat. Clim. Chang. **3**(9), 843–847 (2013)
55. Kinzey, D., Watters, G., Reiss, C.S.: Effects of recruitment variability and natural mortality on generalised yield model projections and the CCAMLR decision rules for Antarctic krill. CCAMLR Sci. **20**, 81–96 (2013)
56. Krupnik, I., Allison, I., Bell, R., Cutler, P., Hik, D., López-Martinez, J., Rachold, V., Sarukhanian, E., Summerhayes, C.: Understanding Earth's Polar Challenges: International Polar Year 2007–2008, vol. 1. University of the Arctic. CCI Press, Rovaniemi (2011)
57. Laws, R.M.: Seals and Whales of the Southern Ocean. Philos. Trans. R. Soc. Lond. B **279**, 81–96 (1977)
58. Lincoln, F.C.: Calculating waterfowl abundance on the basis of banding returns, vol. 118. Cir. U.S. Department of Agriculture (1930)
59. Link, J.S., Ihde, T.F., Harvey, C.J., Gaichas, S.K., Field, J.C., Brodziak, J.K.T., Townsend, H.M., Peterman, R.M.: Dealing with uncertainty in ecosystem models: The paradox of use for living resources. Prog. Oceanogr. **102**, 102–114 (2012)
60. Link, W.A., Barker, R.J.: Bayesian Inference with Ecological Applications. Academic, London (2010)
61. Magnusson, A., Punt, A., Hilborn, R.: Measuring uncertainty in fisheries stock assessment: the delta method, bootstrap, and MCMC. Fish Fish. **14**(3), 325–342 (2012)
62. Martinez, I.: Emperor penguin. In: Handbook of the Birds of the World, vol. 1. Lynx Edicions, Barcelona (1992)
63. McLeod, K.L., Leslie, H.M.: Why ecosystem-based management? In: McLeod, K., Leslie, H. (eds.) Ecosystem-Based Management for the Oceans, pp. 3–12. Island Press, Washington (2009)
64. Meredith, M.P., King, J.C.: Rapid climate change in the ocean west of the Antarctic Peninsula during the second half of the 20th century. Geophys. Res. Lett. **32**, L19604 (2005)
65. Montes-Hugo, M., Doney, S.C., Ducklow, H.W., Fraser, W., Martinson, D., Stammerjohn, S.E., Schofield, O.: Recent changes in phytoplankton communities associated with rapid regional climate change along the Western Antarctic Peninsula. Science **323**, 1470–1473 (2009)
66. Murphy, E.J., Hofmann, E.E.: End-to-end in Southern Ocean ecosystems. Curr. Opin. Environ. Sustain. **4**, 264–271 (2013)
67. Murphy, E.J., Morris, D.J., Watkins, J.L., Priddle, J.: Scales of interaction between antarctic krill and the environment. In: Sahrhage, D. (ed.) Antarctic Ocean and Resources Variability, pp. 120–130. Springer, Berlin (1988)
68. Murphy, E.J., Cavanagh, R.D., Hofmann, E.E., Hill, S.L., Constable, A.J., Costa, D.P., Pinkerton, M.H., Johnston, N.M., Trathan, P.N., Klinck, J.M., Wolf-Gladrow, D.A., Daly, K.L., Maury, O., Doney, S.C.: Developing integrated models of Southern Ocean food webs: including ecological complexity, accounting for uncertainty and the importance of scale. Prog. Oceanogr. **102**, 74–92 (2012)
69. Murphy, E.J., Hofmann, E.E., Watkins, J.L., Johnston, N.M., Piñones, A., Ballerini, T., Hill, S.L., Trathan, P.N., Tarling, G.A., Cavanagh, R.A., Young, E.F., Thorpe, S.E., Fretwell, P.: Comparison of the structure and function of Southern Ocean regional ecosystems: The Antarctic Peninsula and South Georgia. J. Mar. Syst. **109–110**, 22–42 (2013)

70. Nevitt, G.A., Losekoot, M., Weimerskirch, H.: Evidence for olfactory search in wandering albatross, Diomedea exulans. Proc. Natl. Acad. Sci. **105**, 4576–4581 (2008)
71. Nicol, S., Foster, J., Kawaguchi, S.: The fishery for Antarctic krill-recent developments. Fish Fish. **13**, 30–40 (2012)
72. Orr, J.C., Fabry, V.J., Aumont, O., Bopp, L., Doney, S.C., Feely, R.A., Gnanadesikan, A., Gruber, N., Ishida, A., Joos, F., Key, R.M., Lindsay, K., Maier-Reimer, E., Matear, R., Monfray, P., Mouchet, A., Najjar, R.G., Plattner, G.-K., Rodgers, K.B., Sabine, C.L., Sarmiento, J.L., Schlitzer, R., Slater, R.D., Totterdell, I.J., Weirig, M.-F., Yamanaka, Y., Yool, A.: Anthropogenic ocean acidification over the twenty-first century and its impact on calcifying organisms. Nature **437**(7059), 681–686 (2005)
73. Parmelee, D.F., Parmelee, J.M., Fuller, M.R.: Ornithological investigations and Palmer Station: the first long-distance tracking of seabirds by satellites. Antarct. J. USA **20**, 162–163 (1985)
74. Parnell, A.C., Inger, R., Bearhop, S., Jackson, A.L.: Source partitioning using stable isotopes: coping with too much variation. PLoS One **5**, e9672 (2010)
75. Peck, L.S.: Prospects for surviving climate change in Antarctic aquatic species. Front. Zool. **2**, 2–9 (2005)
76. Peck, L.S., Barnes, D.K.A., Cook, A.J., Fleming, A.H., Clarke, A.: Negative feedback in the cold: ice retreat produces new carbon sinks in Antarctica. Glob. Chang. Biol. **16**(9), 2614–2623 (2010). doi:10.1111/j.1365-2486.2009.02071.x
77. Phillips, R.A., Xavier, J.C., Croxall, J.P.: Effects of satellite transmitters on albatrosses and petrels. The Auk **120**(4), 1082–1090 (2003)
78. Phillips, R.A., Silk, J.R.D., Croxall, J.P., Afanasyev, V., Briggs, D.R.: Accuracy of geolocation estimates for flying seabirds. Mar. Ecol. Prog. Ser. **266**, 265–272 (2004)
79. Phillips, R.A., Croxall, J.P., Silk, J.R.D., Briggs, D.R.: Foraging ecology of albatrosses and petrels from South Georgia: two decades of insights from tracking technologies. Aquat. Conserv. Mar. Freshwat. Ecosyst. **17**, S6-S21 (2008)
80. Pinkerton, M.H., Bradford-Grieve, J.M., Hanchet, S.M.: A balanced model of the food web of the Ross Sea, Antarctica. CCAMLR Sci. **17**, 1–31 (2010)
81. Plagányi, É.E., Butterworth, D.S.: The Scotia Sea krill fishery and its possible impacts on dependent predators: modeling localized depletion of prey. Ecol. Appl. **22**, 748–761 (2012)
82. Polovina, J.J.: Model of a coral reef ecosystem I. The ECOPATH model and its application to French Frigate Shoals. Coral Reefs **3**, 1–11 (1984)
83. Prince, P.A., Wood, A.G., Barton, T.R., Croxall, J.P.: Satellite-tracking wandering albatrosses Diomedea exulans in the South Atlantic. Antarct. Sci. **4**, 31–36 (1992)
84. Punt, A.E., Smith, A.D.M.: The gospel of maximum sustainable yield in fisheries management: birth, crucifixion and reincarnation. In: Reynolds, J.D., Mace, G.M., Redford, K.H., Robinson, J.G. (eds.) Conservation of exploited species, pp. 41–66. Cambridge University Press, Cambridge (2001)
85. Rademeyer, R.A., Plagányi, É.E., Butterworth, D.S.: Tips and tricks in designing management procedures. ICES J. Mar. Sci. **64**, 618–625 (2007)
86. Ramos, R., Gonzalez-Solis, J.: Trace me if you can: the use of intrinsic biogeochemical markers in marine top predators. Front. Ecol. Environ. **10**(5), 258–266 (2012)
87. Ratcliffe, N., Trathan, P.N.: A review of the diet and foraging movements of penguins breeding within the CCAMLR area. CCAMLR Sci. **18**, 75–114 (2011)
88. Reid, K.: A Guide to the Use of Otoliths in the Study of Predators at South Georgia. British Antarctic Survey, Cambridge (1996)
89. Rignot, E., Bamber, J.L., Van Den Broeke, M.R., Davis, C., Li, Y.H., Van De Berg, W.J., Van Meijgaard, E.: Recent Antarctic ice mass loss from radar interferometry and regional climate modelling. Nat. Geosci. **1**, 106–110 (2008)
90. Rodhouse, P.G., Xavier, J.C., Griffiths, H.: Southern Ocean squid. In: De Broyer, C., Koubbi, P., Griffiths, H., Danis, B., David, B., Grant, S., Gutt, J., Held, C., Hosie, G., Huettmann, F., Post, A., Raymond, B., Ropert-Coudert, Y., van de Putte, A. (eds.) The CAML/SCAR-MarBIN Biogeographic Atlas of the Southern Ocean, pp. 284–289. Scientific Committee on Antarctic Research, Cambridge (2014)

91. Sarmiento, J.L., Gruber, N., Brzezinski, M.A., Dunne, J.P.: High-latitude controls of thermocline nutrients and low latitude biological productivity. Nature **427**(6969), 56–60 (2004)
92. Smetacek, V., Nicol, S.: Polar ocean ecosystems in a changing world. Nature **437**, 362–368 (2005)
93. Smith, A.D., Brown, C.J., Bulman, C.M., Fulton, E.A., Johnson, P., Kaplan, I.C., Lozano-Montes, H., Mackinson, S., Marzloff, M., Shannon, L.J., Shin, Y.J., Tam, J.: Impacts of fishing low-trophic level species on marine ecosystems. Science **333**, 1147–1150 (2011)
94. Stock, C.A., Alexander, M.A., Bond, N.A., Brander, K.M., Cheung, W.W.: On the use of IPCC-class models to assess the impact of climate on living marine resources. Prog. Oceanogr. **88**, 1–27 (2011)
95. Sumner, M.D., Wotherspoon, S.J., Hindell, M.A.: Bayesian estimation of animal movement from archival and satellite tags. PLoS One **4**, e7324 (2009)
96. Tavares, S., Xavier, J.C., Phillips, R.P., Pereira, M.E., Pardal, M.A.: Influence of age, sex and breeding status on mercury accumulation patterns in wandering albatrosses *Diomedea exulans*. Environ. Pollut. **181**, 315–320 (2013)
97. Turner, J., Bindschadler, R., Convey, P., di Prisco, G., Fahrbach, E., Gutt, J., Hodgson, D., Mayewski, P., Summerhayes, C.: Antarctic Climate Change and the Environment. Scientific Committee for Antarctic Research, Cambridge (2009)
98. Turner, J., Barrand, N.E., Bracegirdle, T.J., Convey, P., Hodgson, D.A., Jarvis, M., Jenkins, A., Marshall, G., Meredith, M.P., Roscoe, H., Shanklin, J., French, J., Goosse, H., Guglielmin, M., Gutt, J., Jacobs, S., Kennicutt, M.C.I., Masson-Delmotte, V., Mayewski, P., Navarro, F., Robinson, S., Scambos, T., Sparrow, M., Summerhayes, C., Speer, K., Klepikov, A.: Antarctic climate change and the environment: an update. Polar Record First View, pp. 1–23 (2013)
99. Turner, J., Maksym, T., Phillips, T., Marshall, G.J., Meredith, M.P.: The impact of changes in sea ice advance on the large winter warming on the western Antarctic Peninsula. Int. J. Climatol. **33**(4), 852–861 (2013)
100. Turner, J., Thomas, J.B., Phillips, T., Marshall, G.J., Hosking, J.S.: An initial assessment of antarctic sea ice extent in the cmip5 models. J. Clim. **26**, 1473–1484 (2013)
101. Ulanowicz, R.E., Puccia, C.J.: Mixed trophic impacts in ecosystems. Coenoses **5**, 7–16 (1990)
102. Walton, D.: Antarctica: Global Science from a Frozen Continent. Cambridge University Press, Cambridge (2013)
103. Walton, D., Xavier, J.C., May, I., Huffman, L.: Polar Educators International - a new initiative for schools. Antarct. Sci. **25**, 473 (2013)
104. Waluda, C.M., Hill, S.L., Peat, H.J., Trathan, P.N.: Diet variability and reproductive performance of macaroni penguins (*Eudyptes chrysolophus*) at Bird Island, South Georgia. Mar. Ecol. Prog. Ser. **466**, 261–274 (2012)
105. Watters, G.M., Hill, S.L., Hinke, J., Matthews, J., Reid, K.: Decision making for ecosystem based management: evaluating options for a krill fishery with an ecosystem dynamics model. Ecol. Appl. **23**, 710–725 (2013)
106. Webber, D.N., Parker, S.J.: Estimating unaccounted fishing mortality in the Ross sea region and Amundsen sea (CCAMLR subareas 88.1 and 88.2) bottom longline fisheries targeting Antarctic toothfish. CCAMLR Sci. **19**, 17–30 (2012)
107. Weimerskirch, H., Bonadonna, F., Bailleul, F., Mabille, G., Dell'Omo, G., Lipp, H.-P.: GPS tracking of foraging albatrosses. Science **295**, 1259–1259 (2002)
108. Wilson, R.P., Grémillet, D., Syder, J., Kierspel, M.A.M., Garthe, S., Weimerskirch, H., Schafer-Neth, C., Scolaro, J.A., Bost, C.-A., Plotz, J., Nel, D.C.: Remote-sensing systems and seabirds: their use, abuse and potential for measuring marine environmental variables. Mar. Ecol. Prog. Ser. **228**, 241–261 (2002)
109. Xavier, J.C.: Predator-Prey Interactions Between Albatrosses and Cephalopods at South Georgia. University of Cambridge, Cambridge (2003)
110. Xavier, J.C., Cherel, Y.: Cephalopod Beak Guide for the Southern Ocean. British Antarctic Survey, Cambridge (2009)

111. Xavier, J.C., Rodhouse, P.G., Trathan, P.N., Wood, A.G.: A Geographical Information System (GIS) atlas of cephalopod distribution in the Southern Ocean. Antarct. Sci. **11**(1), 61–62 (1999)
112. Xavier, J.C., Croxall, J.P., Reid, K.: Inter-annual variation in the diet of two albatross species breeding at South Georgia: implications for breeding performance. Ibis **145**, 593–610 (2003)
113. Xavier, J.C., Croxall, J.P., Cresswell, K.A.: Boluses: an effective method to assess the proportions of cephalopods in the diet of albatrosses. Auk **122**, 1182–1190 (2005)
114. Xavier, J.C., Geraint, G.A., Croxall, J.P.: Determining large scale distribution of pelagic cephalopods, fish and crustaceans in the South Atlantic from wandering albatross (*Diomedea exulans*) foraging data. Ecography **29**, 260–272 (2006)
115. Xavier, J.C., Clarke, M.R., Magalhães, M.C., Stowasser, G., Blanco, C., Cherel, Y.: Current status of using beaks to identify cephalopods: III international workshop and training course on cephalopod beaks, Faial Island, Azores, April 2007. Arquipélago Life Mar. Sci. **24**, 41–48 (2007)
116. Xavier, J.C., Phillips, R.A., Cherel, Y.: Cephalopods in marine predator diet assessments: why identifying upper and lower beaks is important. ICES J. Mar. Sci. **68**, 1857–1864 (2011)
117. Xavier, J.C., Barbosa, A., Agusti, S., Alonso-Sáez, L., Alvito, P., Ameneiro, J., Avila, C., Baeta, A., Canário, A., Carmona, R., Catry, P., Ceia, F., Clark, M.S., Cristobo, F.J., Cruz, B., Duarte, C.M., Figuerola, B., Gili, J.-M., Gonçalves, A.R., Gordillo, F.J.L., Granadeiro, J.P., Guerreiro, M., Isla, E., Jiménez, C., López-González, P.J., Lourenço, S., Marques, J.C., Moreira, E., Mota, A.M., Nogueira, M., Núñez-Pons, L., Orejas, C., Paiva, V.H., Palanques, A., Pearson, G.A., Pedrós-Alió, C., Peña Cantero, A.L., Power, D.M., Ramos, J.A., Rossi, S., Seco, J., Sañe, E., Serrão, E.A., Taboada, S., Tavares, S., Teixidó, N., Vaqué, D., Valente, T., Vázquez, E., Vieira, R., B., V.: Polar marine biology science in Portugal and Spain: Recent advances and future perspectives. J. Sea Res. **83**, 9–29 (2013)
118. Xavier, J.C., Cherel, Y., Roberts, J., Piatkowski, U.: How do cephalopods become available to seabirds: can fish gut contents from tuna fishing vessels be a major food source of deep-dwelling cephalopods? ICES J. Mar. Sci. **70**, 46–49 (2013)
119. Xavier, J.C., Allcock, L., Cherel, Y., Lipinski, M.R., Gomes-Pereira, J.N., Pierce, G., Rodhouse, P.G.K., Rosa, R., Shea, L., Strugnell, J., Vidal, E., Villanueva, R., Ziegler, A.: Future challenges in cephalopod research. J. Mar. Biol. Assoc. UK (2015)
120. Ziegler, P.E.: Influence of data quality and quantity from a multiyear tagging program on an integrated fish stock assessment. Can. J. Fish. Aquat. Sci. **70**, 1031–1045 (2013)

Appendix A: CIM International Planet Earth Events MECC I, 2013

In 2013 the CIM organized the International Conference on the Mathematics of Planet Earth: MECC I, 2013—International Conference Planet Earth, Mathematics of Energy and Climate Change, 25–27 March 2013. Furthermore, the CIM organized the following Advanced School Planet Earth directly before and after the International Conference: School MECC I, 2013—Advanced School Planet Earth, Mathematics of Energy and Climate Change, 21–23 and 27–28 March 2013.

The CIM Mathematics of Planet Earth events stemmed from the CIM's role as a partner institution of the International Program Mathematics of Planet Earth 2013 (MPE 2013). We were pleased that the CIM-MPE events were announced, for example, in the ICIAM newsletter for January 2013 and the EMS newsletter for March 2013.

These events were enthusiastically supported by many Portuguese institutions, including: the SPM; SPE; APDIO; CEMAPRE; CEAUL; CMA-UNL; CMAF-UL; CMUP; INESCTEC; ISR; IT; UECE FCUL; ISEG; Calouste Gulbenkian Foundation (FCG) and Ciência Viva (CV).

The International Conference MECC I, 2013 was hosted by the Calouste Gulbenkian Foundation.

The Advanced School Planet Earth, Mathematics of Energy and Climate Change was hosted by the Faculdade de Ciências, Universidade de Lisboa (FCUL).

In addition, the CIM would especially like to thank Irene Fonseca for her scientific guidance, João Paulo Almeida for his guidance and coordination of the events, Antónia Turkman for her assistance in coordinating with the Calouste Gulbenkian Foundation, Telmo Parreira for organizing and compiling the proceedings, and Paulo Mateus, Pedro Baltazar and Telmo Parreira for developing and maintaining the conference website. The CIM would like to thank the CGF staff and members of the local organizing committee as well as the Calouste Gulbenkian Foundation for their incredible hospitality throughout the event and for providing to speakers and participants the opportunity to experience the beautiful city of Lisbon in a friendly ambiance.

J.-P. Bourguignon et al. (eds.), *Mathematics of Energy and Climate Change*,
CIM Series in Mathematical Sciences 2, DOI 10.1007/978-3-319-16121-1

The CIM would like to thank the following keynote speakers of MECC I, 2013 for their insightful lectures:

- Inês Azevedo, Carnegie Mellon University, USA
- Richard James, University of Minnesota, USA
- Christopher K.R.T. Jones, University of North Carolina, USA
- Pedro Miranda, Universidade de Lisboa, Portugal
- Keith Promislow, Michigan State University, USA
- Richard L. Smith, University of North Carolina, USA
- José Xavier, Universidade de Coimbra, Portugal
- David Zilberman, University of California, Berkeley, USA

The CIM's thanks also go to the 60 invited speakers for their insightful presentations, and to the 17 session organizers, whose energy and commitment were so vital to the success of the events:

- João Paulo Almeida, IPB
- Paulo A.V. Borges, Universidade dos Açores
- Margarida Brito, FCUP
- Miguel Centeno Brito, Universidade de Lisboa
- José Luís dos Santos Cardoso and Mário Gonzalez Pereira, UTAD
- Maria da Conceição Carvalho, FCUL
- Stéphane Louis Clain, Universidade do Minho
- João Gama, Universidade do Porto
- Sílvio M.A. Gama and João Emílio Almeida, FCUP
- Ivette Gomes, Universidade de Lisboa
- Patrícia Gonçalves, Universidade do Minho
- Raquel Menezes, Universidade do Minho
- Alberto Adrego Pinto, Universidade do Porto
- António Pacheco Pires, UTL
- Carlos Ramos, Universidade de Évora
- Delfim F.M. Torres, Universidade de Aveiro
- Tânia Pinto Varela, UTL

The CIM would like to thank the members of the local organizing committee of MECC I, 2013 for their remarkable professionalism: Alberto Pinto (FCUP); Paulo Mateus, (IST); Pedro Baltazar (IST); João Paulo Almeida (IPB); Abdelrahim Mousa (FCUP); Renato Soeiro (FCUP); Bruno Neto (FCUP); Filipe Martins (FCU)P; João Coelho, (FCUP) and Joana Becker (FCUP).

The book of abstracts of MECC I, 2013 can be found in the link:

http://sqig.math.ist.utl.pt/cim/mpe2013/docs/bookMECC2013.pdf

Porto, Portugal Alberto Adrego Pinto

Appendix B: Interviews MPE: MECC I

CIM thanks the participants Margarida Brito (Universidade do Porto), João Coelho (LIAAD-INESC TEC, Universidade do Porto), José Cardoso (Universidade de Trás-os-Montes e Alto Douro), Ricardo Cruz (Universidade do Porto), João Gama (LIAAD-INESC TEC, Universidade do Porto), Ivette Gomes (CEAUL and DEIO/FCUL, Universidade do Lisboa), Richard James (University of Minnesota, USA), Carlos Ramos (Centro de Investigação em Matemática e ções, Universidade de Évora), Andrew Schmitz (Universityof Florida, USA), and Ana Soares (Universidade do Minho) of the International Conference and Advanced School Planet Earth, Mathematics of Energy and Climate Change MECC 2013, Portugal, 21–28 March 2013, for sharing their ideas and points of view with us in this interview.

The questions presented here are based on several interviews; in particular, the interviews published in previous CIM bulletins. CIM thanks Renato Soeiro and Alberto Pinto for organizing this interview (see also CIM Bulletin 35).

On the meeting

What was your general impression of the MECC 2013 meeting?

Margarida Brito: In a word, the meeting, due to its interdisciplinary character and the outstanding quality of the participants was a success. The exchange was very prolific, in a purely scientific sense as well as with regard to possible institutional developments and the social impact in general.

José Cardoso: My overall impression of the meeting was very positive. It joined in the same space researchers from different areas with one important link between them: the planet earth. With talks involving important issues in the everyday life of all living beings of our planet, such as climate, energy, and sustainability, the researchers did not just focus on mathematics as an end in itself. Rather, they discussed, with a pragmatic approach to the implications of the new results, new ideas, and, consequently, new materials and new technologies, whether the

J.-P. Bourguignon et al. (eds.), *Mathematics of Energy and Climate Change*,
CIM Series in Mathematical Sciences 2, DOI 10.1007/978-3-319-16121-1

participating community in this meeting, scientific and non-scientific, could become aware of the vast array of problems and challenges that nature incessantly provides us and that, in our own interests, we seek to solve to improve our well-being.

João Coelho: This meeting was fabulous. It provided a general view about what research areas the mathematical society is working on.

Ricardo Cruz: The meeting combined researchers from a wide range of intersectional mathematical areas. It was a great opportunity for M.Sc. and Ph.D. students to meet researchers in several fields, and a good opportunity for collaboration among the researchers.

João Gama: Conferences are meeting places and opportunities to present and discuss our work. In a conference we need to organize and explain in a coherent and comprehensive way the main ideas behind our results. However, sometimes the most relevant aspect comes from the informal contacts that the coffee breaks promote. The offline discussions and the personal contacts with authors whose work we are interested in allow us to enlarge our scientific network, leading us to other scientific experiences.

Ivette Gomes: I organized a session on Statistics of Extremes in Society at CIM International Conferences and Advanced Schools Mathematics of Planet Earth 2013 (CIM-MPE 2013), and due to my schedule I could only attend two other organized sessions and two plenary talks. My overall impression was quite positive.

Richard James: I enjoyed it very much.

Carlos Ramos: The general impression was very good.

Andrew Schmitz: The meeting was excellent.

Ana Soares: Very good.

Something you would like to highlight?

Margarida Brito: It is difficult to choose. The meeting as such was extremely pleasant, with a nice atmosphere, partly due to the conference location, the Calouste Gulbenkian Foundation in Lisbon, which provided an ambience favorable to prolific interchange, not only during the sessions, but during the intervals and at the end of the sessions as well. It was also remarkable to see the great engagement of postgraduate students at the meeting.

José Cardoso: Just to mention a few examples and not pretending to be exhaustive, one heard interesting ideas and new background on the conversion of heat into electricity, the specific mathematics involved in extreme conditions such as in the polar zones, some issues related to photovoltaic dye sensitized solar cells, the relation between technology and bioeconomy, energy conversion on the nanoscale, some topics on biofuels for food crops, as well as more general and well-known questions such as how to reduce CO_2 emissions, and wind power prediction, and also global questions related to climate change such as the role played by internal waves in the surface-atmosphere interface. Beyond all of this, anyone could find in the thematic sessions a variety of subjects where mathematics plays a crucial role.

João Coelho: I would like to highlight the quality of the speakers and the relevance of their research.

João Gama: MECC 2013 was an amazing multidisciplinary meeting. Conversations were more difficult due to the different languages of the attendees, but much richer for those who participate in the game of talking with people outside their borders.

Ivette Gomes: The large variety of topics presented.

Richard James: It was diverse and fascinating, and the venue of the Calouste Gulbenkian Foundation was superb.

Carlos Ramos: The place—the Calouste Gulbenkian Foundation—and the diversity of researchers and communications. The location is fantastic with very good conditions for communications and most of all for informal talks between researchers and students.

Andrew Schmitz: For me, a highlight was the in-depth questions and answers in the sessions I attended.

Ana Soares: The presence of a significant number of Portuguese researchers representing almost all areas of research.

How important do you think that events like this are for students and researchers?

Margarida Brito: In general, meetings like this are important for researchers to develop their ideas through exchange, especially in fields in which interdisciplinarity proves to be essential. Students who participate in these events are exposed to different approaches, open problems, and questions, which encourage and develop their own capacity of research. In particular, the Conference on Mathematics of Energy and Climate Change stands out due to its intrinsic interdisciplinarity, providing researchers with an absolutely necessary platform of exchange and discussion and providing a challenge for participating students.

José Cardoso: One important consequence of this type of meeting is that the general public will be aware of the fundamental role played by mathematics in nature and in the endless attempts to control it. Furthermore, it enables each researcher not only to display their own results and ideas but also to acquire a global overview of many interesting areas of research, and, possibly, to establish new links with other researchers.

João Coelho: They are very important because students and researchers can increase their knowledge and find new ideas and topics to work on.

Ricardo Cruz: The strong adhesion to the event shows there was a growing demand for a conference providing this spectrum of research fields.

Ivette Gomes: The talks I attended were indeed essentially devised for researchers or students at a Ph.D. level, and not for students at an M.Sc. level.

Richard James: These meetings with an intentional flavor, and with a broad collection of viewpoints, are particularly valuable, because they introduce to students a variety of viewpoints that can never be represented in any single institution.

Carlos Ramos: These events are very important for providing a survey and a broad perspective of the area of dynamical systems and its applications within and outside mathematics. I think this is an appropriate type of conference to initiate advanced students in scientific communication and to provide a good opportunity for the students to meet very good active researchers.

Ana Soares: Very important for students in the sense that the meeting represents an opportunity to follow different topics and approaches.

How do you see the impact of this meeting on your field and outside of your field?

Margarida Brito: Well, it was in fact an interdisciplinary meeting, bringing together researchers in mathematics and science working in different fields. By this, I do not refer specifically to mathematical fields, but to different fields of science. Keeping in mind that mathematics, applied to a specific domain, does not mean just using a tool but rather reflecting this domain and its problems in mathematical terms, which may lead to the development of new mathematical methods or even theories, it becomes evident that the exchange which is promoted and facilitated by a congress such as this one is of great importance to the progress of scientific research. This meeting thus emphasizes the decisive role of mathematics in science. We can't overestimate the impact in the scientific field of research. Moreover, the meeting highlights the importance of mathematics in addressing planetary problems. The scientific fields in question are fields with direct connection to problems of humanity, and as these problems are the sort of problems that demand rapid solutions, we can't overestimate the impact of the meeting on society, as well.

Ricardo Cruz: Beyond the meeting itself, participants were invited to submit papers for a volume published by Springer, and the response was overwhelming.

Ivette Gomes: Mathematics is the sharp tool that allows us to describe, to understand, to forecast and to a certain extent to control all phenomena in the world, and even in the universe. Unfortunately, this idea is left behind in the formal teaching of mathematics, and there is the general misleading opinion that mathematics is an abstract science, and that beyond some elementary algebra, analysis, and differential equations used by engineers, it is a kind of useless puzzle. Therefore, periodic meetings on how mathematics intervenes in our way of dealing with reality are a very welcome initiative. I hope they will continue and attract an even wider audience and diversity of active participants.

Richard James: It is particularly valuable for people to see that mathematics has a lot to offer in the study of energy and the environment.

Carlos Ramos: What is, generally, the impact of these events on specific areas, areas they relate to, and on the interplay between different areas or fields of knowledge? The main impact is on the relation between subjects—some very applied—and the possibility of future work it opens.

Andrew Schmitz: The impact of this meeting is positive from a worldwide perspective.

Ana Soares: The impact is relevant on the field because several experts get together and discuss ideas and new problems. Outside the field, it is important because it shows the interdisciplinary character of mathematics.

What would you say is, generally, the impact of these events on specific areas, as they relate to and on the interplay between different areas or fields of knowledge?

Margarida Brito: Let us briefly look at just one problem as an illustration, taken from the main topics from this conference. The reliability of climate previsions is of high importance for a great number of decisions. Previsions depend on a large number of data. We need, besides other things, knowledge about the surface of the earth, which means the earth's crust and the oceans. We have to consider as well the respective consequences of a variety of possible political decisions, which will possibly interfere. So the model on which we elaborate is very complex. Currently, climate researchers know that the actual available data from geology and oceanology is still far from sufficient and that from sociology is minimal. Furthermore, to establish the theoretical bases for prevision, one must take into account that, vice versa, climate interferes at least on the development of the behavior of the oceans. The fast development of electronic data processing in the last decades of the last century motivated the idea of the development of complete models, inducing a tendency to neglect a reflection of the specificity of models, the methods and forms of simplification. This was accompanied by a pushback of theoretical and analytical reflection of the observed phenomena. And, mainly due to mathematicians, the conscience of the inherent interdisciplinary approach was developed, as well as the conscience of the importance of the quality and quantity of data in order to achieve climate research progress. International meetings of this type are fundamental to identify the relevant questions and the different areas or fields involved.

Ivette Gomes: I have a very favorable and positive opinion on all these issues. The impact of the meeting on the broad area of mathematics, including statistics, is high. And due to the interdisciplinary character of the meeting, the impact of the talks is surely also high outside the field of mathematics.

Andrew Schmitz: At least in our session, additional knowledge was obtained from the impact of the US Ethanol Policy.

On your research:

Did you always want to be a mathematician?

João Coelho: Yes.

Ivette Gomes: Indeed, I wanted to study architecture and not mathematics. But my marks in history at the secondary school were not high enough for a candidacy to architecture. Mathematics, a discipline where I had always had very high marks was thus my choice, and today I think this was the most sensible decision.

Richard James: Not at all. As an undergraduate, I was a biomedical engineer—it was a very broad program (at Brown University) that began with basic cellular and molecular biology and ended near physiology and medicine, and the engineering side included a particular focus on mathematics, mechanics, and thermodynamics. Though I was headed for a medical career, I fortunately realized at some point that I liked the quantitative, mathematical part much better, and I turned in that direction. I was (and still am) fascinated by the idea that, by purely mathematical reasoning, one can understand profound things about nature.

Carlos Ramos: I have always wanted to be a scientist (with mathematics).

Ana Soares: Yes, I did.

How did you start working in this area? What was the motivation? Could you tell us about your mathematical beginnings and subsequent career development?

João Coelho: Earlier in my life I started loving math. I liked to study the properties of the numbers and also to discover the methods of solving problems using mathematics. Now, I have a job in stock management, and I use mathematical methods to optimize the management. In the future, my ambition is to obtain a Ph.D. degree. And, who knows, perhaps I will present my future work at future editions of these meetings.

João Gama: My first research experience was in the context of an interdisciplinary European project. I learned a lot from the long discussions on problem formulation using different languages and approaches. The diversity of methods, assumptions, limitations, algorithms, and interpretations was fundamental in my obtaining a much deeper understanding about my own area. We know this to be true: multiple views are always a plus.

Ivette Gomes: I got a degree in Pure Mathematics at the Faculty of Science of Lisbon (FCUL), and my major topic was algebra. I almost went to the USA to work for a Ph.D. in Goldie's ring theory or some similar topic. Indeed, by the end of my 5th year, Professor Almeida Costa was able to provide me with a grant from Gulbenkian Foundation and all the facilities to go abroad immediately after finishing my degree in Pure Mathematics. At the time I chose pure mathematics, and after getting my B.Sc. in Mathematics, I was absolutely sure about this choice. But in my 5th year I had to choose a few optional courses in the area of applied mathematics, and as far as I remember I have chosen courses in probability theory, mathematical statistics, and stochastic processes. Then, my field of interest changed, since dealing with uncertainty and risk is surely the ultimate challenge for a mathematician. I immediately decided not to go to the USA but to stay in Lisbon in order to get a degree in Applied Mathematics. I even found a job as a teacher at a secondary school. But Professor Tiago de Oliveira got to know this through some of my friends in applied mathematics, and he immediately offered me a position at FCUL, in the Department of Applied Mathematics. It was really a tough but gratifying experience. I had to teach courses like Monte Carlo simulation and population dynamics, and I had to use the computer intensively, something that I had never done before. Tiago de Oliveira helped us in the decision of going to Sheffield for the Ph.D. Indeed, Tiago de Oliveira was a very good friend of Joe Gani, the founder of The Probability Trust in Sheffield. But Joe Gani was no longer at Sheffield when we arrived there in September 1975—I only met him 30 years later, in 2005, at the ISI meeting in Sydney, and it was indeed very gratifying talking with him at the time. In Sheffield, I first began my M.Sc. study in Probability and Statistics. I had courses in probability, statistics, weak convergence theory, and data analysis, among others. But as both Dinis and I had Gulbenkian grants and got very high marks in the first term, they thought it sensible to transfer us immediately to the Ph.D. degree in January 1976. I had already had some exposure in Lisbon to statistics of extremes, indeed in the area of bivariate extremes and dependence function estimation, through the reading of an article by Tiago de Oliveira on the subject. I enjoyed the topic very much, but in order to diversify the topics under research at our university, Tiago thought it sensible and I agreed that it would be better to get a specialization in another area, like density estimation, non-parametric statistics, or inference on stochastic processes. But Clive Anderson was a lecturer there and was working in extreme value theory, and he invited me to work under his supervision in the area of extremes. Clive then provided me with several topics of research beginning with rates of convergence and penultimate approximations, extremes of random fields, concomitants of order statistics, and maxima of different types of weak dependent structures, among others. I am deeply indebted to Clive, a person who served as a thesis supervisor and has often helped me with suggestions but given me a lot of freedom, letting me go my own way. Indeed, I almost always followed this path with my Ph.D. students. If a student is bright enough to make his own way, I think we have no right to impose much on him. Back in the University of Lisbon, I started courses in computational statistics, order statistics, and also in applied areas such as statistical quality control. Although I enjoy teaching, my main interest has been research (and family life).

Carlos Ramos: I started with physics and naturally arrived to dynamical systems.

Ana Soares: I loved fluid mechanics and all mathematical problems motivated in physical and engineering applications. My Master's supervisor proposed that I study shock wave problems and combustion problems. I accepted and I am still working in mathematical physics.

How would you describe the essence of your own research to a young student?

Ivette Gomes: The majority of decisions can be made in terms of averages and their fluctuations, and thus with the "middle" observations, when we order the data available (something we could describe as central order statistics). A few, exceedingly important problems deal with extreme order statistics, either maxima or minima, since extreme down-crossings or up-crossings of thresholds can result in very severe losses (for instance floods, droughts, wild fires, and bankruptcy). Models for extreme events have been developed under a wide variety of assumptions, but the basic models are important guidelines in terms of successfully choosing shape, scale, and location. In the last few years, the focus of my research has been on strategies to choose the most reliable models to deal with concrete situations, working essentially under a semi-parametric framework.

Carlos Ramos: I work with the analogy between mathematical structures and other concepts from outside mathematics.

Ana Soares: I study mathematics which help to understand and explain many applied problems arising in real-world applications mainly related to physics and engineering.

Which would you say are the most interesting/challenging open (or recently solved) problems in your area, and what do you think the future holds in your area and in your line of research?

Ivette Gomes: Although computational statistics has been used to "let the data speak for themselves," I strongly believe that science does not deal with singular data. In fact, what is useful is to abstract the characteristic features of the problem, and try to develop a general theory for that class of problems. One of the ways to do that is to fit useful models—useful because they are general, or mathematically tractable, or have simple characterizations. One way of doing this is to think on a large scale, in the sense that we try to devise what would be good for a large dataset (and indeed in data analysis we may simulate pseudo-observations to observe the behavior of larger datasets than the one at hand). In other words, we develop asymptotic approximations. This requires a much deeper study on the rate of convergence towards these asymptotic behaviors. Much has been done in this field, but there is room for further developments. There is also a need to build up models under more realistic assumptions than the commonly used ones that in general do not go beyond weak convergence hypotheses and some mild form of parental homogeneity. On the other hand, as in many situations data gathering is drastically limited, behavior with small samples is also a crucial area of research. And the analysis of spatial and big data is also quite challenging.

Richard James: My area (applied mathematics) is not so much driven by longstanding hard problems that famous mathematicians could not solve. Rather, it is driven by the ideas that precede the problem. The formulation of the problem is typically the most fascinating and challenging part. This does not imply that the solution is easy! Some of the problems on the theme of the Advanced School on providing alternative methods of producing energy that do not rely on burning fossil fuels, and the reliable, accurate prediction of climate change are challenging. But simple, classical problems, like, "why are the planets of the solar system where they are?" also fascinate me.

Carlos Ramos: One of the biggest challenges is to develop mathematics taking into account biology (natural sciences generally speaking) and the social sciences. Reflecting on how mathematics has been developed since Newton, taking into account mainly physics. This process can help to theorize in the referred sciences.

How do you see your area in terms of its importance in mathematics and in other fields of knowledge, the impact on and from other areas, and how do you expect this interplay to develop further?

Ivette Gomes: Extreme value theory is an important area of probability/statistics, both because of its intrinsic beauty and inspirational value for emerging areas (for instance, stability in generalized convolution algebras) and because of its outstanding performance in dealing with extreme risks—for instance, the use of extreme high quantiles, known as value at risk (VaR) in finance. Statistics blends mathematics with the taming of uncertainty, it deals with using the rigor of deductive reasoning, applying it to uphold the use of induction in knowledge building, and I wouldn't agree with the view that statistical reasoning is no more than a subarea of mathematics. But a large share of statistical research, either in probability or stochastic processes, and is traditionally called mathematical statistics, uses deep results from many areas of mathematics, like numerical methods, analysis, algebra, functional analysis, and many others, to construct new deep rigorous knowledge on how to transform information in knowledge, and how to use randomness as an ally. Under this specific perspective, I feel that my field is a sophisticated and challenging area of mathematical research.

Richard James: Mathematics is the language of science. I always inherently liked mathematics, but, as an undergraduate, I also thought that it would certainly be a good idea to learn the language well, because of the inseparable relation between ideas and language. I'm now even more convinced. I suspect that the importance of mathematics in science will grow.

Carlos Ramos: In my opinion dynamical systems will become a cornerstone in mathematics, influencing all mathematics, conceptually, structurally and from a practice point of view. The area as a pure area will be maintained and will develop itself slowly, the interplay between other mathematical areas will explode, and regarding the scientific applications it will develop tools "ready to use" in a similar way as has happened with statistics. The most important thing is that conceptually DS can furnish the correct concepts and tools for the advance and effective synthesis in science.

Do you have a favorite result, your own and/or from others?

Ivette Gomes: This is a difficult question, since I am convinced that in general we are "infatuated" with our more recent results. So, I could answer that I am proud of my recent work showing that by simply using general definitions of a mean, the Hill estimator of the extreme value index can be much improved. But looking back to more ancient results, I like what I have done on pre-asymptotic approximations and domains of attraction of extreme stable models. Indeed, among the articles I read during my stay in Sheffield, UK (1975–1978), for my Ph.D., the one that influenced me most was possibly the article by Fisher and Tippett (1928), on rates of convergence and penultimate approximations. And indeed I still think there is some kind of magic in this topic, because this, my first passion, has been intermittently revisited after my Ph.D. thesis, either individually or in co-authorship, first with Dinis Pestana, next with Laurens de Haan, and more recently with Luisa Canto e Castro, Sandra Dias, and Paula Reis, in a topic relating pre-asymptotic approximations and reliability of large and coherent systems. But in fact my main reward along my professional life has always been the continued pleasure provided by my research activity. Concerning favorite results from others and outside the field of extremes, I think Jacques Bernoulli was right in naming his law of large numbers "his gold theorem." Indeed, the core of simulation is a clever use of the law of large numbers. And, also because of its many uses, from simulation to meta-analysis, the probability transform theorem, bringing the uniform to the limelight of probability, is also one of my favorite results. On the other hand, the very bright total probability theorem, which is Descartes's method translated into probability language, is a foremost result, and I would be happy to discover who deserves the credit to have first used it and understood its universal value. Mathematical statistics is a recent field, and the pioneering achievements, K. Pearson's chi square criterion, Student's illuminating study on the error of the mean (which contains a lucid view of the uses of simulation), and Fisher's ANOVA and all its ensuing creation of experimental design are landmarks, and not only in the history of statistics, since they played a central role in changing the paradigm of scientific research.

Is it difficult to get funding for research in your area?

Ivette Gomes: Yes, indeed. In the preface of his book on probability, Kallenberg states that while circa 1950 Loève's book on probability covered the main results in the field, by the time he wrote his book, several shelves in a library were needed to provide a fair account of the field. I think that it was in a very interesting book by Ian Hacking that I read that per year more than 600,000 new theorems were published, and that a first-rate mathematician was able to incorporate around 100 of them in his toolbox. This difference between the advancement of science and the filtering of its essentials has a perverse effect on the understanding of the relevance of alien work, and the fact that in Portugal evaluation panels seldom have statisticians has had a very negative impact in funding probability and statistics research.

On research, more generally:

What would you say are the most important things to keep a research group going?

Ivette Gomes: New scientists are trained by the example of the senior way of solving problems, so proximity and facilities for exchange of ideas are important assets for the future of science. Guidance in documentation is also an important step in educating young researchers. Incentives for the group, including funding for presenting and discussing ideas in workshops and seminars and for inviting researchers from other groups that are tackling similar problems, are also important. A peaceful life at the research unit is also something invaluable.

Richard James: It is not so easy in the US to achieve long-term continuity of a group, and this presents distractions and difficulties. But it should be appreciated that this has been true for the whole history of the mathematical sciences, as one can see from the letters of Euler and Newton. From the perspective of individual countries, the percentage of GDP spent on scientific research correlates extremely well with every measure of quality of life.

Andrew Schmitz: The importance of the subject and the competency of the researchers.

Ana Soares: The leadership and the team.

How do you see the relation between traveling and research?

Ivette Gomes: The capacity for imagining new problems and having inspirational ideas when listening to ideas that seem very far from the actual problems the group (or individual) is dealing with is one of the important assets in scientific life. The opportunity to contact others, to listen to their problems and methods, and to extract from this new, path-breaking ways of dealing with problems is something invaluable, and travel is one of the most direct ways of achieving it.

Richard James: I am a huge proponent of sabbatical leaves.

Andrew Schmitz: To carry out research, traveling to conferences along with giving papers is a must.

Ana Soares: It is important to leave, for short periods, the activity related to courses and administrative issues. Sometimes, it is easier to concentrate on a problem and to have new ideas.

Onteaching :

What do you think about the relation between teaching and researching?

Margarida Brito: Teaching at the university level without researching seems problematic to me. We only really understand things if and when we are in a productive relation with them, I think. And what is more, if we want to motivate the students to do their own research, it helps to confront them with working problems. Also from the point of view of research, the relation of teaching and research persists. Teaching clarifies one's own thoughts.

José Cardoso: The relation between teaching and researching is sometimes difficult, but most of the time it is mutually beneficial for the student and for the researcher: the former can realize better the way science works, the latter can have the opportunity to clarify to himself the importance of his own research for other people as well as their utility.

João Coelho: It is fundamental. It is the way to guide students to success.

Ivette Gomes: This is one of the most difficult questions. I have met excellent researchers who are boring speakers. And one of my best professors at the Faculty of Science was a fine scholar, with a superb critical knowledge of many fields of mathematics, and as far as I am aware he did not publish many results in international journals. But in his classes, we were shown brilliantly how it was necessary to alter hypothesis to be able to prove statements, and hence the core of research activity in mathematics. A deep knowledge of the field is an important asset to alter the syllabus of basic courses to accommodate new knowledge (directly, or by preparing students to do it in more advanced courses). Providing appropriate documentation is essential to curtail the exposition of matters in the classroom, leaving to the students the "burden" of completing proofs and solving exercises. In tutorials, it is important to discuss strategies to solve the problem at hand, and to enlighten how a knowledge of the theoretical background is essential to gain from a singular problem the ability to solve many more of its class. Teaching at a more advanced level is simpler, both because the students have chosen this path of study because they have an interest in it, and because the emphasis can be placed almost exclusively on the mathematical explanation. And advanced courses can be much more gratifying when the lecturer has contributed to the field, and can give a lively explanation on how he developed his ideas and got the results, and whether there are open issues that need further developments.

Carlos Ramos: It is natural that they can develop simultaneously.

Andrew Schmitz: Those of us who are fortunate draw strong connections between teaching and research, especially if your research can be tied directly to your teacher. So often people teach classes that bear little relationships to the subjects they teach.

Ana Soares: It depends on the course level. In general, for basic courses, the research can help in finding pertinent examples or to show the students new streamlined methods related to some topics. For advanced courses, it is crucial to be updated and really involved in research activities.

Any thoughts on what's crucial for a university teacher and/or student?

Ivette Gomes: For a university teacher: to have a deep knowledge of the field, to be inventive by using well-chosen examples, to provide adequate documentation and guidelines for further reading, to listen to the students, to be fair. For a student: to understand that she or he is in the university to learn both in and out of class. To realize that it is necessary to quickly develop the capability of making hierarchies in knowledge, discerning what is essential and what is accessory, for the present time, but at the same time to respect all knowledge as a treasure, an asset that can be invaluable in the future.

Andrew Schmitz: An excellent teacher must have both knowledge of the subject as well as interest in the field.

What are your thoughts on the relation between high school and university in terms of education?

Ivette Gomes: High school should be a right for everyone, and hence the teaching there should emphasize what is useful for everyone. But as a large share of students will progress to university courses, and the time frames are shorter and shorter (in my time graduation took 5 years, now it has been reduced to three), there is plainly the need to adapt the syllabus so that students leave with some operational capabilities in basic matters.

Do you have any advice for students starting their research?

João Coelho: Please don't give up, and always believe that success comes from work.

Ivette Gomes: Work hard, read a lot, ask questions to others but mainly to yourselves, when you cannot solve a problem try solving something similar, perhaps weaker in the sense that you either assume more hypotheses or reduce the scope of what you are trying to prove. Using simple examples to start with is a good choice.

Andrew Schmitz: Pick a subject that is of current interest and that you are keen about.

Ana Soares: Yes, please do not concentrate on only one problem. Do not leave important tasks for the last moment.

And for the ones who are hesitating between pursuing a Ph.D. and looking for a different job?

João Coelho: Look for a job, get experience (and money, of course), and then pursue a Ph.D. I will do the same.

Ivette Gomes: I listen and I ask questions, but I do not give answers, since in this matter I feel that the only plausible conduct is to help them to find their own answers, like in Socrates' maieutic method.

Ana Soares: If you like to investigate problems, if you like to develop understanding and to contribute to finding solutions of problems, if you like to do solitary work, pursue a Ph.D. If you like to obtain quick results, if you do not like to invest in studying problems, try another job.

Have all of your research students chosen academic careers?

Ivette Gomes: A great majority of my Ph.D. and M.Sc. research students are in academia. But some of them are also in Brazil, Canada, ...for their choices, but essentially due to the crisis in Portugal, and to the fact that universities are not recruiting new people.

Carlos Ramos: Yes.

Andrew Schmitz: About 60 % of my students have chosen academic careers.

Ana Soares: No.

On other issues:

Do you have hobbies?

João Coelho: Yes, photography, swimming, and agriculture.

Ivette Gomes: I collect owls, coins, and stamps. I enjoy traveling. I love swimming, cycling, and playing table tennis. I also love music, and occasionally I like to do embroidery and knitting.

Andrew Schmitz: My major hobby is farming.

Ana Soares: Yes, music, dancing, swimming.

Do you have a connection to Portugal? How do you see its development?

Ivette Gomes: Yes, I have a strong connection to Portugal. I am Portuguese, I live in Portugal, I felt the happiness of watching the rise of democracy, and now I feel with discomfort all the misfortunes caused by the abuses of some politicians whom we do not respect but that our constitutional laws, judicial power, and even the power of the media seem unable to control. Concerning research, after a favorable period, namely inspired by the late minister Veiga Simão, now there seem to exist guidelines to destroy whole areas of research. Concerning teaching, in my opinion there has been a general decline, mainly as a consequence of the implementation of what is called the Bologna agreement. The democratic regiment of our universities also changed drastically, and for the worse. Sincerely, I am a bit frightened about the developments in the last few years.

Ana Soares: The development of research in Portugal has been notable, but recently the researchers have fewer opportunities and so some notable researchers have had to leave Portugal.

MIX
Papier aus verantwortungsvollen Quellen
Paper from responsible sources
FSC® C105338

If you have any concerns about our products,
you can contact us on
ProductSafety@springernature.com

In case Publisher is established outside the EU,
the EU authorized representative is:
Springer Nature Customer Service Center GmbH
Europaplatz 3, 69115 Heidelberg, Germany

Printed by Libri Plureos GmbH
in Hamburg, Germany